American Warplanes
1908-1988:
A Bibliography

Meckler's Bibliographies of Battles and Leaders

Series Editor: Myron J. Smith, Jr.
Series ISBN 0-88736-517-5

1. The Battle of Antietam and the Maryland Campaign of 1862: A Bibliography
   D. Scott Hartwig
   ISBN 0-88736-321-0 CIP 1990

2. The Central Pacific Campaign, 1943 - 1944: A Bibliography
   James T. Controvich
   ISBN 0-88736-325-3 CIP 1990

3. American Warplanes 1908 - 1988: A Bibliography
   Myron J. Smith, Jr.
   ISBN 0-88736-383-0 CIP 1991

4. Pearl Harbor: A Bibliography
   Myron J. Smith, Jr.
   ISBN 0-88736-305-9 CIP 1991

5. The Battles of Coral Sea and Midway, May - June 1942: A Bibliography
   Myron J. Smith, Jr.
   ISBN 0-88736-683-X CIP 1991

6. Julius Caesar: A Bibliography
   Diane R. Gordon
   ISBN 0-88736-693-7 CIP *forthcoming*

7. The Battle of Jutland: A Bibliography
   Eugene Rasor
   ISBN 0-88736-669-4 CIP *forthcoming*

8. The Falklands/Malvinas Conflict: A Bibliography
   Eugene Rasor
   ISBN 0-88736-668-6 CIP *forthcoming*

9. The Normandy Invasion, 1944: A Bibliography
   Colin F. Baxter
   ISBN 0-88736-557-4 CIP *forthcoming*

10. Military Fortifications: A Bibliography
    Dale E. Floyd
    ISBN 0-88736-307-5 CIP *forthcoming*

11. The War of Spanish Succession: A Bibliography
    W. Calvin Dickinson
    ISBN 0-88736-694-5 CIP *forthcoming*

American Warplanes 1908-1988:
A Bibliography

Myron J. Smith, Jr.

Meckler
Westport • London

Library of Congress Cataloging-in-Publication Data

Smith, Myron J.
American warplanes 1908-1988 : a bibliography / Myron J. Smith, Jr.
p. cm. -- (Meckler's bibliographies of battles and leaders)
Includes bibliographical references and index.
ISBN 0-88736-383-0 (alk. paper) : $
1. Aeronautics, Military -- United States -- History -- Bibliography.
2. Airplanes, Military--United States--History--Bibliography.
I. Title. II. Series.
Z6725.U5S57 1991
[UG633]
016.3584'183'0973--dc 20 90-48967
CIP

British Library Cataloguing in Publication Data

Smith, Myron J.
American warplanes, 1908-1988 : a bibliography, -
(Meckler's bibliographies of battles and leaders).
1. United States. Military aircraft, history -
Bibliographies
I. Title
016. 6237460973

ISBN 0-88736-383-0

Meckler Publishing, the publishing division of Meckler Corporation,
11 Ferry Lane West, Westport, CT 06880.
Meckler Ltd., Grosvenor Gardens House, Grosvenor Gardens,
London SW1W 0BS, U.K.

Printed on acid free paper.
Printed and bound in the United States of America.

Contents

North American P-51 Mustang. Courtesy of Rockwell International.

Preface and Acknowledgments

ON January 10, 1908, the U.S. Army signed a contract with the Wright Brothers for a pusher biplane flying machine, which was tested later that year at Fort Myer, Virginia. The American military has since tested and sometimes purchased hundreds of warplanes (both Fixed- and Rotary-winged) which have grown in sophistication from the Wright Flyer to the Stealth Bomber. An enormous body of data concerning these various flying machines has been assembled in repositories across the land, much of it published far and wide over the past 81 years. In uniform for only 80 years, the warplane has become one of the most written of equipments in our entire military history.

THE literature on individual warplanes continues to grow more voluminous with every passing year; even if one discounts the thousands of newspaper articles penned about military aircraft since 1908, the sheer mass is daunting. In years past, pioneer bibliographers from Paul Brockett to August Hanniball (to name only two) have attempted to provide references to specific warplanes within their volumes. In my own military aerospace guides, considerable space has been allotted to works on named and numbered aircraft. Nowhere, however, has anyone ever concentrated on the preparation of a commercially available compilation devoted strictly to the various types of printed literature devoted to the U.S. military's different airplanes and helicopters. For the first outing in Meckler's new BATTLES AND LEADERS BIBLIOGRAPHY SERIES, I wanted to commence with such a working tool, built from many sources and aimed at a wide audience. We hope this book may prove useful at various levels of interest to military personnel, scholars, students, teachers, librarians, journalists, government policymakers, museum/restoration workers, those within the aerospace community, aviation modellers and enthusiasts.

Objectives/Selection Criteria

DESPITE its several thousand citations, this bibliography is not definitive; to provide truly complete coverage for all of the works penned on specific American warplanes conceived, tested, or purchased since 1908 would require an opus of many volumes. Nevertheless, what is offered here provides coverage for 525 varied fixed- and rotary-wing models, including prototypes.

AS a reference tool, the book will permit users to quickly determine what kinds of materials are available on individual warplanes of interest and help to establish a basis for further research on those types/models not specifically chronicled, weakly covered, or found only in warplane/manufacturer overviews. In general, the items cited are those the user might reasonably expect to find in large university, public, or government libraries and those specialized repositories noted for their acquisition of aerospace resources and listed, beginning on page 1 below. If not immediately available in local libraries, many of the references cited (especially non-reference books and photocopies of periodical articles) can be obtained in America (and a few other nations) through interlibrary loan, details of which service can be provided by the libraries approached. Runs of certain journals like Aerospace Historian are available in microform from such firms as University Microfilms while bookstores and mail-order distributors remain an excellent source for specific titles.

THE criteria for selection in this volume is broad. The following types of published materials covering individual aircraft are represented: books and monographs; periodical articles exceeding a page in length; and documents. A considerable amount of "space work" has also been done in the garden of printed, if not strictly "published," materials. Although it would be impossible to list from all of these extant, a generous selection of the following examples are included: doctoral dissertations and masters theses from civilian and military institutions; "think tank" reports and studies such as those from the RAND Corporation; government departmental/agency/war college histories, reports, and studies; and company publications from aerospace manufacturers. A list of the 334 journals consulted forms the Appendix. All citations in this guide are to unclassified items.

ALTHOUGH much has been included, several categories of aerospace literature have purposefully been omitted, due largely to the space citation would take. Excluded are most works designed expressely for young children; exceedingly technical agency publications (e.g., NASA studies of various rotorcraft); newspaper articles (unless reprinted in anthologies); and general aircraft histories and overviews (unless presented as edited anthologies which could be dissected), especially of the large variety referred to as "coffee table" books. Several of these categories are worthy of special bibliographic coverage outside the scope of this work and, indeed, some are examined in certain of the specialized tools listed in Chapter I. For example, a particularly helpful guide to some of the "coffee table" works is John M. G. Emory's <u>The Source Book of</u>

World War II Aircraft.

DESPITE vigorous research, it is probable that a number of sources published even prior to our January 1989 deadline have been missed. In the event that users have or find non-included material filling our criteria, it would be helpful if forwarded to the compiler for use in any second edition. To compensate in a minor fashion, articles from several of the aviation journals are included as late as March when the manuscript was mailed to the publisher.

Arrangement

THE three main chapters, I through III in the Contents, with their subsections, form something of a classified index to this work, providing the key to its layout. To keep abreast of new works and current events as these relate to certain warplanes and to locate repositories of information, users will find it useful to continuously consult many of the libraries and on-going titles presented in Chapter I. Chapter II, the largest in the work, is divided by types of Fixed-Wing Aircraft (e.g., Fighter-Attack Aircraft), with citations following to individual aircraft models (e.g., F-4, F-5, F6F). The latter are ordered not by manufacturer, but by numerical designation. Although that arrangement is somewhat non-traditional/non-chronological, it offers the advantage of a single-order and the quick access. Except for the Pitcairn Autogiro, all of the Rotary-Wing Aircraft covered in Chapter III are helicopters and thus the type subdivisions are there eliminated in favor of a straight numerical model listing. In both Chapters II and III, cross references between model designations are provided to avoid wherever possible multiple presentations of the same material (e.g., OA-10 see PBY Catalina).

THE bibliographic listings herein are not offered so much as an aid in correcting deficiencies in our knowledge of certain warplane sources, though as such they may serve, but primarily as a means of establishing literature control over references to individual flying machines. Toward that end and given the interest shown in some aircraft at the expense of others (for whatever reason), some listings here will be significantly longer than others. To assist users in reviewing the material provided for the more popular models, all warbirds with 10 or more citations are subjected to two major source breakdowns: Book/Document/Report Literature and Periodical/Anthology Literature. Non-critical annotations are occasionally supplied in Chapters II and III to clarify the contents, acronmyms, or titles of citations while references in Chapter I are liberally analyzed in order to extol their potential to current and future users. An alphabetically arranged Model/Manufacturer/Popular Name index completes the work.

Acknowledgments

FOR their advice, assistance, or encouragement in the formulation, research, and completion of this endeavor, several people and organizations deserve thanks:

Jay Miller, Aerofax, Inc.

Robert B. Lane, Air University

Janet Streek, Air International

Bob Krenkel, Squadron/Signal

Dean C. Allard, Naval Historical Center

Jim Controvich, Springfield, Mass.

Lester L. Miller, Jr., Army Field Artillery School

Murray L. Bradley, Naval War College

J. Thomas Russell, National Defense University

Emil L. Havack, Military Review

B. Franklin Cooling, AFCHO

Mike Miller, Salem College

John J. Sloanaker, Army Military History Institute

Stanley Kalkus, Navy Department Library

E. H. Simmons, Marine Corps History and Museums Division

Betty F. Kemlein, The Tailhook Association

Rosario Rausa, Wings of Gold

Ray Merriam, Bennington, Vt.

John T. Correll, Air Force Magazine

McDonnell-Douglas Corporation

Lockheed Corporation

Sikorsky Aircraft

Boeing Corporation

Boeing-Vertol Corporation

Fairchild
Bell Helicopter Textron
Rockwell International
Grumman Aerospace Corporation
Northrop Corporation
British Aerospace Corporation
General Dynamics Corporation
Piper Aircraft

FINALLY, appreciation is extended to Alan Meckler and Tony Abbott at Meckler Corporation for their spirited assistance and many kindnesses, large and small.

I. Reference Works and Libraries

A. Libraries and Research Centers

A significant number of private and public United States libraries hold published and unpublished materials of potential interest to users of this guide. Indeed, the national **infatuation** with aerospace literature has resulted in at least some aerospace (and related military history) collection development in even the smallest public and school libraries. Additionally, over 100 colleges and universities now offer aviation/aerospace courses and have assembled library collections to support them. Museums and historical societies with even partial aerospace themes sometimes have research libraries while aerospace manufacturers and others concerned with defense matters maintain corporate, professional, or technical libraries. The base and flight or technical facilities of the armed forces maintain libraries while every state has one library designated as a regional depository for all Federal documents.

THE amount and kind of material which might be made available to qualified researchers and other students as well as possible problems of access, particularly to private, professional, technical, or government libraries, should be resolved in advance of visitation either by letter or by telephone. Such contact should result in any required permissions and appointment times, access documents or arrangements, and the delivery of any localized instructions. For additional insight, users are urged to consult the references provided below under the subtopic Published Document/Manuscript Guides as well as the latest edition of the following:

The American Library Directory. 2 vols. New York: R. R. Bowker, 1923--.

Directory of Special Libraries and Information Centers. 3 vols. Detroit, Mich.: The Gale Research Co., 1963--.

The Official Museum Directory. Skokie, Ill.: National Register Publishing Co. for the American Museum Association, 1956--.

Schukert, Michael A. Collegiate Aviation Directory: A Guide to College-Level Aviation/Aerospace Study. Dubuque, Iowa: Kendall/Hunt Publishing Co. for the University Aviation Association, 1982. 120p.

THE extensive list which follows is arranged by state, and then by type of institution, civilian or military. Library names are arranged alphabetically, with addresses and telephone numbers provided.

Alabama

Civilian:

Alabama Aviation and Technical College Learning Resources Center. 1439 South Union Ave., Ozark, Ala., 36360. 205-744-5113, ex. 30.

Alabama Space and Rocket Center von Bruan Library and Archives. Tranquility Base, Huntsville, Ala., 35807. 205-837-3400.

Auburn University Ralph Brown Draughton Library. Auburn, Ala., 36849-3500. 205-826-4500.

Museums of the City of Mobile Reference Library. 355 Government St., Mobile, Ala., 36602. 205-438-7569.

NASA-George C. Marshall Space Flight Center Marshall Library. Code CN-24L, Huntsville, Ala., 35812. 205-453-1880.

Troy State University Library. Troy, Ala., 36802. 205-566-3000, ex. 266.

Tuskegee University Hollis Burke Frissell Library. Tuskegee Institute, Ala., 36088. 205-727-8894.

University of Alabama Amelia Gayle Corgas Library. Bx. S, University, Ala., 35486. 205-348-6010.

Military:

U.S. Air Force

Air University Library. Maxwell AFB, Ala., 36112. 205-293-2888.

Gunter AFS Library. Fl 3370, Gunter AFS, Ala., 36114-5000. 205-279-3179.

USAF Historical Research Center. Maxwell AFB, Ala., 36112-6678. 205-293-5958.

U.S. Army

Abrams Library. Bldg. 2102, Fort McClellan, Ala., 36205-5000. 205-238-3715.

Army Aviation Center Library. Novosel Rd. and 5th Ave., Bldg. 212, Fort Rucker, Ala., 36362. 205-255-5010.

Army Aviation Museum Library. Fort Rucker, Ala., 36360. 205-255-4507.

Army Aviation Technical Library. Fifth Ave. and 23rd St., Bldg. 5906/5907, Fort Rucker, Ala., 36362. 205-255-4591.

Alaska

Civilian:

Alaska Historical Library. State Office Bldg., Pouch G, Juneau, Alaska, 99811. 907-465-2925.

University of Alaska, Juneau Library. 11120 Glacier Hwy., Juneau, Alaska, 99801. 907-789-4466.

Military:

U.S. Air Force

Eielson AFB Library. Fl. 5004, 343 CSG/SSL, Eielson AFB, Alaska, 99702-5000. 907-377-3174.

Elmendorf AFB Library. Alaska Air Command. Fl. 5000. Elmendorf AFB, Alaska, 88506-5000. 907-552-3787.

U.S. Army

Fort Wainwright Post Library. Bldg. 3717, Fort Wainwright, Alaska, 99703-5100. 907-353-5194.

Arizona

Civilian:

Arizona Department of Library, Archives, and Public Records. Third Fl., State Capitol, 1700 W. Washington, Phoenix, Ariz., 85007. 602-255-4121.

Champlain Museum Library. 4636 Fighter Aces Dr., Mesa, Ariz., 85205. 602-830-4540.

Pima Air Museum Library. 6400 South Wilmot Rd., Bx. 17298, Tucson, Ariz., 85731. 602-889-0462.

University of Arizona Library. Tucson, Ariz., 85721. 602-621-2101.

Military:

U.S. Air Force

Davis-Montham AFB Library. Fl. 4877, Davis-Montham AFB, Ariz., 85707. 602-748-4381.

Luke AFB Library. Fl. 4887, Luke AFB, Ariz., 85309-5725. 602-856-6301.

Williams AFB Library. Fl. 3044, Williams AFB, Ariz., 85240-5575. 602-988-2611, ex. 5279.

U.S. Army

Yuma Proving Ground Post Library. Yuma, Ariz., 85365. 602-328-2545.

U.S. Marine Corps

Yuma MCAS Library. Bldg. 633, Yuma Marine Corps Air Station, Ariz., 85369-5000. 602-726-2785.

Arkansas

Civilian:

Arkansas State Library. One Capitol Mall, Little Rock, Ark., 72201. 501-371-2326.

Military:

U.S. Air Force

Blytheville AFB Library. Fl. 4634, Bldg. 555, Blytheville AFB, Ark., 72317-5225. 501-762-7286.

Little Rock AFB Library. Fl. 4460, Bldg. 976, Little Rock AFB, Ark., 72099-5000. 501-988-6979.

California

Civilian:

Aerospace Corporation Charles C. Lauritsen Library. 2350 E. El Segundo Blvd., Bx. 92957-MS M1-199, Los Angeles, Calif., 90009. 213-648-6738.

Aircraft Technical Publishers Library. 101 S. Hill Dr., Brisbane, Calif., 94005. 415-468-1705.

American Aviation Historical Society Library. 2333 Otis, Santa Ana, Calif., 92704. 714-549-4818.

California Institute of Technology Aeronautics Department Branch Library. 101 Guggenheim, Pasadena, Calif., 91125. 213-356-4521.

California Museum of Science and Industry Library. 700 State Dr., Los Angeles, Calif., 90037. 213-749-0101.

California State Library. Bx. 2037, Sacramento, Calif., 95809. 906-324-4863.

California State University, Northridge, Delmar T. Oviatt and South Libraries. 18111 Nordhoff St., Northridge, Calif., 91330. 818-885-2297.

General Dynamics Corporation, Convair Division Library. 5001 Kearny Villa Rd., Bx. 85386, San Diego, Calif., 92138-5386. 619-573-4876.

Hoover Institution on War, Revolution and Peace Library. Stanford, Calif., 94305-2323. 415-497-2058.

Hughes Aircraft Company, Information Services Department Technical Library. Bx. 902, Bldg. E-1, MS E-117, 2000 E. El Segundo Blvd., Los Angeles, Calif., 90245-0902. 213-616-3333.

Lockheed Corporation International Marketing Library. Bx. 551, Burbank, Calif., 91520. 818-847-7454.

McDonnell-Douglas Corporation, Douglas Aircraft Company Technical Library. 3855 Lakewood Blvd., Long Beach, Calif., 90846. 213-593-9541.

McDonnell-Douglas Helicopters Library. 6/A69, Centinela at Teale Sts., Culver City, Calif., 90230. 213-305-3514.

N. Paul Whittier Historical Aviation Library of the San Diego Aerospace Museum. 2001 Pan American Plaza, Balboa Park, San Diego, Calif., 92101.

NASA-Ames Research Center Technical Library. N 202-3, Moffett Field, Calif., 94035-4000. 415-694-5157.

NASA-Dryden Flight Research Facility Library. Bx. 273, Edwards AFB, Calif., 93523. 805-258-3702, ex. 334.

Northrop Corporation, Aircraft Division Library Services. One Northrop Ave., Hawthorne, Calif., 90250-3277. 213-970-4136.

RAND Corporation Library. 1700 Main St., Santa Monica, Calif., 90406. 213-393-0411.

Rockwell International Corporation, North American Aircraft Operations Technical Information Center Library. 201 North Douglas St., Box 92098, Los Angeles, Calif., 90009. 213-647-2961.

Teledyne Ryan Aeronautical Technical Information Services Library. 2701 North Harbor Dr., Bx. 80311, San Diego, Calif., 92138. 619-260-4458.

University of California at Los Angeles Library. 405 Hilgard Ave., Los Angeles, Calif., 90024-1302. 213-825-1201.

Military:

U.S. Air Force

Beale AFB Library. Fl. 4686, Beale AFB, Calif., 95903. 916-634-2706.

Castle AFB Library. Fl. 4672, Bldg. 422, Castle AFB, Calif., 95342. 209-726-2630.

Edwards AFB Library. Fl. 2805, ABG/SSL Stop 115, Edwards AFB, Calif., 93523. 805-277-2375.

Edwards AFB Test Center Library. 6520 TEST G-ENXL, Fl. 2806, Edwards AFB, Calif., 83523-5000. 805-277-3606.

George AFB Library. Fl. 3812, George AFB, Calif., 93394-5000. 619-269-3228.

McClellan AFB Library, Fl. 2040, 2852 ABG/SSL, McClellan AFB, Calif., 95652. 916-643-4640.

March AFB Library. Fl. 4664, Bldg. 938, March AFB, Calif., 92518-5000. 714-655-2203.

Norton AFB Library. Fl. 4448, Norton AFB, Calif., 92409-5985. 714-382-7119.

Travis AFB Mitchell Memorial Library. First St. and Travis Ave., Travis AFB, Calif., 94535-5000. 707-438-3279.

Vandenberg AFB Library. Bldg. 10343-A, Vandenberg AFB, Calif., 93437-5000. 805-866-6414.

Vandenberg AFB Western Space and Missile Center Technical Library. PMET (FL. 2827), Vandenberg AFB, Calif., 93437-6021. 805-866-9745.

U.S. Army

Presidio of San Francisco Post Library. Bldg. 386, Presidio of California, 94129. 415-561-3448.

U.S. Marine Corps

Camp Pendleton Base Library. Bldg. 1122, Camp Pendleton, Calif., 92055. 619-725-5104.

U.S. Navy

Alameda NAS Library. Bldg. 2, Wg. 3, NAS Alameda, Calif., 94501-5051. 415-869-3504.

El Cajon NAF General Library. NAF El Cajon, Calif., 92243. 619-399-2461.

Lemoore NAS Library. Bldg. 821, NAS Lemoore, Calif., 932-646-0001.

Long Beach NAS Library. Bldg. 398, NAS Long Beach, Calif., 90822. 213-437-7349.

Miramar NAS Library. Bldg. M-305, NAS Miramar, Calif., 92145. 619-271-3557.

Moffett Field NAS Library. Bldg. 14, NAS Moffett Field, Calif., 94035. 415-966-5455.

Naval Postgraduate School Dudley Knox Library. Monterey, Calif., 93943-5002. 408-646-2341.

North Island NAS Library. Bx. 29, NAS North Island, Calif., 92135-5000.

Colorado

Civilian:

Martin Marietta Aerospace Research Library. Bx. 179, Denver, Colo., 80201. 303-977-5512.

University of Colorado Library. Bx. 2037, Boulder, Colo., 80309. 303-492-8834.

Military:

U.S. Air Force

USAF Academy Library. USAF Academy, Colo., 80840-5721. 303-472-2590.

Lowry AFB Library. Fl. 3059, 3415 ABG/SSL, Lowry AFB, Colo., 80230. 303-370-3093.

Petersen AFB Library. Fl. 2500, Stop 60, Petersen AFB, Colo., 80914-5000. 303-554-7462.

Connecticut

Civilian:

Bradley Air Museum of the Connecticut Aeronautical Historical Association Library. Bx. 44, Hebron, Conn., 06248.

Connecticut State Library. 231 Capitol Ave., Hartford, Conn., 06106. 203-566-7029.

Kaman Corporation Library. Old Windsor Rd., Bx. 2, Bloomfield, Conn., 06002. 203-243-7305.

United Technologies Corporation, Sikorsky Aircraft Division Library. North Main St., Stratford, Conn., 06601. 203-386-4713.

Military:

U.S. Coast Guard

USCG Academy Library. New London, Conn., 06320-4195. 203-444-8512.

Delaware

Civilian:

Hagley Museum and Library. Bx. 3630, Greenville, Del., 19807-0630. 302-658-2400.

Military:

U.S. Air Force

Dover AFB Library Fl. 4497, 436 ABG/SSL, Bldg. 443, Dover AFB, Del., 19902-5225. 302-678-6246.

District of Columbia

Civilian:

Aerospace Corporation Washington Library. 955 L'Enfant Plaza, SW, Suite 4000, Washington, D.C., 20024. 202-488-6154.

Aerospace Industries Association of America Library. 1725 DeSales St., NW, Washington, D.C., 20036. 202-429-4653.

Center for Defense Information Library. 1500 Massachusetts Ave., NW, Washington, D.C., 20024. 202-862-0700.

Library of Congress. Independence Ave. at First St., NE, Washington, D.C. 20540. 202-287-5000.

NASA-Headquarters Library. 600 Independence Ave., SW, Code NHS-4, One Massachusetts Ave., NW, Washington, D.C., 20001. 202-789-0031.

National Archives. Pennsylvania Ave. and 8th St., NW, Washington, D.C., 20408. 202-423-3218.

Smithsonian Institution Libraries, National Air and Space Museum Branch Library. NASM Rm. 3100, Independence and 7th St., SW, Washington, D.C., 20560. 202-357-3133.

U.S. Government Accounting Office Library. 441 G St., NW, Washington, D.C., 20548. 202-275-3691.

Military:

U.S. Air Force

Air Force Systems Command Headquarters Technical Information Center Library. Fl. 2800, AFSC/MPSLT, Andrews AFB, D.C., 20334-5000. 301-981-3551.

Andrews AFB Library. Fl. 4425, Bldg. 1642, Andrews AFB, D. C. , 20331-5984. 301-981-6454.

USAF Office of Air Force History Library. AF/CHO, Bldg. 5681, Bolling AFB, D. C., 20332-6098. 202-767-0421.

Department of Defense

National Defense University Library. Fort Lesley J. McNair, D. C., 20319-6000. 202-475-1905.

U.S. Army

The Pentagon Library. Rm. 1A518, the Pentagon, D.C., 20310-6000. 202-697-4301.

U.S. Marine Corps

Marine Corps Historical Center Library. Code HDS-4, Bldg. 58, Washington Navy Yard, Washington, D. C., 20374-0580. 202-433-3447.

U.S. Navy

Naval Air Systems Command Technical Information and Reference Center Library. Jefferson Plaza, No. 2, Rm. 278, Washington, D.C., 20361. 202-692-9006.

Navy Department Library. Bldg. 44, Washington Navy Yard, Washington, D.C., 20374-0571. 202-433-4131.

Florida

Civilian:

Emory-Riddle Aeronautical University. Jack R. Hunt Memorial Library. Regional Airport, Daytona Beach, Fla., 32014-3891. 904-252-5561.

University of Florida Libraries. 210 Liberty W., Gainesville, Fla., 32611. 904-392-0342.

Military:

U.S. Air Force

Eglin AFB Library. Fl. 2823, Eglin AFB, Fla., 32542. 904-882-2460.

Homestead AFB Library. Fl. 4829, Homestead AFB, Fla., 33039-5000. 305-257-8184.

Hurlburt AFB Library. Fl. 4417, Hurlburt AFB, Fla., 32544-5000. 904-884-6947.

MacDill AFB Library. Fl. 4814, MacDill AFB, Fla., 33608-5000. 813-830-3607.

Tyndall AFB Library. Fl. 4819, Bldg. 966, Tyndall AFB, Fla., 32403-5715. 904-283-4287.

U.S. Navy

Cecil Field NAS Library. Bx. 104, NAS Cecil Field, Fla., 32215-0104. 904-778-5311.

Key West NAS Library. Bldg. 514, NAS Key West, Fla., 33040-5000. 305-292-2116.

Mayport NAS Library. Bx. 52, Bldg. 930, NAS Mayport, Jacksonville, Fla., 32212. 904-772-3415.

Milton NAS Whiting Library. Bldg. 1417, NAS Milton, Fla., 32570. 904-623-7502.

Naval Aviation Museum Library. NAS Pensacola, Fla., 32508. 904-452-3604.

Pensacola NAS Library. Bldg. 633, NAS Pensacola, Fla., 32508. 904-452-4362.

Georgia

Civilian:

Lockheed-Georgia Company, Technical Information Department Library. Dept. 72-34, Zone 7, 86 South Cobb St., Marietta, Ga., 30063. 404-424-2928.

University of Georgia Libraries. Athens, Ga., 30602. 404-542-8949.

Military:

U.S. Air Force

Moody AFB Library. Moody AFB, Ga., 31699-5725. 912-333-3539.

U.S. Army

Fort Stewart-Hunter Army Airfield Library. Bldg. P-4111, Lundquist Rd., Ft. Stewart, Ga., 31314-6085. 912-767-2260.

Infantry School Donovan Technical Library. Fort Benning, Ga., 31905-5452. 404-544-4053.

Hawaii

Civilian:

University of Hawaii Library. 2550 The Mall, Honolulu, Hawaii, 96882. 808-948-8230.

Military:

U.S. Air Force

Wheeler AFB Library. Fl. 5296, ABG/SSL, Wheeler AFB, Hawaii, 96854-5000. 808-655-1867.

U.S. Marine Corps

Kaneohe MCAS Library. MCAS Kaneohe, Hawaii, 96863-5010. 808-257-3583.

U.S. Navy

Pearl Harbor Naval Base Library. 1514 Makalpa Dr., Honolulu, Hawaii, 96818. 808-471-8238.

Idaho

Civilian:

University of Idaho Library. Moscow, Idaho, 83843. 208-885-6344.

Military:

U.S. Air Force

Mountain Home AFB Library. Fl. 4897, Mountain Home AFB, Idaho, 83648-5000. 208-828-2544.

Illinois

Civilian:

Aerophilatelic Federation of America Earl H. Wellman Sr. Memorial Library. 100 E. Chicago St., Bx. 1239, Elgin, Ill., 60121-1239. 312-742-3328.

Defense Research Institute Library. 750 N. Lakeshore Dr., Suite 5000, Chicago, Ill., 60606. 312-944-0575.

Illinois State Library. Centennial Bldg., Springfield, Ill., 62756. 217-782-5185.

Museum of Science and Industry Library. 57th St. and Lakeshore Dr., Chicago, Ill., 60637. 312-684-1414, ex. 429.

Parks College of St. Louis University Library. Cahokia, Ill., 62206. 618-337-7500, ex. 244.

University of Illinois at Urbana-Champagne Libraries. 1408 W. Gregory Dr., 230 Library UIUC, Urbana, Ill., 61801. 217-333-0790.

Military:

U.S. Air Force

Chanute AFB Library. Fl. 3018, Bldg. 95, Chanute AFB, Ill., 61868. 217-495-3191.

Military Airlift Command Library. Headquarters, MAC/DPSRL, Fl. 4401, Scott AFB, Ill., 62225-5001. 618-256-3228.

Scott AFB Library. Bldg. 1940, 375 ABG/SSL, Scott AFB, Ill., 62225. 618-256-5100.

Indiana

Civilian:

Indiana State Library. 140 N. Senate Ave., Indianapolis, Ind., 46204. 317-232-3686.

Purdue University Aviation Technical Library. Purdue Airport Hanger I, Rm., 225, W. Lafayette, Ind., 47907. 317-494-7640.

Military:

U.S. Air Force

Grissom AFB Library. Fl. 4654, Bldg. 575, Grissom AFB, Ind., 46971-5000. 317-689-2056.

Iowa

Civilian:

Airpower Museum Library. Antique Airfield, Rte. 2, Bx. 172, Ottumwa, Iowa, 52501. 515-938-2773.

Iowa State University Library. Ames, Iowa, 50011-2140. 515-294-1442.

Kansas

Civilian:

Kansas State University Farrell Library. Manhattan and Anderson Aves., Manhattan, Kans., 66506-7166. 913-532-6516.

Military:

U.S. Air Force

McConnell AFB Library. Fl. 4621, McConnell AFB, Kans., 67221-5000. 316-681-5414.

U.S. Army

Combined Arms Research Library. Command and General Staff College, Fort Leavenworth, Kans., 66027-6900. 913-684-4035.

Kentucky

Civilian:

University of Kentucky Library. Lexington, Ky., 40506-0039. 606-257-3139.

Military:

U.S. Army

R. F. Sink Memorial Library. Bldg. 38, Fort Campbell, Ky., 42223-5000. 502-798-5729.

Louisiana

Civilian:

Louisiana State University Troy H. Middleton Library. Baton Rouge, La., 70803. 509-388-2217.

Military:

U.S. Air Force

England AFB Library. Fl. 4805, Bldg. 1213, England AFB, La., 71311. 318-448-5621.

Maine

Civilian:

University of Maine Raymond H. Fogler Library. Oronto, Maine, 04469. 207-581-1680.

Military:

U.S. Air Force

Loring AFB Library. Fl. 4678, Loring AFB, Maine, 04751-5000. 207-999-2416.

U.S. Navy

Brunswick NAS Library. Bx. 21, NAS Brunswick, Maine, 04011-5000. 207-921-2639.

Maryland

Civilian:

NASA-Goddard Space Flight Center Library. Code 252, Greenbelt, Md., 20771. 301-344-7000.

University of Maryland McKeldin Library. College Park, Md., 20742. 301-454-3034.

U.S. Naval Institute Reference and Photographic Library. Annapolis, Md., 21402. 301-268-6110.

Military:

U.S. Air Force

See listing under District of Columbia, Above.

U.S. Army

Fort George G. Meade Post Library. Bldg. 4418, Ft. George G. Meade, Md., 20755-5068. 301-677-4509.

U.S. Navy

Patuxent River NAS Central Library. Bldg. 407, NAS Patuxent River, Md., 20670. 301-863-1927.

U.S. Naval Academy Nimitz Library. Annapolis, Md., 21402-1390. 301-267-2194.

Massachusetts

Civilian:

Boston Public Library. Boston, Mass., 02117. 617-536-5400.

Massachusetts Institute of Technology Aeronautics and Astronautics Department Library. Rm. 33-316, Cambridge, Mass., 02139. 617-253-5665.

Military:

U.S. Air Force

Hanscom AFB Library. Fl. 2835, Hanscom AFB, Mass., 01731. 617-861-2177.

Michigan

Civilian:

Detroit Public Library. 5201 Woodward Ave., Detroit, Mich., 48202-4093. 313-833-1000.

The University of Michigan Library. Ann Arbor, Mich., 48109-1205. 313-764-9356.

Military:

U. S. Air Force

K. I. Sawyer AFB Library. Fl. 4515, K. I. Sawyer AFB, Mich., 49843. 906-346-2864.

Selfridge AFB Library. Bldg. 169, Wagner and George Sts., Selfridge AFB, Mich., 48045. 313-466-5088.

Wurtsmith AFB Library. Fl. 4585, Wurtsmith AFB, Mich., 48753-5000. 517-747-6597.

Minnesota

Civilian:

University of Minnesota Library. 309 19th Ave., S., Minneapolis, Minn., 55455. 612-373-7870.

Mississippi

Civilian:

University of Mississippi Library. University, Miss., 38677. 601-232-7091.

Military:

U.S. Air Force

Columbus AFB Library. Fl. 3022, Columbus AFB, Miss., 39701-5000. 601-434-7762.

Kessler AFB McBridge Library. 3380 MSS/SSL, Kessler AFB, Miss., 39534-5225. 601-377-2181.

U.S. Navy

Meridian NAS Library. Bx. 7-L, NAS Meridian, Miss., 39309-5000. 601-679-2623.

Missouri

Civilian:

McDonnell-Douglas Corporation Corporate Library. McDonnell Blvd., St. Louis, Mo., 63166. 314-232-8515.

University of Missouri Elmer Ellis Library. Columbia, Mo., 65201-5149. 314-882-4701.

Military:

U.S. Air Force

Whiteman AFB Library. Fl. 4625, Whiteman AFB, Mo., 65305-5000. 816-687-3089.

U.S. Army

Army Aviation Systems Command Library and Information Center. 4300 Goodfellow St., AMSAV/DIL, St. Louis, Mo., 63120-1798. 314-263-2345.

Montana

Civilian:

University of Montana Mansfield Library. Missoula, Mont., 59812. 406-243-6700.

Military:

U.S. Air Force

Malmstrom AFB Library. Fl. 4626, Malmstrom AFB, Mont., 59402-5000. 406-731-2748.

Nebraska

Civilian:

Strategic Air Command Museum Library. 2510 Clay St., Bellevue, Neb., 68005. 402-292-2001.

University of Nebraska Love Library. Lincoln, Neb., 68588-0410. 402-472-2562.

Military:

U.S. Air Force

Offutt AFB Library. Fl. 4600, Offutt AFB, Neb., 68113-5000. 402-294-2533.

Strategic Air Command Library. SAC/DPSOL, Offutt AFB, Neb., 68113. 402-294-2367.

Nevada

Civilian:

University of Nevada Library. Reno, Nev., 89557-0044. 702-784-6579.

Military:

U.S. Air Force

Nellis AFB Library. Fl. 4852, Nellis AFB, Nev., 89191-5000. 702-643-2280.

U.S. Navy

Fallon NAS Library. Barracks 2, NAS Fallon, Nev., 89406. 702-426-5161.

New Hampshire

Civilian:

Daniel Webster College Library. University Dr., Nashua, N.H., 03063. 603-883-3556.

Military:

U.S. Air Force

Pease AFB Library. Fl. 4623, Pease AFB, N. H., 03803. 603-430-3734.

New Jersey

Civilian:

Newark Public Library. 5 Washington St., Newark, N.J., 07101-0630. 201-733-7812.

Military:

U.S. Air Force

McGuire AFB Library. Fl. 4484, McGuire AFB, N.J., 08641-5225. 609-724-2079.

U.S. Navy

Naval Air Engineering Center Technical Library. Code 1115, NAS Lakehurst, N.J., 08733. 201-323-2893.

Naval Air Propulsion Library. 1440 Parkway Ave., Bx., 7176, Trenton, N.J., 08628. 609-896-5609.

New Mexico

Civilian:

University of New Mexico Library. Albequerque, N.M., 87131. 505-277-4241.

Military:

U.S. Air Force

Cannon AFB Library. Fl. 4855, Cannon AFB, N. M., 88103. 505-784-3311.

Holloman AFB Library. Fl. 4801, Holloman AFB, N.M., 88310. 505-429-6511.

Kirtland AFB Library. Fl. 4469, Kirtland AFB, N.M., 87117. 505-844-0795.

New York

Civilian:

Academy of Aeronautics Library. 86-01, 23rd Ave., La Guardia Airport Station, Flushing, N.Y., 11371. 718-420-6600.

American Institute of Aeronautics and Astronautics Technical Information Center Library. 555 W. 57th St., New York, N.Y., 10019. 212-247-6500.

Bell Aerospace Textron, Larry D. Bell Memorial Library, Bx. 1, Buffalo, N.Y., 14240. 716-297-1000, ex. 7011.

Fairchild Industries, Fairchild-Republic Company Technical Information Center Library. Conklin St., Farmingdale, N.Y., 11735. 516-531-3497.

Glenn H. Curtiss Museum of Local History Library. Lake and Main Sts., Hammondsport, N.Y., 14840. 607-569-2160.

Grumman Aerospace Corporation Technical Information Center Library. Plant 35 & Stuart Ave., Bethpage, N.Y., 11714. 516-575-3912.

NASA-Goodard Institute for Space Studies Library. Rm. 710, 2880 Broadway, New York, N.Y., 10025. 212-678-5613.

New York Public Library. Fifth Ave. and 42nd St., New York, N.Y., 10018. 212-930-0800.

Military:

U.S. Air Force

Griffiss AFB Library. Fl. 4616, Griffiss AFB, N.Y., 13441. 315-330-7030.

Plattsburgh AFB Library. Fl. 4615, Plattsburgh AFB, N.Y., 12903 518-565-7046.

U.S. Army

U.S. Military Academy Library. West Point, N.Y., 10996-1799. 914-938-2230.

U.S. Coast Guard

USCG Support Center Library. Bldg. 251, USCG Station, Governors Island, New York, N.Y., 10004. 212-668-7394.

North Carolina

Civilian:

University of North Carolina Walter Royal Davis Library. Chapel Hill, N.C., 27514-6080. 919-962-1301.

Wright Brothers National Memorial Library. Cape Hatteras National Seashore, Rte. 1, Bx. 675, Manteo, N.C., 27954. 919-473-2111.

Military:

U.S. Air Force

Pope AFB Library. Fl. 4488, Pope AFB, N.C., 28308-5225. 919-394-2791.

Seymour Johnson AFB Library. Fl. 4809, Seymour Johnson AFB, N.C., 27531-5725. 919-736-5707.

U.S. Army

Fort Bragg Library. Bldg. 2-1256, Knox and Macomb Sts., Ft. Bragg, N.C., 28307-5000. 919-396-3526.

U.S. Marine Corps

Camp Lejeune Base Library. Bldg. 63, Camp Lejeune USMC Base, N.C., 28542. 919-451-5724.

Cherry Point MCAS Library. Bldg. 298, MCAS Cherry Point, N.C., 28533. 919-466-3571.

North Dakota

Civilian:

University of North Dakota Chester Fritz Library. Grand Forks, N.D., 58202. 701-777-4629.

Military:

U.S. Air Force

Grand Forks AFB Library. Fl. 4659, Grand Forks AFB, N.D., 58205. 701-594-6725.

Minot AFB Library. Fl. 4528, Bldg. 138, Minot AFB, N.D., 58705-5000. 701-727-3344.

Ohio

Civilian:

Cleveland Public Library. 325 Superior Ave., Cleveland, Ohio, 44114-1271. 216-623-2800.

Goodyear Aerospace Corporation Library. 152-G, 1210 Massilon Rd., Akron, Ohio, 44315. 216-796-2557.

NASA-Lewis Research Center Library. 21000 Brookpark Rd., Cleveland, Ohio, 44135. 216-433-4000, ex. 419.

Neil Armstrong Air and Space Museum Library. I-75 and Bellfontaine Rd., Bx. 1978, Wapakoneta, Ohio, 45895. 419-738-8811.

The Ohio State University Library. 1858 Neil Ave. Mall, Columbus, Ohio, 43210-1286. 614-422-6151.

Military:

U.S. Air Force

USAF Institute of Technology Academic Library. AFIT/LD, Bldg. 640, Area B., Wright-Patterson AFB, Ohio, 45433. 513-255-5894.

USAF Museum Library. Wright-Patterson AFB, Ohio, 45433. 513-255-3284.

Wright-Patterson AFB Library. Fl. 2300, WG/SSL, Wright-Patterson AFB, Ohio, 45433. 513-257-4815.

Oklahoma

Civilian:

University of Oklahoma Library. 401 W. Brooks, Norman, Okla., 73019. 405-325-2611.

Military:

U.S. Air Force

Altus AFB Library. Fl. 4419, Altus AFB, Okla., 73523-5985. 405-481-6302.

Tinker AFB Library. Fl. 2030, Bldg. 5702, Tinker AFB, Okla., 73145. 405-734-3083.

Vance AFB Library. Fl 3029, Vance AFB, Okla., 73705-5000. 405-249-7368.

Oregon

Civilian:

Portland State University Library. Bx. 1151, Portland, Oreg., 97207. 503-229-3673.

Pennsylvania

Civilian:

Boeing Vertol Company Lydia Rankin Technical Library. Bx. 16858, Philadelphia, Pa., 19142. 215-522-2536.

Franklin Institute Library. 20th St. and Parkway, Philadelphia, Pa., 19103. 215-448-1227.

Pennsylvania State University Fred Lewis Pattee Library. University Park, Pa., 16802. 814-865-0401.

Military:

U.S. Army

Army War College Library. Carlisle Barracks, Pa., 17013-50509. 717-245-4319.

U.S. Army Military History Institute Library. Carlisle Barracks, Pa., 17013-5008. 717-245-3611.

U.S. Navy

Naval Air Development Center Technical Information Branch Library. Warminster, Pa., 18974. 215-441-2918.

Rhode Island

Military:

U.S. Navy

Naval War College Library. Newport, R. I., 02841. 401-841-2641.

South Carolina

Civilian:

The Citadel Daniel Library. Charleston, S.C., 29409. 803-792-5116.

Florence Air and Missile Museum Library. Bx. 1326, Florence, S.C., 29503. 803-665-5118.

Patriots Point Naval and Maritime Museum Library. Patriots Point, Bx. 986, Mt. Pleasant, S.C., 29464. 803-884-2772.

Military:

U.S. Air Force

Charleston AFB Library. Fl. 4418, Charleston AFB, S.C., 29404. 803-554-3134.

Myrtle Beach AFB Library. Myrtle Beach AFB, S.C., 29579-5000. 803-238-7211.

Shaw AFB Library. Fl. 4803, Shaw AFB, S.C., 29152. 803-668-3084.

U.S. Marine Corps

Beaufort MCAS Library. MCAS Beaufort, S. C., 29904-5000. 803-846-7682.

South Dakota

Military:

U.S. Air Force

Ellsworth AFB Library. Fl. 4690, Ellsworth AFB, S. D., 57706-5000. 605-399-7965.

Tennessee

Civilian:

Middle Tennessee State University Andrew L. Todd Library. Bx. 13, Murfreesboro, Tenn., 37132. 615-898-2772.

Tusculum College Tate Library. Bx. 5087, Greeneville, Tn., 37743. 615-639-1111.

Military:

U.S. Air Force

Arnold Engineering Development Center Technical Library. Fl. 2804, M.S.-100, Arnold AFB, Tenn., 37389-9998. 615-454-4430.

U.S. Navy

Memphis NAS Library. Bldg. S-78, NAS Memphis, Tenn., 38054. 901-872-5683.

Texas

Civilian:

Admiral Nimitz State Historical Park and Museum of the Pacific War Library. Bx. 777, 340 E. Main St., Fredericksburg, Tex., 78624. 512-997-4379.

Museum of Aviation Group Library. 300 N. Spur 341, Fort Worth, Tex., 76108. 817-244-1123.

NASA-Johnson Space Flight Center Technical Library. Houston, Tex., 77058. 713-483-4048.

Texas A & M University Sterling C. Evans Library. College Station, Tex., 77843-5000. 409-845-8111.

University of Texas Library. Bx. P, Austin, Tex., 78713-7330. 512-471-3811.

Military:

U.S. Air Force

Bergstrom AFB Library. Fl. 4857, Bergstrom AFB, Tex., 78743-5725. 512-479-3739.

Brooks AFB Library. Fl. 2857, ABG/SSL, Brooks AFB, Tex., 78235-5000. 512-536-2634.

Carswell AFB Library. Fl. 4689, Carswell AFB, Tex., 76127-5000. 817-735-5230.

Dyess AFB Library. Fl. 4661, Dyess AFB, Tex., 79607. 915-696-2618.

Goodfellow AFB Library. Fl. 3030, Goodfellow AFB, Tex., 76908-5000. 915-657-3045, ex. 3232.

Kelly AFB Library. Fl. 2050, 2851 SSL, Kelly AFB, Tex., 78241-5000. 512-925-3214.

Randolph AFB Library. Fl. 3060, Randolph AFB, Tex., 78150-5000. 512-652-2617.

Reese AFB Library. Fl. 3060, Reese AFB, Tex., 79489-5438. 806-885-3344.

Sheppard AFB Library. Fl. 3020, Sheppard AFB, Tex., 76311-5000. 817-851-2687.

U.S. Navy

Beeville NAS Library. Bldg. 1039 Chase Field, NAS Beeville, Tex., 78103. 512-354-5301.

Corpus Christi NAS Library. Bldg. 5, NAS Corpus Christi, Tex., 78419. 512-939-3574.

Dallas NAS Library. NAS Dallas, Tex., 75211. 217-266-6132.

Kingsville NAS Library. Bldg. 1781, NAS Kingsville, Tex., 78363-5000. 512-595-6271.

Utah

Civilian:

Utah State University Merrill Library. UMC 30, Logan, Utah, 84322. 801-750-2682.

Military:

U.S. Air Force

Hill AFB Garrity Memorial Library. Fl. 2020, Hill AFB, Utah, 84056-5000. 801-777-2533.

Vermont

Civilian:

Norwich University Henry Prescott Chaplin Memorial Library. Northfield, Vt., 05663. 802-485-5011, ex. 48.

Virginia

Civilian:

Air Force Association Research Library. 1501 Lee Hwy., Arlington, Va., 22209-1198. 703-247-5800.

Boeing Company Congressional Affairs Library. 1700 N. Moore St., Arlington, Va., 22209. 703-558-9677.

The George C. Marshall Research Foundation Library. Bx. 1600, Lexington, Va., 24550-1600. 703-463-7103.

NASA-Langley Research Center Technical Library. MS 185, Bldg. 1194, Langley, Va., 23665-5225. 804-865-2630.

Virginia Military Institute J. T. L. Preston Library. Lexington, Va., 24450. 703-463-6228.

War Memorial Museum of Virginia Research Library. Huntington Park, 9285 Warwick Rd., Newport News, Va., 23607. 804-247-8523.

Military:

U.S. Air Force

Langley AFB Library. Fl. 4800, Langley AFB, Va., 23665-5725. 804-764-2906.

U.S. Army

Army Transportation and Aviation Logistics Schools Information Center Library. Washington Blvd., Bldg. 705, Ft. Eustis, Va., 23604-5450. 804-878-5563.

Department of Defense

Armed Forces Staff College Library. 7800 Hampton Blvd., Newport News, Va., 23511-6097. 804-444-5155.

U.S. Marine Corps

Marine Corps Aviation Museum Library. USMC Base, Quantico, Va., 22134. 703-640-2606.

U.S. Navy

Norfolk NAS Library. Bldg. U-16, NAS Norfolk, Va., 23511-5117. 804-444-3583.

Washington

Civilian:

Boeing Company Technical Library. MS-74-60, Bx. 3707, Seattle, Wash., 98124-2207. 206-237-8314.

Museum of Flight Foundation Library. PNAHF 2325, Financial Center Bldg., Seattle, Wash., 98161. 206-622-3972.

University of Washington Library. FM-25, Seattle, Wash., 98195-0001. 206-543-9153.

Military:

U.S. Air Force

Fairchild AFB Library. Fl. 4620, Fairchild AFB, Wash., 99011-5000. 509-247-5556.

McChord AFB Library. Fl. 4470, McChord AFB, Wash., 98438. 206-984-3454.

U.S. Navy

Whidbey Island NAS Library. Code ADRL, NAS Whidbey Island, Wash., 98278-2200. 206-257-2702.

West Virginia

Civilian:

West Virginia University Library. Morgantown, W.V., 26506-6069. 304-293-4040.

Wyoming

Civilian:

University of Wyoming Library. Bx. 3334 University Station, Laremie, Wyo., 82071. 307-766-3279.

Military:

U.S. Air Force

Francis E. Warren AFB Library. Fl. 4613, Bldg. 213, Francis E. Warren AFB, Wyo., 82005-5225. 307-775-3416.

B. Published Reference Works

1. Bibliographies

WITH the world of aerospace about to enter its ninth decade, the control of literature relating to it has become a major struggle. The sources here offer significant assistance to those who would attempt to keep abreast of both current and historical writings. Readers should note the bibliographies contained in many of the books, scholarly papers, documents, and journal articles cited in other sections may be particularly useful as well.

Book/Document/Report Literature:

Aerospace Education Association of America. Aerospace Bibliography. Reston, Va., 1982. 140p.
Covers all aspects of **aerospace** science and aircraft, with émphasis on those materials suitable for use by teachers K-12.

Aimone, Alan C. Bibliography of Military History. USMA Library Bulletin no. 14B. West Point, N.Y.: U.S. Military Academy Library, 1982. 152p.

______.______. USMA Library Bulletin no. 14C. West Point, N.Y.: U.S. Military Academy Library, 1987. 167p.
Annotated; particularly helpful for reference sources, including aerospace-aircraft materials.

American Book Publishing Record. New York: R. R. Bowker, 1960. v. 1--. Monthly.
New books arranged by Dewey classification number, with author/title indexes.

American Historical Association. Recently Published Articles. Washington, D.C.: AHA, 1976--. v. 1--. Triannual.
Appearing in February, June, and October, this essential tool averages 18,000 citations per issue from 4,000 journals in 36 languages, although many of the popular aviation titles are not covered; limited by lack of indexes.

_____. Writings on American History. Washington, D.C.: GPO for AHA, 1921-1961; Milwood, N.Y.: Kraus Reprint for AHA, 1962--.
A major bibliographic tool not published in 1941-1946 with books dropped after 1962; surprisingly poor aviation coverage prior to the late 1940s. Annual except for a **1962-1973 cumulation.**

Atkins, Charles M. List of Academic Theses Since 1961 Related to the History of Aeronautics and Astronautics. NASA-HHN-61. Washington, D.C.: History Office, National Aeronautics and Space Administration, 1966. 11p.

This mimeographed tool is now dated, but is helpful for the years covered.

Booth, Robert E. Bibliography of Aerospace Bibliographies: A Keyword-in-Context Index. Detroit, Mich.: Gale Research Company, 1965. Unpaged.

Also dated; also useful.

British Books in Print: The Reference Catalogue of Current Literature. London, Eng.: J. Whitaker, 1879--. Annual.

Published simultaneously in the U.S. since 1967 by the New York firm of R. R. Bowker. As many of the authors of aerospace book titles are British and many of the book titles (especially the large "coffee-table" variety) available from mail order or second outlets are likewise, this is a good source for early publication data.

British Paperbacks in Print. London, Eng.: J. Whitaker, 1960--. Annual.

Especially helpful for the many small aerospace titles that reach America, if at all, later than publication year.

Brockett, Paul, comp. Bibliography of Aeronautics. Smithsonian Miscellaneous Collections, v. 55. Washington, D.C.: GPO for the Smithsonian Institution, 1910; Detroit, Mich.: Gale Research Co., 1966. 940p.

The basic source guide for materials published prior to 1909; Brockett continued his work as chief bibliographer for the NACA works cited below.

Champion, Brian. Advanced Weapons Systems: An Annotated Bibliography of the Cruise Missile, MX Missile, Laser and Space Weapons, and Stealth Technology. Garland Reference Library of Social Science, v. 285. New York: Garland, 1985. 100p.

Contains 647 chronologically-arranged entries.

Coletta, Paolo E. An Annotated Bibliography of United States Marine Corps History. Lanham, Md.: University Press of America, 1986. 417p.

_____. A Bibliography of American Naval History. Annapolis, Md.: U.S. Naval Institute Press, 1981. 453p.

Both largely unannotated, both cover aviation, and the naval guide was slightly updated by the Lanham, Maryland, firm in 1988.

Cumulative Book Index. New York: H. W. Wilson, 1898--. v. 1--. Monthly.

An international bibliography of English-language titles essential for warplane coverage; annual cumulation.

Dickson, Katherine M., ed. History of Aeronautics and Astronautics: A Preliminary Bibliography. NASA-HHR-29. Washington, D.C.: History Office, National Aeronautics and Space Administration, 1968. 420p.

Emory, John J. M. G. The Source Book of World War II Aircraft. New York: Sterling, 1987. 160p.
First published in Great Britain, this work is essentially a guide to the contents of large "coffee book" pictorials; as such, it is very helpful for entry into those kinds of sources.

Estep, Raymond. An Aerospace Bibliography, 1962. Maxwell AFB, Ala.: Documentary Research Division, Aerospace Studies Institute, Air University, 1962. 158p.
Covers air power and aircraft references from 1957 through 1961.

____. An Aerospace Bibliography, 1965. Maxwell AFB, Ala.: Documentary Research Division, Aerospace Studies Institute, Air University, 1965. 99p.
Citations to books and articles published 1962 through 1965.

____. An Air Power Bibliography. Maxwell AFB, Ala.: Documentary Research Division, Aerospace Studies Institute, Air University, 1956. 200p.
Citations to books and articles published 1950 through 1954.

____. An Air Power Bibliography, 1955-1956. Maxwell AFB, Ala.: Documentary Research Division, Aerospace Studies Institute, Air University, 1957. 273p.
Coverage for the years indicated.

Forthcoming Books. New York: R. R. Bowker, 1966--. v. 1--. Bimonthly.
This tool, with its companion Subject Guide to Forthcoming Books (1967--), will give the user advance information on new titles soon to be available in the U.S.

Gamble, William B. History of Aeronautics: A Select List of References to Material in the Library. New York: New York Public Library, 1938; Ayer Co., 1971. 325p.
Especially helpful for citations to pre-World War II items.

Hallion, Richard P. The Literature of Aeronautics, Astronautics, and Air Power. Washington, D.C.: GPO for the Office of Air Force History, 1984. 66p.
Brief but insightful bibliography/bibliographic essay.

Hanni all, August. Aircraft, Engines and Airmen: A Selective Review of the Periodical Literature, 1930-1969. Metuchen, N.J.: The Scarecrow Press, 1972. 849p.
Covers the period 1932-1969, with emphasis on those journals with wide distribution for non-technical readers. Exact titles to articles and authors are not given in a work arranged in sections (aircraft, aircraft engines, biography, and air forces) and then by name, e.g., aircraft type, fighter pilot ace, a process which makes interlibrary loan requests and other uses demanding exact citations all but impossible. Supplemented by the more exact NASM The Aerospace Periodical Index, 1973-1982 (below).

Higham, Robin, ed. A Guide to the Sources of United States Military History. Hamden, Conn.: Archon Books, 1975. 559p.

____. ___: Supplement I. Hamden, Conn.: Archon Books, 1981. 416p.

____.____: Supplement II. Hamden, Conn.: Archon Books, 1986. 332p.
The 1975 master guide contains critical bibliographical essays in 19 chapters covering books, archival materials, some articles and dissertations, and various primary sources. Supplements (scheduled for every five years) add to the value of the work; the 1981 issue contains 23 essays and the 1986 the same.

International Bibliography of Military History. Berne, Switzerland: International Commission for Military History, 1974--. v. 1--. Annual.
An annotated bibliography divided into sections on general, documentary, and bibliographic works; published, with indexes, in German, French, Russian, and English.

Miller, Samuel D. An Aerospace Bibliography. Washington, D.C.: GPO for the Office of Air Force History, 1979. 289p.
A revised and expanded updating of United States Air Force History: An Annotated Bibliography, compiled by Carl Berger and Mary A. Cresswell. The next edition, by this compiler, should be available in the early 1990s.

Neufeld, Jacob, Kenneth Schaffel, and Anne H. Shermer. Guide to Air Force Historical Literature. Washington, D.C.: GPO for the Office of Air Force History, 1985. 174p.
Reports generated by the USAF internally.

New York Public Library. Research Libraries. Subject Catalog of the World War I Collection. 4 vols. Boston, Mass.: G. K. Hall, 1961.
Contains 61,300 entries, including those on aircraft.

___.___. Subject Catalog of the World War II Collection. 3 vols. Boston, Mass.: G. K. Hall, 1977.
Cites in excess of 50,000 sources, including those on aircraft.

Noffsinger, James P. World War I Aviation Books in English: An Annotated Bibliography. Metuchen, N.J.: The Scarecrow Press, 1987. 331p.

Paperbound Books in Print. New York: R. R. Bowker, 1955. v. 1--. Annual.

Pisano, Dominick A. and Cathleen S. Lewis. Air and Space History: An Annotated Bibliography. Garland Reference Library in the Humanities, no. 834. New York: Garland Publishing, 1988. 571p.
Prepared by curators at the National Air and Space Museum, this work contains over 2,000 citations (some repeated) and is most valuable for its body of critical annotation.

Smith, Myron J., Jr. Air War Bibliography, 1939-1945: English-Language Sources. 5 vols. in 7 pts. Manhattan, Kans.: Military Affairs/Aerospace Historian, 1978-1982.
Contains 12,172 references, including those to aircraft.

_____. Air War Southeast Asia, 1961-1973: An Annotated Bibliography and 16mm Film Guide. Metuchen, N.J.: The Scarecrow Press, 1979. 298p.
Citations to books, articles, documents, and studies.

_____. The Airline Bibliography: The Salem College Guide to Sources on Commercial Aviation. 2 vols. West Cornwall, Conn.: Locust Hill Press, 1986-1988.
Vol. II is particularly helpful for citations to the "airliners," including those which did or would serve in military capacities.

_____. The American Navy, 1865-1918: A Bibliography. American Naval Bibliography Series, v. 4. Metuchen, N.J.: The Scarecrow Press, 1974. 372p.
Includes references to early USN aircraft through World War I.

_____. The American Navy, 1918-1941: A Bibliography. American Naval Bibliography Series, v. 5. Metuchen, N.J.: The Scarecrow Press, 1974. 429p.
Includes citations to USMC and USCG aircraft as well as those of the USN.

_____. United States Navy and Coast Guard, 1945-1983: A Bibliography of English-Language Works and 16mm Films. Jefferson, N.C.: McFarland, 1984. 539p.
The final Smith USN bib covers aircraft of all the sea services.

_____. World War I in the Air: A Bibliography and Chronology. Metuchen, N.J.: The Scarecrow Press, 1977. 291p.
Unlike Noffsinger, above, includes periodical article citations.

_____. World War II at Sea: A Bibliography of Sources in English. 3 vols. Metuchen, N.J.: The Scarecrow Press, 1976.
Contains references to U.S. sea service aircraft, 1941-1946.

_____. World War II: The European and Mediterranean Theaters--An Annotated Bibliography. Wars of the United States, no. 2. New York: Garland Publishing, 1983. 450p.
Includes annotations with the aircraft entries.

Strobridge, Truman R. United States Coast Guard Annotated Bibliography. CG-230. Washington, D.C.: GPO for the Public Information Division, U.S. Coast Guard, 1972. 35p.
A 148-page update was made available in 1982.

Subject Guide to Books in Print. New York: R. R. Bowker, 1957--. v. 1--. Annual.
Index to the author/title volumes of Books in Print; not always helpful for works published by lesser-known U.S. firms.

United States. Air Force. Strategic Air Command. Directorate of Information Services, History Division. Bibliography of SAC Sources. Offutt AFB, Neb.: History Division, Directorate of Information Services, Strategic Air Command, 1959. 7p.

____. Army. Military Academy, Library. Subject Catalog of Selected Works Pertaining to the Military Arts and Sciences, With Selected Articles and Added Entries, Including a Preliminary Guide to the Ms. Collection. 4 vols. Westport, Conn.: Greenwood Press, 1970.

____. Library of Congress. Library of Congress Catalog, Books-Subjects: A Cumulative List of Works Represented by Library of Congress Printed Cards. Washington, D.C.: GPO, 1950-1983.

____. ____. The National Union Catalog: A Cumulative Author List. Washington, D.C.: GPO, 1956-1983.

____. ____. Pre-1956 Imprints. 754 vols. London, Eng.: Manswell, 1968-1981.

____. ____. The National Union Catalogs in Microfiche. Andover, Mass.: Advanced Library Systems, 1983--. Monthly/quarterly cumulations.
The complete NUC, 1898-1982, and Subject Catalog, 1950-1982, are also available from ALS; researchers using this guide may prefer to employ the hardback copies held by some larger libraries.

____. National Advisory Committee for Aeronautics. Bibliography of Aeronautics, 1909-1932. 14 vols. Washington, D.C.: GPO, 1921-1936.
A continuation of Brockett's earlier work (above); still an extremely important contribution for the years covered.

____. National Aeronautics and Space Administration. Aeronautical Engineering: A Continuing Bibliography, With Indexes. NASA-SP-7037. Washington, D.C.: GPO, 1970--. Monthly/annual index.
Citations from the STAR Reports (below) plus periodical articles.

____. ____. Scientific and Technical Aerospace Reports (STAR). Washington, D.C.: GPO, 1961--. Bimonthly.

____. Navy. Naval Air Systems Command. Technical Information Indexes: Naval Carrier Aviation. Washington, D.C.: GPO, 1974--. Monthly.

____. ____. Naval History Division. United States Naval History: A Bibliography. 6th ed. Washington, D.C.: GPO, 1972. 91p.
Publication of a new edition is imminent.

____. War Department. Office of the Chief of Air Service. Aeronautical Book and Magazine List, with References to Principal Documents and Papers of Historical Interest. Air Service Information Circular (Heavier-than-Air), v. 1, no. 21. Washington, D.C.: GPO, 1920. 45p.

Werrell, Kenneth P. 8th Air Force Bibliography: An Extended Essay and Listing of Published and Unpublished Materials. Manhattan, Kans.: Military Affairs/Aerospace Historian, 1981. 291p.
Includes aircraft of the force.

Wynar, Bohdan S., ed. American Reference Books Annual. Littleton, Colo.: Libraries Unlimited, 1970--. v. 1--. Annual.
Especially helpful for tracking new bibliographies.

Periodical/Anthology Literature:

Armstrong, William J. "United States Naval Aviation History: A Guide to the Sources." Aerospace Historian, XXVII (June 1980), 109-112.

Estep, Raymond. "Notes on Air Force Bibliography, with a List of Basic Reference Guides." Air University Quarterly Review, XI (Spring 1959), 113-120.

Gorn, Michael and Charles J. Gross. "Published Air Force History: Still on the Runway." Aerospace Historian, XXXI (March 1984), 30-37.

Hallion, Richard P. "Source Guide to the History of Aeronautics and Astronautics." American Studies International, XX (Spring 1982), 3-50.

Hansen, Chuck. "U.S. Military Aircraft and Equipment Research Sources." American Aviation Historical Society Journal, XXII (Fall 1977), 193-197.

Ottaviano, Doris B. "Modern Weapon Systems: A Selective Bibliography." Science and Technology Libraries, V (Summer 1985), 55-81.

Pisano, Dominick A. "40 Years of Jet Aviation: A Selective Bibliography." In: Walter J. Boyne and Donald S. Lopez, eds. The Jet Age: 40 Years of Jet Aviation. Washington, D.C.: Smithsonian Institution Press, 1979. pp. 185-190.

_____. "A Selective Bibliography and Research Guide." In: Walter J. Boyne and Donald S. Lopez, eds. Vertical Flight: The Saga of the Helicopter. Washington, D.C.: Smithsonian Institution Press, 1984. pp. 251-257.

Reynolds, Clark G. "Writing on Naval Flying." Aerospace Historian, XXXI (March 1984), 21-29.

2. Abstracts and Indexes

ABSTRACTING services provide the reader not only with publication citations, but with synopsis of material covered. In some instances, abstracts may also be labelled as bibliographies. Such, for example, would be NASA's Aeronautical Engineering: A Continuing Bibliography, cited above. Abstract users should employ this type of source material in conjunction with bibliographies and indexes. Indexes provide the key to current sources of information. However, as the monthly or quarterly paper indexes are cumulated into bound annual volumes, these become tools for historical research as well and in the same fashion as annual bibliographies or abstracts. The indexes examined here fall into two categories, newspaper and periodical. Both are vital for those who would stay abreast of the freshest reporting and both are becoming increasingly available in the CD-ROM computer-based format.

Abstracts of Military Bibliography. Buenos Aires, Argentina: Reuben A. Ramires Mitchell, 1967--. Quarterly.

Aerospace Defense Markets and Technology. Cleveland, Ohio: Predicasts, Inc., 1982--. Monthly/annual cumulation.
Provides in excess of 20,000 abstracts per year.

America: History and Life--a Guide to Periodical Literature. Santa Barbara, Calif.: ABC-CLIO, 1964--. v.1--. Quarterly/annual and 5-year cumulations.
Classified arrangement providing over 2,000 abstracts per year; also available on-line as a DIALOG database.

Applied Science and Technology Index. New York: H. W. Wilson, 1959--. Monthly/annual cumulation.
Formerly the Industrial Arts Index; covers many aerospace-related periodicals.

Biography Index: A Cumulative Index to Biographical Material in Books and Magazines. New York: H. W. Wilson, 1949--. Quarterly/annual cumulation.

Bogolubsky, Irene W., ed. International Aerospace Abstracts. New York: Technical Information Service, American Institute for Aeronautics and Astronautics, 1961--. Bimonthly.
Multi-language service covering all aspects of its topic; should be the user's first stop among abstracting services.

Botlorff, Robert M. Popular Periodical Index. Wayne, N.J., 1973--. Semiannual.

British Technology Index. London: Library Association Publishing, Ltd., 1889-1961.
Now Current Technology Index. An important source for citations in such well-known publications as The Aeroplane.

Burke, John G. The Access Index. Evanston, Ill., 1975--. 3/yr.
Known as Access, this tool is helpful for little magazines.

Business Periodicals Index. New York: H. W. Wilson, 1958--. Monthly/annual cumulation.
Very helpful for citations to the various aspects of aerospace manufacturing and marketing.

California News Index. Claremont, Calif.: Center for California Public Affairs, 1970--. v. 1--. Monthly.
Newspaper index.

Combined Retrospective Index to Journals in History, 1838-1974. 11 vols. Arlington, Va.: Carrollton Press, 1977.
Indexes 928 periodicals by keyword.

Cumulated Magazine Subject Index, 1907-1949. 2 vols. Boston, Mass.: G. K. Hall, 1964.
Helpful for covering periodicals not indexed 1941-1945 by Writings on American History (above).

Current Technology Index. London, Eng.: Library Association Publishing, 1962--. Monthly/annual cumulation.
Formerly British Technology Index (above); examines a number of British aerospace journals.

Dissertation Abstracts. Ann Arbor, Mich.: University Microfilms, 1945-1968.

Dissertation Abstracts International. Ann Arbor, Mich.: University Microfilms, 1969--. Monthly.

Engineering Index Monthly and Author Index. New York: Engineering Information, Inc., 1884--. Monthly.
Formerly Engineering Index.

General Science Index. New York: H. W. Wilson, 1978--. Monthly/annual cumulation.
Includes aerospace materials.

Index to U.S. Government Periodicals. Chicago, Ill.: Infordata International, 1975--. v. 1--. Quarterly/annual cumulation.
Indexes a number of aerospace/defense-related periodicals not covered in such tools as the Air University Library Index to Military Periodicals (below); examples include Naval Aviation News and NASA Report to Educators.

Industrial Arts Index. New York: H. W. Wilson, 1913-1958.
Superceded by Applied Science and Technology Index (above).

Masters Abstracts. Ann Arbor, Mich.: University Microfilms, 1962--. Monthly.

New York Times Index. New York: New York Times Co., 1851--. Monthly/quarterly/annual cumulation.

Old Line Index to Military and Defense-Related Periodicals. Rockville, Md.: Old Line Publishing Co., Inc., 1986--. Quarterly.
Covers approximately 300 journals.

Owen, J. I. H., ed. Current Military Literature: Comments and Abstracts of Important Articles from International, Strategic, and Defence Periodicals. Oxford, Eng.: Military Press, Ltd., 1982--. 6/yr.

Public Affairs Information Service. Bulletin. New York, 1914--. Monthly/annual cumulation.
Useful for breadth of coverage, which includes not only periodical articles, but entries from newspapers, documents, and books as well.

RAND Corporation. Selected RAND Abstracts. Santa Monica, Calif.: RAND Corporation, 1964--. Quarterly/annual cumulation.
Presents abstracts of the Air Force think tank's research reports, notes, and memorandum, including aerospace-related affairs.

The Readers' Guide to Periodical Literature, 1900--. New York: H. W. Wilson, 1905--. Monthly/annual cumulation.
The most common of U.S. periodical indexes; helpful for coverage of such popular periodicals as Aviation Week and Space Technology and Flying.

Smithsonian Institution Libraries. National Air and Space Museum Branch Library. Aerospace Periodical Index, 1973-1982. Smithsonian Institution Libraries Research Guide, no. 3. Boston, Mass.: G. K. Hall, 1984. 660p.
Supplements Hannibal's guide (above); now itself in need of updating.

Social Sciences and Humanities Index. New York: H. W. Wilson, 1916-1974.
Included coverage from major historical and political science journals until superceded by Hunanities Index and Social Sciences Index (below).

Social Sciences Index. New York: H. W. Wilson, 1972--. Monthly/annual cumulation.
Still quite helpful for political science aspects of our topic, e.g., B-1B bomber.

United States. Air Force. Air University. Library. Abstracts of Student Research Reports. Maxwell AFB, Ala., 1957--. Annual.
Reviews those papers prepared by students/faculty at Air University.

_____._____._____._____. Air University Library Index to Military Periodicals. Maxwell AFB, Ala., 1949--. Quarterly/annual cumulation.
The oldest and most important English-language military periodical index; arranged by subject, it annually indexes some 25,000 articles, news stories, book reviews, and editorials from approximately 70 journals.

_____. Army. Command and General Staff College. Abstracts of Master of Military Art and Science Theses and Special Studies, 1964-1976. Fort Leavenworth, Kans., 1977. Unpaged.
Now listed in the December issue of Military Review.

_____. Department of Commerce. National Technical Information Service. Government Reports--Announcements. Springfield, Va., 1946--. v. 1--. Biweekly/cumulative annual index.
Arranged by topic, this tool now lists almost all important defense studies from think tanks and government agencies, including papers from the services' post-graduate schools.

_____. Navy. Naval War College. Library. Summary Record of Research and Publication. Newport, R. I., 1977--. Annual.
Performs the same service as the AUL Abstracts (above).

United States Naval Institute. United States Naval Institute Proceedings Cumulative Index, 1874-1977. Annapolis, Md., 1982. 377p.
The easiest form of access to this important naval journal for those years prior to the beginning of the AUL Index to Military Periodicals (above).

The Wall Street Journal Index. New York: Dow Jones, 1958--. Monthly.

3. Book Review Media

REVIEWS of new books (and in some cases, periodicals) have been a staple of the publishing world for years. Many of these reviews are cited in such periodical indexes as The Reader's Guide to Periodical Literature and the AUL Index to Military Periodicals. Entire periodicals have been developed to just furnish reviews, two examples being the New York Review of Books and the American Library Association's Choice. The newsletters of such aviation associations as the American Aviation Historical Society provide thoughtful coverage while those from such specialized groups as the American Committee for the History of the Second World War often cover the more important titles as appropriate. Users should note that most monthly and almost all scholarly military journals provide reviews of varying quality; among the more helpful are those contained in the following: U.S. Naval Institute Proceedings; Air Classics; Air Force Magazine; Marine Corps Gazette; Military Review; Naval War College Review; Air Power History (formerly Aerospace Historian); Air Power Journal (formerly Air University Review) and The Journal of Military History (formerly Military Affairs). Major commercial review media include:

Book Review Digest. New York: H. W. Wilson, 1905--. Monthly/annual cumulation.
One of the smaller lists; includes excerpts from reviews.

Book Review Index. Detroit, Mich.: Gale Research Co., 1965--. Bimonthly with annual cumulation.
A 10-volume 1965-1984 cumulation is available; cites many more titles than Book Review Digest, but does not provide excerpts.

Combined Retrospective Index to Book Reviews in Scholarly Journals, 1886-1974. 10 vols. Arlington, Va.: Carrollton Press, 1973.

National Library Service Company. Cumulative Book Review Index, 1904-1974. 6 vols. Princeton, N.J., 1975.

4. Annuals/Weekly Services

ANNUALS and weekly services are designed to provide ready reference and overviews to current events, including aerospace. These works can vary in size and frequency from such weekly sources as Facts on File to almanacs like the World Almanac and yearbooks like the International Security Yearbook. Several military periodicals annually provide service updates (often called "almanacs") in specific issues; examples include those provided in the May issues of Air Force Magazine and U.S. Naval Institute Proceedings. A number of aerospace-oriented journals such as Aviation Week and Space Technology offer various special "almanacs" and updates in different issues throughout the year.

Aerospace Industries Association of America, Inc. Aerospace Facts and Figures. New York: McGraw-Hill, 1919--. Annual.
This organization was formerly known as the Aircraft Industries Association of America (1945-1959) and, before that, the Aeronautical Chamber of Commerce of America (1922-1945). The annual's title has varied.

_____. Aerospace Yearbook. Washington, D.C.: American Aviation Publishing, Inc., and various other publishers, 1923--. Annual.
Formerly called the Aircraft Year Book.

Armaments Year-Books: General and Statistical Information. Geneva, Switzerland: The League of Nations, 1924-1938.

Blechman, Barry M. and Edward N. Luttwak, eds. International Security Yearbook. New York: St. Martin's Press, 1984--. Annual.

Brassey's Annual: The Armed Forces' Yearbook. New York: Various publishers, 1886--. Annual.
This noted British source was renamed Brassey's Defence Yearbook in 1971.

Facts on File Weekly World News Digest with Cumulative Index. New York: Facts on File, Inc., 1941--. Weekly/annual index.

Facts on File Yearbook: The Indexed Record of World Events. New York: Facts on File, Inc., 1941--. Annual.
Annual workup of the weekly news digest.

International Institute for Strategic Studies. The Military Balance. London, Eng., 1958--. Annual.

Jane's All the World's Aircraft. London and New York: Jane's, 1909--. Annual.
The basic source for data on world aircraft, arranged by country and manufacturer; earlier issues should be consulted for aircraft no longer manufactured. Bimonthly supplements published in Air Force Magazine.

Taylor, Michael J. H., ed. Jane's Aviation Review. New York: Jane's, 1981--. Annual.

United States. Joint Chiefs of Staff. Statement by Chairman--to the Congress on the Defense Posture of the United States for Fiscal Year 19--. Washington, D.C.: GPO, 1977-1981.

_____._____. United States Military Posture for Fiscal Year 19--. Washington, D.C.: GPO, 1982--. Annual.
Continuation of the last entry.

_____. President. Aeronautics and Space Report of the President. Washington, D.C.: GPO, 1975--. Annual.

World Almanac and Book of Facts. New York: World Telegram, 1868-1966; Garden City, N.Y.: Doubleday, 1960--. Annual.

5. Encyclopedias/Handbooks/Directories

WHILE the encyclopedia is, perhaps, a more discursive source of background information while the handbook (generally a collection of miscellaneous facts built on a central theme) serves as a working tool, both are similar, so much so that it is often difficult to find the thin line which separates them. Users of this guide will find that encyclopedias might more often be employed for retrospective research while handbooks might be of greater value for ongoing guidance. Indeed, some of the general works cited below might, if retitled, qualify for inclusion in either category. Directories often share similar systematic arrangement with encyclopedias and handbooks, but their function is primarily to list organizations, firms, or persons. Very much current working tools, directories can be an important source for addresses, affiliations, etc., not provided by the other two kinds of works addressed here.

Bowell, Daniel. Encyclopedia of War Machines: An Historical Survey of the World's Great Weapons. London, Eng.: Octopus Books, 1977. 368p.

Chant, Christopher, et al. The Encyclopedia of Air Warfare. New York: T. Y. Crowell, 1970. 256p.

Copley, Gregory L. Defense & Foreign Affairs Handbook. Washington, D.C., 1977--. Annual.

Dean, Donald W., ed. World Aviation Directory. Washington, D.C.: Ziff-Davis, 1940--. Semiannual.

Didelot, J., ed. Interavia ABC 19--: Aerospace Directory. Geneva, Switzerland: Interavia S/A, 1952--. Annual.
A very important source.

Encyclopedia of Aviation. New York: Scribners, 1977. 218p.

Gunston, William T. ("Bill")., ed. The Encyclopedia of World Air Power. London, Eng.: Temple Press, 1986. 319p.

Hartline, David L. Military Dealers and Collectors Directory and Handbook. Worthington, Ohio: Haas Publications, 1981--. Annual.

Keegan, John. Rand McNally Encyclopedia of World War II. New York: Rand McNally, 1977. 256p.

The Marshall Cavendish Illustrated Encyclopedia of Aviation. 20 vols. Hicksville, N.Y.: Marshall Cavendish Corp., 1978.

Mondey, David C., ed. The International Encyclopedia of Aviation. New York: Crown, 1977. 480p.

Norris, Eric, et al. Weapons and Warfare of the 20th Century: A Comprehensive and Historical Survey of Modern Military Methods and Machines. 20 vols. London and New York: Octopus Books, 1976.

Taylor, Michael J. H. Jane's Encyclopedia of Aviation. 5 vols. London and New York: Jane's, 1980.

Watkins, Mary H., et al. Research Centers Directory. 10th ed. 2 vols. Detroit, Mich.: Gale Research Co., 1985.

6. Published Document/Manuscript Guides

SOME of the most important information for the study of aerospace affairs and vehicles is found in public documents and those corporate records of private concerns which may be open for research. Private papers and the materials of public figures together with the artifacts of aerospace can also be of great value. The guides cited here are to sources of public documents and private collections as reported in published sources. Some of these may be indexes, others finding aids. All are important for those engaged in new research.

Book/Document/Report Literature:

Allard, Dean C., Martha L. Crawley, and May W. Edmison. U.S. Naval History Sources in the United States. Washington, D.C.: GPO for the Naval Historical Center, 1979. 235p.

The Declassified Documents Quarterly Catalog. Washington, D.C.: Carrollton Press, 1976--. Quarterly.
Updates Wile (below).

Frank, Benis M. Marine Corps Oral History Collection Catalog. Rev. ed. Washington, D.C.: GPO for the Office of History and Museums, Headquarters, U.S. Marine Corps, 1979.

Hamer, Philip H. A Guide to Archives and Manuscripts in the United States. New Haven, Conn.: Yale University Press, 1961. 775p.

Morse, Richard E. and Thomas C. Lobenstein. Personal Papers in the Albert F. Simpson Historical Research Center. Maxwell AFB, Ala.: Albert F. Simpson Historical Research Center, 1980. 74p.

Paszek, Lawrence J. United States Air Force History: A Guide to Documentary Sources. Washington, D.C.: GPO for the Office of Air Force History, 1973. 225p.
Soon to be updated.

Sommers, Richard J. Manuscript Holdings of the United States Army Military History Institute. 3 vols. Carlisle Barracks, Pa.: U.S. Army Military History Institute, 1972.

Stillwell, Paul. U.S. Naval Institute Oral History Collection Catalog of Transcripts. Annapolis, Md.: U.S. Naval Institute, 1983. 65p.

Sunderman, James F. Documentary Collections Related to the U.S. Air Force: A Guide for Military Writers. Washington, D.C.: USAF Book Program, Office of Information, Office of the Secretary of the Air Force, 1961. 17p.

Taylor, Charles E. and Richard E. Spur. Aerial Photographs in the National Archives. Special List, no. 25. Washington, D.C.: National Archives and Records Service, 1973. 106p.

United States. Air Force. Air Force Historical Research Center. U.S. Air Force Oral History Catalog. Maxwell AFB, Ala.: U.S. Air Force Historical Research Center, 1982. 762p.

_____. Library of Congress. The National Union Catalog of Manuscript Collections, 1959-1969. Ann Arbor, Mich.: Edwards, 1969. 1,061p. Supplements issued after 1964.

_____. National Aeronautics and Space Administration. Index to NACA Technical Publications, 1915-1960. Washington, D.C.: GPO, 1949-1960.

_____. Superintendent of Documents. Monthly Catalog of United States Government Publications. Washington, D.C.: GPO, 1895--. Monthly.

Wile, Anadel, ed. Declassified Documents: Retrospective Collection. 4 vols. Arlington, Va.: Carrollton Press, 1976-1983.

Wooldridge, E. T., comp. Directory of Aerospace Resources: A Guide to Special Collections in College, University, Public, and Special Libraries, Museums, Historical Societies, and the National, State, and Local Archives of the United States. Smithsonian Institution Libraries Research Guides, no. 1. Washington, D.C.: Smithsonian Institution Press, 1984. 112p.
Perhaps the single most important citation in this section.

Periodical/Anthology Literature:

Armstrong, William J. "United States Naval Aviation History: A Guide to the Sources." Aerospace Historian, XXVII (June 1980), 109-112.

Bilstein, Roger E. "Sources in Aerospace History: The Oral History Collection at Columbia University." Aerospace Historian, XXII (March 1975), 46-47.

Casari, Robert B. "A Bibliography of Federal World War I Aviation Agencies and Their Records, 1917-1921." American Aviation Historical Society Journal, X (Spring 1965), 62-63.

Hasdorff, James C. "Sources in Aerospace History: The U.S. Air Force Oral History Collection." Aerospace Historian, XXII (June 1976), 103-104.

Maryanow, Maurice. "Oral History and the Vietnam War: The Air Force Experience." International Journal of Oral History, VII (June 1986), 125-132.

Sunderman, James F. "Documentary Collections Related to the U. S. Air Force: A Guide for Military Writers." Air University Quarterly Review, XII (Spring 1961), 110-126.

II. Fixed-Wing Aircraft

THE major tool of air warfare for over 70 years, fixed-wing air vehicles have become the subject of an enormous literature. This section, the largest in the guide, will examine certain of this material as it relates to specific kinds of warplanes. To this end, our section on fixed-wing aircraft is subdivided by airplane type with citations to individual aircraft models. The latter are ordered not by manufacturer, but by numerical designation. Although this arrangement is somewhat non-traditional/non-chronological, it offers the advantage of a single order and thus quick access. For the most part, fixed-wing operational and training variants are separated; however, details on specific electronic suits/maintenance packages/armaments specifically designed for a given aircraft are entered with that machine.

A. Fighter-Attack Aircraft

Unnumbered Pursuit/Attack

Breguet 14

"The Breguet and Salmson Biplanes." The Aeroplane, XV (October 23, 1918), 1489-1496, 1513.

Bruce, John M. and Jean Noll. Breguet 14. Aircraft in Profile, no. 157. Windsor, Eng.: Profile Publications, 1968. 15p.

Bristol Fighter

Bruce, John M., *et al*. "The American Bristol Fighter Program, World War I." *American Aviation Historical Society Journal*, XVII (Summer 1972), 85-89.

Whitehouse, Arthur G. J. ("Arch"). "The Fabulous Bristol Fighter." *Air Force Magazine*, LVII (October 1974), 64-69.

Nieuport Series

Andrews, C. F. *Nieuport 17*. Aircraft in Profile, no. 49. Windsor, Eng.: Profile Publications, 1966. 12p.

Bowers, Peter M. *Nieuport N.28C-1*. Aircraft in Profile, no. 79. Windsor, Eng.: Profile Publications, 1966. 12p.

_____. "Nieuport 28." *American Aviation Historical Society Journal*, X (Summer 1965), 108-109.

Bruce, John M. "Those Classic Nieuports." *Air Enthusiast Quarterly*, no. 2 (1976), 137-152.

Connors, John F. "Nieuport Chronicles." *Airpower*, III (September 1973), 24-31, 62-65.

Long, W. R. "35 Minutes in a Nieuport 17." *Interceptor*, X (November 1968), 19-21.

"Nieuport 17." *Scale Models*, XII (August 1981), 404-405.

"Nieuport 28." *Scale Models*, XII (October 1981), 532-536.

"The Nieuport 28 Described." *Cross and Cockade Journal*, III (Summer 1971), 117-128.

Salmson

"The Breguet and Salmson Biplanes." *The Aeroplane*, XV (October 23, 1918), 1489-1496, 1513.

"The Salmson Type 2 Military Aeroplane." *The Aeroplane*, XV (October 9, 1915), 1292-1298, 1315-1316.

S.E. 5

Bruce, John M. *S.E. 5*. Aircraft in Profile, no. 1. Windsor, Eng.: Profile Publications, 1965. 12p.

_____. *S.E. 5A*. Aircraft in Profile, no. 74. Windsor, Eng.: Profile Publications, 1967. 15p.

_____. "S.E. 5: Fighter Supreme." Aeroplane Monthly, V (May, July, September-November 1977), 264-269, 355-361, 493-498, 552-558, 608-611.

Connors, John F. "Hun Hunter: Britain's S. E. 5." Airpower, IV (March 1974), 19-24, 30-35.

Hardesty, Bergen. "S.E. 5A Tested in U.S.A." American Aviation Historical Society Journal, VIII (Spring 1963), 38-39.

Hoyt, Ross G. "Flying the Early Birds: The S. E. 5." Air Force Magazine, LIX (September 1976), 110-111.

Rogers, F. D. and E. C. Howes. "The S. E. 5 Story." Aircraft, XXXIII (January 1955), 30-31.

"The S. E. 5A." Air Classics, III (July 1966), 38-42.

"The S. E. 5A Pursuit Plane." Aviation, IV (June 1918), 598-600.

"The S. E. 5A Single-Seater Fighter." Automotive Industries, XXXIX (August 1918), 315-317.

Sopwith Camel

Bowyer, Chaz. Sopwith Camel: King of Combat. Falmouth, Eng.: Glasney, 1978. 192p.

Bruce, John M. Sopwith Camel. Aircraft in Profile, no. 31. Windsor, Eng.: Profile Publications, 1966. 13p.

Holder, William G. "Reborn Camel." TAC Attack, XV (March 1975), 14-15.
That at the USAF Museum.

"The Hun Killer." Air Classics, IV (June 1968), 54-66.

King, H. F. Sopwith Aircraft, 1912-1920. London, Eng.: Putnam, 1981. 322p.
Includes extensive Camel discussion.

Rogers, F. D. and E. C. Howes. "The Sopwith Camel." Aircraft, XXXIII (February 1955), 52-53.

Veale, Sydney E. "Sopwith Sequence." Aircraft Illustrated, VI (November 1973), 448-453.

SPAD Series

Book/Document/Report Literature:

Andrews, C. F. SPAD XIIIC-1. Aircraft in Profile, no. 17. Windsor, Eng.: Profile Publications, 1967. 12p.

Bruce, John M. SPAD Scouts S-VIII to S-XIII. Aircam Aviation Series, no. 9. New York: Arco, 1969. 50p.

Periodical/Anthology Literature:

Bruce, John M. "The SPAD Story." Air International, X (May-June 1976), 237-238, 289-296, 310.

___. "SPAD VII." Air Pictorial, XVIII (July 1956), 231; XXVIII (March-April 1966), 100-103, 144-146.

"The Coloring of the Smithsonian SPAD XIII." Cross and Cockade Journal, II (Winter 1961), 355-358.

Irving, Jimmy. "SPAD 12½: A Tale of Confusion (1918)." TAC Attack, XIII (December 1973), 16-18.

Richardson, Roland W. "SPAD XIII." Aerospace Historian, XXVII (December 1980), 257-261.
Reprinted in Robin Higham and Carol Williams, eds., Flying Combat Aircraft of the USAAF-USAF (Manhattan, Kans.: Sunflower University Press, 1981), pp. 105-109.

Robinson, Neil. "SPADs on the Western Front." Scale Models, XI (May 1980), 249-251.

Rogers, F. D. and E. C. Howes. "The SPAD VII." Aircraft, XXXIV (September 1956), 26-27.

"SPAD S-VII: Original Bleriot Aeronautique General Arrangement Drawings." Cross and Cockade Journal, XV (Winter 1974), 376-380.

"The SPAD Two-Seater Biplane." Automotive Industries, XXXVIII (June 20, 1918), 1212-1214.
Reprinted from Aviation, IV (June 15, 1918), 674-675.

Trimble, Robert. "The Curious Case of the Carbon-Copy Fighters." Air Classics, XI (June 1975), 38-48.

Thomas-Morse Series

Book/Document/Report Literature:

Strnad, Frank. Thomas-Morse Scouts. Aircraft in Profile, no. 68. Windsor, Eng.: Profile Publications, 1966. 12p.

United States. Army Air Service. Performance Test of Thomas-Morse MB-3 With 300 hp Hispano-Suiza Engine. Air Service Information Circular (Heavier-than-Air), v. 1, no. 40. Washington, D.C.: GPO, 1920. 8p.

Periodical/Anthology Literature:

Andrews, Hal. "Naval Aircraft: Thomas-Morse MB-3." Naval Aviation News, (May-June 1987), 18-19.

Begge, A. H. **"Thomas-Morse Mail Plane."** Sibley Journal, XXXIV (April 1920), 44-45.
The MB-4.

Bowers, Peter M. "First Pursuit: The Life and Generally Hard Times of the MB-3." Wings, VI (December 1976), 8-23.

Kucera, J. M. "The Original Tommy Morse." Aeroplane Monthly, V (January 1977), 41-42.

Strnad, Frank. "The Counterfeit Camel: Thomas-Morse Scouts in the Movies." American Aviation Historical Society Journal, XIV (Summer 1969), 83-84.

_____. "Thomas-Morse's Forgotten Fighters." Air Classics, VII (December 1970), 44-61.

"The Thomas-Morse MB-4 Mail Plane." Aviation, VIII (March 1, 1920), 99-100.

"The Thomas-Morse Scout 4-B." Air Progress, XIX (August-September 1963), 52-57.

"Thomas-Morse Single- and Two-Seater Airplanes." Aviation, VI (April 1919), 278-280.

Numbered Fighter-Attack

(Douglas) A-1 Skyraider

Book/Document/Report Literature:

Birdsall, Steve. The A-1 Skyraider. Famous Aircraft Series. New York: Arco, 1970. 64p.

Donovan, James A. The A-1 Aircraft in Southeast Asia. Research Report. Maxwell AFB, Ala.: Air War College, Air University, 1971. 19p.

Dorr, Robert F. Skyraider. Illustrated History of the Vietnam War, no. 17. New York: Bantam Books, 1988. 160p.

Gann, Harry. The Douglas Skyraider. Aircraft in Profile, no. 60. Windsor, Eng.: Profile Publications, 1971. 12p.

Jackson, B. R. Douglas (A-1) Skyraider. Fallbrook, Calif.: Aero Publishers, 1969. 144p.

Rausa, Rosario. Skyraider: The Douglas A-1 "Flying Dump Truck." Annapolis, Md.: Nautical and Aviation Publishing Company of America, 1982., 1987. 239p.

Sullivan, Jim. Skyraider in Action. Aircraft, no. 60. Carrollton, Tex.: Squadron/Signal Publications, 1983. 50p.

Periodical/Anthology Literature:

Andrews, Hal. "Naval Aircraft: AD Skyraider." *Naval Aviation News*, (January 1972), 20-21.

"Another Chapter in the AD Story." *Naval Aviation News*, (July 1957), 18-19.

Bolt, J. F. "Goodbye, Able Dog." *Marine Corps Gazette*, XLI (October 1957), 28-31.

Close, C. C. "Broadside From the Sky." *Skyways*, I (April 1951), 14-15, 42.

____. "The Douglas Skyraider: The Navy's Pedigreed Pulverizer." *Flying*, XLVI (April 1950), 18-19+.

"The Douglas Skyraider." *Aero Digest*, LXIII (November 1951), 26-32+.

Franz, Bob. "A-1 SPAD Goes Way of Sopwith Camel." *Navy*, XII (December 1969), 32-33.

Hail, B. C. "Friendly Versatility." *Naval Aviation News*, (November 1957), 14-15.

Harris, Hap. "Prop Pushers of Bien Hoa." *Airman*, X (January 1965), 34-37.

Heinemann, Edward H. "Skyraider's High Performance Stems From Pinpoint Designing." *Aviation*, XLV (April-May 1946), 55-58, 77-79.

Jones, Donald E. "A-1 Skyraider." In: Robin Higham and Carol Williams, eds. *Flying Combat Aircraft of the USAAF-USAF.* Ames, Iowa: Iowa State University Press, 1978. pp. 3-10.

Klym, Julie O. "The Flying Monument." *Air Classics*, XVII (May 1981), 34-41, 70-74.

McFall, A. Dodge. "Farewell to SPADs." *U.S. Naval Institute Proceedings*, XCI (April 1965), 54-59.

"Operations From a Carrier." *Skyways*, XI (July 1952), 14-15, 41-43.

Pace, Steve. "Skyraider." *Wings*, XVIII (October 1988), 22-51.

Price, David H. "Skyraider Support." *U.S. Army Aviation Digest*, XII (January 1966), 8-11.

Reinburg, J. H. "Trials of a Close Air Support Aircraft." *Army*, XVI (March 1966), 67-68.

Remmel, Carl G. "Able Dog." *Wings*, VIII (October 1978), 46-57.

"Skyraiders." *Joint Services Recognition Journal*, X (May 1955), 132-134.

"Soliloquy." *Naval Aviation News*, (December 1978), 14-19.

Trimble, Robert L. "Skyraider." Air Classics, X (December 1974), 28-37, 56-61; XI (January-February 1975), 40-53, 74-75, 32-37.

"The Venerable 'Able Dog.'" RAF Flying Review, XVIII (December 1962), 18-21.

(Douglas) A-2 Skyshark

Close, G. C. "Turboprop Skyshark." Skyways, IX (October 1950), 10-11.

"Douglas A2D-1 Skyshark." Air Pictorial, XLIV (September 1952), 271-272.

Markgraf, Garry. "Skyshark, son of SPAD." American Aviation Historical Society Journal, XXII (Fall 1977), 200-215.

Self, Thomas M. "Douglas Skyshark Makes First Flight." Aviation Week, LII (June 12, 1950), 13-14.

Trimble, Robert L. "Shark of the Sky." Air Progress Aviation Review, VI (Summer 1982), 6-17.

(Curtiss) A-3 Falcon

"The Curtiss Attack Plane: A-3 Falcon." Aviation, XXIII (July 25, 1927), 200-202.

"Curtiss Attack Plane: Falcon." Aviation, XXIII (December 12, 1927), 1404-1405.

Westburg, Peter. "The Unforgettable Curtiss Falcon." Air Classics, XI (November 1975), 66-81.

(Douglas) A-3 Skywarrior

Book/Document/Report Literature:

Francillon, Rene J. and Edward H. Heinemann. Douglas A-3 Skywarrior. Aerograph, no. 5. Arlington, Tex.: Aerofax, Inc., 1987. 136p.

Periodical/Anthology Literature:

Anderson, J. L. "AD-3 Skywarrior on the Production Line." Automotive Industries, CXVII (August 15, 1957), 64-66.

Anderton, David A. "How Douglas Designed the Skywarrior." Aviation Week, LII (December 13, 1954), 28-30.

Andrews, Hal. "Naval Aircraft: A-3 Skywarrior." Naval Aviation News, (August 1973), 20-21; (October 1982), 24-25.

"The Douglas Skywarrior." Joint Services Recognition Journal, X (April 1955), 94-95.

"Douglas Skywarrior." RAF Flying Review, XI (July 1956), 48-49.

"For Naval Attack: An Analysis of the Douglas A3D Skywarrior of the U.S. Navy." Flight, LXXI (February 15, 1957), 211-214.

Heinemann, Edward H. "A Whale of an Airplane." Naval Aviation News, (November-December 1987), 18-22.

Huff, W. H. "A Prop Pilot Looks at the A3D." Naval Aviation News, (August 1957), 12-14.

Jacobs, Roger P. "Tale of a Whale." Naval Aviation News, (October 1982), 18-23.

"The Mighty Skywarrior." RAF Flying Review, XII (June 1957), 24-27.

Nichols, H. A. "Design Features of the Douglas A3D Skywarrior." Society of Automotive Engineers Journal, LXIII (December 1955), 23-26.

Paul, Lionel. "Skywarrior: Salute to a 20-Ton Whale." Air Classics, VIII (November 1972), 56-63.

Polmar, Norman. "Thar She Blows." U.S. Naval Institute Proceedings, CX (December 1984), 77-89.

Schrader, Richard K. "Skywarrior." Air Classics, XXIV (January 1988), 14-23, 66-67.

"Skywarrior." Flight, LXVII (February 18, 1955), 205-208.

Stone, Irving. "Skywarrior Customed-Designed for Attack." Aviation Week and Space Technology, LXV (August 20, 1956), 66-68+.

(McDonnell-Douglas) A-4 Skyhawk

Book/Document/Report Literature:

Chesneau, Roger. Skyhawk: McDonnell-Douglas A-4M. Aeroguides, no. 14. Vista, Calif.: Aerolus, 1986. 36p.

Gann, Harry K. The Douglas A-4 Skyhawk. Aircraft in Profile, no. 102. Windsor, Eng.: Profile Publications, 1965. 12p.

Kilduff, Peter. Douglas A-4 Skyhawk. Osceola, Wis.: Motorbooks International, 1983. 200p.

Ward, Richard and Ernest R. McDowell. McDonnell-Douglas A-4A/L Skyhawk in USN-USMC-RAN-RNZAF Service. Aircraft Series, no. 27. New York: Arco, 1972. 62p.

Periodical/Anthology Literature:

"A4D Skyhawk." Flight, LXXIII (March 7, 1958), 308-311.

"A-4 Spoilers." Approach, XXI (May 1976), 24-25.

Andrews, Hal. "Naval Aircraft: A-4 Skyhawk." Naval Aviation News, (August 1971), 20-21.

"Bantam Bomber." RAF Flying Review, XI (March 1956), 26-29.

Beach, John. "Bargain-Basement Fighters." Defense Electronics, XII (June 1980), 59-60+.
A-4 retrofits.

Bearden, Bill. "The Last A-4 (#2960)." Naval Aviation News, (May 1979), 8-17.

Brady, James T. "Skyhawks." American Aviation Historical Society Journal, XV (Winter 1970), 217-221.

Butz, A. S., Jr. "Skyhawk: A Proud Past and a Promising Future." Air Force and Space Digest, XLVIII (November 1965), 42-48.

Coopman, Rob. "Transoceanic Ordeal." Combat Crew, XXXVII (June 1987), 10-13.

"Disconnected Skyhawks." Approach, XXII (February 1977), 1-4.

"The Douglas A-4 Skyhawk." Naval Aviation News, (March 1968), 6-13.

Foxworth, Thomas G. "A4D Skyhawk." Aero Album, V (February 1972), 40-43.

Freeman, Edgar. "The Douglas A-4 Skyhawk." Air Classics, VIII (April 1972), 26-35.

Gumble, Bruce. "New Life for a Vintage Fleet." Defense Electronics, XIX (April 1987), 74+.

"Hotting Up the Douglas Hot-Rod." Flying Review International, XXI (March 1966), 389-392.

Key, John H. "More CAS [Close Air Support] for Your Money." Marine Corps Gazette, LVIII (April 1974), 39-40.

"The Last of the Skyhawks." Air Classics, XV (July 1979), 74-79.

Liang, Gerald. "'Scooter' is 23." Air Classics, XIII (December 1977), 50-51.

"The McDonnell-Douglas A-4M: The New Skyhawk." Interavia, XXV (December 1970), 1483-1485.

"More Power for the Skyhawk." Interavia, XXI (April 1966), 497-498.

Rausa, Rosario. "From Rivets to Runway: Building the A-4." Naval Aviation News, (October 1977), 8-17.

Sapp, C. N., Jr. "Skyhawks and Wet Runways." Approach, XX (March 1975), 6-7.

Saw, David. "Something Old, Something New--Aircraft Retrofit." Military Technology, XII (March 1988), 26-28+.

Schaefer, Will F. "A-4 Skyhawk Retrofit and Modernization Extend Service Life." Asian Defence Journal, no. 5 (May 1986), 60+.

Smith, Gene. "Born Again Skyhawk." Air Progress, XLII (January 1980), 48-53.

Stamble, Irwin. "A-4D Simplified Design Proves Out in Navy Tests." Aviation Age, XXVI (December 1956), 26-31.

Stone, Irving. "A-4D Design Opposes Complexity Trend." Aviation Week and Space Technology, LXVII (August 26, 1957), 50-51+.

"Two to Go." Approach, XVI (July 1970), 26-32.

"The U.S. Navy's Mighty Midget." RAF Flying Review, XVII (June 1962), 28-31.

"The Versatile Skyhawk." Aircraft Illustrated, VI (October 1973), 400-401.

Warwick, Grahame. "Skyhawk's Quarter Century." Flight International, CXV (April 14, 1979), 1050-1056.

Wilson, Gill R. "The Skyhawk." Flying, LVIII (April 1956), 36-37.

Wood, R. H. "Heinemann's Bantam Bomber." Aviation Week, LX (June 14, 1954), 106-107.

(North American) A3J/A-5 Vigilante

"A3J Vigilante: A Supersonic Carrier-Based Bomber of the U.S. Navy." Flight, LXXIII (May 30, 1958), 731-732.

Andrews, Hal. "Naval Aircraft: A-5 Vigilante." Naval Aviation News, (October 1973), 20-21.

Coughlin, William J. "A-5D Converts From Bomber to Transport." Aviation Week, LIX (August 24, 1953), 24-26.

"Death at the Ramp." Approach, XXI (February 1976), 26-27.

"Fleet Introduction of the A3J." Naval Aviation News, (November 1961), 19-22.

Golding, D. M. "Painting the A-5 Vigilante Attack Plane." Industrial Finishing, XXXIX (June 1963), 24-26+.

Grove, Michael. "Vigilante--Beauty, Beast, or Both?" The Hook, VIII (Winter 1980), 19-27.

Gunston, William T. ("Bill"). "Vigilante: Tactical Strike and Reconnaissance." Flight International, LXXXVI (August 13, 1964), 251-257.

Heath, Leroy A. "The A3J-1 Vigilante: High or Low, Fast or Slow." U.S. Naval Institute Proceedings, LXXXVII (September 1961), 170-172.

Heinemann, Edward H. "The World's Most Versatile Airplane, A-5D." Aero Digest, LXVIII (April 1954), 50-52+.

Herron, Edward A. "Suddenly, the World Below." Skyline, XXII (Fall 1964), 30-41.

Hillbrink, Will. "Desert Test." Skyline, XVIII (Spring 1960), 18-25.

_____. "Vigilante Attack." Skyline, XVII (Spring 1959), 40-43.

_____. "The Vigilantes are Coming." Skyline, XIX (Fall 1961), 44-45.

Mersky, Peter B. "Vigilante Fadeout." U.S. Naval Institute Proceedings, CVI (June 1980), 63-69.

Saw, David. "Something Old, Something New--Aircraft Retrofits." Military Technology, XII (March 1988), 26-28+.

Siuru, William D. "Vigilante: Farewell to the Fleet's Last Strategic Bomber." Airpower, XI (January 1981), 12-23.

"Versatile Vigilante: A Sophisticated Supersonic U.S. Shipboard Attack Aircraft." RAF Flying Review, XXVII (August 1962), 32-35.

"The Vigorous Vigilante." RAF Flying Review, XV (February 1960), 16-17.

(Grumman) A-6 Intruder

Book/Document/Report Literature:

Chesneau, Roger. Intruder. Aeroguides, no. 15. Vista, Calif.: Aerolus, 1986. 36p.

Dorr, Robert F. Grumman A-6 Intruder. Osprey Air Combat Series. Osceola, Wis.: Motorbooks International, 1987. 200p.

Drendel, Lou. A-6 Intruder in Action. Warren, Mich.: Squadron/Signal Publications 1975. 50p.

Kinzey, Bert. A-6 Intruder in Detail and Scale: Part I, Bomber and Tanker Versions. Detail and Scale Series, no. 24. Blue Ridge Summit, Pa.: TAB, 1987. 64p.

_____ and Ray Leader. A-6 Intruder. Color and Markings Series, v. 5. Blue Ridge Summit, Pa.: TAB, 1987. 64p.

Miska, Kurt H. Grumman A-6A/E Intruder; EA-6A; EA-6B Prowler. Aircraft in Profile, no. 252. Windsor, Eng.: Profile Publications, 1973. 25p.

Thornborough, Anthony M. and Peter E. Davies. Grumman A-6 Intruder/Prowler. Modern Combat Aircraft, no. 26. New York: Scribners, 1987. 112p.

Periodical/Anthology Literature:

Anderton, David A. "Master of Night Attack." Grumman Aerospace Horizon, XXII (Fall 1986), 2-9.

Andrews, Hal. "Naval Aircraft: A-6 Intruder." Naval Aviation News, (October 1971), 20-21.

Andrews, Walter. "New Life For an Old Bird." Armed Forces Journal, CVI (February 8, 1969), 14-15.

Bateman, Kent C. "All-Weather CAS: Fact or Fancy?" Marine Corps Gazette, LVI (June 1972), 41-43.

Brown, David A. "Navy to Deploy A-6A Squadron Next Year." Aviation Week and Space Technology, LXXXI (August 17, 1964), 40-47+.

Brownlow, Cecil. "A-6 Changes Planned to Boost Reliability." Aviation Week and Space Technology, LXXXIX (November 4, 1968), 101+.

Christmann, Timothy J. "The A-6F." Naval Aviation News, (September-October 1986), 22-25.

Dubac, Carl H. "A-6A/RABFAC." Naval Aviation News, (May 1972), 18-19.
RABFAC = Radar Beacon, Forward Air Controller.

Emmons, Randy. "Intruders in the Desert." Naval Aviation News, (April 1977), 16-19.

Greeley, Brendan M., Jr. "Congressional Clash Threatens A-6F, A-6E Compromise Effort." Aviation Week and Space Technology, CXXVIII (January 11, 1988), 18-19.

"The Grumman A-6 Intruder/EA-6B Prowler." Defence Update International, no. 48 (1984), 49-56.

Hays, R. J. "The A-6A Intruder in Vietnam." U.S. Naval Institute Proceedings, XCIII (January 1967), 120-123.

"Interdiction: All Conditions--Grumman's Flexible Intruder." Flying Review International, XXIII (July 1968), 367-369.

"Intruders and Prowlers." Air International, XXXI (1986), 227+.

Johnson, Scott C. "Grumman's Goliath." Aeroplane Monthly, VII (July 1979), 367-370.

Kandebo, Steven W. "Grumman A-6F Will Include Engine and Avionics Upgrades." Aviation Week and Space Technology, CXXIV (January 20, 1986), 104-105.

"Low-Level Intruder." RAF Flying Review, XVII (April 1962), 34-35.

Lynch, Paul K. "The Dream Machine: A-6 Intruder." Navigator, XXVIII (Summer 1981), 28-31.

Miska, Kurt H. "An Introduction to the Intruder." American Aviation Historical Society Journal, XIV (Summer 1969), 112-114.

_____. and Steve J. Kiss. "USN and USMC Intruder Markings." American Aviation Historical Society Journal, XVI (Summer 1971), 79-85.

Mulquin, James J. "The Amazing Intruder Aircraft Proves Itself in Combat." Naval Aviation News, (June 1967), 6+.

"Night Warfare A-6s." Armed Forces Journal, CVI (February 8, 1969), 14-15.

North, David M. "Naval Air Reserve Will Add A-6Es as Part of Modernization Program." Aviation Week and Space Technology, CXXVI (March 2, 1987), 64-65+.

_____. "Navy Proposes Ending A-6E/F Production; Congressional Approval of Plan in Doubt." Aviation Week and Space Technology, CXXVIII (January 25, 1988), 22-23.

Peacock, Lindsay. "Grumman A-6 Intruder." Aeroplane Monthly, IV (November 1976), 579-585, 611.

Rodems, J. D. "A-6E TRAM." Military Electronics/Countermeasures, III (October 1977), 15-18.
TRAM = Target Recognition Attack Multisensor.

Rosenblatt, A. "Electronics Rejuvenate Aging A-6." Electronics, XLV (May 22, 1972), 71-72.

"Target Recognition Attack Multisensor Makes A-6E World's Most Advanced EO Aircraft." Electronic Warfare, XXII (September 1976), 16-17.

(Vought) A-7 Corsair II

Book/Document/Report Literature:

Bear, James T. The Testing of the A-7D. Edwards AFB, Calif.: Air Force Flight Test Center, Air Force Systems Command, 1973. 185p.

Dorr, Robert F. Vought A-7 Corsair II. Osprey Air Combat Series. Osceola, Wis.: Motorbooks International, 1985. 200p.

Drendel, Lou. The A-7 Corsair II in Action. Warren, Mich.: Squadron/Signal Publications, 1976. 50p.

Fiorello, Marco. Estimating Life-Cycle Costs: A Case Study of the A-7D. RAND Report R-1518-PR. Santa Monica, Calif.: RAND Corporation, 1975. 71p.

Head, Richard G. Decision-Making in the A-7 Attack Airplane Program. PhD. Dissertation. Syracuse, N.Y.: Syracuse University, 1971.
Previously prepared in 1970 as a 636-page document for the School of Systems and Logistics, Air Force Institute of Technology.

Kinzey, Bert. A-7 Corsair II in Detail and Scale. Detail and Scale Series, no. 22. Rev. and expanded. Fallbrook, Calif.: Aero Publishers, 1986. 72p.

_____. A-7E Corsair II: U.S. Navy Atlantic Coast Post-Vietnam Markings. Color and Markings Series, v. 9. Blue Ridge Summit, Pa.: TAB, 1987. 64p.

Nelson, J. R., et al. A Weapon System Life-Cycle Overview: The A-7D Experience. RAND Report R-1452-PR. Santa Monica, Calif.: RAND Corporation, 1974. 47p.

Periodical/Anthology Literature:

"A-7 Barricade Arrestments." Approach, XIX (March 1974), 16-17.

"A-7D: A Good, Honest Bird." Aerospace Safety, XXVI (August 1970), 20-24.

"A-7D Corsair II." TAC Attack, XXIII (December 1983), 16-17.

"A-7s Deploy to United Kingdom in European Tactical Exercise." Aviation Week and Space Technology, CXXIII (July 8, 1985), 55+.

Ameel, F. D. "Corsair II Gets a Lift." Approach, XXVI (November 1980), 18-21.

Andrews, Hal. "Naval Aircraft: Corsair II." Naval Aviation News, (May 1970), 20-21.

Barrineau, William E. "SLUF--To Improve Its Accuracy." USAF Fighter Weapons Review, (Summer 1972), 1-6+.

Barringer, L. E. and Ray Thomas. "The Navy Flies the SLUF, Too." USAF Fighter Weapons Review, (Winter 1972), 28-30+.

Basel, G. I. "Flying the SLUF Bird." USAF Fighter Weapons Review, (Spring 1972), 4-7.

Bauernfind, W., et al. "Designing Strike Aircraft: The A-7." Astronautics and Aeronautics, V (June 1967), 30-37.

Bentley, John. "The Corsair Concept." Flight International, XCII (September 21, 1967), 495-497.

Braybrook, Roy M. "Born-Again Combat Aircraft." Pacific Defense Reporter, XI (May 1985), 43-44+.

____. "New Equipment for the 'Mud Movers.'" Pacific Defense Reporter, XIV (November 1987), 47-48.

Bulban, Eric J. "Growing Export Market Seen for A-7s." Aviation Week and Space Technology, CXV (September 28, 1981), 65-67.

Butz, J. S., Jr. "A-7 Development: A Status Report." Air Force and Space Digest International, II (March 1966), 34-35.

Carson, Douglas M. "A-7." Flying Safety, XXXIX (April 1983), 8-9; XL (August 1984), 6-8; XLI (March 1985), 5-8; XLII (March 1986), 3-5.

Cole, Dave. "War Paint on a Hornet Bird." Airman, XV (September 1971), 3-8.

"A Corsair by Any Other Name--Sandy, SLUF, and Little Hummer." Air International, XXII (April 1982), 169-176, 203.

"Corsair II." Flying Review International, XXI (January 1966), 265+.

"The Corsair II Programme Accelerates." Interavia, XXI (January 1966), 89-91.

"The Corsairs are Coming." Flying Review International, XXIII (January 1969), 7-10.

Coyle, J. J. "A-7A Maintainability Program." U.S. Naval Institute Proceedings, XCI (May 1965), 142-145.

Davis, Harold E. "The A-7D Corsair II: An Old Airplane in a New Package." Army, XX (July 1970), 20-21.

English, Malcolm. "Corsair II." Scale Models, X (November 1979), 747-748.

Felker, Randolph B. "The A-7 Challenge." Air Reservist, XXV (December 1973-January 1974), 8-9.

Frisbee, John L. "How the A-7D Rewrote the Book in Southeast Asia." Air Force Magazine, LVI (August 1973), 30-36.

Fuchs, Ron. "Guarding the Canal." Airman, XXVIII (August 1984), 17-21.

Gaines, Mike. "Corsair With More: A-7 Plus." Flight International, CXXXII (August 22, 1987), 21-24.

Geddes, J. Philip. "Avionics and the A-7." Interavia, XXIX (October 1974), 1015-1018.

Geisenheyner, Stefan. "A-7: The Super-Accurate 'Bomb Truck.'" Asian Defence Journal, no. 1 (January 1982), 52-57.

George, J. A. "Heads Up in the Cockpit." Airman, XIII (February 1969), 27-29.

Greene, Smokey. "SLUF Survival: A-7D Bomb Load in a High-Threat Arena." USAF Fighter Weapons Review, (Fall 1975), 14-17+.

Greene, Wiley A. "Hot Pilots." Flying Safety, XXXIX (July 1983), 2-4.

Hanley, A. "A-7 Upgrade for Close Air Support." Aerospace America, XXV (January 1987), 8-11.

Head, Richard G. "The Air Force A-7 Decision: The Politics of Close Air Support." Aerospace Historian, XXI (December 1974), 218-223.

Hembrough, Mike. "A-7 Baro Bombing." USAF Fighter Weapons Review, (Fall 1974), 8-12.

Hill, James C. "The Corsair II: 'As I See It.'" U.S. Naval Institute Proceedings, XCIV (November 1968), 38-42.

Huertas, Salvador M. "Corsair From the Lone Star State." Defence Update International, no. 75 (September 1986), 13-23.

Hutcheson, K. C. "So How Do You Like the A-7 So Far?" TAC Attack, XXIV (November 1984), 18-19.

Kranish, Arthur. "The VAL Program." Western Aerospace, XLIV (March 1964), 3-4.

Lawson, Bob. "Light Attack and the A-7: They Ain't Dead and They Ain't Dyin'." The Hook, XIII (Summer 1985), 57-67.

Leonberger, Loren B. "Season of the Shimmering SLUF." Airman, XVII (June 1973), 24-28.

Letzer, George J. "Corsair II De-Tails." American Aviation Historical Society Journal, XVII (Winter 1972), 244-254.

McClarren, Charles. "A-7 Update." USAF Fighter Weapons News Letter, no. 1 (March 1969), 2-9.

March, D. F. "Full Systems at Launch: A New Concept in Fleet Aviation Operations." U.S. Naval Institute Proceedings, C (January 1974), 34-41.

Mayhew, David L. "The Big D." Airman, XIV (March 1970), 4-8.

Mersky, Peter. "Corsair Relight." Approach, XXXIII (March 1988), 48-49.

Miller, John F., Jr. "SLUF--Improving Your Accuracy." USAF Fighter Weapons Review, (Winter 1975), 28-31.

Miller, Mack. "Wings Over Windmills." Air Force Magazine, LXI (January 1978), 19-43.

"Mission Capability of the Corsair II." Interavia, XXIV (June 1969), 740-743.

Mohamed, Rezh M. "A-7D Notes." USAF Fighter Weapons News Letter, nos. 1-3 (March-September 1970), 33-35, 14-19, 14-16.

Morrissey, John C. "A-7D: Weapon Delivery in the Manual Mode." USAF Fighter Weapons Review, (Spring 1971), 34-45.

Mulquin, James J. "The Corsair II in Combat." Ordnance, LIV (March-April 1970), 506-509.

New, H. L. "Flying the A-7E Corsair II." Marine Corps Gazette, LVI (November 1972), 68-70.

North, David M. "Advances Stretch Service Life of A-7." Aviation Week and Space Technology, CVI (January 31, 1977), 87-89.

"One Pilot's Lament." Approach, XIX (March 1974), 1-3.

"Pick Up the Right One." Approach, XXIII (January 1978), 1-3.

Plattner, C. M. "Navy Phasing A-7Es Into Operation." Aviation Week and Space Technology, XCII (February 16, 1970), 38-39+.

"A Plus for the Corsair." Air International, XXXIII (1987), 61+.

Polmar, Norman. "Corsairs for the Air Force." Air Force and Space Digest, LI (February 1968), 32-39.
Reprinted in Air Force and Space Digest International, IV (March-April 1968), 21-22+.

Ryan, Thomas G. "The A-7D Corsair II: That Super-Accurate SLUF (Short Little Ugly Fellow)." Air Force Magazine, LV (March 1972), 27-32.

Sanger, Ken. "A-7: One Aircraft or Two?" Approach, XXI (June 1976), 10-11.

Schaefer, Will F. "A-7 Strikefighter: Rebirth of the Corsair II." Asian Defence Journal, no. 12 (December 1986), 32-34+.

____. "LTV YA-7F: Updated CAS/BAI Aircraft Nearing Reality." Asian Defence Journal, no. 5 (May 1988), 82-86.
BAI = Battlefield Air Interdiction.

Schemmer, Benjamin F. "Two-Seat Corsair II Works, But the Ride is Painful." Armed Forces Journal International, CXII (May 1975), 8-10.

Schoeni, Arthur L. "New Role for an Old Plane: Remanufacturing the A-7 in Texas." All Hands, no. 734 (March 1978), 4-9.

____. "Short Little Ugly Fella: LTV A-7 Corsair II." Aeroplane Monthly, II (August 1974), 797-803.

Schultz, Wayne. "Navigation/Attack A-7 Style." USAF Fighter Weapons Review, (Fall 1981), 26-30.

Shifrin, Carole A. "LTV Receives Air Force Contract to Modernize, Reengine A-7Ds." Aviation Week and Space Technology, CXXVI (June 1, 1987), 85+.

Smith, Gerald H. "Big D School." Air Reservist, XXXI (May 1979), 12-13.

Stockton, Douglas W. "The SLUF in a Different Sky." Aerospace Safety, XXXIII (April 1977), 12-15.

Taylor, Kelvin. "A-7D Corsair II." TAC Attack, XXV (September 1986), 16-17.

Tomlinson, Frank J. "The Great 7." Aerospace Safety, XXIV (June 1968), 2-6.

"USAF Considers LTV A-7 Corsair for Interim Close Air Support." Aviation Week and Space Technology, CXXV (July 21, 1986), 67+.

"USAF Displays A-7D Ground Support Abilities." Aviation Week and Space Technology, XCV (August 30, 1971), 44-45.

Van der Veen, Linn L. "A-7." Flying Safety, XLIII (March 1987), 2-3; XLIV (January 1988), 2-3.

"The Vought A-7 Joins the USAF." Air Enthusiast, I (January 1971), 128+.

West, Frederick J. "The A-7E Aircraft/Pilot Team: One Man, One Machine, and 101 Missions." U.S. Naval Institute Proceedings, CIV (April 1978), 102-105.

Wetterhahn, R. F. "Bounding the Error Bombing--A-7D." USAF Fighter Weapons Review, (Winter 1972), 1-4.

Wilson, D. R. "Here is the Corsair II." Approach, XIV (September 1968), 16-20.

Witze, Claude. "Air Force Pilots Go to Sea in Carrier Tests of A-7." Air Force and Space Digest, LII (January 1969), 32-35.

(Curtiss) A-8 Shrike

"Curtiss A-8 Shrike." Air Classics, IV (September 1967), 8-15.

Jefferies, Walter M. and Kenn C. Rust. "The Curtiss Shrike." American Aviation Historical Society Journal, X (Summer 1965), 129-139.

Kane, John R. ("Killer"). "Flying the Shrike." American Aviation Historical Society Journal, XI (Spring 1966), 57-59.

Westburg, Peter. "Curtiss A-8 Shrike." Air Classics, XI (December 1975), 14-25.

(Fairchild-Republic) A-10 Thunderbolt II

Book/Document/Report Literature:

Bell, Dana. A-10 Thunderbolt II. Warbirds Illustrated, no. 40. New York: Sterling, 1986. 72p.

____. A-10 Warthog in Detail and Scale: Close Air Support Tank Killer. Details and Scale Series, no. 19. Fallbrook, Calif.: Aero Publishers, 1986. 72p.

Bennett, Russell H. Can the A-X Adequately Replace the A-1 in the COIN and SAR Roles? Research Paper. Maxwell AFB, Ala.: Air Command and Staff College, Air University, 1973. 50p.

Burke, David K. A-10 Effectiveness Against Soviet Offensive Operations in Central Europe. Thesis. Fort Leavenworth, Kans.: Army Command and General Staff College, 1984. 50p.

Carleton, Roger E. The A-10 and Design to Cost: How Well Did It Work? Thesis. Fort Leavenworth, Kans.: Army Command and General Staff College, 1979. 52p.

Drendel, Lou. A-10 Warthog in Action. Aircraft, no. 49. Carrollton, Tex.: Squadron/Signal Publications, 1982. 50p.

Duehring, Craig W. The A-10 and Battlefield Air Interdiction in Europe. Research Paper. Maxwell AFB, Ala.: Air Command and Staff College, Air University, 1981. 76p.

Furtak, Ronald E. The A-10 in the Rescue Escort Role: Tactics and Techniques. Research Paper. Maxwell AFB, Ala.: Air Command and Staff College, Air University, 1975. 74p.

Kasbeer, Charles I., Jr. and Thomas J. Lyon. A-10 Pilot SAR Handbook. Research Paper. Maxwell AFB, Ala.: Air Command and Staff College, Air University, 1981. 52p.

Lake, Lewis V. The Next FAC Aircraft. Research Paper. Maxwell AFB, Ala.: Air Command and Staff College, Air University, 1983. 47p.
FAC = Forward Air Control.

Pulham, Thomas D. and Robert C. Mason. New Concepts for the A-10. Research Report. Maxwell AFB, Ala.: Air War College, Air University, 1975. 91p.

Roper, Robert W. and Gail F. Tatum. The A-10 Close Air Support Mission in the NATO Environment: Is It a Viable Mission? Research Paper. Maxwell AFB, Ala.: Air Command and Staff College, Air University, 1980. 36p.

Spick, Mike. A-10 Thunderbolt II. Modern Combat Aircraft, no. 28. Osceola, Wis.: Motorbooks International, 1987. 112p.

Sweetman, Bill. A-10 Thunderbolt II. Modern Fighting Aircraft, v. 6. New York: Arco, 1984. 64p.

Terry, J. G. A-10 Operations and the Battle for North Norway. RAND Report R-3439-AF. Santa Monica, Calif.: RAND Corporation, 1988. 39p.

Watson, George M. The A-10 Close Support Aircraft: From Development to Production, 1970-1975. Andrews AFB, Md.: History Office, Air Force Systems Command, 1979. 136p.

Periodical/Anthology Literature:

"A-10: A NATO Preview." NATO's Fifteen Nations, XXII (April-May 1977), 90-92+.

"A-10 Survivability in Attack Role Shown During Simulated Combat." Aviation Week and Space Technology, CVI (June 20, 1977), 88-89+.

"A-10 Warthog." TAC Attack, XXV (November 1985), 16-17.

Allward, Maurice. "Fairchild's Close Support A-10." Design Profile, no. 15. Air Pictorial, XXXVII (August 1975), 300-304.

Anderton, David A. "USAF A-10 Thunderbolt II: Tactical Support for the Tankbuster." Grumman Aerospace Horizons, XXIV (Fall 1988), 2-7.

Aronson, R. B. "A-10: Lost Cost, Lethal, and Rugged." Machine Design, XLV (November 15, 1973), 30-32.

Beavers, Jim. "The A-10: Monstrosity With a Mission." Air Force Magazine, LXIII (February 1980), 66-70.

Branby, Harlan E. "A-10: Tank Killer." USAF Fighter Weapons Review, (Winter 1975), 1-7.

Brindley, John F. "Fairchild's A-10 is Entering Service: The Return of the Flying Can-Opener." Interavia, XXXI (November 1975), 1089+.

Brown, David A. "A-10 Crews Sharpen Air Support Skills." Aviation Week and Space Technology, CX (April 2, 1979), 37-38+.

Carter, Jimmy. "A Boot Full of Rudder." TAC Attack, XX (July 1980), 4-6.
The author was a Warthog pilot, not U.S. President.

Cutler, Katie. "Blue Four at the Fore." Airman, XXIV (August 1980), 20-23.

Dilger, Robert G. "Tank Killer Team." National Defense, LXI (November-December 1976), 190-191.

"Fairchild A-10." Flight International, CVII (March 20, 1976), 707-717.

"Fairchild A-10: A Remarkable New Close Air Support Weapon System." Aircraft Illustrated, VII (April 1974), 147-152.

"The Fairchild A-10A." Scale Models, VIII (October 1977), 506-509.

"The Fairchild Can-Opener: Shturmovik of the 80s." Air International, XVI (June 1979), 267-272, 287.

Ferguson, Michael L. "A-10: Close Air Support Stone." Infantry, LXVI (May-June 1976), 24-28.

Freeman, Roger A. "Whispering Warthog." Air Pictorial, XL (May 1978), 192-193.

Frye, Wayne T. "The A-10 Aircraft: A Case History." Defense Management Journal, X (September 1974), 32-33.

Geddes, J. Philip. "A-10: The U.S. Air Force's Choice for the Close Air Support Role." International Defense Review, VII (February 1974), 70-76.

_____. "The Fairchild Industries A-10: Designed for Close Air Support." Interavia, XXIX (February 1974), 119-122.

_____. "Production of the Fairchild A-10 Begins." Interavia, XXX (September 1975), 975-977.

Gilson, Charles. "Can the A-10 Thunderbolt II Survive in Europe?" International Defense Review, XII (February 1979), 184-189.

Givans, David W. "Here in Cajun Country, They're...'High on the Hog.'" Sergeants, XXVI (April 1987), 10-11.

Godfrey, David W. H. "All-Weather Tacticians." Air International, X (January 1976), 7-14.

Ground, James H. "A-10." Flying Safety, XXXVIII (May 1982), 6-9; XL (April 1984), 9-12.

Gulick, John F. "The A-10 Does It Better." Air Force Magazine, LIX (July 1976), 75-79.

Hansen, W. "A-10 Prototype Designed for Production." Aviation Week and Space Technology, XCVI (June 26, 1972), 114-115+.

Harris, Lorenzo D. "Warthogs." Airman, XXIX (April 1985), 36-39.

Harvey, David. "Dangerous Midnight: Delay, Confusion on Road to a Night Tank Killer." Defense and Foreign Affairs, VIII (November 1980), 12-15+.

Houle, Edward H. "A-10 Level Bombing." USAF Fighter Weapons Review, (Summer-Fall 1983), 2-5.

_____. "Fight Early: An A-10 Perspective." USAF Fighter Weapons Review, (Fall 1982), 20-22.

Jacobus, Tom. "Cope Elite." Airman, XXIII (March 1979), 21-28.

Kieffer, Gary L. "Sky Warriors." Soldiers, XXXVI (August 1981), 12-14.

Lambert, Mark. "A-10s in the German Front Line." Interavia, XXXVI (November 1981), 1143-1146.

Lopez, Ramon. "Night/Adverse Weather A-10 at the Crossroads." Interavia, XXXV (August 1980), 707-709.

Lutz, Joe. "The A-10 Fly-by-Wire: F-16s, Eat Your Heart Out!" TAC Attack, XXI (February 1981), 22-24.

McDermont, Alexander V. "The Avenger Cannon." USAF Fighter Weapons Review, (Fall 1978), 14-15.

Major, Michael G. "The Winner: A-10." USAF Fighter Weapons Review, (Summer 1973), 4-9.

Neal, Willaim D., Jr. "The A-X." TAC Attack, XII (July 1972), 16-18.

O'Leary, Michael. "A-10: It Sure Comes on Strong." Air Progress, XL (March 1978), 73-75.

Patton, Orin C. and Kenneth W. Engle. "Basing the A-10 in Europe: Strategic and Mobility Problems." Air University Review, XXXI (April-May 1980), 82-85.

Pennington, Rana. "Yukon Lightning." Air Force Magazine, LXX (March 1987), 102-106.

Pentland, Doc. "A-10 Completes Forward Operating Location Validation Test." TAC Attack, XVIII (December 1978), 18-20.

Philipp, J. W. "The A-10 and Fire-Support Co-Ordination." Field Artillery Journal, XLVI (January-February 1978), 30-33.

Phillip, John S. "A-10: Built for Close Support." Asian Defence Journal, no. 2 (February 1983), 56-62.

Powell, Craig. "The A-10." Army, XXVI (March 1976), 43-47.

Quinn, Jimmie B. "Deadly as a Praying Mantis." Armor, LXXXVI (July-August 1977), 12-15.

Rasmussen, Robert D. "The A-10 in Central Europe: A Concept of Deployment." Air University Review, XXX (November-December 1978), 26-44.

Rhea, John. "A-10 Ready for Attack." Flight International, CX (October 22, 1977), 1191-1194.

____. "Tank Killer of the Future." NationalGuardsman, XXXII (April 1978), 14-16.

Robinson, Clarence A., Jr. "A-10 Future Keyed to Gun Development." Aviation Week and Space Technology, XCVIII (January 29, 1973), 16-17.

____. "USAF Plans A-10 Deployment to Europe." Aviation Week and Space Technology, CIV (March 1, 1976), 44-47.

Ropelewski, Robert R. "Night/Adverse Weather Versions of A-10 Provides Confident Low-Level Operations." Aviation Week and Space Technology, CXI (September 17, 1979), 46-48+.

Rubenstein, Michael. "The Ugliest Plane in the Air Force: A-10 Tank-killer." Mechanics Illustrated, LXXIII (February 1977), 64-65+.

Ruddell, David A. "Travelling With the Warthog." Maintenance, VI (Spring 1981), 32-38.

Schlitz, William P. "Getting A-10 Firepower Forward." Air Force Magazine, LXVI (July 1983), 44-47.

____. "Thunderbolt II: Flying From the Folta and the Dols." Air Force Magazine, LXIV (January 1981), 84-88.

Scholin, Allan R. "The A-10s Join the Air Guard Fleet." National Guard, XXXIII (September 1979), 18-21.

Sharp, Nigel P. "'Charcoal Lizard' Thunderbolts." Scale Models, XI (May 1980), 238-240.

Shawler, Wendell H. "A-10 Canopy-Off Tests." TAC Attack, XIX (October 1979), 10-12.

Sheehan, William G. "A-10 Turkey Shoot." Air Reservist, XXXIII (December 1980-January 1981), 16-17.

Sherman, Ron, et al. "The Warthog." NATO's Fifteen Nations, XXV (June-July 1980), 69-70+.

Singer, Mike. "Off Line, On Line, First for Reserve." Air Reservist, XXXII (August 1980), 23-24.

Smith, Gerald H. "Above All Else, a Tank Killer." Air Reservist, XXXII (December 1979-January 1980), 12-14.

Spurlock, Kenneth M. "A-10." Flying Safety, XLI (March 1985), 8-12; XLII (March 1986), 6-7.

Stitzel, Jack. "Say Hello to Warthog." TAC Attack, XVI (April 1976), 26-28.

Sweetman, Bill. "A-10 Thunderbolt II." In: William T. ("Bill") Gunston, ed. The Great Book of Modern Warplanes. New York: Crown, 1987. pp. 9-72.

Tablor, Dale. "Open Letter to A-10 Drivers." TAC Attack, XVI (August 1976), 4-7.

Tarter, Bob. "A-10 Low-Level Attack." USAF Fighter Weapons Review, (Spring 1984), 5-6.

Taylor, John B. "The Mean Machine." Airman, XXI (September 1977), 20-28.

Taylor, Michael B. and Tony Field. "Rendezvous With a Thunderbolt." Airfax Magazine, XXI (June 1980), 494-498.

Ulsamer, Edgar. "The A-10 Approach to Close Air Support." Air Force Magazine, LVI (May 1973), 41-45.

____. "New Muscle for the Tactical Arm." Air Force Magazine, LVII (October 1974), 22-28.

Van der Veen, Linn L. "A-10." Flying Safety, XLIII (March 1987), 4-6; XLIV (January 1988), 4-6.

Verdi, John M. "An Advanced Attack Airplane Design." U.S. Naval Institute Proceedings, CIV (December 1978), 130-135.

Wacker, Rudolph F. "A-10 Close Support Operations in Europe." Air Force Magazine, LXII (August 1979), 73-77.

"Warthog in the Wild: A-10 Thunderbolt II." Defence Update International, no. 72 (June 1986), 2-7+.

Warwick, Graham. "A-10: Europe's Antitank Vanguard." Flight International, CXV (February 24, 1979), 523-524.

____. "All-Weather A-10." Flight International, CXVI (December 1, 1979), 1844-1855.

Wetmore, Warren C. "Adverse Weather A-10 to Be Evaluated." Aviation Week and Space Technology, CX (April 30, 1979), 157-158.

Whitcomb, Darrel D. "Hogs in the Rear: A-10s in Rear Operations." Air Power Journal, I (Winter 1987-1988), 16-21.

A-11 see YF-12A

(Curtiss) A-12 Shrike

Dean, Jack. "Secret of the Shrike." Airpower, IX (September 1979), 8-17.

Jefferies, Walter M. and Kenn C. Rust. "The Curtiss Shrike." American Aviation Historical Society Journal, X (Summer 1965), 129-139.

Rust, Kenn C. and Walter M. Jefferies. Curtiss A-12 Shrike. Aircraft in Profile, no. 128. Windsor, Eng.: Profile Publications, 1971. 25p.

(Douglas) A-20 Havoc/Boston

Book/Document/Report Literature:

Gann, Harry K. Douglas A-20 (7A to Boston III). Aircraft in Profile, no. 202. Windsor, Eng.: Profile Publications, 1971. 15p.

Hess, William N. A-20 Havoc at War. New York: Scribners, 1980. 128p.

Mesko, Jim. A-20 Havoc in Action. Aircraft, no. 56. Carrollton, Tex.: Squadron/Signal Publications, 1983. 50p.

Periodical/Anthology Literature:

"A-20 Production and Assignment." American Aviation Historical Society Journal, XV (Summer 1970), 99-101.

Alcorn, John S. "Douglas A-20 Boston." Scale Models, IX (January-February 1978), 17-22, 82-85.

Andrews, Paul. "The Douglas A-20 Havoc." AirTech, III (October 1943), 20-27.

Boyne, Walter J. "Cry Havoc." Wings, III (June-July 1976), 22-37, 50-66.

"Cry Havoc!" Air Classics, III (July 1966), 42-45.

"Douglas (A-20) Havoc Attack Bomber." Aero Digest, XLII (May 1943), 180-181.

Ellmore, Howard S. "Coconut Bombing." Air Force, XXVII (October 1944), 56-57.

Gault, Owen. "The Amazing 'Coconut Bomber.'" Air Classics, IX (July 1973), 26-37.

Halley, J. J. "Cry Havoc!" Air Pictorial, XXXIX (November 1977), 434-439.

Harald, Eric. "Evolution of the Havoc." Skyways, II (August 1944), 18-23+.

Lodge, J. E. "Name Your Job, the A-20 Does It: The Boston, Renamed the Havoc." Popular Science, CXLV (July 1944), 77-80+.

Powell, Hickman. "Porcupine Squadron: The A-20 Attack Bomber, Famous Abroad as the Boston and Havoc." Popular Science, CXL (May 1942), 82-87.

Smith, T. R. "A-20 Havoc." Aviation, XLIII (January 1944), 125-146.

(Martin) A-22 Maryland

"Martin Maryland A-22." Air Classics, III (July 1967), 36-43.

A-24 see SBD Dauntless

A-25 see SB2C Helldiver

(Douglas) A-26/B-26 Invader

Book/Document/Report Literature:

Mesko, Jim. A-26 Invader in Action. Carrollton, Tex.: Squadron/Signal Publications, 1982. 50p.

United States. Army Air Forces. Air Technical Service Command. A-26, November 1940-October 1945. ATSC Historical Study, no. 48. Wright Field, Ohio: USAAF Air Technical Service Command, 1945. 33p.

Periodical/Anthology Literature:

"The A-26 Invader Goes to War." Aero Digest, XLVII (December 1, 1944), 72-74.

Bangs, Stephen. "The Lifting of Ban Reveals the A-26 Invader as the Fastest All-Purpose Bomber." Aviation News, II (November 6, 1944), 12-13.

Boone, Andrew R. "Meet the Invader, Our Fastest Bomber." Popular Science, CXLV (December 1944), 80-81+.

Carigan, William. "B-26 Invader." In: Robin Higham and Carol Williams, eds. Flying Combat Aircraft of the USAAF-USAF. Ames, Iowa: Iowa State University Press, 1978. pp. 65-74.

"Douglas A-26 Invader." AirTech, VI (January 1945), 46-47, 68.

"Douglas A-26 Invader." Air Pictorial, XV (January 1953), 17-18.

"The Douglas A-26 Proves Hottest Attack Bomber." Aviation, XLIV (April 1945), 175-176+.

"The Douglas Invader Story." Air Enthusiast, no. 7 (1978), 215-223.

Grinsell, Bob. "Invader." Wings, IV (June 1974), 32-45.

Johnston, Robert D. "The Invader Returns." Air University Review, XV (November-December 1963), 9-22.

Kimura, H. "The Flying Fuel Tank." Flying Review International, XIX (March 1964), 43-45.

"Medium Bombers Standardized in the A-26 Invader." Aero Digest, XLIX (May 1, 1945), 104+.

Morton, Joseph. "They Call It the Streaming Demon: Our New Invader Plane." Saturday Evening Post, CCXVI (March 4, 1944), 17+.

Norton, Jay. "The A-26 Invader." TAC Attack, XV (June 1975), 16-17.

"Plane of Many Faces: The A-26 Invader." Popular Science, CXLVII (July 1945), 100-105+.

Roth, Michael J. C. "Nimrod: King of the [Ho Chi Minh] Trail." Air Force Magazine, LIV (October 1971), 30-34.

Ventolo, Joseph A. "A-26: Three-Decade Workhorse." Air Force Magazine, LXIV (July 1981), 92-97.

(Lockheed) A-28/A-29 Hudson

Bowers, Peter M. "The Makeshift Armada." Wings, XIV (April 1984), 26-41.
Continued in Airpower, XIV (May 1984), 10-17.

"Flying the Harassing Hudson." Air Enthusiast, III (February 1972), 292+.

"The Handy Hudson." Air International, XXIX (1985), 240+.

Shores, Christopher F. Lockheed Hudson Mks. I to IV. Aircraft in Profile, no. 254. Windsor, Eng.: Profile Publications, 1972. 24p.

(Vultee) A-31/A-35 Vengeance

Marshall, James L. "Flying Hell: Our New Vultee Vengeance." Collier's, CIX (May 9, 1942), 13+.

Self, Mary and C. Gilbert. V-72, A-31, A-35, July 1940-September 1944. ATSC Historical Study, no 10. Wright Field, Ohio: USAAF Air Technical Service Command, 1944. 47p.

Smith, Peter C. Vengeance: The Vultee Dive Bomber. Washington, D.C.: Smithsonian Institution Press, 1988. 188p.

"The Vultee A-31 Vengeance." Air News, V (June 1943), 44-48.

"The Vultee A-35 Vengeance." Flight, XXXVII (July 12, 1945), 36-37.

A-34 see SB2A

(North American) A-36 Mustang
(see also P-51 Mustang)

Everett, Arthur W., Jr. "Those 'Screaming' Invaders." Air Force, XXVII (February 1944), 35-36.

Holder, William G. "The A-36 Dive-Bombing Mustang." TAC Attack, XIV (August 1974), 24-27.

(Cessna) A-37 Dragonfly
(see also T-37 Tweety Bird)

"The A-37 Close-Support Attack Plane Gave a Superb Account of Itself in a Four-Month Vietnam Test Called 'Combat Dragon.'" Aerospace International, IV (March-April 1968), 48-49.

"A-37 Proves Effective in Vietnam Tests." Aviation Week and Space Technology, LXXXVIII (February 19, 1968), 38-39+.

"A Bold New Look for the '70s." Air Reservist, XXIII (March 1971), 6-7.

Brown, David A. "YAT-37 Shows Agility in COIN Role." Aviation Week and Space Technology, LXXXIII (July 12, 1965), 66-67+.

"The Cessna A-37." Air Progress, XXIII (December 1967), 36-37+.

"Cessna A-37B Includes Probe for Refueling." Aviation Week and Space Technology, LXXXVIII (June 17, 1968), 40-41.

Hart, John W. "They Call It 'Super Tweet.'" Airman, XII (September 1968), 42-45.

Johnston, Robert L. "Combat Evaluation of the A-37." Ordnance, LII (March-April 1968), 424+.

Kronenwatt, Horst K. "A/T-37." Flying Safety, XLII (March 1986), 8-9; XLIII (March 1987), 6-7.

Martinez, Marty. "Newest Viet Cong Foe." Airman, XI (December 1967), 10-11.

"OA-37 Dragonfly." TAC Attack, XXVI (January 1986), 16-17.

Ponert, Horst. "A/T-37." Flying Safety, XL (April 1984), 12-14; XLI (March 1985), 25-27.

Pugh, Craig. "Still Leading the Way." Airman, XXIV (August 1980), 2-6.

"A Singular Aircraft: The A-37 in Vietnam." Air Force and Space Digest, LII (June 1969), 49-51.

Swift, Jack. "The World's Littlest Fighter." Airman, XIII (August 1969), 40-44.

Wright, Tom. "To Ride a Dragonfly." Air Reservist, XXIX (December 1977-January 1978), 10-11.

(Douglas) AC-47 Shadow
(see also C-47 Skytrain/Dakota)

Book/Document/Report Literature:

Maltais, Richard E. Warner Robins Air Materiel Area's Role in Southeast Asia: AC-47 Gunship (Project Big Shoot). AFLC Historical Study, no. 484. Robins AFB, Ga.: History Office, Warner Robins Air Materiel Area, 1970. 25p.

Periodical/Anthology Literature:

Bowman, Alfred C. "AC-47 Navigation." Navigator, XV (Summer 1968), 22-24.

Brownlow, Cecil. "AC-47 Broadens Viet Attack Envelope." Aviation Week and Space Technology, LXXXVI (April 17, 1967), 54-55+.

Cole, J. L. "An Old Airplane in a New War." Aerospace Historian, XVIII (June 1971), 71-74.

Engle, Eloise. "An Agressive 'Gooney Bird' and the Man Behind It." Data, XIV (January 1969), 20-21.

Glines, Carroll V. "Tribute to the Goon." Airman, X (March 1966), 18-21.

Knopf, Robert P. "The AC-47." Airman, XI (June 1967), 44-46.

"'Puff the Magic Dragon': The AC-47 Gunship in Vietnam." Air Classics, VII (December 1970), 62-65.

Rodwell, Robert R. "A Night With 'Spooky.'" Flight International, XCI (January 26, 1967), 124-128.

Thompson, Leroy. "Dragon of the Vietnamese Skies." Air Classics, XVII (May 1981), 56-61.

(Fairchild) AC-119 Spooky
(see also C-119 Provider)

Book/Document/Report Literature:

Bartrand, Louis E. The AC-119K Impact on Close Air Support/Interdiction Mission of the U.S. Air Force. Research Paper. Maxwell AFB, Ala.: Air Command and Staff College, Air University, 1971. 51p.

Maltais, Richard E. Warner Robins Air Materiel Area's Role in Southeast Asia: AC-119G/K Gunship Program (Project Combat Hornet)--A Summary Report, 1967-1970. AFLC Historical Study, no. 485. Robins AFB, Ga.: History Office, Warner Robins Air Materiel Area, 1970. 30p.

Periodical/Anthology Literature:

"AC-119 Gunship Readied for Deployment." Aviation Week and Space Technology, LXXXIX (September 23, 1968), 40-41.

Allen, Robert J. "Navigating the AC-119G." Navigator, XVII (Fall 1969), 6-7.

Casey, William S. "The AC-119: The U.S. Air Force's Flying Battleship." Air Force and Space Digest, LIII (February 1970), 48-50.

Everett, Robert P. "Just a Shadow of Its Former Self." Airman, XIV (February 1970), 11-14.

Lessels, Robert J., Jr. "AC-119 Gunships in Southeast Asia." Air Force Magazine, LIV (November 1971), 38-40.

Reynolds, Donald E. "Gunship Story." Air Reservist, XX (October 1968), 12-13.

Rodwell, Robert E. "A Night With 'Spooky.'" Flight International, XCI (January 26, 1967), 124-128.

"The Story of 883." Airman, XVI (January 1972), 35-36.

Thompson, Leroy. "The Flying Battleships of Vietnam." Chronicle of War: Vietnam, I (Fall 1988), 38-45.

(Lockheed) AC-130 Spectre
(see also C-130 Hercules)

Book/Document/Report Literature:

Aiken, Gerald G. U.S. Air Force Gunships in Future Conflicts. Research Paper. Maxwell AFB, Ala.: Air Command and Staff College, Air University, 1973. 52p.

Blatchley, P. A. AC-130U Gunship Warranty Analysis: Lessons Learned. Research Paper. Maxwell AFB, Ala.: Air Command and Staff College, Air University, 1988. 38p.

Fuss, John R. Effectiveness of the AC-130 Gunship in Southeast Asia. Research Paper. Maxwell AFB, Ala.: Air Command and Staff College, Air University, 1985. 38p.

Spicka, Charles F. Eight Years of Serendipity: The U.S. Air Force Gunship Program. Research Report. Maxwell AFB, Ala.: Air War College, Air University, 1973. 30p.

Periodical/Anthology Literature:

"AC-130 Gunship Upgrades Will Improve Combat Lethality." Aviation Week and Space Technology, CXXVII (December 14, 1987), 51-57.

Colucci, Frank. "Night Spectre." Wings, X (April 1980), 30-37.

Correll, John T. "From a Bucket and a Rope." Airman, XIX (July 1975), 24-29.

Douglas, Kent. "The Ghostriders Ride Again." Airlift, VI (Fall 1984), 21-24.

Fuchs, Ron. "The Ghostriders." Airman, XXVII (June 1983), 42-47.

Humphries, James F. "Spectre in the Night Sky." Airman, XVII (July 1973), 2-6.

Lehnert, Richard A. "The Ghostrider in the Sky." Air Force Magazine, LVI (September 1973), 118-122.

O'Connor, John O. "Hours of Terror." MAC Flyer, XXIII (February 1976), 3-5.

"Pave Pronto." Air Reservist, XXVII (November 1975), 10-11.

Pittman, Benjamin C. "Gunship Navigation." Navigator, XXV (Winter 1978), 25-27.

Rucker, James. "Son of Spooky." Air Combat, VIII (May 1980), 16-19.

Smith, Gerald H. "Eyes of Spectre." Air Reservist, XXX (August 1978), 12-13.

Thompson, Leroy. "The Flying Battleships of Vietnam." Chronicle of War: Vietnam, I (Fall 1988), 38-45.

Zeybel, Henry. "Gunships at Grenada." National Defense, LXVIII (February 1984), 53-56.

____. "Spectre." National Defense, LXVII (October 1983), 47-50.

(Grumman) AF Guardian

Andrews, Hal. "Naval Aircraft: AF-2S Guardian." Naval Aviation News, (November 1972), 20-21.

"Grumman Guardians." American Aviation Historical Society Journal, IX (Fall 1964), 219-221.

"Grumman Guardians Designed for Seeking Out and Destroying Enemy Submarines." Aero Digest, LXII (May 1951), 26-28+.

"Guardians." Joint Services Recognition Journal, X (April 1955), 104-105.

Johnson, Scott O. "Hunter/Killer." Airpower, X (March 1980), 30-34, 51.

Kowalski, Robert J. "Grumman AF-2 Guardian." The Hook, XIII (Fall 1985), 34-50; XIV (Summer 1986), 34-37.

Miska, Kurt H. "Development of the Grumman XTB3F and the AF Guardian." American Aviation Historical Society Journal, XVIII (Winter 1973), 218-225.

Trimble, Robert L. "Bird of Prey." Air Classics, XVIII (November 1982), 14-33, 74-75.

(North American) AJ Savage

Collins, Neil. "An Eye for the Job." Skyline, XI (February 1953), 18-21.

"North American Savage." Aero Digest, LXIV (February 1952), 20-27.

"North American Savage." Joint Services Recognition Journal, IX (January 1954), 18-19.

Pace, Steve. "Shortlived Savage." Airpower, XVII (July 1987), 12-29.

Redman, Rod. "Turbine Savage." Air Classics, X (July 1974), 20-23.

"VC-6's Heavyweight Champions." Naval Aviation News, (January 1955), 1-4.

(Martin) AM-1 Mauler

Kowalski, Robert J. "Mable Wasn't Very Able: The Martin AM-1." The Hook, IX (Spring 1981), 26-34.

Trimble, Robert L. "Martin AM-1 Mauler." Air Classics, XVIII (September 1982), 16-27.

(BAe/McDonnell-Douglas) AV-8 Harrier

Book/Document/Report Literature:

Braybrook, Roy M. British Aerospace Harrier and Sea Harrier. Osceola, Wis.: Motorbooks International, 1984. 192p.

Chant, Christopher F. Harrier. Super Profile Series. Osceola, Wis.: Motorbooks International, 1988. 56p.

Fozard, J. W. The British Aerospace Harrier: Case Study in Aircraft Design. New York: American Institute of Aeronautics and Astronautics, 1978. 190p.

Goza, Joel L. The AV-8B Decision. Thesis. Monterey, Calif.: Naval Postgraduate School, 1982. 96p.

Gunston, William T. ("Bill"). Harrier. Modern Fighting Aircraft, no. 5. New York: Arco, 1984. 64p.

Jones, William R. AV-8B in Defense of the Amphibious Task Force. Research Paper. Newport, R. I.: Department of Naval Operations, Naval War College, 1985. 31p.

Linn, Don. AV-8 Harrier in Detail and Scale. Detail and Scale Series, v. 28. Blue Ridge Summit, Pa.: TAB, 1987. 72p.

____. Harrier in Action. Aircraft, no. 58. Carrollton, Tex.: Squadron/Signal Publications, 1982. 50p.

McDonnell-Douglas Corporation. Public Information Office. AV-8B: A Success Story. St. Louis, Mo., 1981. 13p.

____.____. AV-8B Harrier II: A Closer Look. St. Louis, Mo., 1987. 20p.

____.____. AV-8B Operations in NATO: An Evaluation of the U.S. Marine Corps AV-8B V/STOL Light Attack Aircraft in a NATO Environment. St. Louis, Mo., 1979.

____.____. AV-8B Operations Summary. St. Louis, Mo., 1980.

Mason, Francis K. Harrier. 2nd ed. Annapolis, Md.: Naval Institute Press, 1983. 200p.

Myles, Bruce. Jump Jet: The Revolutionary V/STOL Fighter. 2nd ed. New York: Pergamon Press, 1986. 300p.

Orr, James W. Feasibility of Tactical U.S. Marine Corps V/STOL Aircraft Operations Aboard Merchant Ships, 1990-2000. Advanced Research Paper. Newport, R. I.: Center for Advanced Research, Naval War College, 1978. 109p.

Siuru, William D. British Aerospace and McDonnell-Douglas Harrier AV-8A/B. Aero Series, no. 31. Fallbrook, Calif.: Aero Publishers, 1985. 163p.

Sweetman, Bill and James Goulding. Jane's Aircraft Spectaculars: Harrier. New York: Jane's, 1984. 52p.

United States. Government Accounting Office. Mission Analysis and Acquisition Division. Mission Effectiveness of the AV-8B Harrier 2 Could Be Improved If Actions are Taken Now. Washington, D.C., 1982. 28p.

_____._____. Procurement and Systems Acquisition Division. Is the AV-8B Advanced Harrier Aircraft Ready for Full-Scale Production? Washington, D.C., 1979. 29p.

Wogstad, James and Jay Miller. AV-8 Harrier. San Antonio, Tex.: Aerophile, 1982. 83p.

Periodical/Anthology Literature:

"Advanced Harrier Pushed for Fleet Use." Aviation Week and Space Technology, CVII (November 28, 1977), 55+.

Anderson, Lance D. "Piggyback Technology." U.S. Naval Institute Proceedings, CXII (November 1986), 53-60.

Andrews, Hal. "Naval Aircraft: Harrier." Naval Aviation News (February 1971), 20-21.

Aronson, R. B. "Brown-Wing V/STOLs on Trial." Machine Design, XLIX (October 6, 1977), 26-28.

Ashley, Steven. "Jump Jet." Popular Science, CCXXXIII (October 1988), 56-59, 112-114.

"AV-8A." Approach, XVII (July 1971), 18-23.

"AV-8B Harrier II Moves Into the Field With the Marines." Defense Electronics, XVI (December 1984), 84+.

"AV-8B: The Marine Machine." Surface Warfare, VII (October 1982), 5-6.

"AV-8B: Second Generation V/STOL." Automotive Engineering, LXXXVIII (January 1980), 61-67.

Bingham, Price T. "Increasing Force Flexibility Through V/STOL." Air University Review, XXXVI (January-February 1985), 72-87.

Birch, Steven. "Harrier's Slow Takeoff." Engineer, CCLVIII (May 31, 1984), 26-27+.

Blot, Harry W. "Fly, Hover, and Fight: A Pilot's Report." Air Force Magazine, LXI (February 1978), 49-54.

Boquet, A. H. "A Step Forward With the Harrier." Marine Corps Gazette, LX (April 1976), 59-61.

Braddon, John R. "V/STOL and Marine Aviation." Marine Corps Gazette, LIX (May 1975), 47-51.

____. "What Every Marine Should Know About the Harrier." Marine Corps Gazette, LV (May 1971), 20-24.

Braybrook, Roy M. "Harrier Saga, Phase Two." Pacific Defense Reporter, XII (March 1986), 41-42+.

____. "Military V/STOL: What is the Future?" Air Progress Aviation Review, IV (August 1980), 54-68.

Brophy, Bernard. "Flight of the Harrier." Forbes, CXXVII (June 8, 1981), 38+.

Brown, David A. "Harrier Program Seeks Broad VTOL Data." Aviation Week and Space Technology, XCV (October 25, 1971), 47-48.

____. "Harrier Ski Jump Takeoff Ramp Tested." Aviation Week and Space Technology, CVII (December 5, 1977), 39-41.

_____. "Harriers Exceed Sortie Goal in Exercise." Aviation Week and Space Technology, XCVI (May 1, 1972), 40-42.

_____. "Marines Fit Harrier to Own Tactics." Aviation Week and Space Technology, XCV (October 18, 1971), 38-44.

Browne, Peter. "Jump Jet--The Shape of Things to Come: Harrier." Reader's Digest, XCIX (October 1971), 174-176+.

Coleman, Herbert J. "Harrier Seen Shifting Concept of Ground-Air Warfare." Aviation Week and Space Technology, LXXXVIII (June 24, 1968), 204-207.

Collins, Helen. "The Harrier: Battle Proven and Getting Better." Naval Aviation News, (May-June 1984), 12-13.

Copley, Gregory. "No Ten Commandments." Defense and Foreign Affairs, III (May 1975), 20-23+.

Corbett, David C. "What V/STOL Really Is: A Pilot's View." U.S. Naval Institute Proceedings, CIV (June 1978), 111-114.

Cordier, Sherwood S. "Command of the Air at Sea: V/STOL and Small Carriers." Naval War College Review, XXXIV (July-August 1981), 93-99.

Covault, Craig. "Harrier Breaks Path in Navy V/STOL." Aviation Week and Space Technology, CVI (January 31, 1979), 99-101+.

Cummings, Edward A. "Landing That Isn't Harrier." Naval Aviation News, (December 1982), 38-40.

Cunningham, J. L. "Can the Marine Corps Really Use V/STOL Aircraft?" Marine Corps Gazette, LXI (January 1977), 18-24.

DeMais, Richard. "Designing a V/STOL Fighter: McDonnell-Douglas AV-8B Harrier II." Aerospace America, XXIII (May 1985), 88-91.

"Development AV-8B to Fly by October." Aviation Week and Space Technology, CXV (September 21, 1981), 59-60.

Dodt, Robert C., Jr. "Harrier in a Direct Support Role." Marine Corps Gazette, LXVII (May 1983), 41+.

Elliott, Jim and Tom Bartlett. "Harrier: AV-8A." Leatherneck, LIV (May 1971), 24-29.

Ethell, Jeffrey L. "Harrier: The Marines' V/STOL Baby." Air Progress, XXXIX (February 1977), 32+.

Fozard, John. "Wind, Seapower, and Jet V/STOL." U.S. Naval Institute Proceedings, CIII (September 1977), 109-114.

Fricker, John. "The Incomparable Harrier." Flying, XCIX (September 1976), 86-89+.

Gaines, Mike. "AV-8B Means Marine V/STOL Machine." Flight International, CXXVII (June 1, 1985), 148-151.

Greeley, Brendan M., Jr. "AV-8B Night Attack Advances." Aviation Week and Space Technology, CXXIX (August 8, 1988), 34-36.

Griffiths, David R. "AV-8B Passes Flight Test Milestone." Aviation Week and Space Technology, CXI (August 20, 1979), 49-50.

Griggs, Tom. "Harrier Goes to the Bush." Profile, XXI (December 1977), 16-19.

____. "Harriers in Force." Leatherneck, LXIV (July 1981), 16-23.

Gunston, William T. ("Bill"). "Harrier." In: William T. ("Bill") Gunston, ed. The Great Book of Modern Warplanes. New York: Crown, 1987. pp. 329-392.

Gustafson, Richard A. "The Marine Corps View." Armed Forces Journal International, CXVIII (April 1981), 58-61.

____. "Why V/STOL for the Marines?" Shipmate, XLI (November 1978), 23-25.

"Harrier in the USMC." Interavia, XXVI (September 1971), 1035-1036.

"Harrier Shows Operational Capabilities." Aviation Week and Space Technology, LXXXIX (October 14, 1968), 38-39.

"Harrier In-Flight Thrust Vectoring Honed." Aviation Week and Space Technology, C (June 10, 1974), 32-35.

"Harrier II." Air International, XXXI (1986), 175+.

"Harrier II: Successful Linkup of American and British Skills." Asian Defence Journal, no. 2 (February 1986), 124+.

"Harrier II AV-8B." Defence Update International, no. 53 (1984), 30-32+.

"The Harrier: V/STOL Vindicated." Air University Review, XXXIII (September-October 1982), 71-79.

Harvey, David. "Harrier: The Cherry Point Connection." Defense and Foreign Affairs, VIII (September 1980), 17-20+.

Hazen, D. C. "Breaking V/STOL Free of Catch 22." Astronautics and Aeronautics, XVIII (March 1980), 24-30.

Hieronymus, W. S. "Marines' Plan for Harrier Introduction." Aviation Week and Space Technology, XCIII (September 14, 1970), 55-56.

Hite, Kenneth F. "Why the V/STOL Fighter?" Air University Review, XIX (July-August 1968), 86-92.

Holloway, James L., III. "The Transition to V/STOL." U.S. Naval Institute Proceedings, CIII (September 1977), 18-24.

Iles, J. E. "Training to Fly the Harrier." Approach, XXIII (March 1978), 1-6.

Kiffer, Jim. "Harrier in the Groove." Approach, XXII (January 1977), 6-8.

Kinnucan, P. "Supersonic Jump Jets." High Technology, V (September 1985), 38-43.

Kitfield, James. "Harrier: A Happy Transatlantic Marriage." Military Logistics Forum, IV (November-December 1987), 29-32.

Kuszewski, Michael D. "AV-8B: The Only Jet We Need?" Marine Corps Gazette, LXXII (July 1988), 22-25.

Lambert, Mark. "AV-8B/Harrier GR.5: Range, Payload, and V/STOL." Interavia, XXXVII (March 1982), 224-227.

Lehman, J. R. "Here Comes the AV-8B." Marine Corps Gazette, LXVIII (May 1984), 64-68.

Leng, John. "Alpha, Beta, and Theta Jay." Approach, XXI (November 1975), 1-3.

Lewis, Stanley P. "V/STOL Close Air Support in the U.S. Marine Corps." U.S. Naval Institute Proceedings, CII (October 1976), 113-116.

Lillie, Jay C. "Seabasing the Harrier." Marine Corps Gazette, LXXI (May 1987), 78-83.

Linn, Don. "AV-8B: The Once and Future Harrier." Air Progress Aviation Review, LIV (January 1980), 32-38.

Lowe, W. W. "AV-8B Flying Qualities and Performance Integration From Design to Flight Test." Cockpit, XVIII (July-September 1983), 5-11.

McCalla, John C. "AV-8B: Two Meanings to Air Support." Marine Corps Gazette, LXXI (May 1987), 72-76.

Manning, Harry. "USMC Harrier--Base to Battlefield in a Flash." International Combat Arms, VI (July 1988), 20-26.

"Marines' Bad-Luck Plane: Harrier." Time, CX (August 15, 1977), 15-16.

Mehta, K. "The AV-8B Light Attack Aircraft." Asian Defense Journal, no. 12 (December 1982), 84-86+.

Mersky, Peter. "The AV-8B on the JFK." Approach, XXXIV (October 1988), 24-26.

Miller, Thomas H., Jr. "Fixed-Wing V/STOL for the Marines." Grumman Aerospace Horizons, XIV (Spring-Summer 1978), 16-21.

____ and C. M. Baker. "The AV-8A Harrier in USMC Service." Interavia, XXVIII (February 1973), 150-151.

Morgan, David H. S. "Harrier the Viffer." Aerospace America, XXIV (August 1986), 34-35.

New, Noah C. and John R. Braddon. "Harrier: A New Dimension in Close Air Support." Marine Corps Gazette, LIV (May 1970), 26-30.

Nihart, Brooke. "USMC/DOD/STOL Exercise 'Warrior" Validates AV-8A Harrier Goals." Armed Forces Journal, CIX (May 1972), 52-53.

O'Rourke, Gerald G. "V/STOL and the U.S. Navy." Marine Corps Gazette, LXIII (January 1979), 43-47.

____. "The Wonderous World of Jet V/STOL." U.S. Naval Institute Proceedings, XCVIII (November 1972), 33-41.

Orr, James W. "What It's All About." Approach, XIX (March 1974), 18-20.

____ and R. E. O'Dare. "V/STOL Will Do the Job." Marine Corps Gazette, LXI (May 1977), 39-46.

Petersen, F. S. "Facing Up to the Military V/STOL Challenge." Astronautics and Aeronautics, XV (November 1977), 16-21.

Petersen, Stefan. "Harrier II: The AV-8B in USMC Service." Defence Update International, no. 84 (December 1987), 48-57.

Phillip, John S. "V/STOL Goes to Sea." Aerospace International, XII (July-August 1976), 24-26+.

Plank, Thomas H. "Jump Jet." USAF Fighter Weapons Review, (Fall 1973), 18-20.

Plummer, C. A., Jr. "AV-8B Technical Update." Society of Experimental Test Pilots Technical Review, XVI (Spring 1981), 203-209.

____. "YAV-8B/AV-8B Advanced Harrier Program." Society of Experimental Test Pilots Technical Review, XIII (Fall 1977), 43-65.

____. "YAV-8B Status Report." Society of Experimental Test Pilots Technical Review, XIV (Fall 1979), 52-58.

Priest, Richard H. "AV-8B: The New Generation of V/STOL." Rolls Royce Magazine, (December 1982), 7-120.

Rhea, John. "Marine Aviation: Looking Up." Sea Power, XIX (November 1976), 9-13.

Riley, B. L. "AV-8B/GR Mk. 5 Airframe Composite Applications." Proceedings of the Institute of Mechanical Engineers, CC (1986), 167-183.

Robinson, Clarence A., Jr. "AV-8B Pivotal in Naval Aviation Fight." Aviation Week and Space Technology, CVII (October 31, 1977), 18-20.

Salamon, George. "From Pilot to Grunt: The AV-8B Bridges the Gap in Marine Corps Close Air Support." Defense Systems Review, II (April 1984), 13-15.

Scharfen, John C. "Barges, Ramps, and V/STOL." Marine Corps Gazette, LXIV (September 1980), 20-21.

Schier, J. F. and R. J. Juergens. "Design Impact of Composites on Fighter Aircraft." Aerospace America, XXI (September-October 1983), 44-49+, 41-45.

Scott, L. and R. W. Morrissey. "USMC AV-8A Maintenance Experience." Journal of Aircraft, XIX (August 1982), 694-696.

"Second Generation V/STOL: AV-8B." Interavia, XXXII (January 1977), 33-34.

Showers, Robert G. "V/STOL for Close Air Support." National Defense, LXIII (September-October 1978), 34-37.

Smith, G. M. "Harrier Flies at Quantico." Marine Corps Gazette, LIV (September 1970), 10-11.

Stromberg, Russ. "The AV-8B Opeval: Mission Complete--The Best Ever!" Wings of Gold, X (Summer 1985), 36-39, 42.

Sweetman, Bill. "From AV-8B to the ASTOVL." Interavia, XXXIX (December 1984), 1324-1327.
ASTOVL = Advanced Short Take Off, Vertical Landing.

Taylor, Peter P. W. "The Impact of V/STOL on Tactical Air Warfare." Air University Review, XXIX (November-December 1977), 65-81.

Tegler, John. "Premier V/STOL Flying." U.S. Naval Institute Proceedings, CXII (October 1986), 161-163.

Thompson, Pat. "AV-8B Harrier 2 Unveiled at Cherry Point." Marines, XIII (March 1984), 6-7.

Thoms, Wayne. "We Fly the Remarkable Harrier Jet." Plane and Pilot, XVI (December 1980), 54-57.

Tierney, G. P. "V/STOL Night Approach." Approach, XXIII (March 1978), 28-29.

Urbanik, Theodore J. "Fiber-Optic Technology Takes to the Air." Military Electronics/Countermeasures, VIII (October 1982), 32+.

Verdi, John M. "The Advanced Attack Airplane Design." U.S. Naval Institute Proceedings, CIV (December 1978), 130-135.
AV-8B Harrier II.

Warwick, Graham. "AV-8B Advanced Harrier." Flight International CXVI (December 29, 1979), 2127-2142.

Watson, J. C. "AV-8B Composite Fuselage Design." Journal of Aircraft, XIX (March 1982), 235-238.

Weatherup, Robert A. "Missions for V/STOL Aircraft." U.S. Naval Institute Proceedings, CVI (July 1980), 111-112.

Wilson, Michael. "Supersonic V/STOL Prospects Brighten." Interavia, XXXVI (March 1981), 219-220.

Wood, Derek. "Big Wing Harrier." Interavia, XXXIV (May 1979), 447-448.

____. "Harrier: The Horizon Widens." Interavia, XXX (November 1975), 1175-1179.

Yackle, A. R. "V/STOL: A Practical Weapon System." Astronautics and Aeronautics, XXI (January 1983), 18-25, 33.

(Curtiss) C-18 Triplane

Trimble, Robert L. "Triplanes for the Navy." Air Classics Quarterly Review, VI (Fall 1979), 72-79.

F-1 see FJ Fury

(Brewster) F-2A Buffalo

"Brewster Buffalo Fighter." Aero Digest, XXXIX (September 1941), 185-186.

"Brewster F2A Buffalo." Air Classics, III (July 1967), 16-23.

"Brewster XF2A-1 Single-Seater Fighter." Aero Digest, XXXII (June 1938), 83-84.

"Brewster's Benighted Buffalo." Air Enthusiast, VII (Spring 1975), 66-96.

Green, William. "Peanut Special." RAF Flying Review, XV (May 1960), 16-18.

Maas, Jim. F2A Buffalo in Action. Carrollton, Tex.: Squadron/Signal Publications, 1987. 50p.

"Naval Aircraft: Brewster Buffalo." Naval Aviation News, (August 1976), 20-21.

Sanders, James V. "The Brewsters of Fighting Two [VF-2]." American Aviation Historical Society Journal, XIV (Fall 1969), 153-158.

Shores, Christopher F. The Brewster Buffalo. Aircraft in Profile, no. 217. Windsor, Eng.: Profile Publications, 1970. 24p.

(Curtiss) F2C-1 Hawk

Andrews, Hal. "Naval Aircraft: BF2C-1 Hawk." Naval Aviation News, (January-February 1986), 16-17.

(Grumman) F2F

Andrews, Hal. "Naval Aircraft: F2F-1." Naval Aviation News, (January 1976), 20-21.

"The Grumman F2F-1 Single-Seater Fighter." Aero Digest, XXIX (February 1936), 56-57.

Reams, Benton E. "The Displaced F2F-1." American Aviation Historical Society Journal, XIV (Summer 1969), 97-98.

Westburg, Peter. "The Grumman F2F-1." Model Builder, X (August 1980), 42-45.

Wixey, Kenneth E. "Grumman's Biplane Family." Aircraft Illustrated, XII (November 1979), 540-545.

(Vought) F2G Corsair

Daniels, C. M. "Too Late for Glory." Air Classics, VIII (April 1972), 54-63.

(McDonnell) F2H Banshee

Braitsch, Fred G., Jr. "Photos by Banshee." Leatherneck, XXXV (October 1952), 26-30.

"Death of an Airplane." Interavia, VII (September 1952), 496-497.

Ginter, Steven J. McDonnell F2H Banshee. Naval Fighters Series, no. 2. Simi Valley, Calif.: Naval Fighters, 1980. 78p.

Gunston, William T. ("Bill"). "Fighters of the Fifties: McDonnell Banshee." Aeroplane Monthly, III (September 1975), 442-447.

"Let's Land One." Interavia, X (April 1955), 262-263.

Linn, Don. "Banshee: McDonnell's Flying Banjo." Airpower, IX (November 1979), 18-35.

"McDonnell's Banshee." RAF Flying Review, XV (August 1960), 16-17.

McLarren, Robert. "Developing the McDonnell Banshee." Aviation Week, LI (October 31, 1949), 22-25.

Silsbee, Nathaniel F. "Navy's F2H by McDonnell." Aviation Op, XIII (February 1950), 29, 54-55.

(Boeing) F3B

Andrews, Hal. "Naval Aircraft: F3B." Naval Aviation News, (October 1978), 22-23.

(Douglas) F3D Skynight

Book/Document/Report Literature:

Ginter, Steven J. Douglas F3D Skynight. Naval Fighters Series, no. 4. Simi Valley, Calif.: Naval Fighters, 1982. 82p.

Periodical/Anthology Literature:

Andrews, Hal. "Naval Aircraft: F3D Skynight." Naval Aviation News, (May 1977), 20-21.

"Deceptive Cat: Flight Impressions of the New Douglas Skynight." Interavia, V (November 1950), 576-578.

"The Douglas Skynight." Aero Digest, LXIV (April 1952), 24-31.

Ginter, Steve. "Willie the Whale." Airpower, XVI (January 1986), 28-43.

Heinemann, Edward H. "First Details of the F3D Skynight." Aviation Week, LII (March 6, 1950), 23-24.

Hotz, Robert. "F3Ds Outfly Red Night Fighters in Korea." Aviation Week, LVIII (February 9, 1953), 13-14.

"Navy Reveals New Night Fighter: Douglas Twin-Jet Skynight." Aviation Week, XLIX (October 11, 1948), 12-13.

"Skynight Rider." Skyways, IX (September 1950), 12-13.

"Skynight: The Navy's Forgotten Night Fighter." Air Combat, X (May 1982), 8+.

Trousdale, J. H., Jr. "Birth of the Navy's Night Fighters." U.S. Naval Institute Proceedings, LXXVIII (June 1952), 620-623.

Williams, R. E. "Skynight." Naval Aviation News, (January 1983), 34-37.

(Grumman) F3F

Andrews, Hal. "Naval Aircraft: F3F." Naval Aviation News, (September 1977), 20-21.

"The Grumman F3F Biplane Fighter." Air Classics, VII (May 1971), 20-35.

"Last U.S. Fighting Biplane: Grumman F3F-3." RAF Flying Review, XVII (October 1961), 50-51.

Reams, Benton. "F3F-2, BuNo. 1031, Neo-XF3F-3." American Aviation Historical Society Journal, XVIII (Fall 1973), 181-183.

____. Grumman F3F Series. Aircraft in Profile, no. 92. Windsor, Eng.: Profile Publications, 1966. 12p.

Tillman, Barrett. "Grumman's Flying Beer Barrels." Wings, IV (February 1974), 32-47.

Westburg, Peter. "Grumman F3F-2." Model Builder, XI (June-August 1981), 29, 57, 101, 36-37.

Wixey, Kenneth E. "Grumman's Biplane Family." Aircraft Illustrated, XII (November 1979), 540-545.

(McDonnell) F3H Demon

Book/Document/Report Literature:

Ginter, Steven J. McDonnell F3H Demon. Naval Fighters Series, no. 12. Simi Valley, Calif.: Naval Fighters, 1985. 150p.

Periodical/Anthology Literature:

"All-Weather Demon: Single-Jet Carrier-Based Fighter." Mechanical Engineering, LXXVII (August 1955), 691-692.

Andrews, Hal. "McDonnell F3H Demon." Aeroplane Monthly, VI (March 1978), 143-149.

____. "Naval Aircraft: F3H Demon." Naval Aviation News, (March 1973), 20-21.

"Demon." American Aviation, XIX (May 26, 1954), 22-24.

Kennan, F. J. "Probers Find No Fraud in F3H Procurement." American Aviation, XIX (November 7, 1955), 26-27.

"Navy Blames F3H-1 Difficulties on Former Procurement Policy." Aviation Week, LXIII (October 10, 1955), 16-17.

"Navy Gets New Demon." Aviation Week, LX (January 18, 1954), 19+.

"New Fighter Joins U.S. Navy." RAF Flying Review, IX (March 1954), 18-19.

"Who's to Blame for Shortcomings of F3H?" American Aviation, XX (October 10, 1955), 23-24.

(McDonnell-Douglas) F-4 Phantom II

Book/Document/Report Literature:

Anderson, Kenneth C. and Ronald B. Nenner. A Wild Weasel Penetration Model. Thesis. Wright-Patterson AFB, Ohio: School of Engineering, Air Force Institute of Technology, 1982. 198p.
F-4G.

Barber, Thomas J. F-4 Strike/Attack Operations in NATO's Central Region: Is USAFE Prepared to Face the Threat? Research Paper. Maxwell AFB, Ala.: Air Command and Staff College, Air University, 1975. 57p.

Boyne, Walter J. Phantom in Combat. Washington, D.C.: Smithsonian Institution Press, 1985. 176p.

Buck, Michael L. The Future Use of the F-4E as an Operational Air-to-Ground Fighter. Research Paper. Maxwell AFB, Ala.: Air Command and Staff College, Air University, 1984. 49p.

Burns, Michael. McDonnell-Douglas F4K and F-4M Phantom II. Osceola, Wis.: Motorbooks International, 1984. 186p.

Corder, John A. F-4D Radar Bombing Tactics and Techniques. Research Paper. Maxwell AFB, Ala.: Air Command and Staff College, Air University, 1971. 71p.

Davis, Larry. F-4 Phantom II in Combat. Aircraft, no. 65. Carrollton, Tex.: Squadron/Signal Publications, 1984. 58p.

Dorr, Robert F. McDonnell-Douglas F-4 Phantom II. Osprey Colour Series. Osceola, Wis.: Motorbooks International, 1984. 192p.

____. Phantoms Forever. Osceola, Wis.: Motorbooks International, 1987. 128p.

Drendel, Lou. F-4 Phantom II in Action. Warren, Mich.: Squadron/Signal Publications, 1976. 50p.

____. Phantom II: A Pictorial History of the McDonnell-Douglas Phantom II. Warren, Mich.: Squadron/Signal Publications, 1977. 62p.

____. U.S. Air Force Phantoms in Combat. Carrollton, Tex.: Squadron/Signal Publications, 1987. 63p.

F-4 Phantom II Aircraft and Ogden Air Logistics Center: 30 Years of Pride and Progress. Washington, D.C.: GPO for the OALC, 1988. 58p.

Gill, Anthony B. and Michael C. Short. Implications of the F-4 Phase-out on the Tactical Air Forces: Modernization Program. Thesis. Washington, D.C.: Industrial College of the Armed Forces, 1985.

Gray, David L. An Analysis of the F-4 Weapon System in Out-Country Counterair Operations, 1965 to the Present. Research Report. Maxwell AFB, Ala.: Air War College, Air University, 1968. 125p.

Gunston, William T. ("Bill"). F-4 Phantom II. New York: Scribners, 1977. 112p.

Holmes, David T. Wild Weasel Tactical Employment Guide. Research Paper. Maxwell AFB, Ala.: Air Command and Staff College, Air University, 1983. 9p.

Honlahan, Robert F. Evaluation of the F-4 Weapon System in Out-Country Interdiction Operations, 1965-1967. Research Report. Maxwell AFB, Ala.: Air War College, Air University, 1968. 87p.

Jabara, James. An Analysis of the Requirements for an Air Superiority Aircraft to Establish Air Superiority in Limited War. Research Report. Maxwell AFB, Ala.: Air War College, Air University, 1961. 62p.

Kelly, John J. and Thomas E. McNiff. Foreign Military Sales: The F/RF-4 Weapon System as a Special Case Study. Research Report. Maxwell AFB, Ala.: Air War College, Air University, 1975. 96p.

Kinzey, Bert. F-4C Phantom II; Pt. 1, Post-Vietnam Markings, 1974-1984. Color and Markings Series, v. 3. Fallbrook, Calif.: Aero Publishers, 1985. 64p.

____. F-4 Phantom II in Detail and Scale. Detail and Scale Series, v. 6-7, 12. 3 vols. Fallbrook, Calif.: Aero Publishers, 1981-1983.

___ and Ray Leader. F-4E Phantom II. Color and Markings Series, v. 13. Blue Ridge Summit, Pa.: TAB, 1989. 64p.

Lindsay, William E. F-4 Phantom Aircrew Survival Equipment Evaluation. Thesis. Fort Leavenworth, Kans.: Army Command and General Staff College, 1979. 120p.

Mason, Francis K. Phantom: A Legend in Its Own Time. Osceola, Wis.: Motorbooks International, 1984. 192p.

Middleton, Larry J. Advanced Wild Weasel System. Research Paper. Maxwell AFB, Ala.: Air Command and Staff College, Air University, 1975. 40p.

O'Rourke, Gerald G. The F-4 Phantom II. Famous Aircraft Series. New York: Arco, 1969. 64p.

Richardson, Doug. F-4 Phantom II. Modern Fighting Aircraft, v. 4. New York: Arco, 1984. 64p.

Russell, Jan M. Handbook on Reconnaissance to Fighter Conversions: Signficant Requirements and Problems. Research Paper. Maxwell AFB, Ala.: Air Command and Staff College, Air University, 1982. 43p.
Converting the RF-4 into an F-4.

Stern, Robert C. F-4 Phantom II Part I. Warbirds Illustrated, no. 27. Harrisburg, Pa.: Stackpole Books, 1984. 72p.

_____. F-4 Phantom II, Part II. Warbirds Illustrated, no. 46. New York: Sterling, 1987. 72p.

Sweetman, Bill. Phantom. Boston, Mass.: Jane's, 1984. 52p.

United States. Marine Corps. "Here are the Tigers": F-4B Phantom II Fighter-Bomber. Washington, D.C.: GPO, 1970. 64p.

Periodical/Anthology Literature:

Abler, Pete. "The Hardwing F-4." TAC Attack, XVIII (November 1978), 8-9.

Adee, Donald P. and David L. McFarland. "Phantom Profile." Navigator, XV (Summer 1968), 12-15.

Alder, Konrad. "F-4G Phantom II." Armanda International, X (April 1986), 41-42+.

Allen, R. C. "The F-110A." Fighter Weapons News Letter, no. 2 (June 1962), 20-22.
F-4C.

Allsup, Dan. "Those Wild, Wild Weasels." Airman, XXVIII (November 1984), 20-27.

Allward, Maurice. "McDonnell's Fantastic Phantom." Air Pictorial, XL (May 1978), 168-174.

Alquist, Jon. "The Back-Seat Driver." Airman, X (June 1966), 22-23.

"Anatomy of Phantom." Flight, LXXXVII (March 4, 1965), 330-331.

Andrews, Hal. "Naval Aircraft: F-4 Phantom II." Naval Aviation News, (March 1971), 20-21.

Baxter, Robert H. "Let's Get Serious About Bomb Toss." USAF Fighter Weapons Review, (Summer 1976), 28-34.

Beck, Michael O. "The Agile Eagle Becomes a Reality." USAF Fighter Weapons Review, (Winter 1973), 20-29.

____. "The New Phantom Switchology." USAF Fighter Weapons Review, (Summer 1973), 26-30.

____. "Super Phantom." USAF Fighter Weapons Review, (Winter 1972), 18-24.
F-4E.

Bielinski, Henry E. "The F-4E--A Pilot's View." Air Force Magazine, LV (November 1972), 34-39.

Blesse, Frederick C. "One Candle for a Piece of Cake." Aerospace Safety, XX (July 1964), 10-13.
On the F-4's two-engine powerplant.

Boyne, Walter J. "The Fabulous Phantom." Wings, XV (December 1985), 12+.

Butz, J. S., Jr. "F-4 Phantom II, Versatile and Long-Lived." Air Force and Space Digest, XLIX (August 1966), 30-36.

____. "How the Phantom II Got That Way." Aerospace International, II (November 1966), 16-18+.

Carroll, Edward G. "Deployment to England." National Guard, XXXVII (November 1983), 24-27.

Cleaver, Thomas M. "Wild Weasels Smash the Vermin." Air Force Magazine, LXV (July 1982), 80-82.
F-4G.

Coble, Donald W. "The World's Best Fighter Aircraft." Airman, VII (June 1963), 1+.

Coyne, James P. "Burn Up the Sky." Airman, VIII (April 1964), 23-26.

Davis, Harry F. F-4 WRCS/CAS." USAF Fighter Weapons Review, (Spring 1978), 7-14.

"The Development of the McDonnell Phantom." Interavia, XXI (May 1966), 737-741.

Ellis, Joe. "Bringing It Home." Airman, XIX (February 1975), 8-9.

Enman, Ann. "Whizzo of a Day." Air Progress, XLVIII (October 1986), 50-54.

"F-4 Phantom." Interceptor, XX (November 1978), 20-23.

"F-4 Trend Analysis." Approach, XVIII (July 1972), 22-23.

"F-4D." TAC Attack, XXVI (June 1986), 16-17.

"F-110, USAF's Newest Tactical Fighter." Air Force and Space Digest, XLV (May 1962), 36-37.
F-4C.

Feighny, James P., Jr. "Back to Basics." TAC Attack, XVI (December 1976), 4-9.

"Focus on Phantoms." Airfax Magazine, XXII (July 1981), 506-509.

Frisbie, Norman H. "Never Excelled." Aerospace Safety, XXII (April 1966), 8-11.

Gardner, Guy S. "F-4 AIM-7 Boresight Shots." USAF Fighter Weapons Review, (Fall 1979), 20-24.

"Get the Facts Straight." TAC Attack, XXIII (July 1983), 4-7.

Gilchrist, John R. and Thomas S. Dewberry. "Forging F-4 Combat Crews." USAF Instructors Journal, V (Summer 1967), 67-71.

Gish, Donald L. "F-4 Air-to-Air Training." USAF Fighter Weapons Review, (Fall 1975), 1-5.

Gregory, William H. "Advances May Extend Life of F-4 Program." Aviation Week and Space Technology, LXXXII (February 15, 1965), 67+.

Griffiths, David R. "F-4Es Deploying to Egypt for Training." Aviation Week and Space Technology, CXII (June 23, 1980), 20-21.

Gunston, William T. ("Bill"). "McDonnell Phantom II." Aeroplane Monthly, II (February 1974), 497-503.

Hankin, Raymond. "The Phantastic Phantom." Flying Review International, XXV (July 1969), 34-41+.

Harris, Hap. "The Phantom's Men." Airman, XI (April 1967), 4-7.

Hart, Charles R. and John J. Harty. "F-4G Advanced Wild Weasel." Navigator, XXVI (Summer 1979), 16-18.

Harvey, Frank L. "I Rode Our Hottest Jet." Popular Science, CLXXX (March 1962), 53-56+.

Harwick, Alexander H. C. "F-4 Phantom II." In: Robin Higham and Carol Williams, eds. Flying Combat Aircraft of the USAAF-USAF. Ames, Iowa: Iowa State University Press, 1978. pp. 113-120.

Henderson, Gordon. "Here's the F-110." Aerospace Safety, XVIII (June 1962), 2-4.
F-4C.

Hirst, Mike. "5,000 Phantoms." Flight International, CXIV (May 27, 1978), 1621-1633.

Johnson, Jerome L. "F-4." Flying Safety, XLIII (March 1987), 8-11.

Kessler, M. J. "Super Phantom." Defence Update International, no. 58 (1985), 15-23.

"King of the Road." Approach, XXI (August 1975), 1-3.

Kinnan, Timothy A. "All New Digital F-4E." USAF Fighter Weapons Review, (Winter 1980), 2-8.

Kinzey, Bert. "Wild Weasel SAM Killer." Air International, XX (January 1981), 11-14.
F-4G.

Kline, Don. "The Long-Haul Champ." USAF Fighter Weapons Review, (Fall 1984), 12-13.

Koehnke, Richard K. "Improved Capability." USAF Fighter Weapons News Letter, no. 1 (March 1970), 29-32.

Kostelnik, M. C. "The Phantom Flies UHF." TAC Attack, XVII (March 1977), 26-28.

Kronenwett, Horst K. "F/RF-4." Flying Safety, XLII (March 1986), 10-12; XLIV (January 1988), 7-9.

Lefavor, Marshall W. "Subject: End of an Era." Marine Corps Gazette, LXXII (May 1988), 78-82.

Letzer, George J. "Phantom II De-Tails." American Aviation Historical Society Journal, XII (Summer 1967), 114-119; XV (Spring 1970), 49-55.

McCoy, Neil. "Until the Aggressors Come." USAF Fighter Weapons Review, (Fall 1976), 9-13.

"McDonnell-Douglas F-4E Phantom II." Aireview, X (October 1980), 88-89.

"The McDonnell-Douglas Phantom II: An Outstanding All-Rounder." Interavia, XVIII (November 1963), 1698-1701.

"McDonnell-Douglas Phantom II: Profile of a Two-Seat Fighter." Interavia, XX (February 1965), 259-263.

McDowell, Donald E. "Five Years in a Phantom." Air Classics, IX (January-February 1973), 25-31, 54-55, 40-49, 64-65.
A USAF appreciation.

McGraw, Karen L. and Bruce A. "The Phantom Crew: A-1 in the Cockpit." Defense Science and Electronics, IV (November 1985), 44-45+.

McInerney, Thomas G. "Dive Toss With the F-4D." USAF Fighter Weapons News Letter, no. 1 (March 1967), 10-13.

______. "F-4 Dive Toss." USAF Fighter Weapons News Letter, no. 1 (March 1968), 7-11.

Mackay, William A. "Flying the Phantom." Flying Review International, XIX (July 1964), 25-28.

Martin, Richard W. "Introducing the Phantom: Newest Member of the Air Force Fighter Family." Aerospace Accident and Maintenance Review, XVII (October 1962), 2-5.

Mason, E. P. "Phantom Alert." Wings, XIV (October 1984), 26-45.

Meschko, E. T. "The F-4E Austers HUD/Gunsight." Society of Experimental Test Pilots Technical Review, XIV (Winter 1978), 25-30.

"Mighty Spectre: First Two-Seat Dual Mission Shipboard Fighter With Mach 2 Capability." RAF Flying Review, XVII (January 1962), 30-33.

Miller, Thomas B. "Phantom II: 500-km Speed Record." Naval Aviation Museum Foundation, V (Fall 1985), 2-9.

Morphew, Gary R. "F/RF-4." Flying Safety, XL (April 1984), 14-16; XLI (March 1985), 10-12.

"New Weasel Aircraft Will Play Vital Role." Electronic Warfare/Defense Electronics, X (February 1978), 75-77.
F-4G.

Ohlgren, Arthur. "Go It Alone With Phantom II." Marine Corps Gazette, XLVII (October 1963), 26-27.

O'Rourke, Gerald G. "Flying the Fabulous Phantom." Reader's Digest, XCII (February 1968), 132-135.

Owen, Kenneth. "Big Day at St. Louis." Flight, XC (July 7, 1966), 5-7.

Peterson, Chester. "Ride of a Lifetime." Air Progress, XLVII (March 1985), 28+.

"Phantom--Fighter Fantastic." Flying Review International, XIX (July 1964), 21-25.

"Phantom II, Fastest Shipboard Fighter." RAF Flying Review, XVI (March 1961), 38-39.

Pratt, Bob. "The GI and the F-4." USAF Fighter Weapons Review, (Summer 1977), 33-36.

Pulin, W. D. "F-4 Functional Modernization." Journal of Engineering for Gas Turbines and Power, CVII (October 1985), 815-820.

Reitmaier, L. W. "Phantom II, Record-Breaking Fighter." Flying, LXX (May 1962), 62-63+.

"Retrofits Keep Navy F-4Js Combat Ready." Electronic Warfare/Defense Electronics, XI (April 1979), 62-64.

Richardson, Doug and Mike Spick. "F-4 Phantom II." In: William T. ("Bill") Gunston, ed. The Great Book of Modern Warplanes. New York: Crown, 1987. pp. 521-584.

Riley, Stephen. "Black Lions and Green Phantoms." American Aviation Historical Society Journal, XIX (Spring 1974), 29-32.

Robinson, Clarence A., Jr. "Defense Department Backs F-4 Upgrade Program." Aviation Week and Space Technology, CXX (January 9, 1984), 16-18.

_____. "F-4G Boosts Wild Weasel Capabilities." Aviation Week and Space Technology, CVIII (February 6, 1978), 181-185.

Robinson, Dennis E. "SAM Suppression: Our Profession." Navigator, XXXIII (Summer 1986), 16-17.
F-4G.

Rodwell, Robert R. "Test Flight in a -4B: Mach 2 in a U.S. Navy Version." Flight, LXXXIX (June 30, 1966), 1100-1101.

Romaglis, Frank J. "Flying the F-4 Leading Edge Slat." TAC Attack, XVII (March 1977), 22-25.

Rucker, Cindy. "Pounce!" Air Progress, XXXVIII (November 1976), 6+.

Ryckman, Kenneth R. "Threshold of Sin." USAF Fighter Weapons News Letter, no. 1 (March 1970), 18-21.

Scholin, Allan R. "New 'Phlyers' for the Phantom." Air Force and Space Digest, LI (January 1968), 46-50.

____. and Edward C. Tripp. "F-4 Phantom:Deadliest Fighter of Them All." Air Progress, XXXII (May 1975), 46-53.

Scott, William B. "Wild Weasel Aircraft Being Upgraded." Aviation Week and Space Technology, CXVI (June 14, 1982), 115-116.
F-4G.

Shannon, Richard H. and Wayne L. Waag. "Prediction of Pilot Performance in the F-4 Aircraft." Aerospace Medicine, XLV (February 1974), 167-170.

____. "Toward the Development of a Criterion for Fleet Effectiveness in the F-4 Fighter Community." Aerospace Medicine, XLIV (April 1975), 453-455.

Sheffield, Dick. "On a Sacred Mission." Airman, XXIII (July 1979), 21-28.
F-4G.

Sims, Donald. "New Weasel in the Sky." Air Classics Quarterly Review, VI (Fall 1979), 62-71.
F-4G.

Smith, Dave, et al. "F-4, 1980-1985." USAF Fighter Weapons Review, (Fall 1975), 6-13.

Spick, Mike. "2001 Phantom Odyssey." Air International, XXII (December 1985), 287-292.

Stiles, Gerald. "F-4G Wild Weaseldom: Fact and Fancy." Journal of Electronic Defense, VI (February 1983), 47-51.

Stricklin, Charles R. "Eyes of the Phantom." Airman, XVII (May 1973), 2-7.

Studdard, Gary L. "F-4." Flying Safety, XXXIX (April 1983), 10-12.

Swalm, Tom. "YF-4E: Project Agile Eagle." USAF Fighter Weapon News Letter, no. 3 (September 1969), 31-33.

Sweetman, Bill. "F-4G Wild Weasel." International Defence Review, no. 16 (November 1983), 1585-1588.

"USAF F-4C Carries Varied Ordnance Loads." Aviation Week and Space Technology, LXXXVI (April 24, 1967), 75-78.

"U.S. Navy Phantoms." Aircraft Illustrated, XI (August 1978), 384-387.

Vanasdlen, Ron. "End of an Era." All Hands, no. 829 (April 1986), 14-15.
USN F-4 phaseout.

"The Very Tangible Spectre." Flying Review International, XXI (September 1966), 5-10.

Vion, Jerry E. "The Digital F-4E." TAC Attack, XVIII (March 1978), 18-21.

Wagner, William. "Brute Force." Combat Crew, XXII (August 1971), 18-20.

Wayne, Lloyd G. "Flying the Phantom." Aerospace Safety, XXIV (April 1968), 16-21.

"Wild Weasels Prepare to Fly the F-4G." Electronic Warfare/ Defense Electronics, X (February 1978), 70-73.

Wilson, Michael. "Preparing for the Phantom." Flight International, XCII (September 21, 1967), 483-487.

Winkler, John L., Sr. "F-4 Harmonization." "USAF Fighter Weapons News Letter, nos. 3-4 (September and December 1967), 18-24, 24-33; no. 3 (June 1968), 18-30.

Wiseman, Charles. "Upgrading Today's Wild Weasel System." Defense Electronics, XX (September 1988), 62+.

(Boeing) F4B Hawk

Andrews, Hal. "Naval Aircraft: F4B." Naval Aviation News, (January 1978), 20-21.

Barrow, Jess C. "The Last of a Breed: A Synopsis of the World's Only Surviving Boeing F4B-4." American Aviation Historical Society Journal, XXIV (Fall 1979), 235-238.

____. "Teething Troubles With F4B-1: A New Fighter Joins the Fleet." American Aviation Historical Society Journal, XVI (Summer 1971), 98-104.

"Boeing F4B-4." Air Classics, III (September 1966), 19-23.

Bowers, Peter M. The Boeing F4B-4 Hawks. Aircraft in Profile, no. 27. Windsor, Eng.: Profile Publications, 1966. 15p.

"Evolution of the Boeing F4B-4." Air Classics, I (Fall 1964), 7-12.

O'Leary, Michael. "Devildogs' Biplane." Air Classics, XII (January 1976), 40-44.

Westburg, Peter. "Boeing F4B-4." Model Builder, IX (December 1979), 29-32.

(Douglas) F4D Skyray

Book/Document/Report Literature:

Francillon, Rene J. Douglas F4D, Aerofax Minigraph, no. 4. Arlington, Tex." Aerofax, Inc., 1985. 40p.

Williams, Nick and Steven J. Ginter. Douglas F4D Skyray. Naval Fighter Series, no. 13. Simi Valley, Calif.: Naval Fighters, 1986. 189p.

Periodical/Anthology Literature:

"All For the Record." Flight Magazine, XL (November 1953), 20-21.

Anderton, David A. "Ten Minute Killer Mission Guided F4D Skyray Design." Aviation Week, LXII (June 6, 1955), 40-42.

Andrews, Hal. "Douglas F4D Skyray." Aeroplane Monthly, V (May 1977), 255-261.

_____. "Naval Aircraft: Skyray." Naval Aviation News, (April 1975), 20-21.

"Closeup: Skyray." Joint Services Recognition Journal, X (July 1955), 176-177.

"Douglas XF4D-1 Skyray." Air Pictorial, XIV (June 1952), 173-174.

"Navy Plane Sets Record in Climbing." U.S. Naval Institute Proceedings, LXXXIV (August 1958), 127+.

O'Rourke, Gerald G. "The Douglas Navy F4D 'Ford.'" American Aviation Historical Society Journal, XXIV (Summer 1979), 125-132.

____. "F4D 'Ford': A Better Idea." U.S. Naval Institute Proceedings, CXII (June 1986), 123-126.

Phillips, Ward. "Bat Out'a Hell." Air Classics, XIV (June 1978), 36-41, 70-73.

Rahn, Robert O. "XF4D Skyray Development: Now It Can Be Told." Society of Experimental Test Pilots Technical Review, XIV (Spring 1979), 147-161.

Smartt, C. H. "New Final Assembly Methods Used to Build F4D-1 Delta-Type Fighter." Automotive Industries, CXV (August 15, 1956), 48-52.

Smith, John F. "Shaped for Speed." BeeHive, XXX (January 1955), 3-6.

"The Sting Ray." RAF Flying Review, XV (November 1959), 30-33.

Trimble, Robert L. "El Segundo's 10-Minute Killer." Air Progress Aviation Review, VI (Winter 1982), 24-37.

Williams, Nicholas M. "The Douglas Skyray in Fleet Colors." American Aviation Historical Society Journal, XXIII (Summer 1978), 129-140; XXVI (Summer 1981), 98-113; XXX (Summer 1985), 82-93.

____. "Douglas F4D-1 Skyray Record Flights." American Aviation Historical Society Journal, XXI (Spring 1976), 50-60.

____. "Firebird I: The Highway Skyray." American Aviation Historical Society Journal, XXII (Winter 1977), 261-262.

____. "Mayday Skyrays." American Aviation Historical Society Journal, XXXI (Summer 1986), 82-91.

____. "Silver Eagle Skyrays." The Hook, VI (Spring 1978), 14-17.

____. "The X-Rays: Douglas' XF4D-1 Skyrays." American Aviation Historical Society Journal, XXII (Winter 1977), 242-260.

(Grumman) F4F Wildcat

Book/Document/Report Literature:

Greene, Frank L. Grumman F4F-3 Wildcat. Aircraft in Profile, no. 49. Windsor, Eng.: Profile Publications, 1966. 15p.

____. The Wildcat Story: History of the Grumman F4F Wildcat. Bethpage, N.Y.: Grumman Aircraft Engineering Corp., n.d.

The Grumman Wildcat. Appleton, Wis.: Aviat Publications, 1978. 63p.

Kinzey, Bert. F4F Wildcat in Detail and Scale. Detail and Scale Series, v. 30. Blue Ridge Summit, Pa.: TAB, 1988. 64p.

Linn, Don. F4F Wildcat in Action. Aircraft, no. 84. Carrollton, Tex.: Squadron/Signal Publications, 1984. 50p.

Tillman, Barrett. Wildcat in World War II. Annapolis, Md.: Nautical and Aviation Publishing Company of America, 1983. 270p.

Periodical/Anthology Literature:

Andrews, Hal. "Naval Aircraft: Wildcat." Naval Aviation News, (December 1971), 20-21.

Aubuchon, Nathan. "Wildcat: The Lethal Loser." Flying, XCII (January 1973), 36-43+.

Bowers, Alan F. "Grumman Wildcat." Air Pictorial, XXIII (December 1961), 380-381.

Copeland, Gizella. "Second Life for a Wildcat." Aeroplane Monthly, XII (November 1975), 560-564.
With the Confederate Air Force.

Greene, Frank L. "Addendum to the Wildcat Story." American Aviation Historical Society Journal, VII (Spring 1962), 46-47+.

"Grumman's Willing Wildcat." Air Enthusiast, III (1976), 49-69.

Heuer, Mike. "Wildcat." Sport Aviation, XXIV (December 1975), 50-54.

Lehardt, Jack. "Wildcat Dream." Sport Aviation, XXIV (June 1975), 38-47.

Mendenhall, Charles. "Taming the Wildcat." Air Classics, XX (May 1984), 22-31.

Miller, Jay. "Confederate Wildcat." Air Classics, XIII (May 1977), 42-51.

Semmes, Raphael. "Wildcat: The First Navy Cat." Aerospace Historian, XXVIII (March 1981), 2-9.

Trimble, Robert L. "First of the Bethpage Cats." Air Classics, XIV (August-September 1978), 34-41, 64-67, 36-41, 70-74.

"Wildcat." Air Classics, IV (June 1968), 26-39.

(Chance-Vought) F4U Corsair

Book/Document/Report Literature:

Abrams, Richard. F4U Corsair at War. New York: Scribners, 1981. 160p.

Chance-Vought F4U Corsair. Appleton, Wis.: Aviat Publications, 1973. 67p.

Dial, Jay F. The Chance-Vought F4U-1 Corsair. Aircraft in Profile, no. 47. Windsor, Eng.: Profile Publications, 1965. 11p.

____. Chance-Vought F4U-4 to F4U-7 Corsair. Aircraft in Profile, no. 150. Windsor, Eng.: Profile Publications, 1972. 25p.

Johnsen, Frederick A. F4U Corsair. New York: Crown, 1983. 56p.

Maloney, Edward T. Chance-Vought F4U Corsair. Aero Series, v. 11. Rev., 2nd ed. Fallbrook, Calif.: Aero Publishers, 1984. 68p.

Musciano, Walter A. Corsair Aces: The Bent-Wing Bird Over the Pacific. New York: Arco, 1978. 136p.

O'Leary, Michael, ed. Fighting Corsairs. War Eagle Special, no. 1. Canoga Park, Calif.: Challenge Publications, 1987. 64p.

Rice, Michael, ed. Pilots Manual for the Chance-Vought F4U Corsair. Glendale, Calif.: Aviation Book Company, 1973. 72p.

Sullivan, Jim. F4U Corsair in Action. Aircraft, no. 29. Warren, Mich.: Squadron/Signal Publications, 1977. 49p.

____. F4U Corsair in Color. Fighting Colors Series. Carrollton, Tex.: Squadron/Signal Publications, 1981. 32p.

Tillman, Barrett. Corsair: The F4U in World War II and Korea. Annapolis, Md.: Naval Institute Press, 1979. 219p.

Periodical/Anthology Literature:

Ahnstrom, D. Newell. "Evolution of the Corsair." Skyways, III (February 1944), 19-23+.

Andrews, Hal. "Naval Aircraft: F4U Corsair." Naval Aviation News, (May-June 1986), 28-29.

"Another Corsair Sweeps the Skies." Aviation Week, LVI (April 7, 1952), 32+.
F4U-7.

Baxter, Gordon. "Corsair: Old Hoss-Nose Becomes a TV Star." Flying, C (June 1977), 46-52.
In the series "Baa Baa Black Sheep."

____. "Pride of the Corps." Flying, CXI (October 1984), 110-111.

Bibee, A. J. "The Long-Flying Corsair." American Aviation Historical Society Journal, IX (Winter 1964), 294-295.

Bordelon, Guy P. "A Corsair Pilot Reminisces." Approach, XVI (May 1971), 40-43.

Brown, Eric. "Viewed From the Cockpit." Air International, XVI (May 1979), 230-238.

Caldwell, George E. "FrU-1D." Scale Modeler, XIV (June 1977), 41-45.

Cleaver, Tom. "Back From the Graveyard." Air Enthusiast, XVII (1981), 43-46.

"The Corsair." Air Classics, III (November 1966), 4-17.

"Corsair: Exploded Sketches Show How the Parts Fit to Make the Airplane." Life, XVI (May 8, 1944), 41-44.

Daniels, C. M. "Too Late for Glory." Air Classics, VIII (April 1972), 54-61, 75, 82.
F4U-7.

Ebersacher, Warren A. "XF4U-1, First of the Bent-Wing Birds." Air Classics, X (September 1974), 14-18.

"End of the Line, 1938-1953." Flight Magazine, XXXIX (February 1953), 20-21.

"F4U U-Birds." Naval Aviation News, (July-August 1984), 12-13.

Folmar, J. G. "The Corsair's First MIG Kill." Air Trails, XXXIX (February 1953), 26-27, 56.

Greenway, Sean. "Leatherneck Corsairs Over Korea." Air Classics, XIV (November 1978), 20-31.

Grinsell, Bob. "Ultimate Corsairs." Airpower, IV (January 1974), 18-27.

____ and Hap Langstaff. "Sweathogs of the Solomons." Airpower, VIII (January 1978), 30-48.

Guyton, Boone T. "The Crash That Threatened the Solomons Campaign." Naval Aviation Museum Foundation, IV (Fall 1984), 6-14.

"Higher, Faster at Chance-Vought." BeeHive, XXVIII (Fall 1953), 30-31.

Hofrichter, Paul. "Television Brought Back Memories of World War II Glory for Old Corsair Pilots." Marine Corps Gazette, LXIII (May 1979), 79-82.

Johnson, J. W. "The Last of the Corsairs." BeeHive, XXVIII (January 1953), 29-31.
F4U-7.

Kuokka, H. "New to the U-bird." Skyways, X (July 1951), 18-19, 48.

Langewieche, William. "Flying the F4U." Flying, C (June 1977), 53+.

"The Last of the Corsairs." Naval Aviation News, (March 1953), 1-5.

Matheson, Bruce J. "The Corsair and Its Contributions." Marine Corps Gazette, LXV (May 1981), 59-63.

Merriam, Ray. "The Chance-Vought F4U Corsair." World War II Journal, II (May-June 1975), 90-101.

Mikesh, Robert C. "The Last Corsair." Airpower, XI (September 1981), 20-36.

"Pilots Knew It as 'Whistling Death': The Corsair." RAF Flying Review, XVI (March 1961), 18-21+.

Sullivan, Jim. "Corsairs in Reserve." American Aviation Historical Society Journal, XVI (Fall 1971), 164-169.

Trickey, Robert. "Corsairs of the U.S.S. Bataan [CVL-29]." Air Classics, XXI (April 1985), 27-37.

"The Vought Corsair." Aero Digest, LXV (December 1952), 26-32.

Walton, Frank E. "Air and Space Museum Inducts Corsair." Marine Corps Gazette, LXV (May 1981), 57-59.

(Northrop) F-5 Freedom Fighter-Tiger II/F-20 Tigershark (see also T-38 Talon)

Book/Document/Report Literature:

Bloss, John N. U.S. Arms Transfer Policy and the F-X: Can Tigershark Survive? Research Report. Maxwell AFB, Ala.: Air War College, Air University, 1983. 30p.

Drendel, Lou. F-5 in Action. Carrollton, Tex.: Squadron/Signal Publications, 1980. 50p.

Kinzey, Bert. F-5E Tiger II in Detail and Scale. Detail and Scale Series. Fallbrook, Calif.: Aero Publishers, 1982. 71p.

Krempel, Donald M. Can the "Aggressors" Continue to Be Effective in the F-5E? Thesis. Fort Leavenworth, Kans.: Army Command and General Staff College, 1984. 50p.

Northrop Corporation. Public Relations Department. F-5 Tactical Fighter Series. Los Angeles, Calif., 1987. 2p.

_____. _____. F-20 Tigershark Multirole Tactical Fighter. Los Angeles, Calif., 1987. 2p.

Scutts, Jerry. Northrop F-5/F-20. Modern Combat Aircraft, no. 25. New York: Scribners, 1986. 112p.

Stuart, W. G. Northrop F-5: Case Study in Aircraft Design. New York: American Institute of Aeronautics and Astronautics, 1978. 219p.

Periodical/Anthology Literature:

Archer, Robert D. "Northrop N-156F: Design and Development of America's First Military Assistance Programme Fighter." Flight, LXXVII (January 8, 1960), 43-50.

Bean, David. "22-Hour Airplane." American Aviation Historical Society Journal, XXVI (Summer 1981), 116-117.

Berent, Mark. "F-20: Tomorrow's Fighter Available Today." Asian Defence Journal, no. 9 (September 1984), 18-20+.

Bowers, Kenneth. "Co-production: The U.S. F-5E in Taiwan and Switzerland." Defense Systems Management Review, II (Spring 1979), 34-45.

Braybrook, Roy M. "From Claws to Jaws: Tiger Into Tigershark." Air International, XXII (March 1982), 111-116, 136-138.

Briggs, Ernest A. "F-5." Flying Safety, XXXIX (April 1983), 13-14.

Butz, J. S., Jr. "F-5: Little Plane With a Big Future." Air Force and Space Digest, XLVII (December 1964), 34-36+.

"A Cat for Dogfighting." Air Enthusiast, III (January 1972), 184+.
F-5E Tiger II.

"A Classic Comes of Age." Air International, XII (February 1977), 67-70.

Copley, Gregory. "A Smile on the Face of the Tiger." Defence and Foreign Affairs, X (November 1982), 6-8+.
F-20 Tigershark.

Coyne, James P. "Trials of the Tigershark." Air Force Magazine, LXVIII (January 1985), 71-76.

Dane, William. "Long Life for F-5 Freedom Fighters." Asian Defence Journal, no. 10 (October 1985), 42+.

DelleCapre, Enrico. "The New/Old Fighter: F-5G." Asian Defence Journal, no. 1 (January 1981), 48-52.
F-20.

"Design for Minimum Maintenance: Northrop N-156F." Interavia, XIV (August 1959), 972-974.

Dunstan, Keith. "Northrop: The Daring Innovator." Pacific Defense Reporter, IX (August 1982), 17+.
F-20.

Easterbrook, Gregg. "The Airplane That Doesn't Cost Enough." Atlantic, CCLIV (August and November 1984), 46-56, 6+.
F-20.

"F-5 Program Ends." Air Progress, XLIX (May 1987), 68-69.

"F-5A." Flight, LXXXVI (December 17, 1964), 1046-1049.

"F-5E Tiger II." TAC Attack, XXV (October 1985), 16-17.

"The F-5F: Northrop's Two-Seat Trainer/Fighter." Interavia, XXIX (December 1974), 1208-1209.

"F-20 Tigershark." Defence Update International, no. 79 (March-April 1987), 18-26.

Fink, Donald E. "F-5F Melds Fighter/Trainer Roles." Aviation Week and Space Technology, CII (May 26, 1975), 35-39.

_____. "Northrop Plans Advanced F-5 Versions." Aviation Week and Space Technology, CV (December 13, 1976), 18+.

"Flight Test Plan for F-5E is Accelerated." Aviation Week and Space Technology, XCVI (June 26, 1972), 120-122.

"Freedom Fighter--Combat Ready." Flying Review International, XXI (December 1965), 197-199.

Frey, William J. "F-20: Saga of an F-X." Air University Review, XXXVII (May-June 1986), 43-51.

Fricker, John. "Supersonic Strike in Miniature." Flying Review International, XXI (September 1965), 13-14+.

Gallagher, J. T. "F-20 Tigershark." Defence Update International, no. 34 (1983), 50-58.

Gilson, Charles. "F-5G Tigershark: In for a Testing Time." International Defence Review, XV (September 1982), 1281-1284.

____. "F-5G Tigershark: Testing Time Lies Ahead." Interavia, XXXVII (October 1982), 1045-1047.

Gordon, Michael R. "Selling the F-20; or, How Northrop Corporation Turned a White Elephant Into a Prize Bull." National Journal, XVII (July 13, 1985), 1608-1613.

Gumble, Bruce. "Will F-20 Sales Finally Get Off the Ground?" Defense Electronics, XVII (September 1985), 64-68+.

Hieronymus, W. S. "Early Study of Users Aided F-5 Program." Aviation Week and Space Technology, XCII (March 16, 1970), 46-47+.

Johansen, H. O. "The F-5: Our Bantam Supersonic Jet." Popular Science, LXXXVIII (March 1966), 77-81.

Kerr, John. "Is There a Future for Northrop's F-20 Fighter?" Electronic Business, XI (April 1, 1985), 28-30.

Labreque, Ron. "Tom Jones and His Amazing Flying Machine." Los Angeles, XXIX (March 1984), 181+.
F-20.

Lambert, Mark. "Is 1984 the Year of the Tigershark?" Interavia, XXXIX (April 1984), 363-364.

____. "Northrop F-5B." Flight, LXXXVIII (July 29, 1965), 174-179.

____. "Northrop F-5G Tigershark Faces the Competition." Interavia, XXXVI (September 1981), 900-901.

"Maintainability and the F-5." Aerospace Maintenance Safety, XIX (September 1964), 10-12.

Mann, Paul. "Reagan's F-5G Decision Pleases No One." Aviation Week and Space Technology, CXVI (January 15, 1982), 16-17.

Miller, Barry. "F-5E Rollout Beats Schedule." Aviation Week and Space Technology, XCVII (July 3, 1972), 19-20.

Mohamed, Rezk M. "Willy and the Dart." USAF Fighter Weapons News Letter, no. 4 (December 1969), 25-27.

Mulvihill, Bob. "F-5." Flying Safety, XLI (March 1985), 13-14; XLII (March 1986), 12-13; XLIII (March 1987), 12-13, XLIV (January 1988), 10-11.

"No. 4 F-20 Configured With Production Changes." Aviation Week and Space Technology, CXXII (February 25, 1985), 54-56.

"Northrop F-5." Air Progress, XXXVIII (December 1976), 62+.

"Northrop F-5: A Different Kind of Fighter." Flying Review International, XIX (August 1964), 21-27.

"The Northrop F-5: A Versatile Fighter for Regional Warfare." Interavia, XVII (November 1962), 1462-1464.

"Northrop F-5G Makes First Flight." Aviation Week and Space Technology, CXVII (September 6, 1982), 48-49.

"The Northrop F-5-21: Study of a Fighter in Evolution." Interavia, XXIV (July 1969), 912-915.

"The Northrop N-156F Weapons System." Interavia, XV (July 1960), 828-835.

"Northrop Rolls Out Company-Funded F-5G Tigershark." Journal of Electronic Defense, V (September 1982), 49-50.

"Northrop's Campaign to Get a New Fighter Flying in the Third World." Business Week, (July 18, 1984), 74+.
F-20.

"Northrop's Muscular Minimums." RAF Flying Review, XV (March 1960), 26-29.

Pochhacker, Christian. "New Fighter for Austria." Defence Update International, no. 56 (1985), 49-51+.

Redman, Charles. "Northrop Aims for a Killing With the Tigershark." Fortune, CXI (June 24, 1985), 58-59+.

Richardson, Doug. "F-5G Tigershark." Flight International, CXX (October 17, 1981), 1144-1150.

Robinson, Clarence A., Jr. "Defense, State Aid F-20 Sales Drive." Aviation Week and Space Technology, CXVII (December 13, 1982), 19-21.

Ropelewski, Robert R. "F-5E Keyed to Agility in Combat." Aviation Week and Space Technology, XCV (July 12, 1971), 41-43.

_____. "F-5G Performance Emphasized." Aviation Week and Space Technology, CXIII (July 28, 1980), 12-14.

_____. "Flight Tests Verify Predictions for F-20." Aviation Week and Space Technology, CXVIII (May 9, 1983), 65-67.

_____. "Northrop F-20 Achieving Flight Test Program Goals." Aviation Week and Space Technology, CXX (June 11, 1984), 56-60.

Shreves, Keith. "The Case for the F-5G Export Fighter." Asian Defence Journal, no. 10 (October 1982), 33+.

"'Skoshi Tiger': Evaluating the F-5 in Combat." Air Force and Space Digest, XLIX (August 1966), 45-48.

Smith, Bruce A. "Northrop Closes Out F-20 Fighter Project." Aviation Week and Space Technology, CXXV (November 24, 1986), 24-25.

_____. "Northrop Cutting Costs on F-5G Line." Aviation Week and Space Technology, CXVII (July 5, 1982), 24-25.

Swift, Jack. "The World's Littlest Fighter." Airman, XIII (August 1969), 40-44.

Taylor, John H. "The F-5E Tiger II--A 'Sports Model' With a Punch: A Pilot's Report." Air Force Magazine, LVIII (January 1975), 54-57.

_____. "Tiger by the Tail." In: Michael O'Leary, ed. Air Progress Pilot Reports. Canoga Park, Calif.: Challenge Publications, 1979. pp. 86-91.

"Tigershark." Asian Defence Journal, no. 11 (November 1982), 50-52.

Vance, C. H. "Gliding Flight." Aerospace Safety, XXVII (March 1971), 6-9.

Warwick, Graham. "F-X: The Export Fighter." Flight International, CXIX (February 14, 1981), 416+.
F-20.

Wisdom, James L. "Both Sides of the Fence." USAF Fighter Weapons Review, XXX (Fall 1982), 13-19.
F-20.

Woellner, Joan C. "Tigers With a Tail." Airman, XIX (September 1975), 40-41.
F-5E.

Wright, Roger. "Tiger II's on Target." Aircraft Illustrated, X (May 1977), 205-207.

F-6A see F4D Skyray

(Curtiss) F6C Hawk

Adams, Joe. "Preserving an Endangered Species." Naval Aviation Museum Foundation, VIII (Spring 1987), 61-63.

Van Deurs, George. "Hawk and the Was." Naval Aviation News, (August 1973), 18-19.

(Grumman) F6F Hellcat

Book/Document/Report Literature:

Anderton, David A. Hellcat. New York: Crown, 1981. 56p.

The Grumman F6F Hellcat. Appleton, Wis.: Aviat Publications, 1973. 60p.

Kinzey, Bert. F6F Hellcat in Detail and Scale. Detail and Scale Series, v. 26. Blue Ridge Summit, Pa.: TAB, 1987. 72p.

Sullivan, Jim. F6F Hellcat in Action. Aircraft, no. 36. Carrollton, Tex.: Squadron/Signal Publications, 1979. 50p.

Tillman, Barrett. Hellcat: The F6F in World War II. Annapolis, Md.: Naval Institute Press, 1979. 265p.

United States. Navy Department. Office of the Chief of Naval Operations. Aviation Training Division. Meet the Hellcat. NAVAER 00-80R-10. Washington, D.C.: GPO, 1944. 97p.

Periodical/Anthology Literature:

"Ace Maker From the Iron Works." Air International, X (April 1976), 76-87.

Andrews, Hal. "Naval Aircraft: F6F Hellcat." Naval Aviation News, (September-October 1988), 16-17.

Colby, Carroll B. "How the Hellcat Got That Way." Popular Science, XCVIII (December 1943), 71-74.

Demme, Robert E. "Evolution of the Hellcat." Skyways, II (December 1943), 19-21+.
Reprinted in William H. Fetridge, ed. The Second Navy Reader (Indianapolis, Ind.: Bobbs-Merrill, 1944), pp. 193-293.

"The Grumman Hellcat." Flying, XXXVI (March 1945), 35-38.

"The Grumman Hellcat." RAF Flying Review, XVII (December 1961), 36-39.

Harp, A. Norman. "The Grumman Hellcat." Aerospace Historian, XXVI (December 1979), 215-220.

Hill, Richard M. "Hellcat 40467: The Story of One F6F." American Aviation Historical Society Journal, XVII (Fall 1972), 201-205.

Lee, George C. "F6F Colors and Insignia." American Aviation Historical Society Journal, XIV (Summer-Winter 1969), 122-124, 185, 248-249.

Miller, Jay. "Saved From a Swamp: The Story of the Confederate Air Force Grumman Hellcat." Air Classics, XIII (June 1977), 42-51.

"The Navy's Newest Fighter, the Hellcat." Life, XV (September 27, 1943), 57-58.

Tillman, Barrett. "Dark Hunter: Grumman's F6F-5N Nightfighter." Wings, VII (August 1977), 10-17.

____. "Hellcat: F6F-3." Wings, IV (October 1974), 8-35, 55.

____. "Homage to the Big Cat." Wings, IX (December 1979), 22-39.

(Chance-Vought) F6U Pirate

Koehnen, Richard. Chance-Vought F6U Pirate. Naval Fighters Series, no. 9. Simi Valley, Calif.: Naval Fighters, 1984. 34p.

____. "Reluctant Pirate." Wings, VII (December 1977), 24-45, 53.

Schoeni, Arthur L. "'You Can't Win 'Em All': Vought's Pirate." The Hook, VIII (Spring 1980), 42-45.

"Vought F6U-1 Pirate." Scale Models, XII (November 1981), 568-569.

(Grumman) F7F Tigercat

Carr, Orrin I. "Fire Cat." Air Classics, XII (September 1976), 38-47.

Daniels, C. M. "On the Track of the Tigercat." Wings, III (August 1973), 20-37.

Gault, Owen. "Grumman's Tiger Twins: Skyrocket and Tigercat." Air Classics, IX (August 1973), 22-27.

"Grumman F7F Tigercat Navy Fighter." Popular Science, CXLVII (September 1945), 96-97.

Gunston, William T. ("Bill"). "Prototype Pursuit: Tigercat." Aeroplane Monthly, VIII (April 1980), 202-204.

Larsen, Jim. "Tigercat Pictorial." Air Classics, XIV (July 1978), 59-69.

Scarborough, Bill. F7F Tigercat in Action. Carrollton, Tex.: Squadron/Signal Publications, 1986. 50p.

____. "Warplanes in Detail: The F7F Tigercat." Air Combat, III (January 1975), 20-31, 68-73.

(Chance-Vought) F7U Cutlass

"Chance-Vought Cutlass." RAF Flying Review, XI (June 1956), 48-49.

"Chance-Vought F7U-3 Cutlass." Air Pictorial, XIV (June 1952), 173-174.

"Crusaders in G-Suits." Air Trails, XXXVIII (September 1952), 26-27.

"The Curious Cutlass." RAF Flying Review, XII (December 1956), 38-41.

"Cutlass Case History." The Aeroplane, XCI (October 5, 1956), 515+.

"Cutlass Cuts." Aeronautics, XXXII (February 1955), 42-43.

"Cutlass Joins the Fleet." Aviation Week, LXIII (September 26, 1955), 32-33.

"The F7U-1." Aviation Age, XV (June 1951), 3-4.

Ginter, Steve. "Cutlass." Airpower, XIII (November 1983), 9-37.

"How Chance-Vought Builds the Cutlass." American Aviation, XVII (May 10, 1954), 26-27.

"Project Cutlass." Naval Aviation News, (June 1954), 1-5.

Schoeni, Arthur L. "Cutlass: The History of Chance-Vought's Unusual U.S. Navy Fighter." Air Classics, XII (June-August 1976), 55-60, 76-78, 36-41, 86-92, 48-56, 84-91.

"Sky Slicers." All Hands, no. 461 (July 1955), 19-20.

"Vought F7U Cutlass." Aero Digest, LXIV (January 1952), 26-32.

(Chance-Vought) F-8 Crusader

Book/Document/Report Literature:

Joos, Gerhard. Chance-Vought F-8A/E Crusader. Aircraft in Profile, no. 91. Windsor, Eng.: Profile Publications, 1966. 15p.

Pace, Steven J. Vought's F-8 Crusader, Pt. One. Naval Fighters Series, no. 16. Simi Valley, Calif.: Naval Fighters, 1988. 114p.

Sullivan, Jim. F-8 Crusader in Action. Carrollton, Tex.: Squadron/Signal Publications, 1985. 50p.

Tillman, Barrett. MIG Master: The F-8 Crusader. Annapolis, Md.: Nautical and Aviation Publishing Company of America, 1980. 224p.

Ward, Richard and Arthur L. Schoeni. Vought F-8 Crusader. Aircam Aviation Series, no. 31. New York: Arco, 1973. 48p.

Periodical/Anthology Literature:

Andrews, Hal. "Naval Aircraft: F-8 Crusader." Naval Aviation News, (April 1972), 20-21.

"Bloodlines of the Champion Crusader." Flying Review International, XIX (July 1964), 30-33+.

Bulban, Eric J. "Afterburner Pushes F-8 to Trans-U.S. Speed Mark." Aviation Week and Space Technology, LXVII (July 22, 1957), 30-32.

____. "F-8s Begin Duty on U.S.S. Saratoga." Aviation Week and Space Technology, LXVIII (January 27, 1958), 70-72.

"Carrier-Based Mach 1.5 Fighter." RAF Flying Review, XI (October 1955), 52-53.

"Carriers Crusade." Aeronautics, XXXVIII (May 1958), 52-53.

Cross, Roy. "F-8 Crusader." Air Pictorial, XIX (September 1957), 312-313.

"Crusader." Flight, LXXI (May 24, 1957), 691-696.

"The Crusader." RAF Flying Review, XII (December 1957), 64-65.

"Crusader Wins Veteran Status." Naval Aviation News, (December 1958), 22-24.

Cunningham, Doug. "The Crusaders." Naval Aviation News, (September 1976), 30-35.

"Design and Production of the Crusader Fighter." Automotive Industries, CXVII (July 1, 1957), 56-59.

"F-8 was Production Engineered." Aviation Week and Space Technology, LXVI (February 4, 1957), 60-61.

Haggerty, James J., Jr. "Collier Trophy Award: Chance-Vought F-8." Look, XXI (December 24, 1957), 36+.

Heumann, Gert W. "The Crusader Family." Air Pictorial, XX (October 1958), 353-355.

Irwin, Thomas C. "Last of the Crusaders." Naval Aviation News, (July 1982), 26-31.

Letzter, George J. "Crusader De-Tails." American Aviation Historical Society Journal, VII (Summer 1962), 115-119; IX (Summer 1964), 104-108.

Lord, Charles, Jr. "Crusaders." Air Classics, VIII (December 1971), 52-56.

"Mark III Crusader Gets a Test Run." All Hands, no. 501 (October 1958), 16-17.

Mersky, Peter S. "The Crusader: Farewell to the Fleet." U.S. Naval Institute Proceedings, CVIII (June 1982), 56-61.

Miller, Jay. "Crusader Without a Cause." Airpower, VII (July 1977), 13-25.

Mormillo, Frank B. "No More Crusaders." U.S. Naval Institute Proceedings, CXIII (August 1987), 80-81.

"Navy Puts F-8 Through Carrier Shakedown Trial." Aviation Week and Space Technology, LXVII (July 8, 1957), 60-61.

"The Navy's Fastest Fighter." Military Review, XXXVI (October 1956), 69-70.

Pace, Steve. "Crusader With a Cause." Wings, XVII (August 1987), 10-37.

Peck, John P. "Farewell to the Crusader." Wings of Gold, XII (Summer 1987), 23-27.

Quaile, John E. "Supersonic Jet." Military Engineer, L (November-December 1958), 450-451.

"Record for Crusader." Flight Magazine, LXV (October 1956), 40+.

Saul, J. R. "Production Planning for the Crusader." American Machinist, CI (February 25, 1957), 126-128.

Schoeni, Arthur L. "Crusader." Air Classics, XIII (November-December 1977), 28-37, 32-43; XIV (January 1978), 28-33, 92-97.

"Souped Up Sailor." RAF Flying Review, XV (September 1960), 22-23.

Witze, Claude. "F-8 Hits 1,000-mph Under Wraps." Aviation Week and Space Technology, LXV (August 27, 1956), 27-28.

F8C see OC Falcon

(Grumman) F8F Bearcat

Book/Document/Report Literature:

Andrews, Hal. The Grumman F8F Bearcat. Aircraft in Profile, no. 107. Windsor, Eng.: Profile Publications, 1965. 12p.

Chant, Christopher F. Grumman F8F Bearcat. Super Profile Series. Newbury Park, Calif.: G. T. Foulis/Interbook, 1985; Osceola, Wis.: Motorbooks International, 1988. 56p.

Jackson. B. R. and Tom E. Doll. Grumman F8F Bearcat. Aircraft, no. 20. Fallbrook, Calif.: Aero Publishers, 1970. 64p.

Periodical/Anthology Literature:

Brown, Eric. "Last of the Wartime Cats: The Bearcat Was No Kitten." Air International, XVIII (May 1980), 233-240.

Davisson, Bud. "A Classic Fighter is Brought Out of Hibernation." Air Progress, XLIV (Feburary 1982), 40-43.

Hanson, N. R. "Last of the Red Hot Cats: Grumman's Bearcat." Flying, LXXIV (February 1964), 44-47+.

"Last of the Line: The Bearcat." Flying Review International, XXI (January 1966), 310-314.

Miller, Jay. "Confederate Bearcats." Air Classics, XIV (June 1978), 28-33, 68.

Pace, Steve. "F8F Bearcat." Airpower, XVII (November 1987), 30-52.

Slaughter, Ted. "Last of the Red Hot Cats." Air Classics, X (July 1974), 38-47, 78.

Tillman, Barrett. "Bearcat." Wings, VI (February 1976), 32-45.

(Curtiss) F9C Sparrowhawk

Book/Document/Report Literature:

United States. Navy Department. Naval Air Station, Anacostia. Report on XF9C-1 Curtiss Single-Seat Fighter. Report, no. 91. Washington, D.C.: GPO, 1931. 69p.

_____._____._____. Report on XF9C-2 Curtiss Airship and Carrier Fighter. Report, no. 122. Washington, D.C.: GPO, 1932. 66p.

Periodical/Anthology Literature:

Andrews, Hal. "Naval Aircraft: Sparrowhawk." Naval Aviation News, (September 1971), 20-21.

____. "Sparrowhawk: The Airship Fighter." Naval Aviation News, (April-May 1980), 22-27, 20-25.

Baskerville, Hugo. "Sparrowhawk." Air Trails, NS III (Summer 1978), 36-43.

Hook, Tom. "The Flying Hookers." Air Classics, XV (February 1979), 72-79, 97-98.

Mackey, Donald. "The Control of Airship Planes." American Aviation Historical Society Journal, VIII (Spring 1963), 13-22.

Mikesh, Robert C. "That Great Hookup in the Sky." Wings, V (February 1975), 16-29, 54.

Mitchell, John C. "A Chronological History of the F9C-2 and Its Use on the Akron and the Macon." American Aviation Historical Society Journal, III (April-June 1958), 100-109.

Nye, William L. "The Genealogy of the Curtiss Sparrowhawk Dirigible Fighters." American Aviation Historical Society Journal, III (April-June 1958), 70-99.

Redman, Rod. "The F9C-2 Sparrowhawk." Air Classics, IX (April 1973), 14-17, 78.

"Sparrowhawk." Air Classics, XI (March 1975), 40-47.

Westburg, Peter. "Curtiss F9C-2 Sparrowhawk." Model Builder, VIII (October 1978), 36-39.

(Grumman) F9F Cougar/Panther

Book/Document/Report Literature:

Kinzey, Bert. F9F Cougar in Detail and Scale. Detail and Scale Series, v. 16. Simi Valley, Calif.: Naval Fighters, 1983. 71p.

_____. F9F Panther in Detail and Scale. Detail and Scale Series, v. 15. Simi Valley, Calif.: Naval Fighters, 1983. 71p.

Sullivan, Jim. F9F Panther/Cougar in Action. Aircraft, no. 51. Carrollton, Tex.: Squadron/Signal Publications, 1982. 50p.

Periodical/Anthology Literature:

Anderton, David A. "New Tiger is Supersonic in Level Flight." Aviation Week, LXI (August 23, 1954), 26-27+.
Cougar.

Andrews, Hal. "Grumman Panther." Aeroplane Monthly, IV (August 1976), 407-412.

_____. "Naval Aircraft: F9F-2/5 Panther." Naval Aviation News, (July 1978), 28-29.

_____. "Naval Aircraft: F9F-6/8 Cougar." Naval Aviation News, (September 1972), 24-25.

"The Big Transition Has Its Problems." Naval Aviation News, (February 1954), 17-19.
F9F-6.

Brewer, R. P. "And Then There Were None." Approach, XV (May 1970), 1-5.

"A Different Breed of Cat." Wings, VIII (June 1978), 20-39, 57.

Foxworth, Thomas G. "F9F Cougar." Aero Album, V (January 1972), 30-37.

"Grumman Cougar." Aero Digest, LXIX (September 1954), 32-36.

"Grumman Cougar." RAF Flying Review, XII (February 1957), 26-29.

"Grumman F9F Panther." Aero Digest, LXI (December 1950), 24-27.

"Grumman F9F-6 Cougar." Air Pictorial, XIV (June 1952), 173-174.

"Grumman F9F-8 Cougar." Air Pictorial, XVIII (September 1956), 310-311.

Hawkes, Robert. "Cougar Doubles as Fighter-Trainer." Aviation Week and Space Technology, LXVI (January 28, 1957), 66-67.

Jones, Cecil B. "F9F-2 Panther." In: Robin Higham and Carol Williams, eds. Flying Combat Aircraft of the USAAF-USAF. Ames, Iowa: Iowa State University Press, 1978. pp. 137-148.

McLarren, Robert. "Navy Reveals Grumman Jet Fighter." Aviation Week, XLVII (December 19, 1947), 12-13.

_____. "Panther 650-mph. Fighter Solves Jet Carrier Problem." Automotive Industries, XCVIII (January 1, 1948), 42+.

McLaughlin, G. F. "Grumman F9F." Aero Digest, LXIX (September 1954), 26-27.

Perkins, Kenny. "F9F-8 Cougar." Model Airplane News, CI (November 1980), 24-28.

Pitiky, Mark. "16,000-Pound Gorilla." Air Progress, XLVIII (October 1986), 6-11+.

Silsbee, Nathaniel F. "Grumman Panther." Aero Digest, LVII (July 1948), 28-30.

___. "Panther Performance." Air Op, XIII (June 1950), 32-33.

"Those Prowling Panthers." Naval Aviation News, (July 1974), 40-49.

(Curtiss) F11C Goshawk
(see also F2C Hawk)

Larkins, William T. "The Curtiss Goshawk." American Aviation Historical Society Journal, IX (Spring 1964), 58-63.

(Grumman) F11F Tiger

Book/Document/Report Literature:

Ginter, Steven J. Grumman F11F-1 Tiger. Naval Fighters Series, no. 1. Simi Valley, Calif.: Naval Fighters, 1980. 64p.

Kinzey, Bert. F11F Tiger in Detail and Scale: The Navy's First Supersonic Fighter. Detail and Scale Series, v. 17. Fallbrook, Calif.: Aero Publishers, 1984. 72p.

Periodical/Anthology Literature:

Anderton, David A. "New Tiger is Supersonic in Level Flight." Aviation Week, LXI (August 23, 1954), 26-27.

Andrews, Hal. "Naval Aircraft: Tiger." Naval Aviation News, (September 1973), 20-21.

Cross, Roy. "Grumman F11F-1F Tiger." Air Pictorial, XIX (April 1957), 126-127.

"F11F Carrier Suitability Tested on Saratoga." Aviation Week and Space Technology, LXVII (July 8, 1957), 82-83.

"The Grumman F11F-1F Super Tiger." Interavia, XIII (May 1958), 439-441.

"The Grumman Tiger." RAF Flying Review, XI (August 1956), 40-41.

Hotz, Robert. "F11F Sets New Speed, Altitude Marks." Aviation Week and Space Technology, LXV (December 24, 1956), 26-29.

"New Design Changes on Navy's Supersonic F11F." Aviation Week, LX (January 2, 1956), 30-31.

Trimble, Robert L. "F11F Tiger." Air Classics, XVI (November 1980), 32-37, 70-71.

Tsakis, Homer U. "Superiority Through Simplicity." Interavia, XIV (March 1959), 282-283.

Wetmore, Warren C. "Thrust Reverser Use Tested for Combat on Navy/Grumman F11." Aviation Week and Space Technology, CI (August 19, 1974), 61-63.

(Grumman) F-14 Tomcat

Book/Document/Report Literature:

Anderson, B. T., R. R. Meyer, and H. R. Chiles. Techniques Used in the F-14 Variable-Sweep Transition Flight Experiment. NASA-TM-100444. Edwards, Calif.: NASA Dryden Flight Research Center, 1988. 25p.

Drendel, Lou. F-14 Tomcat in Action. Carrollton, Tex.: Squadron/Signal Publications, 1977. 50p.

Kinzey, Bert. F-14A/B Tomcat in Detail and Scale: Su-22 Killer. Detail and Scale Series, v. 9. Fallbrook, Calif.: Aero Publishers, 1982, 1987. 72p.

____. F-14 Tomcat, Pt. I: Atlantic Coast Markings, the First Ten Years, 1974-1984. Color and Markings Series, v. 2. Fallbrook, Calif.: Aero Publishers, 1985. 64p.

____. and Ray Leader. F-14 Tomcat, Pt. II: Pacific Coast Squadrons, 1974-1987. Color and Markings Series, v. 8. Blue Ridge Summit, Pa.: TAB, 1988. 64p.

Miller, Jay. Grumman F-14A/B Tomcat. Minigraph, no. 3. Arlington, Tex.: Aerofax, Inc., 1984. 40p.

Reed, Arthur. F-14 Tomcat. New York: Scribners, 1978. 111p.

Spick, Mike. F-14 Tomcat. Modern Fighting Aircraft, v. 8. New York: Arco, 1985. 64p.

Stevenson, James. Grumman F-14 Tomcat. Aero Series, v. 25. Fallbrook, Calif.: Aero Publishers, 1975; Blue Ridge Summit, Pa.: TAB, 1986. 103p.

United States. Congress. House. Committee on Armed Services. Investigating Subcommittee. Crash of the F-14A: Report. 92nd Cong., 1st sess. Washington, D.C.: GPO, 1971. 10p.

____. ____. Senate. Committee on Armed Services. Subcommittee on Tactical Air Power. Navy Department--F-14/Grumman Financing: Hearings. 93rd Cong., 2nd sess. Washington, D.C.: GPO, 1974. 52p.

Periodical/Anthology Literature:

Antal, Jeffrey. "Time to Put the Cats Out." U.S. Naval Institute Proceedings, CXII (April 1986), 90-91.

Benjamin, Dick. "The Navy Takes a Look." Naval Aviation News, (February 1972), 35-39.

Bihrle, William, Jr. "F-14A High Angle of Attack Characteristics." Journal of Aircraft, XIII (August 1976), 576-583.

Blease, Frederick C. "The Tomcat is Purring." Grumman Aerospace Horizons, XIV (Spring-Summer 1978), 3-7.

Booda, Larry. "Faith, Patience, Perserverance, and the Tomcat." Sea Power, XV (March 1972), 13-18.

Bowes, Bill. "F-14D: Tomorrow's Tomcat." Wings of Gold, IX (Fall 1984), 36-38.

Braybrook, Roy M. "Grumman Tomcat: In a Class of Its Own." Air International, XXII (January 1982), 23-30.

____. "When Only the Best Will Do." Pacific Defense Reporter, IX (October 1982), 17-18+.

Brown, Kevin V. "Hot New Fighter for the Navy." Popular Mechanics, CXXXIII (January 1970), 83-86+.

Brown, Michael. "Tomcat Joins the Fleet: F-14A Fully Operational." Interavia, XXIX (December 1974), 1203-1207.

Campbell, J. A. "Tomcat Tailspin: Some Flat Facts." Approach, XXIII (October 1977), 10-13.

Christenson, Chris. "Sea Releases Her Grip." All Hands, no. 721 (February 1977), 4-6.
Salvage of a lost F-14A.

"The Congressional Dogfight Over the F-14." Business Week, (December 16, 1972), 59+.

Connolly, Thomas F. "The Air Combat Navy is in Good Shape." Data, XIV (June 1969), 25+.

Cornell, Douglas. "Grumman Prepares the F-14A for Its First Flight." Interavia, XXV (November 1970), 1384-1387.

"Dogfighter Supreme: The Grumman F-14A--The Tomcat." Air Enthusiast, V (January 1974), 7-16, 46-47.

Ethell, Jeffrey L. "The Modern Dogfight." Air Progress, XLII (June 1980), 41+.

_____. "Tomcat Tangle: We Fly the F-14 Tomcat." Air Progress, XLII (August 1980), 71+.

"The F-14 Cat Roars On." Naval Aviation News, (March-April 1983), 22-25.

"F-14 Adapts to Changing Threats." Defense Electronics, XVI (December 1984), 81-82.

"The F-14A Tomcat Air Superiority Fighter." International Defence Review, VI (August 1973), 455-457.

"F-14D: Tomorrow's Tougher Tomcat." Grumman Aerospace Horizons, XVIII (Fall 1982), 8-15.

Furlong, George M., Jr. "Tomorrow's Tomcat." Grumman Aerospace Horizons, XVIII (Fall 1982), 8-15.

Godfrey, David W. H. "Grumman Improves the Breed." Aircraft, LII (August 1973), 14-19, 48.

Griffin, Larry. "Bend the Sky." Car and Driver, XXVIII (July 1982), 113-114+.

"The Grumman F-14 in Detail." Interavia, XXIV (April 1969), 384-385.

Haering, George. "Flying Backseat in the F-14." Armed Forces Journal International, CXII (October 1974), 38-40.

Hanley, Art. "Sharpening the Claws of the Tomcat." U.S. Naval Institute Proceedings, CXVIII (June 1982), 109-113.

Harvey, Frank L. "Tomcat: The World's Deadliest Fighter Plane." Reader's Digest, C (February 1972), 41-44+.

Haugland, Vern. "New Fighter Aircraft." Ordnance, LIII (May-June 1969), 590-593.

Johnston, Robert L. "Grumman Wins F-14 Competition." Ordnance, LIII (March-April 1969), 448+.

Kandebo, Stanley W. "Grumman Develops Plan to Upgrade F-14D Weapon Sensor Capabilities." Aviation Week and Space Technology, CXXIX (August 29, 1988), 68-69.

____. "Navy/Grumman F-14 Upgrade Improves Range, Radar Avionics." Aviation Week and Space Technology,CXXIII (July 9, 1985), 52-53.

Kolcum, Edward H. "F-14 to Receive Improved TF-30 Engine." Aviation Week and Space Technology, CX (August 17, 1981), 62-63.

Kozicharow, Eugene. "Navy Switches F-14D to F-110 Engine." Aviation Week and Space Technology, CXX (July 16, 1984), 24-25.

Lert, Peter. "Grumman's MIG-Master: The F-14." Air Progress, XXXVI (March 1975), 40+.

Martin, Clark W. "F-14 Shows Dogfight Capabilities." Aviation Week and Space Technology, XCVIII (June 4, 1973), 38-40+.

____. "Navy Fighter to Counter Varied Threats: F-14 Systems Capabilities." Aviation Week and Space Technology, XCVIII (March 12, 1973), 31-35.

Miller, Barry L. "AWG-9 Provides Multi-Target Capability: F-14 Weapons Control System." Aviation Week and Space Technology, XCVIII (March 12, 1973), 36-37.

____. "Engine Woes Nag F-14 Fleet Introduction." Aviation Week and Space Technology, CIII (July 14, 1975), 16-17.

"More Reserve Tomcats." Naval Aviation News, (March-April 1987), 10-13.

Morrocco, John D. "Grumman Offers Tomcat 21 as Alternative to Navy ATF." Aviation Week and Space Technology, CXXIX (November 7, 1988), 18-19.

Myers, Henry E. "Navy F-14: New Fighter's Cost, Sophistication, Stir Controversy." Science, CLXXVI (April 7, 1972), 35-37.

"Navy's Revenge?: VFX, Replacement for F-111B." Forbes, CII (August 1, 1968), 19-20.

Nihart, Brooke. "Navy's F-14: Costly, But Nothing Else Will Do the Job." Armed Forces Journal, CIX (December 1971), 40-41.

Norris, John G. "The Navy's Hot New Fighter Plane." Navy, XII (March 1969), 9-12.

North, David M. "Navy Opts to Cut F-14D Buy, Will Complete F-14A Update." Aviation Week and Space Technology, CXXVI (March 30, 1987), 18-20.

Pace, Steve. "Tomcat." Wings, XVI (April 1986), 10+.

Pearson, J. F. "A Hot Plane Gets Its Sea Legs: F-14." Popular Mechanics, CXLVI (December 1976), 68-69+.

Polmar, Norman. "VFX-1: Successor to the F-111B." Aerospace International, IV (September-October 1968), 22-24.

Powers, William M. "The F-14 Tomcat." U.S. Naval Institute Proceedings, CII (October 1976), 75-87.

Prina, L. Edgar. "New Competition for 'Scaled Down' Fleet Fighter." Sea Power, XVI (July 1973), 34-37.

____. "Tomcat Gets Into Dogfight on Capitol Hill." Sea Power, XVI (April 1973), 43-45.

Robinson, Clarence A., Jr. "Backfire Threat Spurs F-14 Upgrading." Aviation Week and Space Technology, CXVII (August 30, 1982), 49-52.

____. "F-14 Demonstrates Agile Aerial Combat Capabilities." Aviation Week and Space Technology, CV (November 29, 1976), 46-53.

____. "F-14 Procurement to Cost $13.9 Million Per Unit." Aviation Week and Space Technology, XCIX (September 10, 1973), 13-15.

Ropelewski, Robert R. "F-14 Performance Up With F-101 Engine." Aviation Week and Space Technology, CXV (October 19, 1981), 44-45.

Saki, Joe. "Tamiya's Big Cat." Scale Models, XII (September 1981), 460-464.

Sellers, B. R. "Today's Acquisition Strategy for Tomorrow's Tomcat." Program Manager, XIV (September-October 1985), 18-21.

Simonsen, Erik. "Navy's New 'Top Gun' Team." International Combat Arms, VI (January 1988), 48-51+.

Spick, Mike. "F-14 Tomcat." In: William T. ("Bill") Gunston, ed. The Great Book of Modern Warplanes. New York: Crown, 1987. pp. 585-648.

"The Swing-Wing Birds of Peace." Air Enthusiast, I (July 1971), 7+.

"Testing the Tomcat." Interavia, XXVII (January 1972), 35-37.

Tillotson, F. Lee. "The F-14D: The Right Stuff." Wings of Gold, VIII (Fall 1983), 22-26.

Velocci, Tony. "Tomorrow's F-14 Tomcat." National Defense, LXVII (April 1983), 23-24+.

Winnefeld, James A., Jr. "Fresh Claws for the Tomcat." U.S. Naval Institute Proceedings, CXIII (July 1987), 103-107.

Winston, Donald C. "Navy Urges Full F-14 Funding." Aviation Week and Space Technology, XCVI (April 3, 1972), 14-15.

"Working Up for the F-14." Flight International, XCV (March 6, 1969), 370-371.

Yaffee, Michael L. "Combat Agility Marks F-14 Design Concept." Aviation Week and Space Technology, XC (February 24, 1969), 18-21.

Yorke, Duane. "Grumman Couldn't Bear It: The F-14 Story." Washington Monthly, V (July-August 1973), 72-79.

(McDonnell-Douglas) F-15 Eagle

Book/Document/Report Literature:

Beard, Terry L. U.S. Air Force Instructor Upgrade Course: F-15. Research Paper. Maxwell AFB, Ala.: Air Command and Staff College, Air University, 1982. 50p.

Burcham, F. W., Jr., et al. Summary of Results of NASA F-15 Flight Research Program. NASA-TM-86811. Moffett Field, Calif.: NASA Ames Research Center, 1986. 34p.

Campbell, Frank A. and George W. Hawks, Jr. Handbook for F-15 Instructor Pilots in Tactical Training Wings. Research Paper. Maxwell AFB, Ala.: Air Command and Staff College, Air University, 1980. 62p.

Christenson, Gerald A. and James C. Horne. A Study of the F-15 Contract Structure and Its Contribution to Effective Program Management. Thesis. Wright-Patterson AFB, Ohio: School of Systems and Logistics, Air Force Institute of Technology, 1973. 163p.

Dedrick, Craig R. Effect of Current Rated Personnel Management Policies on the Combat Capability of the F-15 Aircraft. Research Paper. Maxwell AFB, Ala.: Air Command and Staff College, Air University, 1986. 54p.

Dickey, Roy S. The Advocacy of the F-15. Research Report. Maxwell AFB, Ala.: Air War College, Air University, 1973. 57p.

Digital Electronic Engine Control (DEEC) Flight Evaluation in an F-15 Airplane. NASA-CP-2298. Edwards, Calif.: NASA Dryden Flight Research Center, 1984. 243p.

Drendel, Lou. Eagle: The McDonnell-Douglas F-15. Carrollton, Tex.: Squadron/Signal Publications, 1986. 64p.

___. F-15 Eagle in Action. Carrollton, Tex.: Squadron/Signal Publications, 1986. 50p.

Ethell, Jeffrey L. F-15 Eagle. Osceola, Wis.: Speciality Press, 1981. 128p.

Foxwell, Mark B. F-15 Capabilities Against the AS-X-15. Research Report. Maxwell AFB, Ala.: Air War College, Air University, 1985. 30p.

Gething, Michael J. F-15 Eagle. Modern Fighting Aircraft, no. 1. New York: Arco, 1983. 64p.

Gordon, William S., III. Developing an Air Combat Formation for the F-15. Research Paper. Maxwell AFB, Ala.: Air Command and Staff College, Air University, 1973. 67p.

Kinzey, Bert. F-15 Eagle in Detail and Scale: Versions A, B, C, D, and E. Detail and Scale Series, v. 14. Fallbrook, Calif.: Aero Publishers, 1984. 72p.

Lennon, Tom and Jim Wray. Bringing the F-15 to Operational Readiness. Langley AFB, Va.: History Office, Tactical Air Command, 1987. 62p.

McDonnell-Douglas Corporation. Public Information Office. The F-15E Dual-Role Fighter. St. Louis, Mo., 1987. 6p.

____.____. The McDonnell-Douglas F-15 Eagle. St. Louis, Mo., 1987. 2p.

____.____. The MSIP F-15. St. Louis, Mo., 1986. 4p.
MSIP = Multi-Stage Improvement Program.

____.____. What If Our Adversaries Had Aircraft Like the Eagle? St. Louis, Mo., 1987. 12p.

Pinter, W. W. Computers in Weapons Systems: A Look at the F-15. Research Paper. Maxwell AFB, Ala.: Air Command and Staff College, Air University, 1988. 28p.

Rogers, Danny R. Effect of the Visual Identification Requirement on Employment of the F-15 Eagle in the NATO Theater. Thesis. Wright-Patterson AFB, Ohio: School of Engineering, Air Force Institute of Technology, 1982. 66p.

Smith, Stanton T., III. F-15 Surface-Attack Systems. Research Paper. Maxwell AFB, Ala.: Air Command and Staff College, Air University, 1985. 48p.

Spick, Mike. F-15 Eagle. Combat Aircraft Series, no. 1. Harrisburg, Pa.: Stackpole Books, 1986. 48p.

Stevenson, James P. McDonnell-Douglas F-15 Eagle. Aero Series, no. 28. Fallbrook, Calif.: Aero Publishers, 1986. 104p.

Westwood, Edward C. The F-15 Eagle Engine Development. Research Report. Maxwell AFB, Ala.: Air War College, Air University, 1975. 154p.

Wetterhahn, Ralph F. F-15 Tactics Against a Look-Down, Shoot-Down Threat. Research Report. Maxwell AFB, Ala.: Air War College, Air University, 1984. 47p.

Periodical/Anthology Literature:

"The ADCOM Eagle Sharpens Its Claws." Interceptor, XVII (October 1975), 8-11.
ADCOM = USAF Aerospace Defense Command.

Alexander, John V. "The Eagles Fly Red Flag: Full Details of the F-15 in Operational Training." Air Combat, V (July 1977), 56-83.

Allward, Maurice. "F-15: America's Superior Eagle." Design Profile, no. 7. Air Pictorial, XXXVI (August 1974), 289-293.

Andrews, Walter. "The F-15: Problems and Prospects." Armed Forces Journal, CVI (May 31, 1969), 15+.

"ASD Pushes Total Management for F-15." Aviation Week and Space Technology, XCVI (June 26, 1972), 66-67.
ASD = USAF Aerospace Systems Division.

Bailly-Cowell, G. M. "NATO Central Europe Gets the F-15." NATO's Fifteen Nations, XX (April-May 1975), 50-58.

Banholzer, Dick. "F-15E: Dual-Role Fighter." Aerospace America, XXII (May 1984), 54-58.

Beason, Robert G. "We Fly in the Hottest Plane of All: The F-15 Eagle." Mechanics Illustrated, LXXV (February 1979), 44-58.

Bellis, Benjamin H. "The Coming Cost Crunch of the F-15." Air Force Magazine, LV (January 1972), 38-43.

____. "For Air Superiority--The F-15." Ordnance, LV (July-August 1970), 64-67.

Berry, F. Clifton, Jr. "Becoming an F-15 Wingman." Air Force Magazine, LXV (March 1982), 116-122.

____. "Flying the Enhanced Eagle." Air Force Magazine, LXIV (May 1981), 44-47.

Blair, Edwin T. "The Fighter's Fighter." Airman, XIII (June 1969), 42-45.

Bowman, Buddy. "To Fly and Fight." USAF Fighter Weapons News Letter, no. 4 (Winter 1970), 10-11+.

Braybrook, Roy M. "New Roles for the F-15 Eagle." Air International, XXI (August 1981), 62-68, 82-83.

Brookes, William R. "Bitburg's Eagles." Air Combat, VIII (May 1980), 38-41.

Brown, Kevin V. "A Fighter Pilot's Fighter Plane, the F-15." Popular Mechanics, CXXXIV (December 1970), 73-77.

Brownlow, Cecil. "Air-to-Air Capability Seen for F-15." Aviation Week and Space Technology, XCVII (July 3, 1972), 18-19.

Buchholz, B. B. "When Your Only Customer is the Federal Government." Marketing and Media Decisions, XIX (April 1984), 58-61.

Burrows, Irv. "F-15 Flight Test Report." USAF Fighter Weapons Review, (Winter 1973), 18-22.

Carson, Don. "Flying the F-15." Air Force Magazine, LVII (February 1974), 23-29.

Reprinted in Aerospace International, IX (July-August 1975), 12-14+.

Cole, Dave. "Battle Birds Ace-High." Airman, XV (March 1971), 12-15.

Correll, John T. "The Fastest Climb." Airman, XIX (May 1975), 30-34.

Cox, Bill. "F-15 Fighter: What It's Like to Fly the Hottest of the Fighters." Plane and Pilot, XV (July 1979), 9-11.

Dennison, John M. "Making the McDonnell-Douglas F-15 Even Better." Interavia, XXXIII (December 1978), 1178-1180.

"Dogfight Over the F-15." Business Week, (December 20, 1969), 96-98.

Durocher, Edward N. "F-15 Aerodynamics." Flying Safety, XXXVII (January 1981), 8-13+.

"The Eagle." TAC Attack, XII (September 1972), 29-30.

"Eagle: McDonnell-Douglas's New Act." Aircraft, LIII (October 1973), 12-17, 40.

Ethell, Jeffrey L. "F-15: World's Best Dogfighter." Air Progress, XLII (March 1980), 64-69, 80.

"The F-15 Competition." Armed Forces Journal, CVII (September 13, 1969), 14-23.

"F-15E." Defence Update International, no. 71 (May 1986), 62-63.

"The F-15 Eagle Air Superiority Fighter." International Defence Review, V (October 1972), 522-524.

"The F-15 Eagle Starts Flight Tests." Interavia, XXVII (August 1972), 813-814.

"F-15 STOL: A New Technology Demonstrator." Defence Update International, no. 79 (March-April 1987), 46-51.
STOL = Short Takeoff and Landing.

"F-15 STOL Technology Demonstrator Passes Airframe, Engine Reviews." Aviation Week and Space Technology, CXXV (August 25, 1986), 97+.

Fallows, James M. "I Fly With the Eagles." Atlantic, CCXLVIII (November 1981), 70-77.
Reprinted in Reader's Digest, CXX (May 1982), 144-148.

____. "The Trouble With the Air Force's Eagle." Atlantic, CCXLIX (January 1982), 64-66.

Ferguson, James. "F-15: Can This Management Prototype Pass the 'Real World' Test?" Armed Forces Management, XVI (November 1969), 41-43.

Foster, Bob. "The Sun Never Sets on World's F-15s." Eagle Flight Leader, VI (September 1987), 1-8.

Furlong, R. D. M. "Operational Aspects of the F-15 Eagle." International Defence Review, VIII (June 1975), 361-371.

Gaines, Mike. "Air Combat: Eagle Style." Flight International, CXVIII (August 30, 1980), 886-890.

____. "STOL Eagle." Flight International, CXXXII (November 8, 1987), 32-34.

Geddes, J. Philip. "High Performance and the F-15." Interavia, XXX (February 1975), 190-195.

Gething, Michael J. "F-15 Eagle." In: William T. ("Bill") Gunston, ed. The Great Book of Modern Warplanes. New York: Crown, 1987. pp. 73-136.

Gillert, Doug. "Eagle's Comeback." Airman, XXX (March, 1986), 9-14.

Gilson, Charles. "Strike Eagle Proposed for Air-to-Ground Missions." International Defence Review, XIII (August 1980), 1188-1189.

____. and Dan Boyle. "The F-15 and the Defence of Europe." Internationl Defence Review, XI (May 1978), 696-700.

Greene, Robert M. "The F-15 and System Safety." Aerospace Safety, XXVI (July 1970), 1-4.

Grimes, Vincent. "F-15 and F-16 First-String Strike Team." International Combat Arms, VI (September 1988), 48-51+.

Guarino, Gilbert B., et al. "Faith Restored: The F-15 Program." Air University Review, XXVII (January-February 1976), 63-77.

Gulick, John F. "The Eagle Makes the Scene." Airman, XVI (October 1972), 24-30.

Gunston, William T. ("Bill"). "More Eagles for the USAF Eyrie." Air International, XXXIII (July 1987), 7-13, 47-48.

Hamilton, Bill. "One Versus Many." USAF Fighter Weapons Review, (Fall 1977), 1-7.

Hardison, Stan. "A Visit With the Eagle: The Maintenance Man's Dream." TAC Attack, XIV (March 1974), 4-6.

Harvey, David. "Eagles at Angels: Flying the McDonnell-Douglas USAF F-15 Eagle." Defense and Foreign Affairs, VII (May 1979), 12-15+.

Hawkes, George W., Jr. "Flying the F-15 Eagle." Aerospace Historian, XXVII (March 1980), 2-8.
Reprinted in Robin Higham and Carol Williams, eds., Flying Combat Aircraft of the USAAF-USAF (Manhattan, Kans.: Sunflower University Press, 1981), pp. 97-104.

Heenan, Mike. "The Eagle is a Pretty Bird." Air Classics Quarterly Review, III (Fall 1976), 66-77.

Hill, Martin V. "F-15." Flying Safety, XLIII (March 1987), 14-16; XLIV (January 1988), 12-15.

____. "Out of Control in the Eagle." Flying Safety, XLIV (August 1988), 8-11, 24-26.

Hoag, Robert J. "It's a Bird--No, It's a Plane--It's--Super Fighter." TAC Attack, XIV (November 1974), 4-10.

____. "Super Fighter." USAF Fighter Weapons Review, (Summer 1974), 18-30.

Jacobs, Gene. "The F-15E Strike Eagle: When Only the Best Will Do." Asian Defence Journal, no. 10 (October 1985), 16+.

Jennings, G. L. and P. Henry. "F-15 SAR." Society of Experimental Test Pilots Technical Review, XVI (Spring 1981), 182-197.
SAR = Synthetic Aperture Radar.

Johnson, Lionel P. "Alert Eagle." Interceptor, XX (August 1978), 8-11.

Jose, John. "F-15E on Track." Defense Electronics, XVII (August 1985), 10-12+.

Kandebo, Stanley W. "First Flight of STOL Performance F-15 Demonstrator Set for July." Aviation Week and Space Technology, CXXVIII (June 20, 1988), 107-109.

Kaye, Michael J. "F-15." Flying Safety, XL (April 1984), 18-19; XLI (March 1985), 15-16; XLII (March 1986), 14-16.

Kozicharow, Eugene. "USAF Selects F-15 as Dual-Role Fighter." Aviation Week and Space Technology, CXX (March 5, 1984), 18-19.

"Landing on Less." Air Force Magazine, LXX (April 1987), 74-78.
F-15 STOL demonstrator.

Larkin, Michael B. "F-15 Gun Employment." USAF Fighter Weapons Review, (Spring 1981), 28-31.

Lay, Theodore W., II. "FLIR, Fast Pac, and the F-15: The Newest Fighter in Town--The F-15E Dual-Role Fighter." USAF Fighter Weapons Review, (Summer 1988), 21-24.
FLIR = Forward Looking Infrared.

Levin, Stuart M. "F-15: The Teething of a Dogfighter." Space/Aeronautics, LII (December 1969), 36-47.

Loosbrock, John P. "F-15 Eagle: U.S. Air Force's Top-Priority Fighter." Air Force Magazine, LV (August 1972), 22-23.

Machat, Mike. "Eagle Flight." Wings, XVIII (February 1988), 24-51.

McNickle, Marvin L., et al. "The F-15: Tomorrow's Champion of the Dogfight." Air Force and Space Digest, LII (July 1969), 28-34.

Malvern, Donald. "The F-15A Eagle Program: A Reliability Case History." Defense Management Journal, XII (April 1976), 40-45.

Martin, Clark W. "F-15: Special Report." Aviation Week and Space Technology, C (March 25-April 8, 1974), 40+, 50-53, 51-55.

____. "Simplified Assembly of F-15 Stressed." Aviation Week and Space Technology, XCIX (October 29, 1973), 47-53.

Martin, Robert P. "The Look of Eagles." Newsweek, XCI (April 24, 1978), 52-53.

Mascot, Thomas K. "Aerial Gunnery: Eagle Style." USAF Fighter Weapons Review, (Summer 1986), 2-7.

Meyer, John C. "Air Superiority Into the 1990s." Air Force Magazine, LIV (August 1971), 42-45.

"New Eagle on the Horizon." Defence Update International, no. 70 (August 1986), 24-33.

Nicholson, Robert W., Jr. "Eagles in Arabia." Airman, XXII (July 1979), 2-6.

O'Keefe, Tim, Jr. "Eagle Eye." USAF Fighter Weapons Review, (Fall 1978), 26-29.

Pluta, John C. "F-15." Flying Safety, XXXIX (March 1983), 9-10.

Powell, Craig. "The FX: Designed to Control the Skies." *Armed Forces Journal*, XV (November 1968), 38-40.

"Predator With a Pedigree." *Air International*, XXVI (1984), 142+.

Prochoroff, Alan. "Eagle Owns the Sky." *Citizen Airman*, XXXVIII (August 1986), 8-11.

Pugh, Craig. "Alert Eagles." *Airman*, XXXI (July 1987), 20-21.

"Quality in the Air: The F-15." *Ms*, I (February 1984), 70-73.

Raymer, Daniel P. "Supercruise for a STOL Fighter." *Aerospace America*, XXIII (August 1985), 72-75.

"Report on the F-15 Programme." *Interavia*, XXVIII (February 1973), 148-149.

Rhodes, Jeffrey P. "Newest Eagle." *Air Force Magazine*, LXX (February 1987), 74-75.

Ritchie, Steve. "An Eagle for All Arenas." *Air Force Magazine*, LXVI (November 1983), 43-49.

Robins, Yves. "Flying the Eagle." *Aerospace International*, XIV (April-May 1978), 24-26+.

Ropelewski, Robert R. "F-15 Fighter Abilities Evaluated." *Aviation Week and Space Technology*, CXVI (April 26, 1982), 39-43.

____. "Integrated Systems Evaluated on F-15." *Aviation Week and Space Technology*, CXVIII (April 11, 1983), 47-50.

Ross, Donald O., Jr. "F-15 Mixed-Force Employment." *USAF Fighter Weapons Review*, (Winter 1980), 9-13.

Schemmer, Benjamin F. "The Eagle is Not a Turkey or a Dove!" *Armed Forces Journal International*, CXII (March 1975), 38-40+.

Schultz, James B. "Production Nears for F-15E Dual-Role Fighter." *Defense Electronics*, XVI (September 1984), 90-92+.

Scott, Phil. "Fighter Country." *Flying*, CXV (February 1988), 76-82.

Scott, William B. "F-15E Achieving Operational Sortie Rate While Still in Test." *Aviation Week and Space Technology*, CXXVIII (May 9, 1988), 59-60.

____. "USAF, McDonnell-Douglas Prepare to Deploy F-15E for Winter Testing." *Aviation Week and Space Technology*, CXXX (January 2, 1989), 79-82.

Skanchy, Thomas C. "Flying the Eagle." *Aerospace Safety*, XXXIII (February 1977), 2-5.

Smith, Gene. "Eagle Driver." *Air Progress*, XLIX (May 1987), 46-53+.

Smith, Roger J. "The Day the Eagle Streaked." Air Force Magazine, LVIII (July 1975), 32-37.

Sweetman, Bill. "F-15E: Heavyweight Eagle." Interavia, XLII (March 1987), 262-263.

____. "Short-Field F-15." Interavia, XLI (July 1986), 791-792.

"The Swing-Wing Birds of Peace." Air Enthusiast, I (July 1971), 7+.

Turk, Paul. "The F-15E." International Defence Review, XVIII (August 1985), 1261-1262.

Ulsamer, Edgar. "The F-15: Tomorrow's Champion of the Dogfight Arena." Air Force and Space Digest, LII (July 1969), 28-34.

"USAFE F-15s Used for Air Defense Intercepts." Aviation Week and Space Technology, CXVI (July 7, 1982), 68-69.

Utz, Terry E. "The Nuts 'n Bolts of F-15 Standardization." Defense Management Journal, IX (April 1973), 26-27+.

Winters, Deacon. "Two, Bingo, May-Day." USAF Fighter Weapons Review, (Fall 1987), 13-21.

Witze, Claude. "Back on Track." Air Force and Space Digest, LIII (February 1970), 16-18.

Wright, Wayne K. "The F-15E: A New Home for the WSO." Navigator, XXXI (Winter 1984), 5-9.

(General Dynamics) F-16 Fighting Falcon

Book/Document/Report Literature:

Abell, J. B., et al. The Costs and Performance Implications of Reliability Improvements in the F-16A/B Aircraft. RAND Note RN-2499-AF. Santa Monica, Calif.: RAND Corporation, 1988. 77p.

Bell, Robert C. F-16 Coproduction Experience. Research Report. Maxwell AFB, Ala.: Air War College, Air University, 1980. 25p.

Dorfer, Ingamar. Arms Deal: The Selling of the F-16. New York: Praeger, 1983. 287p.

Drendel, Lou. F-16 Fighting Falcon in Action. Aircraft, no. 53. Carrollton, Tex.: Squadron/Signal Publications, 1983. 50p.

Gentry, Jerauld. Evolution of the F-16 Multinational Fighter. Thesis. Washington, D.C.: Industrial College of the Armed Forces, 1976. 72p.

Greenlee, Arthur and Michael D. O'Neill. Peace Gate: A Case Study of F-16 FMS Management. Thesis. Wright-Patterson AFB, Ohio: School of Systems and Logistics, Air Force Institute of Technology, 1985. 143p.
Pakistan sale.

Hoff, Larry A. The F-16 Coproduction Program: An Analysis. Research Paper. Maxwell AFB, Ala.: Air Command and Staff College, Air University, 1981. 32p.

Holder, William G. and William D. Siuru, Jr. General Dynamics F-16. Aero Series. Fallbrook, Calif.: Aero Publishers, 1976. 104p.

Huff, Byron L. The F-16 Wild Weasel: A Feasibility Study. Thesis. Fort Leavenworth, Kans.: Army Command and General Staff College, 1979. 63p.

Kinzey, Bert. F-16A/B Fighting Falcon in Detail and Scale. Detail and Scale Series. Fallbrook, Calif.: Aero Publishers, 1982. 72p.

Loy, Noah E. Specialized Aircrews and Prudent Tasking Will Improve the F-16's Combat Potential. Research Report. Maxwell AFB, Ala.: Air War College, Air University, 1979. 96p.

Miller, Jay. General Dynamics F-16 Fighting Falcon. Arlington, Tex.: Aerofax, Inc., 1982. 124p.

Richardson, Doug. F-16 Fighting Falcon. Modern Combat Aircraft, no. 2. New York: Arco, 1983. 64p.

Siuru, William D., Jr. General Dynamics F-16 Fighting Falcon. Aero Series, v. 26. Fallbrook, Calif.: Aero Publishers, 1983. 103p.

Talty, Patrick. F-16 XL Demonstrates New Capabilities in Its Flight Test Program at Edwards AFB, California. Research Paper. Maxwell AFB, Ala.: Air Command and Staff College, Air University, 1986. 36p.

United States. Government Accounting Office. Logistics and Communications Division. F-16 Integrated Logistics Support: Still Time to Consider Economical Alternatives. Washington, D.C., 1980. 56p.

____.____. National Security and International Affairs Division. An Assessment of the Air Force's F-16 Aircraft Multiyear Contract. Washington, D.C., 1986. 26p.

____.____. Procurement and Systems Acquisition Division. The Multinational F-16 Aircraft Program: Its Progress and Concerns. Washington, D.C., 1979. 43p.

____.____.____. Sharing the Defense Burden: The Multinational F-16 Aircraft Program. Washington, D.C., 1977. 46p.

____.____.____. Status of the Air Force's F-16 Aircraft Program. Washington, D.C., 1978. 48p.

Periodical/Anthology Literature:

Adam, J. A. "Fighter Aircraft: Small and Agile." IEEE Spectrum, XXIV (May 1987), 64-65.

"Advanced Fighter Technology Integration/F-16: Flying Testbed for Future Aviation." Defense Electronics, XIV (August 1982), 88-91.

"AFTI F-16: In Search of New Ways to Fly and Fight." Defence Update International, no. 73 (July 1986), 48-52+.

Anderson, Neil R. "F-16 ACF Pilot Report." NATO's Fifteen Nations, XXI (August-September 1976), 24-32.

Andrews, Hal. "Naval Aircraft: F-16N Fighting Falcon." Naval Aviation News, (May-June 1988), 20-21.

Baker, Mark. "F-16 Loft." USAF Fighter Weapons Review, (Winter 1982), 2-12.

Berent, Mark. "F-16 Fighting Falcon: A Phenomenal Fighter." Asian Defence Journal, no. 3 (March 1987), 76+.

Berry, F. Clifton, Jr. "Flying the F-16." Air Force Magazine, LXIV (October 1981), 73-79.

____. "The Revolutionary Evolution of the F-16XL." Air Force Magazine, LXVI (November 1983), 50-55.

Boyle, Dan. "F-16: The Multinational, Multirole Aircraft." Interavia, XXXIV (January 1979), 39-43.

Brake, Mike. "The New Kids on the Block." USAF Fighter Weapons Review, (Fall 1984), 22-29.

Braybrook, Roy M. "Everybody Loves the Falcon." Pacific Defense Reporter, XII (November 1985), 57-58+.

____. "Fighting Falcon vs. Hornet." Air Progress, XLIII (January 1981), 42-47.

Brechner, Berl. "Electric Jet." Flying, CXI (December 1984), 66-71+.

Bulban, Eric J. "F-16 Demonstrator to Test New Combat Technology." Aviation Week and Space Technology, CXVIII (July 19, 1982), 23-24.

____. "F-16 Modifications Authorized." Aviation Week and Space Technology, CXIII (September 1, 1980), 291+.

____. "Studies Pinpoint Cost-Saving F-16 Manufacturing Techniques." Aviation Week and Space Technology, CVI (May 2, 1977), 81-82+.

____. "YF-16 Stresses Advanced Technology." Aviation Week and Space Technology, C (January 7, 1974), 40-44+.

Carroll, Robert G. H. "F-16: Swing-Force Fighter for the '80s." Air Force Magazine, LIX (April 1976), 30-35.

Clostermann, Jacques. "Mirage 2000 and F-16: One Pilot Flies Them Both." Interavia, XLI (September 1986), 975-976.

____. "9G and Other Experiences in the F-16." Interavia, XLI (June 1986), 617-619.

Culkin, Dennis. "New Fight for the Fighting Falcon." Defense and Foreign Affairs, XI (May 1983), 20-21+.

Cutler, Katie. "Five Friendly Forces and Their Fighting Falcons." Airman, XXVII (March 1983), 21-25.

Day, Bonner. "The F-16 Goes Operational." Air Force Magazine, LXII (January 1979), 34-37.

Delle Capre, Enrico. "F-16: A Fighter Pilot's Fighter." Asian Defence Journal, no. 2 (February 1983), 8-12+.

Diskin, George W. "The F-16 Connection." Electronic Warfare, VIII (July-October 1976), 83-85+.

Dornheim, Michael A. "Eidetics Proposes Upgrades to Improve F-16 Performance." Aviation Week and Space Technology, CXXVII (September 28, 1987), 59-61.
Eidetics International, Inc.

Eagle, Dan. "A Two-Edged Sword." TAC Attack, XXIV (June 1984), 18-21.

Edwards, Wayne C. "The F-16: Not Like Any Other." Air Force Magazine, LXIII (August 1980), 34-39.

Ennis, Michael. "Tally Ho!" Texas Monthly, IX (June 1981), 132+.

Ethell, Jeffrey L. "Dog Fighter." Air Progress, XLVII (February 1985), 34+.
F-16B.

Ettinger, R. C. and M. B. Johnston. "F-16 Flight Test Progress Report." Society of Experimental Test Pilots Technical Review, XIII (Fall 1977), 1-13.

"F-16 Air Combat Fighter." NATO's Fifteen Nations, XIX (December 1974-January 1975), 37-40.

"F-16 Avionics Show Revolutionary Progress." Defense Electronics, XVI (December 1984), 92-93.

"F-16 European Test and Evaluation." Interavia, XXXIV (December 1979), 1177-1179.

"F-16 Export Push: Special Report." Aviation Week and Space Technology, CII (February 24, 1975), 14-18.

"The F-16 Fighting Falcon: Simply Lethal." National Guard, XXXVII (September 1983), 20-23.

"F-16 Fighting Falcon: Why Switch When You've Got a Winner?" Defence Update International, no. 69 (March 1986), 6-10+.

"The F-16 Meets NATO Goals." Defense Systems Review, II (February 1984), 52-54+.

"F-16 Multinational Fighter Program: Special Report." Aviation Week and Space Technology, CVI (May 2, 1977), 44-130.

"F-16A." TAC Attack, XXVI (May 1986), 16-17.

"F-16B Wild Weasel Versions Studied." Aviation Week and Space Technology, CVIII (April 17, 1978), 22-23.

"F-16C." TAC Attack, XXV (May 1985), 16-17.

"First of a New Breed." Armed Forces Journal International, CXI (January 1974), 8-10.

"Fly Me: General Dynamics F-16." Newsweek, LXXXV (January 27, 1975), 49-50.

Frantz, Bob. "A New-Generation Adversary Challenges Fleet Fighters." Naval Aviation News, (September-October 1987), 20-22.
F-16N.

Gallagher, S. M. "F-16: The Fighter Everybody Wants." Popular Mechanics, CXLIV (November 1975), 72-73+.

Geddes, J. Philip. "F-16 Coproduction: An American Point-of-View." Interavia, XXXIV (November 1979), 1090-1092.

"General Dynamics F-16: The U.S. Air Force's New Air Combat Fighter." Aircraft Illustrated, VIII (April 1975), 150-153.

"The General Dynamics F-16XL." Asian Defence Journal, no. 10 (October 1982), 102-103.

Gennin, George S. "F-16." USAF Fighter Weapons Review, (Spring 1976), 1-8.

Godfrey, David W. H. "U.S. Air Force Selects the YF-16." Aircraft, LIV (February 1975), 18-20.

Grasset, Phillipe. "The F-16 Choice: A Lesson for the Future." Interavia, XXX (August 1975), 883-886.

Greeley, Brendan M., Jr. "USAF Selects Modified F-16A as Air Defense Fighter." Aviation Week and Space Technology, CXXV (November 10, 1986), 25-26.

Gross, Richard C. "Regaining Lost Agility for the F-16--Agile Falcon." Defense Science and Electronics, VI (October 1987), 7-8+.

Harvey, David. "Another Stab at the F-16." Defense and Foreign Affairs, VI (May 1978), 35-36.

Harvey, Frank L. "The F-16: Super Hot Rod of the Skies." Mechanics Illustrated, LXXII (December 1976), 36-37+.
Abridged in Reader's Digest, CIX (December 1976), 142-145.

Hillaker, H. J. "YF-16 Design Concept and Philosophy." Israel Journal of Aircraft, XIX (Spring-Summer 1981), 3-18.

Hoey, Brian. "Best Airplane on the Block." Airman, XXX (September 1986), 40-41.

___. "Flying Into the Future." Airman, XXIX (July 1985), 21-27.
F-16XL.

Horton, Chuck and Bob Baker. "F-16 Time-on-Target Clock." USAF Fighter Weapons Review, (Spring 1983), 8-11.

Kincaid, C. G. and G. A. Cude. "The F-16 Fighter: Marvel of Maintainability." National Defense, LXI (September-October 1976), 116-117.

Kozicharow, Eugene. "Air Force Considers Using Integrated Avionics in F-16." Aviation Week and Space Technology, CXX (February 13, 1984), 160+.

___. "First European F-16 Delivery Scheduled." Aviation Week and Space Technology, CX (January 22, 1979), 42-45+.

___. "General Dynamics Offers USAF Specially-Configured F-16C." Aviation Week and Space Technology, CXXII (June 24, 1985), 19-20.

___. "Updates Begin on Multinational F-16." Aviation Week and Space Technology, CXVI (April 20, 1982), 157+.

Kraar, Louis. "General Dynamics Struggles to Build a Plane for All Nations: F-16." Fortune, XCV (March 1977), 180-184+.

Lambert, Mark. "F-16 Improvement Program: Making the Digits Count." Interavia, XXXVI (March 1981), 258-259.

___. "F-16: Passing the 2,000 Mark." Interavia, XXXVI (December 1981), 1248-1249.

___. "Multinational Fighter: F-16 Update." Flight International, CXIII (March 11, 1978), 697-699.

Larcom, Craig. "Electric Jet." Wings, XVIII (June 1988), 22+.

Lloyd, Allwyn T. "More Fight for the Fighting Falcon." Air International, XXI (October 1981), 161-170, 202-203.

McAtee, Tuck P. and Loren E. Timm. "F-16 European Test and Evaluation." Society of Experimental Test Pilots Technical Review, XIV (Fall 1979), 37-51.

___. "'Please Tower, Zoom Flight is Number One.'" USAF Fighter Weapons Review, (Summer 1980), 2-13.

McCarthy, Geoffrey W. "Human Factors in F-16 Mishaps." Flying Safety, XLIV (May 1988), 17-21.

McKean, Kevin. "A Voice-Controlled Superplane." Discover, III (October 1982), 78+.

McKenzie, Mark. "F-16A/B Operational Capabilities Upgrade." USAF Fighter Weapons Review, (Winter 1987), 9-13.

McLendon, Michael H., et al. "The F-16: Medical and Industrial Hygiene Aspects of the U.S. Air Force's Newest Air Combat Fighter." Aviation, Space and Environmental Medicine, LII (March 1981), 145-153.

McMahan, Andrew H. "F-16 Program Spurs Modernization." National Defense, LXII (January-February 1978), 312-313+.

Manning, Steven O., III. "Lightweight Classic." Airman, XIX (September 1975), 42-47.

Marsh, Alton K. "F-16C Fixes Key to 1985 Deliveries." Aviation Week and Space Technology, CXXI (November 5, 1984), 22-23.

Mersky, Peter S. "Flying the Fighting Falcon: The Navy's Future Adversary Aircraft." Naval Aviation News, (November-December 1985), 20-21.

Milam, David W. "Automating Tactical Fighter Combat." Aerospace America, XXII (May 1984), 68-70.

Miller, John. "Low Angle, Low Drag." USAF Fighter Weapons Review, (Summer 1981), 26-34.

Morrocco, John D. "USAF May Develop Agile Falcon Without Allied Participation." Aviation Week and Space Technology, CXXVII (October 26, 1987), 23-24.

"New Fighter for Europe." Air International, XIX (September 1980), 126-130, 143.

Newcomb, Harold. "The Combat Coop." Airman, XXII (February 1978), 20-26.

North, David M. "Navy Selects F-16 for Aggressor Mission." Aviation Week and Space Technology, CXXII (January 14, 1985), 25-26.

____. "TAC Team Supports F-16 Joint Tests." Aviation Week and Space Technology, CVIII (February 6, 1978), 219+.

Oestricker, P. F. and R. C. Ettinger. "F-16 High Angle of Attack Testing." Society of Experimental Test Pilots Technical Review, XIV (Spring 1979), 1-10.

Olds, David R. "A Case Study: 'Peace Victor I.'" Air Force Journal of Logistics, IX (Fall 1985), 21-24.
F-16 sale to Egypt.

Panyalev, Georg. "MIG-21bis and F-16A: Air Combat Potential--A Comparison." International Defence Review, XI (September 1978), 1429-1434.

Perini, Michael B. "Coming on Strong in the F-16." Air Force Magazine, LXIII (August 1980), 40-43.

Pomphrey, Michael K. "Flying in the Electric Jet: A Navigator's Perspective." Navigator, XXVII (Summer 1980), 20-22.

Quinlan, J. C. "Innovation Builds the F-16." Tool Production, XLIX (November 1983), 58-62.

Reed, Fred. "The Electric Jet." Smithsonian Air and Space, I (December 1986-January 1987), 42-49.

Rendine, Michael J. "Meeting the Challenge of Multinational Programs." Concepts, IV (Spring 1981), 119-126.

Richardson, Doug. "F-16 Fighting Falcon." In: William T. ("Bill") Gunston, ed. The Great Book of Modern Warplanes. New York: Crown, 1987. pp. 137-200.

Richardson, Michael. "The F-16 for Southeast Asia: Arms Race or Strategic Balance?" Pacific Defense Reporter, XI (May 1985), 17-19.

Rider, James G. "F-16 Pilot Report." Aerospace International, XIII (January-February 1977), 12-18+.

Riemer, Jeff and Terry Tomeny. "Out of Control in the F-16." TAC Attack, XXVII (May 1987), 18-22.

Riggs, Roger. "F-16 Low-Altitude Attack." USAF Fighter Weapons Review, (Fall 1985), 26-32.

Ropelewski, Robert R. "F-16 Displays Combat Capabilities." Aviation Week and Space Technology, CX (May 28, 1979), 16-22.

____. "F-16/79 Offers Export Market New Capability." Aviation Week and Space Technology, CXV (November 23, 1981), 38-39+.

____. "F-16XL Shows Advances in Range, Ride Qualities." Aviation Week and Space Technology, CXIX (September 26, 1983), 62-63+.

Rossetti, Paul. "F-16: Just the Facts." USAF Fighter Weapons Review, (Fall 1978), 3-8.

Rost, Paul. "F-16." Flying Safety, XXXIX (May 1983), 8-11; XL (April 1984), 20-23.

Scarborough, Rowan. "F-16: Promises Kept." Military Logistics Forum, IV (November-December 1987), 14-17.

Schemmer, Benjamin F. "The Airplane Nobody Wanted." Armed Forces Journal International, CXII (October 1974), 34-35+.

Schrader, Steven. "F-16A/B Loss of Control." USAF Fighter Weapons Review, XXXV (Summer 1987), 11-17.

Scott, William B. "Air Force, NASA Conduct Tests to Define Fighter Aircraft Agility." Aviation Week and Space Technology, CXX (January 9, 1989), 45-47.

____. "Combined Test Force Evaluates Upgrades for U.S., Foreign F-16s." Aviation Week and Space Technology, CXXVI (May 18, 1987), 93+.

Shifrin, Carole A. "General Dynamics Will Produce Kits to Upgrade F-16A/B Fighters." Aviation Week and Space Technology, CXXIV (March 31, 1986), 92-93.

____. "USAF Prepares F-16 Modifications for Growth Into 21st Century." Aviation Week and Space Technology, CXXV (July 21, 1986), 48-49+.

Singh, Pushpindar. "F-16s for Pakistan: Reactions and Views." Asian Defence Journal, no. 9 (September 1981), 40-41.

Smith, Gerald. "Some Kinda Wonderful: The F-16." Air Reservist, XXXV (July-August 1983), 18-21.

Snodgras, Mike. "F-16C Avionics Set Up." USAF Fighter Weapons Review, XXXVI (Summer 1988), 25-27.

Stewart, John T., Jr. "Multinational Aircraft: Present and Future Problems." Aerospace International, XVII (November-December 1981), 30+.

Stichell, Dean. "20,000 Pounds of Fighter Performance." TAC Attack, XV (August 1975), 4-8.

Sweetman, Bill. "F-16: Into the 1990s." Interavia, XL (March 1985), 251-253.

____. "F-16XL: General Dynamics Hatches a New Falcon." Interavia, XXXVIII (July 1983), 747-748.

Talley, Dorsey J. "Once Gained, Twice Accomplished: Micromanaging the Perfect Aircraft." Defense Management Journal, XVII (Fall 1981), 8-13.

____. "The Quest for Quality: A Defense Contractor's Story." Defense Management Journal, XXI (Fall 1985), 14-19.

____ and Richard M. Doubt. "Assaulting the Limits." TAC Attack, XVIII (December 1978), 6-8.

Talty, Patrick K. and Donald J. Caughlin. "F-16XL Demonstrates New Capabilities in Flight Test at Edwards AFB." Journal of Aircraft, XXV (March 1988), 206-215.

Ulsamer, Edgar. "YF-16: On Time, On Track, On Budget." Air Force Magazine, LVII (January 1974), 50-54.

United States. Air Force. Directorate of Aerospace Safety. F-16 Safety Task Force. "F-16." Flying Safety, XLIV (January 1988), 16-19.

Walsh, Bill. "Countdown for a New International Deterrent." Countermeasures, III (January 1977), 28+.

Warwick, Graham. "Advanced Fighter Technology F-16." Flight International, CXVII (January 5, 1980), 28-34.

____. and Hugh Lucas. "F-16s Into Service." Flight International, CXVII (January 5, 1980), 28-34.

Watkins, H. D. "The Politics of the F-16." Forbes, CXVIII (December 15, 1976), 61-64.

"Weapon System Series: F-16." NATO's Fifteen Nations, XXIII (December 1978-January 1979), 68-70.

Wise, Clare E. "F-16: First With Fly-by-Wire." Machine Design, XLVII (August 19, 1975), 16-18, 20-21.

Yonay, Ehud. "Living and Dying at 9Gs." California, XII (April 1987), 57+.

(McDonnell-Douglas) F/A-18 Hornet

Book/Document/Report Literature:

Drendel, Lou. Hornet. Modern Military Aircraft. Carrollton, Tex.: Squadron/Signal Publications, 1988. 64p.

Dunn, Joe L. Testing the F-18 at the U.S. Naval Air Test Center. NATC-TM-78-5-SA. Patuxent River, Md.: Naval Air Test Center, 1978. 16p.

Dyer, Joseph W. An Analysis of Cost Growth in the F/A-18 Airplane Acquisition Program. Thesis. Monterey, Calif.: Naval Postgraduate School, 1981. 231p.

Gunston, William T. ("Bill"). F/A-18 Hornet. Modern Combat Aircraft, no. 22. Osceola, Wis.: Motorbooks International, 1985. 111p.

Linn, Don. F-18 Hornet in Detail and Scale, Pt. I: Developmental and Early Production Aircraft. Detail and Scale Series, v. 6. Fallbrook, Calif.: Aero Publishers, 1982. 71p.

McDonnell-Douglas Corporation. Public Information Office. F/A-18 Hornet Strike Fighter: Background Information. St. Louis, Mo., 1987. 3p.

____.____. F/A-18 Strike Fighter. St. Louis, Mo., 1987. 16p.

____.____. The Hornet Strike Fighter. St. Louis, Mo., 1986. 12p.

Miller, Jay. McDonnell-Douglas F/A-18 Hornet. Minigraph, no. 25. Arlington, Tex.: Aerofax, Inc., 1988. 48p.

Naas, C. W. Proposal For a New Aggressor Aircraft. Research Paper. Maxwell AFB, Ala.: Air Command and Staff College, Air University, 1988. 39p.

Nicholson, Peter G. The F/A-18 Fighter/Attack Aircraft Program. Research Report. Maxwell AFB, Ala.: Air War College, Air University, 1985. 65p.

Northrop Corporation. Public Relations Department. F/A-18 Hornet Strike Fighter. Los Angeles, Calif., 1987. 2p.

Peacock, Lindsay. F/A-18 Hornet. Combat Aircraft Series, no. 2. Harrisburg, Pa.: Stackpole Books, 1986. 48p.

Spick, Mike. F/A-18 Hornet. Modern Fighting Aircraft, v. 7. New York: Arco, 1984. 64p.

United States. Congress. Senate. Committee on Armed Services. Subcommittee on Tactical Airpower. F-18 Program: Hearings. 94th Cong., 1st sess. Washington, D.C.: GPO, 1976. 254p.

____. Government Accounting Office. Mission Analysis and Acquisition Division. Navy's F/A-18 Expected to Be an Effective Performer, But Problems Still Face the Program. Washington, D.C., 1982. 39p.

____. ____. Procurement and Systems Acquisition Division. F/A-18 Naval Strike Fighter: Its Effectiveness is Uncertain. Washington, D.C., 1980. 37p.

____.____.____. F/A-18 Naval Strike Fighter: Progress Has Been Made, But Problems and Concerns Continue. Washington, D.C., 1981. 41p.

____.____.____. Operational and Support Costs of the Navy's F/A-18 Can Be Substantially Reduced. Washington, D.C., 1980. 58p.

Vangysel, Gary R. Marine Corps Aviation: Is the Hornet the Answer? Research Report. Maxwell AFB, Ala.: Air War College, Air University, 1985. 62p.

Zimmerman, R. R. F/A-18D Hornet MAGTF Force Multiplier of the 1990s. Research Paper. Maxwell AFB, Ala.: Air Command and Staff College, Air University, 1987. 21p.
MAGTF = Marine Air-Ground Task Force.

Periodical/Anthology Literature:

Andrews, Hal. "Naval Aircraft: F/A-18 Hornet." Naval Aviation News, (November 1980), 8-13; (January 1983), 24-25.

Bartlett, Tom. "The F/A-18 Hornet." Leatherneck, LXIV (September 1981), 22-27.

Bergman, Don. "New F-18 Promises to Be a Fighter/Attack Airplane For All Seasons." Marine Corps Gazette, LXI (May 1977), 31-38.

______. "The Hornets are Here." Marine Corps Gazette, LXVII (May 1983), 56-57.

Bourisaw, David L. and Carl F. Brandenberg. "F/A-18's Landing, Launch, and Recovery System." Astronautics and Aeronautics, XVIII (November 1980), 49-52.

Brinks, W. H. "F-18A." Society of Experimental Test Pilots Technical Review, XIII (Fall 1977), 32-42.

Brown, David A. "F/A-18 Hornet: The Lightweight With the Deadly Sting." Asian Defence Journal, no. 11 (November 1986), 36+.

Bulban, Erwin J. "Reliability Stressed in Navy F-18 Program." Aviation Week and Space Technology, CIV (May 31, 1976), 40-43.

Chakravartz, S. N. "The Flying Station Wagon." Forbes, CXXXI (April 11, 1983), 115+.

Christmann, Timothy J. "F/A-18: Reserve Air in Renaissance." Naval Aviation News, (January-February 1984), 8-11.

Clelland, John P. "Ordnance for the F/A-18 Hornet." Marine Corps Gazette, LXIV (October 1980), 16-18.

Dunn, Richard C. "...and No More Hornets are Needed." Marine Corps Gazette, LXVII (May 1983), 58-59.

Edney, L. A. "F/A-18: Adding Versatility and Punch to Battle Group Options." Wings of Gold, IX (Winter 1984), 18-19, 22-23.

Eggman, Phil. "Hornet--a New Name for Air Superiority." Naval Aviation News, (January 1983), 26-31.

Eustace, Harry F. and Richard T. Davis. "Commonpoly: The Billion Dollar Game." Electronic Warfare, IX (January-February 1977), 35+.

"F-18: Air Power for the Eighties." All Hands, no. 720 (January 1977), 35+.

"The F-18 Multimission Fighter/Attack." NATO's Fifteen Nations, XXI (June-July 1976), 91-92.

"F-18 Navy Air Combat Fighter Awaits Approval." Interavia, XXX (August 1975), 872-873.

"F-18 Special Report." Flight International, CXVIII (July 19, 1980), 167-169.

"The F/A-18: Hornet With Two Stings." Vectors, XXIV (April 1982), 7-11.

Farrow, George. "Interavia Combat Aircraft Design Analysis, no. 2: McDonnell-Douglas/Northrop F-18." Interavia, XXXIII (July 1978), 605-608.

Fink, Donald E. "F-18 Blends Advanced Fighter Concepts." Aviation Week and Space Technology, CVII (September 5, 1977), 38-41.

Foster-Simeon, E. "F/A-18 Hornet: Adding Sting to Naval Aviation." All Hands, no. 824 (November 1985), 28-31.

Geddes, J. Philip. "Two Versions of F-18 in Hot Competition." Interavia, XXXIV (April 1979), 351-356.

____. "The U.S. Navy's View of the F-18 Hornet." International Defence Review, XI (February 1978), 164-168.

Gershanoff, Hal. "The F/A-18 Hornet: NavAir's New Strike Fighter." Journal of Electronic Defense, VII (April 1984), 23-24+.

Gilson, Charles. "F/A-18: A Status Report." Interavia, XXXV (February 1980), 139-146.

____. "F/A-18: One Aircraft, One Man, Multiple Missions." International Defence Review, XIII (February 1980), 211-218.

Griffiths, David R. "F/A-18 Testing Meets Navy's Schedule." Aviation Week and Space Technology, CXI (July 30, 1979), 45-48.

Gustafson, R. A. "The Marine Corps View." Armed Forces Journal International, CXVIII (April 1981), 58-61.

Haliscak, George. "What the Maintenanceman Can Expect From the F/A-18." Mech, (Fall 1980), 26-32.

Halleland, Hank L. "The F-18." U.S. Naval Institute Proceedings, CII (October 1976), 123-127.

Hamill, Tom. "F-18 Hornet." Flight International, CXIV (December 2, 1978), 2021-2028, 2033-2036.

Hammett, Tom. "Hornet's Nest." Flying, CXI (October 1984), 38-42+.

Hanley, Arthur. "The F-18 Hornet: Did the Navy Get Stung?" Sea Classics, XIV (September 1981), 48-55, 69-71.

Harvey, David. "F-18 Hornet and the Grain of Truth Syndrome." Defense and Foreign Affairs, VIII (July 1980), 36-37.

____. "The Sting of the Hornet." Sea Power, XXVII (January 1984), 10-15.

Hopkins, G. E. "Better, Cheaper Plane the Pentagon Didn't Want." Washington Monthly, IX (March 1977), 32-37.

"Hornet Joins the Marines." Design News, XXXIX (March 14, 1983), 20+.

"Hornet Starts to Swarm." Aircraft Illustrated, XV (May 1982), 215-217.

Jenista, James M. "Super Hornet: An F/A-18 Hornet Upgrade Proposal." U.S. Naval Institute Proceedings, CXIII (August 1987), 97-100.

Kendall, Kenneth. "Electronic Advances on the F/A-18 Hornet." National Defense, LXVII (May-June 1983), 22-26.

Krings, J. E. "F-18 Status Report." Society of Experimental Test Pilots Technical Review, XIV (Spring and Fall 1979), 21-25, 1-9.

Lew, H. K. "Generating Innovation for Navy's F-18." Aircraft Engineering and Maintenance, III (October 1979), 22-25.

Lockard, John A. "The F/A-18." Wings of Gold, IX (Fall 1984), 62-66.

Macbain, Merle. "Is Tomorrow's Fighter/Attack Plane Here Now?" Sea Power, XXII (September 1979), 24-26.

McCutcheon, Richard L. "Marine Corps F/A-18s: Now More Than Ever." Marine Corps Gazette, LXXII (October 1988), 31-33.

Mancus, Peter. "It's the Hornet." Wings, XIV (August 1984), 10+.

Morrocco, John D. "Navy to Select Hornet 2000 Design by End of January." Aviation Week and Space Technology, CXXVIII (January 18, 1988), 27+.

____. "U.S. Will Discuss Codevelopment of Updated F/A-18 With Allies." Aviation Week and Space Technology, CXXVII (December 27, 1987), 20-21.

North, David M. "Blue Angels, NASA Moving to Acquire F/A-18 Aircraft." Aviation Week and Space Technology, CXXII (June 10, 1985), 105+.

____. "Single-Pilot Effectiveness of F-18 Tested in Simulator." Aviation Week and Space Technology, CVI (January 31, 1977), 64-69.

____. "Navy F/A-18 Demonstrates Dual-Mission Performance." Aviation Week and Space Technology, CXII (June 23, 1980), 42-55.

O'Mara, K. P. "Past Performance and Mission Dictate That the Corps Use a Two-Seat Fighter." Marine Corps Gazette, LX (August 1979), 60-64.

Philip, John S. "Built for Land and Sea." Aerospace International, XII (September-October 1976), 28-31.

Plimpton, George. "Life Begins at Mach 2." Popular Mechanics, CLXIII (April 1986), 75-79+.

Prina, L. Edgar. "F-18: Slow Race Horse, Pig-in-a-Poke, or Weapons Bargain?" Sea Power, XVIII (June 1975), 27-28.

____. "New Cost Increases Endanger F-18 Program." Sea Power, XXIII (May 1980), 22-24.

Rahal, Bobby. "Trading Places." Car and Driver, XXXII (April 1987), 153+.

Richards, D. and C. D. Pitcher. "F/A-18A Initial Sea Trials." Society of Experimental Test Pilots Technical Review, XVI (Winter 1981), 10-22.

Robertson, James S. "Too Many Roles for the F/A-18?" Marine Corps Gazette, LXXII (July 1988), 27-28.

Robinson, Clarence A., Jr. "Navy Plans Two Distinct F-18 Versions." Aviation Week and Space Technology, CII (May 26, 1975), 14-16.

Roser, Hans. "Thumbs Up for the F/A-18." Pacific Defense Reporter, IX (May 1983), 17-18.

Rucker, Bob. "F/A-18: New Multi-Mission Strike Fighter." All Hands, no. 752 (September 1979), 6-10.

Salamon, George. "More Sound Than Sense: The F/A-18 Controversy." Defense Systems Review, I (June 1983), 29-34.

Scheibla, S. H. "Hornet's Nest: Trouble is Buzzing Around the Navy's New Fighter Plane." Barron's, LX (August 25, 1980), 9+.

Schreiber, Jon A. "Hornets: Ready When Needed." Wings of Gold, XII (Winter 1987), 32-35.

Schultz, James B. "Marines Put F/A-18s to Test." Defense Electronics, XV (November 1983), 104+.

Service, Brian. "Hornets Go to Sea." International Defence Review, XIX (June 1986), 795-797.

Shohat, Murray and F. Clifton Berry, Jr. "F-18: A Special Report." Military Electronics/Defense Countermeasures, V (May 1979), 29-30+.

Simonsen, Erik. "Navy's New 'Top Gun' Team." International Combat Arms, VI (January 1988), 48-51+.

Smith, Gene. "Every Boy Should Have One." Air Progress, XLVIII (August 1986), 42-51.

Spick, Mike. "F/A-18 Hornet." In: William T. ("Bill") Gunston, ed. The Great Book of Modern Warplanes. New York: Crown, 1987. pp. 201-264.

Stevenson, James P. "F/A-18 Hornet." NATO's Fifteen Nations, XXVI (April-May 1981), 88-90.

Stumpf, Robert E. "One Hornet, Two Stings." U.S. Naval Institute Proceedings, CVIII (September 1982), 115-119.

Sullivan, Martin J. "Integration of the F/A-18 Into the FMF." Marine Corps Gazette, LXXII (July 1988), 24-27.

Tegler, John. "The Best Carrier Airplane." U.S. Naval Institute Proceedings, CXII (October 1986), 75-81.

Thach, M. J. "F/A-18 Weapon System Development." Canadian Aeronautics and Space Journal, XXVII (Fall 1981), 242-252.

Thomas, Vincent C., Jr. "A Clean Bill of Health for the F-18?" Sea Power, XXIII (September 1980), 53-54+.

Trotter, T. W. "One Seat vs. Two." U.S. Naval Institute Proceedings, CXII (October 1986), 75-81.

"The U.S. Navy's F-18 Programme Gathers Momentum." Interavia, XXXI (July 1976), 627-629.

"Versatile F/A-18 Hornet Performs Fighter and Attack Missions." Defense Electronics, XVI (December 1984), 71-72.

Waldner, Joseph C. "F-18 Hornet Program Progress Report." U.S. Naval Institute Proceedings, CV (November 1979), 111-114.

Walker, R. W. "New Roles for the F/A-18." Marine Corps Gazette, LXX (May 1986), 90-95.

Warwick, Graham. "F-18s Into Service." Flight International, CXX (November 28, 1981), 1613-1616.

"Weapon System Closeup: F/A-18 Hornet." Defence Update International, no. 39 (1983), 25-38.

White, William J. "The F/A-18 Program." Marine Corps Gazette, LXIV (October 1980), 26-30.

Zadarnowski, J. H. "CADD on the F-18 Program." Astronautics and Aeronautics, XVIII (March 1980), 48-51, 61.
CADD = Computer-Designed Drafting.

Zimmermann, Robert R. "F/A-18 Hornet: An Aircraft For All Missions." U.S. Naval Institute Proceedings, CXIII (November 1987), 108-111.

F-19 see F-117A Stealth

F-20 see F-5 Freedom Fighter/Tiger II

(I.A.F.) F-21 Kfir

Andrews, Hal. "Naval Aircraft: F-21 Kfir." Naval Aviation News, (November-December 1986), 16-17.

Greeley, Brendan M., Jr. "Navy Halts East Coast Kfir Flight Operations." Aviation Week and Space Technology, CXXVIII (April 18, 1988), 21-22.

____. "Navy Upgrades Fighter Training Capability With Israeli F-21A Kfir." Aviation Week and Space Technology, CXXIII (October 21, 1985), 45+.

"Israel's Kfir Joins the U.S. Navy." *Defence Update International*, no. 83 (September-October 1987), 50-57.

Mersky, Peter. "Israel's Kfir Accepted by U.S. Navy." *Naval Aviation News*, (July-August 1985), 14-15.

____. "Kfir Joins the Navy: Such a Nice Plane." *Approach*, XXX (January 1985), C1-C2.

Tegler, John. "The Young Lions of Oceana." *U.S. Naval Institute Proceedings*, CXIII (January 1987), 114-116.

Walker, M. A. "Kfir Fighters Enhance Combat Air-to-Air Training." *Marines*, XVI (October 1987), 24-25.

F-51 see P-51 Mustang

(Lockheed) F-80 Shooting Star
(see also T-33 Silver Star)

Book/Document/Report Literature:

Davis, Larry. *P-80 Shooting Star in Action*. Carrollton, Tex.: Squadron/Signal Publications, 1980. 50p.

Jensen, Edna. *XP-80, XP-80A, YP-80A, P-80A-1, P-80A-5, P-80B-1, P-80B-5, P-80N and P-80R, June 1943-July 1947*. AMC Historical Study, no. 177. Wright Field, Ohio: USAAF Air Materiel Command, 1948. 126p.

Lockheed F-80 Shooting Star. Appleton, Wis.: Aviat Publications, 1974. 53p.

Rhodes, Arnold. *Shooting Star, T-Bird, and Starfire: A Famous Lockheed Family*. Tucson, Ariz.: Aztex Corp., 1981. 128p.

Woolridge, E. T., Jr. *The P-80 Shooting Star: Evolution of a Jet Fighter*. Famous Aircraft of the National Air and Space Museum, no. 3. Washington, D.C.: Smithsonian Institution Press, 1979. 110p.

Periodical/Anthology Literature:

"AAF Unveils Speedy Shooting Star as Answer to Nazi Jets." *Aviation News*, III (March 15, 1945), 8-10.

"Birds of Slight Feather." *Air Force*, XXXIII (February 1950), 40-41.

Casari, Robert. "The Legend of the Shooting Star." *Flying*, LVIII (March 1956), 28-29+.

"Designing the F-80." *Flying*, XLIII (December 1948), 33-35+.

"F-80 Shooting Star." *TAC Attack*, XXV (March 1985), 16-17.

Gunston, William T. ("Bill"). "Lockheed Shooting Star." Aeroplane Monthly, IV (January 1976), 15-20.

Hasskari, Robert A., Jr. "Every Inch a Fighter." Airpower Historian, IX (July 1962), 180-183.

Jessup, Altheus W. "Combat Reports Prove F-80 Can Take It." Aviation Week, LIII (July 31, 1950), 12-14.

Johnson, Clarence L. ("Kelly"). "Development of the Lockheed P-80A Jet Fighter." Journal of Aeronautical Science, XIV (December 1947), 659-679.

____. "How We Engineered the Lockheed Shooting Star." Aviation, XLIV (July 1945), 149-151.

Keeler, John W. "F-80 Shooting Star." In: Robin Higham and Abigail T. Siddall, eds. Flying Combat Aircraft of the USAAF-USAF. Ames, Iowa: Iowa State University Press, 1975. pp. 61-65.

____. "Just Another Airplane?" Airman, IV (September 1960), 14-15.

Klibbe, Frank W. "Fox Able One: The First Transatlantic Jet Deployment." Air Force Magazine, LXIII (October 1980), 72-80.
July 1948.

Leach, Al. "The Acrojets." Skyways, IX (August 1950), 23-24.

Littrell, Gaither. "Shooting Star." Flying, XXXVII (October 1945), 31+.

McLarren, Robert. "P-80 Still Right After Five Years." Aviation Week, XLIX (July 19, 1948), 22+.

"Memorandum on the Shooting Star." Aero Digest, L (August 1, 1945), 55+.

O'Hara, Bob. "Shooting Star." Air Classics, V (August 1969), 50-65.

"P-80: U.S. Army Uses Fast-Flying Jet Planes to Develop a Spectacular New Air Force." Life, XXI (December 9, 1946), 98-103.

"They Call It the Apparition of Death: The Lockheed Jet Fighter, P-80 Shooting Star." Popular Science, CXLVI (April 1945), 92-94+.

Winchester, James H. "The F-80 in Combat." Aviation Age, XIV (December 1950), 17-18.

(North American) F-82 Twin Mustang

Book/Document/Report Literature:

The F-82 Twin Mustang. Appleton, Wis.: Aviat Publications, 1973. 66p.

McLaren, David. North American P-82/F-82 Variants. Minigraph, no. 17. Arlington, Tex.: Aerofax, Inc., 1984. 40p.

Miller, Martin J. and Margaret C. Bagwell. F-82E, F & G Airplanes and Related Engines (V-1710-145), August 1945-November 1950. AMC Historical Study, no. 215. Wright-Patterson AFB, Ohio: USAF Air Materiel Command, 1951. 23p.

Periodical/Anthology Literature:

Daniels, C. M. "The Twin-Tailed Threat." Air Classics, VI (October 1969), 14-18, 60-61.

Hendrix, Lin. "Rangefinder." Wings, VII (April 1977), 58-63.

Miller, Jay. "P-82 Restoration Nears Completion." Air Classics, XII (December 1976), 56-61.

Oatley, D. E. "Taming the Mustang." Skyways, X (October 1951), 18-19, 44-45.

Pape, Garry R. "Escort Fighter: Flying the F-82 Twin Mustang in the Cold War." RC Modeler, VII (November 1977), 48-59.

Sharp, John F. "Flying the Twin Mustang." In: Michael O'Leary, ed. Air Progress Pilot Reports. Canoga Park, Calif.: Challenge Publications, 1979. pp. 72-79.

Thompson, Warren. "Double Trouble." Wings, XIII (August 1983), 22-37.

____. "F-82: Killers Over Korea." Air Enthusiast, VI (1978), 121-138.

____. "North American's Twin Mustang." Aeroplane Monthly, VI (March 1978), 116-123.

Trimble, Robert L. "Twin Mustang." Air Classics, X (March-April 1974), 60-67, 33-39, 54-57.

(Republic) F-84 Thunderjet/F-84F Thunderstreak

Book/Document/Report Literature:

Brassell, John C. and Robert A. Smith. F-84F, 1949-1954. AMC Historical Study, no. 267. Wright-Patterson AFB, Ohio: USAF Air Materiel Command, 1955. 80p.

Davis, Larry. F-84 Thunderjet in Action. Carrollton, Tex.: Squadron/Signal Publications, 1983. 50p.

Keaveney, Kevin. Republic F-84 (Swept-Wing Variants). Minigraph, no. 15. Arlington, Tex.: Aerofax, Inc., 1987. 48p.

McDowell, R. Ernest. Republic F/RF-84F Thunderstreak/Thunderflash. Aircam Aviation Series, no. 14. New York: Arco, 1970. 48p.

Miller, Dorothy. F-84, 1944-1950. AMC Historical Study, no. 203. Wright-Patterson AFB, Ohio: USAF Air Materiel Command, 1950. 78p.

Wagner, Ray. Republic F-84F Thunderstreak. Aircraft in Profile, no. 95. Windsor, Eng.: Profile Publications, 1966. 11p.

Periodical/Anthology Literature:

Baines, T. E. "The 'Hog': A U.S. Air Force Fighter That Never Fought." Aerospace Historian, XIX (September 1972), 132-136.
F-84F.

Beaird, Harry G., Jr. "Driving the Thunderstreak." Combat Crew, VI (September 1955), 8-13.

Boyd, John R. "F-84F Harmonization." USAF Fighter Weapons News Letter, no. 1 (March 1957), 17-21.

Bright, Charles D. "F-84 Thunderjet." In: Robin Higham and Abigail T. Siddall, eds. Flying Combat Aircraft of the USAAF-USAF. Ames, Iowa: Iowa State University Press, 1975. pp. 66-72.

Burchard, A. S. "Destroyers Up." Skyways, IX (July 1950), 18-19.
F-84E.

"Dante's Inferno." Air Training, IV (September 1954), 26-27.

"F-84E Thunderjet." Joint Services Recognition Journal, X (April 1955), 90-91.

"F-84F High- and Low-Angle Strafe." USAF Fighter Weapons News Letter, no. 1 (February 1956), 37-44.

"F-84F Thunderstreak." TAC Attack, XXVI (December 1986), 16-17.

Flomer, Leonard W. and Cosier S. Kline. "Babys, It's Cold Outside." Combat Crew, VII (December 1956), 26-29.
F-84F.

"Grandson of the Thunderbolt." Air Trails, XL (June 1953), 26-27.
F-84.

Green, William. "Thunderstreak." RAF Flying Review, X (January 1955), 22-25.

Greenway, Sean. "The Last Thunderstreak." Air Classics, XV (October 1979), 53-56.

Grimm, Peter. "The Fastest Tricycle in the Air Force: The Republic F-84F Thunderstreak." Aerospace Historian, XXXII (December 1985), 214-225.

Heumann, Gert W. "The Republic Thunderstreak." Air Pictorial, XXII (January 1960), 15-17.

Higgenbotham, R. R. "Engine Installation Problems in the XP-84." Society of Automotive Engineers Journal, LVI (January 1948), 53-57.

Hoehn, Jean-Pierre. "Thunderstreaks and Flashes Over France." American Aviation Historical Society Journal, XVII (Fall 1972), 145-153.

"How to Fly the Thunderjet." Flying, XLIV (March 1949), 22-26.

Johnsen, Frederick A. "Thunder Maker." Wings, III (June 1976), 40-49.

Johnson, Gerald W. "On Emergency Standby." Combat Crew, VI (September 1955), 4-7.
F-84F.

"Long-Range F-84F Strikes With Guns, Bombs, Rockets." Aviation Week, LI (January 15, 1951), 14-15.

McLarren, Robert. "F-84 Thunderjet: A Story of Development." Aviation Week, LI (September 5, 1949), 20-22+.

____. "Republic's P-84 Thunderjet Designed Around Specific Engine Type, Axial-Flow J-35." Aviation Week, XLVIII (May 24, 1948), 18+.

"NATO's First Sting." RAF Flying Review, XVI (August 1961), 22-25.

"Republic F-84F." Aero Digest, LXII (February 1951), 36+; LXIII (July 1951), 25-32.

Roth, R. M. "The Fas in F-A-S-T." Flying Safety, X (February 1954), 2-5.

____. "Flying the 'F.'" Flying Safety, XI (January 1955), 8-10.

Silsbee, Nathaniel F. "The Thunderjet." Aviation Op, XIII (March 1950), 24-25.

____. "Thunderjet, Air Force Ace for '48.'" Aero Digest, LVI (February 1948), 44-48.

Stone, Irving. "Production Details on Republic F-84F." Aviation Week, LVII (December 15, 1952), 31-32+.

Taylor, Norman E. "F-84s Around the World." American Aviation Historical Society Journal, XV (Summer 1970), 126-127.

Thompson, Warren. "Thunder Over Korea." Airpower, XIV (March 1984), 26-41.
F-84 Thunderjet.

"The Thunderstreak." Air Force, XXXVI (February 1953), 29-30.

Trimble, Robert L. "Thunderjet." Air Classics, XII (May-June 1976), 52-63, 28-43.

"U.S. Air Force's New F-84E." Skyways, VIII (October 1949), 14-15.

Vogel, Joseph L. "F-84F Thunderstreak." In: Robin Higham and Carol Williams, eds. Flying Combat Aircraft of the USAAF-USAF. Ames, Iowa: Iowa State University Press, 1978. pp. 121-128.

____. "I Fly the Thunder Maker." Airman, XIII (January 1969), 8-10.

Weldinger, Charles L. "Guardsman Reminisces--Flying the F-84E Thunderjet." Aerospace Historian, XXXIV (Spring 1987), 2-7.

____. "Guardsman Reminisces--Flying the F-84F Thunderstreak." Aerospace Historian, XXXIV (Spring 1987), 8-12.

Woten, Homer G. "F-84F Tips on Harmonization." USAF Fighter Weapons News Letter, no. 2 (May 1956), 58-60.

(North American) F-86 Sabre/FJ Fury

Book/Document/Report Literature:

Childerhose, R. J. F-86 Sabre. Famous Aircraft Series. New York: Arco, 1966. 64p.

Davis, Larry. F-86 Sabre in Action. Aircraft, no. 33. Carrollton, Tex.: Squadron/Signal Publications, 1978. 48p.

____. F-86 Sabre in Color. Carrollton, Tex.: Squadron/Signal Publications, 1981. 32p.

Ginter, Steven J. North American FJ-2 Fury. Naval Fighters Series, no. 10. Simi Valley, Calif.: Naval Fighters, 1984. 46p.

Mason, Francis K. North American FJ Fury. Aircraft in Profile, no. 42. Windsor, Eng.: Profile Publications, 1966. 12p.

Shacklady, E. G. North American F-86A Sabre. Aircraft in Profile, no. 20. Windsor, Eng.: Profile Publications, 1965. 13p.

Wagner, Ray. The North American Sabre. Garden City, N.Y.: Doubleday, 1963, 162p.

Ward, Richard and Ernest McDowell. North American F-86A/L Sabre in USAF and Foreign Service. Aircam Aviation Series, no. 13. New York: Arco, 1970. 48p.

Williams, Fenton L. Project Hop Up: The F-86D Modification Program, 1954-1955. AMC Historical Study, no. 401. McClellan AFB, Calif.: History Office, Sacremento Air Materiel Area, 1956. 29p.

Periodical/Anthology Literature:

"Air Force Sabre." Skyways, XI (January 1952), 42-43.

"All-Weather Sabre K." RAF Flying Review, XVIII (July 1963), 30-31.

Baumgardner, Clyde. "F-86D Sabre." RC Modeler, XIII (February 1976), 24-31, 142.

Beckett, Rod, et al. "Sabre Dart." USAF Fighter Weapons Review, (Winter 1981), 2-7.

Bell, Colin. "F-86D Sabre." Scale Models, XII (September 1981), 473-476.

Blount, Eliot. "New World Speed Record." Skyline, XI (February 1953), 11-15.

____. "Sabrebomber." Skyline, XI (November 1953), 10-13.

Bonnalie, A. F. "Is the MIG 15 Superior to the F-86?" U.S. Air Services, XXXVIII (June 1953), 10-14.

Buerger, John. "Furys on the Forrestal." Skyline, XIV (August 1956), 10-13.

Christian, G. L. "USAF Pilots in Korea Love the F-86." Aviation Week, LVI (June 30, 1952), 52-54.

Close, G. C. "Killer Jet in Korea." Skyways, X (June 1951), 10-12, 46-48.

Cole, R. A. "Sabriety." Aeronautics, XXVII (December 1952), 30-33.

Collins, N. and Earl Blount. "Swept Wings for the Leathernecks." Skyline, XII (May 1954), 4-7.
FJ Fury.

"Combat Ready." Air Reservist, IX (March 1957), 3-4.
F-86L.

Coughlin, William J. "F-86 Gimmick--Improved Wing." Aviation Week, LIX (September 7, 1953), 15-16.

"The 'D' is Different." Flying Safety, IX (August 1955), 12-13.

Darnell, Dan. "The Edge in Performance." Flying Safety, XI (November 1955), 22-24.

"Development Highlights." Air Trails, XXXVIII (August 1952), 24-25.

"Dogs on Watch." The Aeroplane, LXXXVIII (April 8, 1955), 438-439.
F-86D.

Dougherty, A. J. "F-86L Formation Fundamentals." Interceptor, I (November 1959), 22-24.

Ellis, Steve. "Hall's Hot Ship." Aeroplane Monthly, VII (August 1979), 433-436.

Elman, H. L. "Curing Furious Headaches; or, Down to the Sea in Jets." American Aviation Historical Society Journal, XIII (Summer 1968), 116-124.

Evans, Douglas K. "My Love Affair With the Sabre." Air Force Magazine, LXI (October 1978), 78-83.

"The F-86 Sabre." RAF Flying Review, X (July 1955), 30-33.

"The F-86F Sabre." RAF Flying Review, X (July 1955), 30-33.

"F-86Ks Ordered by Air Force From North American Aviation." American Helicopter, XXXVI (September 1954), 13-14.

"First of the Furys." RAF Flying Review, XVII (December 1961), 30-31.

Foster, Cecil G. "F-86 Sabre." In: Robin Higham and Abigail T. Siddall, eds. Flying Combat Aircraft of the USAAF-USAF. Ames, Iowa: Iowa State University Press, 1975. pp. 73-78.

Girando, J. C. and E. P. McNeff. "Rag Drag." Flying Safety, XI (February 1955), 20-23.

Godwin, Orgle D. "The F-86H." USAF Fighter Weapons News Letter, no. 1 (February 1956), 80-81.

____. and Bruno A. Giordano. "F-86H Gunnery." USAF Fighter Weapons News Letter, no. 2 (May 1958), 25-42.

Gunston, William T. ("Bill"). "North American F-86 Sabre." Aeroplane Monthly, V (September and November 1977), 467-474, 580-586.

Harper, John. "George Turns Tiger." Flying Safety, XII (July 1956), 18-21.

Hilbrink, Will. "Buckeye Fury: A Story of Achievement From the FJ-1 to the FJ-4 Fury Jet." Skyline, XIII (November 1955), 12-19.

____. "Furys Fly the Bird." Skyline, XV (Winter 1957), 22-27.

Holm, Skip. "Yalu River Raider." Air Progress, XLVI (September 1984), 41+.

Hoover, Bob. "Off and Running." Flying Safety, XII (March 1956), 22-25.

Horn, Ron. "Dog Sabre." In: Michael O'Leary, ed. Air Progress Pilot Reports. Canoga Park, Calif.: Challenge Publications, 1979. pp. 34-43.

Hotz, Robert. "F-86A Hits Record, Cameras Miss." Aviation Week, XLIX (September 13, 1948), 12-13.

____ "Salute to the Sabre." Aviation Week and Space Technology, LXVII (July 15, August 19, September 2 and 23, 1957), 21, 166, 118, 150.

Ingram, Bill. "I Test Flew the FJ-3 Fury." Naval Aviation News, (July 1955), 32-33.

Jessup, Alpheus W. "F-86F is Top Fighter Bomber." Aviation Week, LIX (October 19, 1953), 21-22.

Lauren, Ward. "Furys Pass the Test." Skyline, XIII (February 1955), 22-27.

___. "Fury With Legs." Skyline, XV (Fall 1957), 36-41.
FJ-4.

Letzter, George J. "Old Sabres Never Die." American Aviation Historical Society Journal, X (Fall 1965), 212-213.

Littrell, Gaither. "F-86: World's Fastest Fighter." Flying, XLIII (November 1948), 15-17+.

McLarren, Robert. "Design Highlights of North American F-86." Aviation Week, LI (November 21, 1949), 18+.

___. "XF-86 Flies Faster Than Sound." Aviation Week, XLVIII (June 14, 1948), 11-12.

"Marine Furys and Their Death Angels." Naval Aviation News, (September 1954), 13-15.

Martin, E. L. "The Soft Touch." Flying Safety, XI (August 1955), 10-11.

Merifield, J. H. R. "Sabre vs. MIG." Royal Air Forces Quarterly, V (July 1953), 251-252.

"North American F-86D Sabre." Flight, LXIX (June 8, 1956), 709-710.

"North American Sabre." Aero Digest, LXIV (June 1952), 22-31.

O'Leary, Michael. "Sabre." Air Classics, XVIII (May 1982), 44-49, 82.

Otto, Richard. "The Sabre Saga." Air Classics, VIII (January 1972), 10-21.

Rice, Robert. "Evolution of the F-86 Sabre Jet Fighter." Automotive Industries, CVIII (April 15, 1953), 45-48+.

___. "The Sabre Story: A Study in Exploitation of a Proven Design." Skyline, XI (February 1953), 4-7.

"Sabre: A Study of a Renowned Fighter." Flight, LXIII (January 30, 1955), 136-147.

"Sabre Dog." RAF Flying Review, X (June 1955), 52-53.
F-86D.

"Sabres are a Prolific Breed." Air Force, XXXVI (January 1953), 36-39.

"Sabres Cross the Atlantic." Interavia, IX (June 1954), 370-371.

Schmidt, E. "Fury for the Navy." Aero Digest, LVII (October 1948), 28-30.

Seltz, E. A. "Flying Sword." Wings, V (October 1975), 8-20, 50-55.

Sixman, Herman J. "Sabre Sign-Off." Aircraft Illustrated, XIII (January 1980), 23-25.

Stroud, William P. and J. W. Thomas. "The F-86 and Its Aces." Talon, XI (December 1965), 4+.

Taylor, Kelvin. "F-86 Sabre." TAC Attack, XXVII (January 1987), 16-17.

Thompson, Warren. "King of MIG Alley." Airpower, XII (September 1982), 14-31.

Trimble, Robert L. "Swept-Wing Furys: An Examination of the Navy's Version of the Sabre." Air Classics, X (October 1974), 38-47, 68-73.

Tuttle, Jim. "The Mystery of the Stuck Tail." American Aviation Historical Society Journal, XXXIII (Summer 1988), 146-147.
An F-86D.

"The U.S. Navy's Latest Carrier-Based Jet Fighter." Air Pictorial, XIV (June 1952), 178-179.

Welch, George. "Sky High Sabre Dance." Skyways, IX (April 1950), 14-16.

(Northrop) F-89 Scorpion

Book/Document/Report Literature:

Schultz, Helen. F-89, 1945-1951. AMC Historical Study, no. 229. Wright-Patterson AFB, Ohio: USAF Air Materiel Command, 1952. 32p.

Periodical/Anthology Literature:

Bratcher, Fred. "Night Watch of the Air Force." Skyways, IX (June 1950), 14-15.

Cleary, Ed. "Great Gobs of Gravel Gobblers." Interceptor, I (September 1959), 21-23.

Close, G. C. "Depth Defense Wolfpack Style." Aviation Age, XXI (April 1954), 17-18.

____. "F-89 All-Weather Interceptor." Skyways, X (December 1951), 14-15, 42-43.

"Closeup: Scorpion." Joint Services Recognition Journal, IX (December 1954), 306-307.

Cuff, Victor D. "Scorpion Stuff." Aerospace Maintenance Safety, XIX (January 1964), 10-11.

"Design of the Scorpion." Aero Digest, LXX (January 1955), 28-31.

"F-89 Scorpion." Joint Services Recognition Journal, X (May 1955), 118-119.

Gatlin, Wayne C. "F-89 Scorpion." In: Robin Higham and Abigail T. Siddall, eds. Flying Combat Aircraft of the USAAF-USAF, Ames, Iowa: Iowa State University Press, 1975. pp. 79-84.

____. "Flying the Scorpion." Air Classics, XIV (October 1978), 29-33, 74-76.

Gunston, William T. ("Bill"). "Fighters of the Fifties: Northrop Scorpion." Aeroplane Monthly, III (May 1975), 227-233.

"Little Auggie and the Scorpion." Honeywell Flightline, IV (November 1953), 10-11.

Miller, Jay. "The Scorpion: A Pictorial Report." Air University Review, XXXI (July-August 1980), 34-45.

Nelson, L. A. "Do's and Don'ts for 'Ds." Flying Safety, XI (January 1955), 2-4.

"New Scorpion Designed for Falcon." Aviation Week, LXIV (March 19, 1956), 32-33.

"Northrop F-89 Scorpion." Aero Digest, LXII (January 1951), 30-32+.

"Northrop Scorpion." RAF Flying Review, XI (May 1956), 48-50.

"Northrop Scorpion: New Night Fighter." Flying, XLVI (February 1950), 14-15+.

Quinn, J. J. "Safety With a Sunday Punch." Flying Safety, IX (November 1953), 2-5.

"Round-the-Clock Scorpion." Interavia, VII (May 1952), 269-271.

Solwold, Don. "The Navigator Heritage: F-89." Navigator, XXVIII (Spring 1981), 14-15+.

"They All Watch Scorpion Weight." Aviation Week, LVII (September 29, 1952), 31-32.

Trimble, Robert L. "Scorpion." Air Classics, XIV (October 1978), 20-28.

(Lockheed) F-94 Starfire

Book/Document/Report Literature:

Francillon, Rene J. and Kevin Keaveney. Lockheed F-94 Starfire, Minigraph, no. 14. Arlington, Tex.: Aerofax, Inc., 1986. 48p.

Rhodes, Arnold. Shooting Star, T-Bird, and Starfire: A Famous Lockheed Family. Tucson, Ariz.: Aztex Corp., 1981. 128p.

Williams, Fenton L. Project Pull Out: The F-94C Modification Program, 1954-1955. AMC Historical Study, no. 401. McClellan AFB, Calif.: History Office, Sacramento Air Materiel Area, 1956. 26p.

Periodical/Anthology Literature:

"All-Weather Jet Fighter: Lockheed's F-94." Aero Digest, LIX (August 1949), 23+.

Amody, Francis J. "We Got Ours at Night: The Story of the Lockheed F-94 Starfire in Combat." American Aviation Historical Society Journal, XXVII (Summer 1982), 148-150.

Anderton, David A. "F-94: Variations on a Lockheed Theme." Aviation Week, LIV (June 25, 1951), 20-23.

Boone, Andrew R. "Flying Bazooka Outsmarts Bombers: Starfire." Popular Science, CLXI (August 1952), 72-73.

Christian, G. L. "Far East Pilots Praise F-94's Durability." Aviation Week, LVII (July 7, 1952), 61+.

Delear, Frank. "Flying the Night Picket." BeeHive, XXIX (Fall 1954), 15-18.

Downie, D. "F-94 Flight to Alaska." Skyways, X (October 1951), 10-12, 40-43.

"F-94C Nearly All-Automatic Interceptor." Aviation Week, LVII (July 7, 1952), 16-17.

Fithian, Ben. "The F-94: First Kill in Korea." Navigator, XXVIII (Winter 1981), 15-18.

Fuller, C. "Killer in the Night: F-94 Jets." Flying, XLIX (August 1951), 22-23+.

Gatlin, Wayne C. "F-94 Starfire." In: Robin Higham and Carol Williams, eds. Flying Combat Aircraft of the USAAF-USAF. Ames, Iowa; Iowa State University Press, 1978. pp. 149-158.

"Guardians of New York." Interavia, VII (July 1952), 390-391.

Gunston, William T. ("Bill"). "Fighters of the Fifties: Lockheed F-94." Aeroplane Monthly, VI (December 1978), 636-642.

"The Lockheed F-94C All-Weather Fighter." Aero Digest, LXIII (August 1951), 20-23.

"The Lockheed F-94C Starfire." Aero Digest, LXIX (November 1954), 32-33.

Mason, S. "The 'C' Around You." Flying Safety, X (May 1954), 16-18; XI (January 1955), 5-7.

"New Weapon in Our Air Defense Arsenal." Air University Quarterly Review, V (Fall 1952), 124-125.

Patton, C. B. "Lockheed F-94C." Skyways, XII (April 1953), 11, 58.

Schrader, Richard K. "Starfire." Air Classics, XXIV (September 1988), 20-33.

"Starfire." Joint Services Recognition Journal, IX (February 1954), 30-31.

Thompson, Warren. "Night Patrol." Wings, XVIII (April 1988), 26+.

(North American) F-100 Super Sabre

Book/Document/Report Literature:

Anderton, David A. North American F-100 Super Sabre. Osprey Air Combat Series. Osceola, Wis.: Motorbooks International, 1987. 200p.

Kinzey, Bert. F-100 Super Sabre in Detail and Scale. Detail and Scale Series, v. 33. Blue Ridge Summit, Pa.: TAB, 1989. 72p.

Memlock, Stanley M. Use of Jet Fighter Aircraft as a Forward Air Control Vehicle in a Non-Permissive Environment. Research Report. Maxwell AFB, Ala.: Air War College, Air University, 1970. 93p.

Sullivan, Farrell J. An Analysis of F-100 Aircrew Training Program. Research Paper. Maxwell AFB, Ala.: Air Command and Staff College, Air University, 1966. 68p.

Periodical/Anthology Literature:

Andersen, Roger E. "F-100A Gunnery." USAF Fighter Weapons News Letter, no. 2 (May 1956), 43-51.

Arnberger, Herbert F. "F-100 Pitch-up Tactics." USAF Fighter Weapons News Letter, no. 4 (December 1964), 8-19.

Barnes, W. F. "F-100 Conversions in the Far East Air Force." Air University Quarterly Review, IX (Spring 1957), 98-105.

Beindorff, William G. "Meet the Super Sabre." TAF Review, II (July 1956), 20-22.

"A Big Noise." Flight, LXIX (June 8, 1956), 696-697.

Blakely, Garth. "The F-100 Super Sabre." Aerospace Historian, XXIV (September 1977), 149-152.
Reprinted in Robin Higham and Carol Williams, eds. Flying Combat Aircraft of the USAAF-USAF (Manhattan, Kans.: Sunflower University Press, 1981), pp. 35-38.

Brooks, Ralph L. "Tips on Turning the F-100." USAF Fighter Weapons Review, no. 3 (September 1961), 28-32.

Brown, Bruce L. "Flight Characteristics of the F-100." USAF Fighter Weapons News Letter, no. 4 (December 1968), 1-6.

"C-Note Series." Air Training, IV (November 1954), 20-21.

"Closeup: Super Sabre." Joint Services Recognition Journal, IX (August 1954), 193-194.

Cornelius, Art. "The Lady From El Segundo." Air Classics, VIII (September 1972), 30-39, 64-68, 81-83.

Coughlin, William D. "F-100 Passes Mach 1 in Transonic Dives." Aviation Week, LIX (October 26, 1953), 12-13.

Doggette, James C. "F-100 Armament Improvement Modification." USAF Fighter Weapons News Letter, no. 2 (June 1967), 1-4.

Edney, Robert. "Bunt Bombing." USAF Fighter Weapons News Letter, no. 1 (March 1969), 16-21.

"F-100 Super Sabre: The Greatest Step in Aviation Since Jet Power." Skyline, XII (May 1954), 16-19.

"F-100D Super Sabre." TAC Attack, XXVII (March 1987), 16-17.

"F-100F Seats Two Without Sacrificing Performance." American Aviation, XXVII (April 1957), 54-57.

Fero, Robert W. "F-100." Flying Safety, XIV (May 1958), 17-19.

Filliaman, Michael G. "Air-to-Ground Century Style." USAF Fighter Weapons News Letter, no. 3 (September 1959), 16-28.

Fricker, John. "Presenting the Super Sabre." The Aeroplane, XC (June 1, 1956), 474-476.

Green, William. "The First Supersonic Fighter." RAF Flying Review, XIII (February 1958), 16-19+.

Guillot, Athos E., Jr. "End of an Era." USAF Fighter Weapons News Letter, no. 1 (March 1970), 22-23.

Gunston, William T. ("Bill"). "Fighters of the Fifties: North American F-100 Super Sabre." Aeroplane Monthly, VIII (June 1980), 299-304.

Hart, John W. "Master of the Century." Airman, XII (July 1968), 30-31.

Hemm, Wayne H. "Old Figher Planes Never Die." Aerospace Safety, XXVI (January 1970), 20-22.

Hillbrink, Will. "World's Shortest Runway." Skyline, XVII (Winter 1958-1959), 24-29.

Hoover, Robert A. "Be an Old Smoothie." Aerospace Safety, XXI (June 1965), 6-7.

Hopkins, Alfred H., Jr. "Air-to-Ground in the F-100D/F." USAF Fighter Weapons News Letter, no. 1 (March 1958), 14-15.

____. "Two-Place F-100F." USAF Fighter Weapons News Letter, no. 3 (November 1957), 9-10.

"How to Sharpen Old Sabres." Skyline, XX (Fall 1962), 26-31.

"Jack High." Skyline, XIX (Spring 1961), 10-19.

Jackson, Bud and Cam Dooley. "Farewell to a Gallant Lady." TAC Attack, XII (August 1972), 8-11.

Jackson, Robert A. "F-100 Ripple Bombing." USAF Fighter Weapons News Letter, no. 1 (March 1968), 1-6.

Koyen, K. "Supersonic Fighter No. 1." BeeHive, XXVIII (Fall 1953), 3-6.

Lauren, Ward. "Jet Strike Force: Long-Range Super Sabres Help Demonstrate TAC's New Global Mobility." Skyline, XV (Spring 1957), 4-11.

____. "March on the Rhine." Skyline, XIV (August 1956), 20-23.

McCarthy, Dan B. "Super Sabre." Wings, IX (October 1979), 24-35, 46.

McLaughlin, George F. "F-100 Super Sabre." Aero Digest, LXVII (November 1953), 17-19.

Miller, Jay. "Super Sabre." Airpower, IX (September 1979), 18-31.

"Most Fire for the Buck." Air Reservist, XXII (August-September 1970), 12-14.

Murata, Stephen K. "When the Thunderbirds Roar." Aerospace Historian, XV (Spring 1968), 16-27.

"New Mission Aircraft Assigned to Three ANG Fighter Units." National Guardsman, XXIV (November 1970), 37-38.

Newton, Bob. "F-100 Retired by Air Guard TAC Units." Air Progress, XLI (August 1979), 76-77.

North, David M. "F-100s to Become Target Drones." Aviation Week and Space Technology, CXV (September 28, 1981), 69-71.

"North American F-100C Super Sabre." Flight, LXIX (June 8, 1956), 710-711.

Rodwell, Robert R. "Supersonic Sabreurs." Aeronautics, XXXVII (September 1957), 140-141.

Scholin, Allan R. "Farewell to the F-100s." National Guard, XXXIV (January-February 1980), 14-17.

Shadbolt, Stuart W. "No Before, No After." Aerospace Historian, XIV (Spring 1967), 50-51.

"Stop of the Century." Flying Safety, XIII (January 1957), 20-23.

"Super Sabre." Air International, XXVIII (1985), 238+.

"Super Sabre Details." Flight, LXVI (July 30, 1954), 137-138.

"Supersonic Fighter: A Critical Examination." Flight, LXVII (May 20, 1955), 680-684.

"TAC Flexes New Super Sabre Muscle." Aviation Week, LXIII (July 26, 1955), 18-19.

"Thoughts on the F-100." Flight, LXIV (November 6, 1953), 612-614.

"To Fly and To Fight." National Guardsman, XXIV (June 1970), 30-32.

Trimble, Robert L. "Salute to the Super Sabre." Air Classics, XI (July-September 1975), 26-37, 26-37, 24-35.

Tucker, Roger D. "Fire Control" USAF Fighter Weapons News Letter, no. 4 (December 1959), 14-23.

Vogel, Joseph L. "The Best Fighter I Ever Saw." Aerospace Historian, XXIV (September 1977), 146-148.

Weidinger, Charles L. "'The Hun': The F-100 Super Sabre." Aerospace Historian, XXIV (March 1987), 13-19.

____. "Test Flying the F-100." Aerospace Historian, XXXIV (March 1987), 20-21.

White, Alvin S. "Stability, Tigers, and the F-100." Flying Safety, XVI (February 1960), 6-8.

Wilson, Tom. "Wild Weasel Legacy: The Early Days." Defense Electronics, XX (September 1988), 52-54+.

(McDonnell) F-101 Voodoo

Book/Document/Report Literature:

Becklund, Thornton E. Air National Guard Increased Weapon System Capability. Research Report. Maxwell AFB, Ala.: Air War College, Air University, 1972. 22p.

Bonnell, Harry L. The F-100 Voodoo, August 1952-June 1962. AFLC Historical Study, no. 359. Wright-Patterson AFB, Ohio: History Office, Air Force Logistics Command, 1962. 161p.

Dorr, Robert F. McDonnell F-101 Voodoo. Osprey Air Combat Series. Osceola, Wis.: Motorbooks International, 1987. 172p.

Drendel, Lou and Paul Stevens. Voodoo. Carrollton, Tex.: Squadron/Signal Publications, 1985. 64p.

Kinzey, Bert. F-101 in Detail and Scale. Details and Scale Series, v. 21. Blue Ridge Summit, Pa.: TAB, 1986. 72p.

Periodical/Anthology Literature:

"Air Force Gives Nod to McDonnell Voodoo." Aviation Week, LV (October 29, 1951), 16-17.

"Air Force's Biggest Fighter." Air Training, IV (December 1954), 8-9.

"Arm Chair Analysis: Voodoo." Interceptor, VII (August 1965), 9-11.

"A Bird's Eye View of the Voodoo." Interceptor, X (April 1968), 14-15.

"F-100." Interceptor, XVI (December 1974), 12-15.

"F-100 Voodoo." Interceptor, XX (November 1978), 24-27.

"F-100 Voodoo." TAC Attack, XXVII (November 1987), 16-17.

"F-100 Voodoos Come to Compete." Interceptor, XIX (January 1977), 8-11.

Francillon, Rene J. "It's Witchcraft: McDonnell's F-101 Voodoo." Airpower, X (May 1980), 10-29, 50.

Green, William. "The Amazing Voodoo." RAF Flying Review, X (February 1955), 40-41.

Griffith, Wayne E. "Some War Story." Interceptor, XXI (March 1979), 5-9.

"How to Brake the Voodoo." Interceptor, IV (December 1962), 8-13.

Jones, J. E. "Up is a Relative Word." Interceptor, I (October 1959),8-9.

Kell, R. J. and D. L. Rodgers. "Automatic Attack." Interceptor, III (September 1961), 5-9.

Little, Bob. "The Voodoo." Combat Crew, VII (April 1957), 4-8.

Lundquist, Tom H. "Voodoos." Interceptor, XIII (January 1971), 7-9.

Newell, Brad. "F-101 Sleigh Ride." Interceptor, XIV (December 1972), 24-25.

Polek, Fran. "Supersonic Voodoo Magic." Air Progress, XL (August 1978), 22-25, 76.

Ruddock, Larry. "Voodoos in England." RAF Flying Review, XIV (December 1958), 37-38.

Shackleton, Dwayne M. "Brakes of the Voodoo Game." Interceptor, XI (August 1969), 8-11.

Spering, Don. "Mounting Guard." Wings, XIII (October 1983), 18-29.

Stevens, Paul D. "Operation Fire Wall." American Aviation Historical Society Journal, XXIII (Spring 1978), 72-76.

Stuck, Donald H. "F-101." Flying Safety, XIV (May 1958), 19-21.

___ and P. A. McWilliams. "Voodoo: Air Defense Style." Interceptor, I (May 1959), 9-13.

"Voodoo Blowouts." Interceptor, XV (July 1973), 25-28.

"Wahoo: Voodoo." Interceptor, XVIII (March 1976), 5-9.

(Convair) F-102 Delta Dagger

Book/Document/Report Literature:

Alling, Frederick. The F-102A Airplane. AMC Historical Study, no. 289. Wright-Patterson AFB, Ohio: USAF Air Materiel Command, 1957. 67p.

LaRosa, Benedict D. Project Pave Deuce: The Droning of the F-102 Interceptor. Kelly AFB, Tex.: History Office, San Antonio Air Logistics Center, 1982. Unpaged.

Lee, Philip M. An Analysis of F-102A Combat Crew Training. Research Paper. Maxwell AFB, Ala.: Air Command and Staff College, Air University, 1967. 62p.

United States. Department of the Air Force. Air Training Command. F-102 Aircrew Standardization/Evaluation Manual: Flying. AFM 60-102. Washington, D.C., 1961. 63p.

Periodical/Anthology Literature:

Anderton, David A. "F-102 Pioneers Production Concept." Aviation Week and Space Technology, LXIV (June 11, 1956), 50-52.

"The Area Rule." Flight, LXVIII (September 30, 1955), 549-552.

Bristow, Dean. "Delts." Talon, IV (November 1958), 24-25.

Burns, Robert W. "F-102 Operational Training." Air University Review, XV (July-August 1964), 9-23.

Carlton, James L. "Do's and Don'ts for Deuces." Interceptor, II (January 1960), 8-11.

Chillo, R. J. "Mating the Pratt and Whitney 257 Turbojet to the F-102." Society of Automotive Engineers Journal, LXV (December 1957), 33-36.

"Convair Converts Test F-102As to Latest Tactical Configurations." Aviation Week and Space Technology, LXVII (August 19, 1957), 63+.

"Convair F-102." Flight, LXIX (June 8, 1956), 707-708.

"Convair F-102: An Analysis of America's Home-Defence Interceptor." Flight International, LXXI (April 19, 1957), 512-518.

"Convair's Delta Defender." Air International, XXX (January 1986), 28+.

"Convariant." Aeronautics, XXXII (March 1955), 44, 50.

Cutler, Katie. "What the Deuce." Airman, XXIII (May 1979), 21-26.

"Delta Daggers and Ground Egresse." Interceptor, XII (December 1970), 14-15.

Driskill, D. E. "Familiarity Breeds Content." Aerospace Maintenance Safety, XXII (March 1967), 8-9.

"F-102." Interceptor, XVI (December 1974), 20-23.

"F-102 Delta Dagger." TAC Attack, XXVII (December 1987), 16-17.

"Future's Baby: Delta F-102 Break's Aviation's Design Barrier." Air Training, IV (February 1955), 2-3.

Gatlin, Wayne C. "Daggers." Interceptor, XIII (January 1971), 10-12.

Green, William. "Delta Dagger." RAF Flying Review, XIII (December 1957), 34-37.

Harvey, Frank L. "Deadliest Fighter in the Air: F-102A." Evening Post, CCXXIX (December 15, 1956), 43+.

Hucker, Robert. "First of the Coke Bottle Fighters." Air Progress Aviation Review, IV (Summer 1980), 18-25.

Hull, Richard. "Deuces for Defense." BeeHive, XXXII (Summer 1957), 19-23.

"The Imperfect Match." Interceptor, VIII (September 1966), 12-15.

Johnson, Richard L. "F-102." Flying Safety, XIV (May 1958), 21-22.

Mathewson, Tracy B. "The 102 Below Zero." Flying Safety, XII (October 1956), 8-13.

Ragay, John D. "The F-102 in European Service." American Aviation Historical Society Journal, XXIX (Fall 1984), 192-201.

Robotti, Aurelio C. "Satellite Launching From an F-102." Missiles and Rockets, I (November 1956), 54+.

Saromines, Sascha. "Last of an Old Friend." Interceptor, XIX (March 1977), 5-6.

"Shaped for Speed." Military Review, XXXV (November 1955), 65-66.

Stambler, Irwin. "F-102 is Built Around Its Electronic System." Aviation Age, XXVII (January 1957), 26-31.

Stone, Irving. "F-102 Fits Into Cook-Cragie Pattern." Aviation Week, LXII (June 13, 1955), 51-52.

"Supersonic Delta Fighter." Air Force, XXXV (December 1952), 31-32.

Vadas, John F. "Eleven Sixty Three." American Aviation Historical Society Journal, XXIX (Summer 1984), 116-117.
F-102 serial number 56-1163.

____. "326: The Phantom Deuce." American Aviation Historical Society Journal, XXXI (Winter 1986), 318-319.

(Lockheed) F-104 Starfighter

Book/Document/Report Literature:

Barondes, Arthur D. F-104G: Sword for the NATO Shield. Research Paper. Maxwell AFB, Ala.: Air Command and Staff College, Air University, 1962. 93p.

Drendel, Lou. F-104 Starfighter in Action. Warren, Mich.: Squadron/Signal Publications, 1976. 50p.

Foster, Peter R. F-104 Starfighter. Warbirds Illustrated, no. 44. New York: Sterling, 1987. 64p.

Joos, Gerhard. Lockheed F-104G/CF-104. Aircraft in Profile, no. 131. Windsor, Eng.: Profile Publications, 1967. 15p.

Reed, Arthur. F-104 Starfighter. New York: Scribners, 1981. 112p.

Periodical/Anthology Literature:

Bond, Robert M. "Turning With the 104." USAF Fighter Weapons News Letter, no. 3 (September 1960), 10-11.

Canan, James W. "Era of the Starfighter." Air Force Magazine, LXX (January 1987), 42-48.

Coble, Donald W. "The World's First Space Trainer." Airman, VII (October 1963), 26-27.

Darge, Thomas F. "Fantastic 104." Flying, LVIII (January 1956), 34+.

"Evolution of the F-104 Starfighter." Interavia, XXII (December 1967), 1917-1920.

"The F-104." Air Pictorial, XVIII (July 1956), 232-233.

"F-104 Starfighter." TAC Attack, XXVIII (January 1988), 16-17.

"F-104G: A Much-Discussed Weapon." Interavia, XVIII (August 1963), 1196-1200.

"F-104G Super Starfighter." Flying Review International, XIX (June 1964), 19-25.

"Fighter for Tomorrow." RAF Flying Review, XII (February 1957), 31-32.

Fried, Jules. "F-104 Run." Air Progress, XXXVIII (April 1976), 67-73.

"Gee-Whizzer." Military Review, XXXIV (November 1954), 34-64.

Green, William. "Super Starfighter." RAF Flying Review, XVI (February 1961), 26-30.

Gunston, William T. ("Bill"). "F-104, Holder of the Absolute Records for Speed and Altitude." Flight International, LXXIII (May 30, 1958), 739-743.

____. "Starfighter: Lockheed's Mach 2 Fighter for the U.S. Air Force." Flight, LXIX (April 20, 1956), 440-443.

Heumann, Gert W. "The Lockheed F-104A Starfighter." Air Pictorial, XXI (August 1958), 277-279.

Jabara, James. "A Fighter Pilot's Airplane." Air Force, XLIII (August 1960), 60-62+.

Kitchens, Charles A. "Spinning the F-104." Flying Safety, XIV (December 1958), 2-6.

____. "Parade of the Centuries: F-104." Flying Safety, XII (November 1956), 10-15.

Lambert, Mark. "F-104G Starfighter." Flight, LXXXIII (March 14, 1963), 368-377.

LeVier, Anthony W. "How It Feels to Fly the World's Fastest Fighter." Popular Science, CLXVIII (June 1956), 74-77.

____. "Day to Remember: Flying the F-104 Starfighter." Flying, LX (April 1957), 47+.

____. "The Thing." Interceptor, XII (May 1970), 8-13.

____ and Louis W. Schalk, Jr. "F-104." Flying Safety, XIV (May 1958), 23-28.

"The Lockheed F-104." Interavia, XIV (January 1959), 43-46.

"Lockheed F-104 Details." Aviation Week and Space Technology, LXV (December 3, 1956), 32-33.

"Lockheed Reveals the F-104A Starfighter." Interavia, XI (May 1956), 360-361.

"Lockheed's Starfighter." The Aeroplane, XC (April 20, 1956), 284-285.

Macho, Dean C. "An Exchange View." Interceptor, I (October 1959), 10-11.

Major, Mike. "Snake-Eyes and Starfighters." USAF Fighter Weapons News Letter, no. 1 (March 1970), 12-16.

Messerli, Robert E. "Slipping the Surly Bonds With a Zip: The F-104." In: Robin Higham and Carol Williams, eds. Flying Combat Aircraft of the USAAF-USAF. Manhattan, Kans.: Sunflower University Press, 1981. pp. 61-68.
Reprinted in Aerospace Historian, XXIX (June 1982), 28-29.

"New Era Fighter." Flight Magazine, XLV (May 1956), 28-29.

Petersen, Stefan. "Starfighter F-104." Defence Update International, no. 80 (May-June 1987), 28-35.

Plattner, C. M. "Mission Variety Demonstrated by F-104G." Aviation Week and Space Technology, LXXVIII (February 11, 1963), 72-73+.

"Programmed for Maintenance." Aerospace Maintenance Safety, XX (February 1965), 6-9.

Rayburn, Scott. "Starfighter." Air Progress, XXXIX (January 1977), 37-39, 74.

"Return of the 'High Fasties." Interceptor, V (April 1963), 20-21.

Rodwell, Robert R. "Pilots Say the Starfighter 'Goes Like a Bomb, Climbs Like a Homesick Angel, and Glides Like a Streamlined Brick.'" Flight International, LXXXVII (June 17, 1965), 979-983+.

Seitz, E. A. "Flying the Winged Missile." In: Michael O'Leary, ed. Air Progress Pilot Reports. Canoga Park, Calif.: Challenge Publications, 1979. pp. 4-11.

Simmons, H. T. "Lockheed Bids for Lead in Jet Fighters." American Aviation, XIX (May 7, 1956), 28-30.

"The Smashing F-104A." RAF Flying Review, XI (April 1956), 17-18.

Stambler, Irwin. "Lockheed F-104: Designed for Mach 2 Plus." Aviation Age, XXVI (July 1956), 26-31.

"Starfighter Status." Interceptor, VI (April 1964), 12-16.

Stone, Irving. "USAF F-104 Completes Test Program, Enters Squadron Use." Aviation Week and Space Technology, LXVIII (February 10, 1958), 66-84.

___. "USAF Shows Unique Details of Lockheed Starfighter." Aviation Week and Space Technology, LXIV (April 23, 1956), 34-37.

Van der Klaauw, B. "The Ten Years of the Starfighter." NATO's Fifteen Nations, XVI (February-March 1971), 76-83.

Vandyk, Anthony. "NATO Firms Tool Up to Produce F-104G Starfighter." Armed Forces Management, VII (August 1961), 28-29.

Watts, A. F. "F-104." Aviation Age, XXIX (May 1958), 60-67.

"A Weapon System in Practice." Interavia, XIII (April 1958), 326-327.

(Republic) F-105 Thunderchief.

Book/Document/Report Literature:

Anderton, David A. Republic F-105 Thunderchief. Osprey Air Combat Series. Osceola, Wis.: Motorbooks International, 1983. 198p.

Archer, Robert D. The Republic F-105. Aero Series. Fallbrook, Calif.: Aero Publishers, 1969. 80p.

Dorr, Robert F. F-105 Thunderchief. Warbirds Illustrated, no. 49. New York: Sterling, 1988. 64p.

Drendel, Lou. F-105 Thunderchief in Action. Aircraft, no. 17. Warren, Mich.: Squadron/Signal Publications, 1974. 49p.

___. Thud. Carrollton, Tex.: Squadron/Signal Publications, 1986. 63p.

Hildreth, James R. Can the F-105D Provide Effective All-Weather Close Air Support? Research Paper. Maxwell AFB, Ala.: Air Command and Staff College, Air University, 1965. 64p.

Kinzey, Bert. F-105 Thunderchief in Detail and Scale. Detail and Scale Series, v. 8. Fallbrook, Calif.: Aero Publishers, 1982. 72p.

Long, Patrick G. *Evaluation of the F-105 Weapon System in Role of Out-Country Interdiction, Southeast Asia, 1965-1967*. Research Report. Maxwell AFB, Ala.: Air War College, Air University, 1968. 103p.

Scutts, Jerry G. *F-105 Thunderchief*. New York: Scribners, 1981. 112p.

Sorlie, Donald M. *An Analysis of the F-105 Weapon System in Out-Country Air Operations, Southeast Asia, 1965-1967*. Research Report. Maxwell AFB, Ala.: Air War College, Air University, 1968. 127p.

Van Geffin, Theodore, Jr. *Republic F-105 Thunderchief*. Aircraft in Profile, no. 226. Windsor, Eng.: Profile Publications, 1972. 24p.

Periodical/Anthology Literature:

"Ballad of a Brake." *Airman*, XIII (July 1969), 10-13.

Baset, Gene I. "Ripple Six." *USAF Fighter Weapons News Letter*, no. 3 (September 1970), 24-26.

Beaird, Henry G., Jr. "F-105." *Flying Safety*, XIV (May 1958), 28-29.

Brownlow, Cecil. "All-Weather F-105s Phase Into Europe." *Aviation Week and Space Technology*, LXXIV (May 22, 1961), 74-75.

____. "F-105 Modified for Viet Role." *Aviation Week and Space Technology*, LXXXVI (May 22, 1967), 16-17.

____. "F-105D's Limited War Capability Boosted." *Aviation Week and Space Technology*, LXXVIII (February 25, 1963), 105+.

____. "USAF to Strengthen F-105 Survivability." *Aviation Week and Space Technology*, LXXXIX (September 23, 1968), 78-79+.

Butz, J. S., Jr. "F-105s in Vietnam: New Lustre for a Tarnished Image." *Air Force and Space Digest*, XLIX (October 1966), 37-43.

Carson, Don. "F-105s Join Air Force Reserve Forces." *Air Force Magazine*, LVII (April 1974), 24-25.

____. "Flying the 'Thud.'" *Air Force Magazine*, LVII (April 1974), 18-23.

Carter, Vic. "Squash Bomber." *AirFix Magazine*, XXIII (June 1982), 486-488.

Cook, Mark D. "F-105D: Curios, Knickknacks, Doodads." *USAF Fighter Weapon News Letter*, no. 2 (July 1961), 5-10; no. 2 (June 1964), 18-19.

Cross, Roy. "F-105 Thunderchief." *Air Pictorial*, XIX (October 1957), 354-355.

Duncan, Scott. "The Combat History of the F-105." *Aerospace Historian*, XXII (September 1975), 121-128.

Everett, R. P. "The 'Thud' is No Dud." Airman, XII (May 1968), 4-8.

"F-105: A Boost to NATO's Defense of Berlin." NATO Journal, I (September 1961), 41-42.

"F-105 Shows Steep Climb, Weapons Load." Aviation Week and Space Technology, LXVII (August 5, 1957), 74-75.

"F-105G Wild Weasel." TAC Attack, XXIV (March 1984), 16-17.

Flynn, George. "No Job for Boys." BeeHive, XXXV (Spring 1960), 28-30.

Glass, James. "Readiness Through Realism." Air Reservist, XXVIII (August-September 1976), 7-9.

Harvey, Frank L. "The F-105 Thunderchief: The World's Deadliest One-Man Plane." Popular Science, CLXXII (June 1958), 64-67.

Hendrix, Lin. "How to Bellyland a Thunderchief." Aeroplane Monthly, V (October 1977), 515-519.

Hoblit, Jerry N. "F-105 Thunderchief." In: Robin Higham and Abigail T. Siddall, eds. Flying Combat Aircraft of the USAAF-USAF. Ames, Iowa: Iowa State University Press, 1975. pp. 85-90.

Hopkins, Alfred H. "F-105D Conversion Course." USAF Fighter Weapons News Letter, no. 1 (March 1961), 9-11.

Interavia Study Group. "The Republic F-105D Weapons System." Interavia, XV (March 1960), 311-317.

Johnson, Harold E. "Of Bees, Weasels, Ferrets, and Eagles." Air University Review, XXXIII (January-February 1982), 86-93.

King, Tom., et al. "People, Let Me Tell You 'Bout My Best Friend." Air Reservist, XXXVI (March-April 1984), 6-7+.

Lambert, Mark. "The Five is the Thunderchief." Flight International, LXXXV (February 6, 1964), 204-212.

Lynn, Tom. "The Weasel Joins the Guard." Navigator, XXVIII (Spring 1981), 8-10.
F-105G.

Macknight, Nigel. "F-105s Over Southeast Asia." Air Pictorial, XXXIX (April 1977), 140-143.

McNally, Frank. "Coronet Fife." Air Reservist, XXVIII (February 1977), 6-7.

Mancus, Peter. "SAM Slayer." Airpower, XIV (January 1984), 24-37.

Mason, Dwight E. "Low Tossing With the F-105D." USAF Fighter Weapons News Letter, no. 2, (June 1963), 7-23.

"The Mighty Thunderchief." RAF Flying Review, XIII (February 1958), 35-36.

Miller, Barry L. "F-105 Modified for Blind-Bombing Role." Aviation Week and Space Technology, XC (January 20, 1969), 64-65.

Mitchell, James R. "Down on the 99th." Air Force Magazine, LVI (September 1973), 112-116.

Morrison, Blake G. "Requiem for a Heavyweight." USAF Fighter Weapons Review, (Summer 1980), 36-37.

Morrissey, John C. "Attack--Thunderstick Style." USAF Fighter Weapons News Letter, no. 4 (December 1967), 1-7.

Murphy, Frank L. "Lead Scout of the Thunderchief." BeeHive, XXXIV (Summer 1959), 26-29.

Myrahi, Joseph V. and Peter Mancus. "Thud, Goes the Weasel." Airpower, XI (January 1981), 24-33.

"The NASARR System for the F-105G." Interavia, XVIII (January 1963), 88-90.

NASARR = North American Search and Ranging Radar.

Noel, Martin A., Jr. "The Heavyweight Champ." TAC Attack, XII (October 1972), 16-19.

Palmer, E. C. and John W. Wiley. "The Extra Margin." Interceptor, IV (April 1962), 12-15.

Reed, William S. "TAC Trains Pilots for F-105D Missions." Aviation Week and Space Technology, LXXIV (April 17, 1961), 88-89.

Roth, Rusty. "Thunderchief." Flying Safety, XV (December 1959), 21-23.

Smith, Gene. "Thud." Air Progress, XLII (April 1980), 24-30, 72.

Soldeus, Lara G. "Thunderchiefs on the Rhine." American Aviation Historical Society Journal, XVIII (Spring 1973), 34-47.

Spering, Don. "The Last Goodbye." Wings, XIII (April 1983), 30-37.

Stambler, Irwin. "F-105 Production Line Stresses Colony System." Aviation Age, XXIX (April 1958), 54-57.

Stanfield, Robert I. "F-105 Spearheads TAC Fighter Forces." Aviation Week and Space Technology, LXXI (December 28, 1959), 37-45.

Taylor, Levin. "F-105F Thunderchief." TAC Attack, XXVII (April 1987), 16-17.

"Thunderchief." RAF Flying Review, XVI (November 1960), 28-31.

"Thunderchief: The Big Stick." Flying Review International, XIX (January 1964), 21-23.

"The Thunderchief: Vietnam's Versatile Heavyweight." Flying Review International, XXII (July 1967), 699-702.

Van Geffin, Theodore, Jr., and Gerald C. Arruda. "Thunderchief." Air University Review, XXXIV (January-February 1983), 46-51+.

Waxman, Saul. "The Thinking Thunderchief." Aerospace Safety, XVII (September 1961), 2-3.

"Wild Weasel." Electronic Warfare, II (February-March 1970), 15+. F-105G.

Wilson, Tom. "Wild Weasel Legacy--The Early Days." Defense Electronics, XX (September 1988), 52-54+.

Wright, Tom. "To Ride a One-Oh-Five." Air Reservist, XXX (July 1978), 15-16.

Yow, Frank L., Jr. "Flying the Five." USAF Fighter Weapons News Letter, no. 3 (September 1962), 5-6.

(Convair) F-106 Delta Dart

Book/Document/Report Literature:

Carson, Don and Lou Drendel. F-106 Delta Dart in Action. Aircraft, no. 15. Warren, Mich.: Squadron/Signal Publications, 1974. 49p.

Flannagan, William R. F-106 Accidents and Instrument Training: Is There a Relationship? Research Paper. Maxwell AFB, Ala.: Air Command and Staff College, Air University, 1975. 59p.

Holder, William G. Convair F-106 Delta Dart. Aero Series. Fallbrook, Calif.: Aero Publishers, 1977. 104p.

Kinzey, Bert. F-106 Delta Dart in Detail and Scale: The Ultimate Interceptor. Detail and Scale Series, v. 13. Fallbrook, Calif.: Aero Publishers, 1983. 72p.

____. F-106 Delta Dart: USAF, ANG, and NASA Markings. Color and Markings Series, v. 1. Fallbrook, Calif.: Aero Publishers, 1984. 64p.

Norris, William D. The F-106 Dilemma. Research Paper. Maxwell AFB, Ala.: Air Command and Staff College, Air University, 1965. 59p.

Scott, Jessie C. The F-106 Fighter-Interceptor, July 1958-March 1960. AFLC Historical Study, no. 451. Kelly AFB, Tex.: History Office, San Antonio Air Materiel Area, 1960. 94p.

___. The Struggle For Survivability: The F-106, April 1960-February 1962. AFLC Historical Study, no. 461. Kelly AFB, Tex.: San Antonio Air Materiel Area, 1963. 107p.

Williams, Fenton L. Sacramento Air Materiel Area's F-106 Experience, 1959-1961. AFLC Historical Study, no. 433. McClellan AFB, Calif.: History Office, Sacramento Air Materiel Area, 1962. 22p.

Periodical/Anthology Literature:

Bach, Richard. "Air Defense and the F-106." Flying, LXIX (August 1961), 46-47+.

Baer, John. "General Aviation Pilot Tries It: Beyond the Speed of Sound." Plane and Pilot, IX (July 1973), 34-37.

Carlton, James L. "Durable Dart." Interceptor, I (September 1959), 8-11.

____. "Fuel for Thought." Interceptor, II (November 1960), 10-12.

Carson, Donald D. "Flying the Six." Air Force Magazine, LVI (October 1973), 38-44.

Crewse, Roger G. "Give It Lots of Lovin' Care." Interceptor, XIV (March 1972), 20-25.

"Delta Dart." Air International, XXXI (1986), 198+.

"Delta Dart Joins Guard in Major Restructuring." National Guardsman, XXVII (May 1972), 28-29.

DeRamus, Dave. "Night Flight in a Delta Dart." Air Progress, XLI (May 1979), 52+.

Dwyer, Tom. "Last Chance on Runway Three Three." Airman, XVIII (May 1974), 32-35.

"F-106." Interceptor, XVI (December 1974), 24-27.

"F-106 Delta Dart." Interceptor, XX (November 1978), 12-17.

"F-106 Flameout." Interceptor, XIX (June 1977), 20-21.

"F-106 Minor Miracle." Interceptor, XV (February 1973), 23-25.

Fairey, William. "Spinning Darts." Interceptor, XI (June 1969), 6-7.

Fitzpatrick, J. M. "Keep Your Nose Up in the Blue." Interceptor, III (May 1961), 10-12.

Gamble, Jack and Patrick. "F-106 Delta Dart." In: Robin Higham and Abigail T. Siddall, eds. Flying Combat Aircraft of the USAAF-USAF. Ames, Iowa: Iowa State University Press, 1975. pp. 91-100.

Golden, Gordon H. "F-106." Flying Safety, XXXIX (March 1983), 11-12.

Gorman, Tom. "Three Minutes After Flameout." Interceptor, XIX (July 1977), 12-15.

Greenfield, William D. "Global Air Defense Through Mobility." Air University Review, XX (November-December 1968), 84-94.

Hennigan, George R. "A New Potential." Interceptor, XII (May 1970), 20-22.

Isham, Marty J. "The F-106: Twenty Years." Interceptor, XXI (May 1979), 12-15.

Johnson, Dick and Maxwell Frank. "The Delta Dart." Interceptor, I (July 1959), 10-14.

Larcom, Craig. "Goodbye to All That." Wings, XVII (October 1987), 24-35.

Luke, Terry R. "Block 'S.'" Interceptor, XV (January 1973), 5-9. Modifications.

McBride, Larry B. "...& Darts." Interceptor, XIII (January 1971), 13-15.

Mack, William M. "Silver Sentinels." Interceptor, XI (June 1969), 12-14.

Mancus, Peter. "Red Alert: The F-106 and the Case for Manned Interceptors." Wings, XI (June 1981), 25-35.

Moutman, James H. and Michael M. Dunn. "Aerial Refueling the Gray Ghost." Interceptor, XIX (December 1977), 5-9.

Normile, Thomas H. "Eleven Inches From Success." Interceptor, XII (September 1970), 5-7.

Palmer, E. C. and John W. Wiley. "The Extra Margin." Interceptor, IV (April 1962), 12-15.

"The Potent Dart." RAF Flying Review, XIII (May 1958), 16-17.

Prochoroff, Alan. "Shooting for a '10.'" Airman, XXV (April 1981), 20-27.

Ratcliff, M. David. "Six Seconds of Terror." Interceptor, XV (July 1973), 20-22.

Rogers, Joseph W. "The Blistering Dart." Interceptor, II (April 1960), 20-23.

Scott, William A. "Down and In: Dead Dart." Interceptor, XVI (February 1974), 26-28.

"Speedline: The Ultimate Weapon is Alive and Well and Recuperating in California." Interceptor, XIV (June 1972), 11-13.

Stanfield, Robert I. "F-106 Hits Mach 2 on Radar Intercept." Aviation Week and Space Technology, LXXI (September 7, 1959), 126-128+.

"A Step Forward." Interceptor, IX (February 1967), 5-7.

Talladay, Keaith R. "Air Superiority Tactics Training." Interceptor, XVIII (January 1976), 20-23.

Taylor, Lelvin."F-106 Delta Dart." TAC Attack, XXVII (May 1987), 16-17.

"That 106 is Tough." Interceptor, III (December 1961), 14-15.

"Updating the Convair F-106." Flight International, XC (September 1, 1968), 341-342.

"Webster, W. F. "Give Me Air!" Interceptor, III (February 1961), 11-13.

Weisenburg, Sandy. "Mission of the Dart." Airman, XIV (June 1970), 8-9.

F-110 see F-4 Phantom II

(General Dynamics) F-111 Aardvark

Book/Document/Report Literature:

Amrhein, Robert L. "Harvest Ranger"/"Combat Lancer." Research Paper. Maxwell AFB, Ala.: Air Command and Staff College, Air University, 1969. 159p.
F-111s in Southeast Asia.

Art, Robert J. The TFX Decision: McNamara and the Military. Boston, Mass.: Little, Brown, 1968. 202p.

Barrett, John G. A Study of F-111A Crew Structure. Research Paper. Maxwell AFB, Ala.: Air Command and Staff College, Air University, 1966. 52p.

Coulam, Robert F. Illusions of Choice: The F-111 and the Problems of Weapons Acquisition Reform. PhD. Dissertation. Cambridge, Mass.: Harvard University, 1978.
Published commercially later that year with the same title by Princeton University Press in a 432-page edition.

Dixon, Ford. McNamara, General Dynamics, and the F-111: A Business and Political History. PhD. Dissertation. Waco, Tex.: Texas Christian University, 1972.

Fincher, Clifton F. F-111 Aircrew Training for Low-Altitude Operations. Research Paper. Maxwell AFB, Ala.: Air Command and Staff College, Air University, 1972. 48p.

Franklin, Carl E. Conventional Employment of the F-111F in NATO. Research Paper. Maxwell AFB, Ala.: Air Command and Staff College, Air University, 1979. 58p.

Gorn, Michael H. TFX: From Conceptual Development to F-111B Termination, 1958-1968. Andrews AFB, Md.: History Office, Air Force Systems Command, 1986. 84p.

Green, Louis A. Considerations for Optimizing F-111A Radar Bombing. Research Paper. Maxwell AFB, Ala.: Air Command and Staff College, Air University, 1973. 76p.

Gunston, William T. ("Bill"). F-111. Modern Fighting Aircraft, no. 3. New York: Arco, 1983. 64p.

____. F-111. New York: Scribners, 1979. 112p.

Hagerty, Richard K. The Administrative Actions of the Secretary of Defense [McNamara] in the Procurement of the TFX. Research Paper. Maxwell AFB, Ala.: Air Command and Staff College, Air University, 1966. 90p.

Hub, Michael G. F-111D Retention Issue. Thesis. Fort Leavenworth, Kans.: Army Command and General Staff College, 1986. 176p.

Jones, Laurence M. F-111A Navigator's Introductory Guide. Research Paper. Maxwell AFB, Ala.: Air Command and Staff College, Air University, 1981. 53p.

Jordan, Larry M. F-111A Evaluation Program: Can It Be Done Without an Operational Flight Trainer? Research Paper. Maxwell AFB, Ala.: Air Command and Staff College, Air University, 1982. 25p.

Kinzey, Bert. F-111 Aardvark in Detail and Scale. Detail and Scale Series, v. 2. Fallbrook, Calif.: Aero Publishers, 1982. 72p.

Larson, R. R. Flight Control System Development and Flight Test Experience With the F-111 Mission-Adaptive Wing Aircraft. NASA-TM-88265. Moffett Field, Calif.: NASA Ames Research Center, 1986. 21p.

Link, Gordon H. F-111A: A Wild Weasel? Research Report. Maxwell AFB, Ala.: Air War College, Air University, 1968. 80p.

Mathie, Lloyd P. An Analysis of Electronic Warfare Training for F-111A Aircrews. Research Paper. Maxwell AFB, Ala.: Air Command and Staff College, Air University, 1973. 34p.

Miller, Jay. General Dynamics F-111. Aero Series, v. 29. Fallbrook, Calif.: Aero Publishers, 1981. 104p.

Olsen, Thomas A. The F-111F: The Multi-Mission Fighter. Research Report. Maxwell AFB, Ala.: Air War College, Air University, 1975. 62p.

Osterholtz, Robert A. FB-111 Pilot Instructor Guide. Research Paper. Maxwell AFB, Ala.: Air Command and Staff College, Air University, 1985. 69p.

Parks, Scott E. Pilot Report: AFTI (Advanced Fighter Technology Integration) F-111. Research Paper. Maxwell AFB, Ala.: Air Command and Staff College, Air University, 1988. 23p.

Patterson, Jock P. The F-111 in Combat. Research Report. Maxwell AFB, Ala.: Air War College, Air University, 1977. 38p.

Savage, James W., Jr. Consolidation of F-111 Test and Evaluation Activities. Research Paper. Maxwell AFB, Ala.: Air Command and Staff College, Air University, 1983. 17p.

Schwerdt, Gary F. A Guide for F-111F Aircrews for Operation of the AGM-65A/B Maverick Missile. Research Paper. Maxwell AFB, Ala.: Air Command and Staff College, Air University, 1982. 37p.

Sullivan, John J. The F-111A Test Program. Research Paper. Maxwell AFB, Ala.: Air Command and Staff College, Air University, 1967. 69p.

Taylor, Harris J. The F-111 in Southeast Asia. Research Report. Maxwell AFB, Ala.: Air War College, Air University, 1971. 17p.

Trout, Gary G. The F-111 Aircraft: So Unwanted, So Misunderstood, So What Now? Research Paper. Maxwell AFB, Ala.: Air Command and Staff College, Air University, 1975. 51p.

United States. Congress. Senate. Committee on Government Operations. Permanent Subcommittee on Investigations. TFX Contract Investigation: Hearings. 88th Cong., 1st sess. 10 pts. Washington, D.C.: GPO, 1963.

Wood, James W. The F-111: Myths and Realities. Research Report. Maxwell AFB, Ala.: Air War College, Air University, 1967. 78p.

Periodical/Anthology Literature:

Alelyunae, Peter. "The F-111 Today." Space/Aeronautics, XLIV (August 1965), 46-54.

Alexander, Tom. "McNamara's Expensive Economy Plane: The F-111." Fortune, LXXV (June 1, 1967), 88-91+.

"All-Purpose Airplane." Machine Design, XXXVI (December 3, 1964), 134-145.

Allen, Nelson. "Put the Ordnance on Target." Aerospace Safety, XXIII (October 1967), 1-3.

Andrews, Hal. "Naval Aircraft: F-111B Tiger." Naval Aviation News, (September 1973), 20-21.

Armbrister, Tom. "Is This Plane a Billion-Dollar Blunder?: The F-111." Saturday Evening Post, CCXL (June 17, 1967), 23-25+.

Arnold, Brian A. "Red Flag: Prep for the FB-111." Combat Crew, XXXIII (September 1983), 14-15.

"Back in the Bomber Business." Flight International, CXII (November 12, 1977), 1413-1415.

Bamford, Hal. "What Managers Can Learn from TFX." Armed Forces Management, IX (January 1963), 15-17.

Bennett, Ralph E. "The F-111: Is Adequate Good Enough?" Data, XIII (May 1968), 10-11.

Blackburn, A. W. "Blackburn Memo Criticizes TFX Decision." Aviation Week and Space Technology, LXXVIII (June 3, 1963), 79-83.

Bolton, Malcolm F. "Stopping the Aardvark." TAC Attack, XVII (February 1977), 4-9.

Boyd, Stuart B. "High Angles of Attack." Combat Crew, XXII (August 1971), 8-11.

____. "High Angles of Attack in the Swingwing." Aerospace Safety, XXVII (August 1971), 1-4.

Braybrook, Roy M. "Variable Sweep." Flying Review International, XXIV (September 1968), 68-73.

Broadway, Cal. "F-111D Mark II Avionics." USAF Fighter Weapons News Letter, no. 4 (December 1969), 2-7.

Brownlow, Cecil. "F-111 Shows Bombing Support Strength in Indochina." Aviation Week and Space Technology, XCVIII (April 30, 1973), 88-89+.

Bulban, Erwin J. "F-111 Lessons Applied to Other Projects." Aviation Week and Space Technology, XCVI (June 26, 1972), 123-125.

____. "General Dynamics Defends F-111 Program." Aviation Week and Space Technology, LXXXVII (November 13, 1967), 34-35.

____. "Malfunction Shortens First Flight of F-111." Aviation Week and Space Technology, LXXXI (December 28, 1964), 18-19.

"Bumpy Grind." Combat Crew, XXIII (July 1972), 4-6.

Bunton, U. D. "Concept and Conduct of Proof Test of F-111 Production Aircraft." Aeronautical Journal, LXXVI (October 1972), 587-598.

Burnham, James. "Why Do They Hate Robert Strange McNamara?" National Review, XVIII (November 15, 1966), 1153-1159.

Butz, J. S., Jr. "F-111: Second Look at the Third Version." Air Force and Space Digest, XLIX (February 1966), 37-38.

____. "FB-111: A Technical Discussion of the Third Version." Air Force and Space Digest International, II (March 1966), 26-27.

Chalmer, Lawrence. "The FB-111 Goes Conventional." Combat Crew, XXX (November 1980), 20-21.

Chandler, John. "Bright Points for the F-111." Air Force and Space Digest, LII (September 1969), 40-41.

"The Combat-Proven F-111." International Defence Review, VIII (June 1975), 348-350.

Correll, John T. "The Swinging Single." Airman, XVIII (June 1974), 10-14.

Coulam, Robert F. "The Importance of the Beginning: Defense Doctrine and the Development of the F-111 Fighter-Bomber." Public Policy, XXIII (Spring 1975), 1-38.

Covault, Craig. "F-111 Fighter Role Being Expanded." Aviation Week and Space Technology, CVIII (February 6, 1978), 119-124.

_____. "F-111's Effectiveness Increased." Aviation Week and Space Technology, CIV (May 10, 1976), 103-113.

Cribbe, Thomas J. "Coming--the Variable Sweep." Aerospace Maintenance Safety, XXI (May 1966), 2-4.

Dawson, Judith. "Cost of the F-111 Fighters." Air Force and Space Digest International, II (May 1966), 41-42+.

"Developing a Multi-Mission Aircraft." Interavia, XXI (February 1966), 166-169.

Dunn, Peter M. "F-111 Aardvark." In: Robin Higham and Abigail T. Siddall, eds. Flying Combat Aircraft of the USAAF-USAF. Ames, Iowa: Iowa State University Press, 1975. pp. 101-107.

Eccles, Henry E. "The TFX F-111 Aircraft: A Perspective in Military Command and Defense Management." Naval War College Review, XXIII (April 1971), 66-87.

Ely, Arven R., Jr. "FB-111 Crew Training." Navigator, XVII (Spring-Summer 1970), 63-64.

"F-111 About to Fly." Flight International, LXXXVI (December 24, 1965), 1093-1095.

"F-111: The Big, Swing-Winger from Texas." Flight International, LXXXIX (February 24, 1966), 301-307.

"F-111A Aardvark." TAC Attack, XXV (December 1985), 16-17.

"F-111A Mission Profile." Air Force Magazine, LXIV (April 1981), 64-68.

"The F-111A Spreads Its Wings." Flying Review International, XX (February 1965), 21-23.

"F-111A Starts Flight Trials." Interavia, XX (February 1965), 264-265.

"F-111 Variants." Air Pictorial, XXIX (March 1967), 79-80.

Farrell, G. Francis. "That Swing-Wing Thing." Combat Crew, XXII (May 1971), 4-7.

Fillgrove, George. "Tactical Jamming: A New Role for the F-111." Military Electronics/Countermeasures, VIII (February 1982), 28+.

Forman, Jim. "Low to the Ground at Night F-111 Style." Combat Crew, XXIX (June 1979), 10-11.

Francis, John, Jr. "The F-111: A Pilot's View." Air Force Magazine, LIV (April 1971), 31-39.

Freeman, Roger A. "Fastest Gun in the West: The General Dynamics F-111 in Britain." Air Pictorial, XXXIII (April 1971), 135-138.

Fricker, John. "The People vs. the F-111." Flying, LXXXIV (May 1969), 64-72+.

"General Dynamics F-111." Air Pictorial, XXVIII (January 1966), 4-7.

George, James A. "The Fantastic F-One-Eleven." Airman, XIII (August 1969), 28-34.

____. "A Report on the F-111." Airman, X (September 1966), 12-15.

Gilmore, C. P. "The Truth About the Amazing F-111 From the Men Who Fly Them." Popular Science, CXCII (May 1968), 61-65+.

Green, William. "F-111 Controversy and Capability." Flying Review International, XXV (May 1969), 34-37+.

Griffiths, David R. "FB-111 Bombers Playing Crucial Role." Aviation Week and Space Technology, CXII (June 16, 1980), 145-146.

Gunston, William T. ("Bill"). "F-111." In: William T. ("Bill") Gunston, ed. The Great Book of Modern Warplanes. New York: Crown, 1987. pp. 265-326.

____. "TFX: A Next Generation Military Aeroplane." Flight, LXXXI (February 8, 1962), 207-208.

Harvey, Frank L. "The Flying Switchblade." Popular Science, CLXXXII (January 1963), 96-98+.

Higgins, Edward J., Jr. "The Navigator's Airplane: FB-111." Navigator, XXXI (Winter 1984), 14-15.

Hornbaker, David M. "The F-111 'Wizzo.'" Navigator, XXI (Fall 1974), 11-15.

Huddleston, James R. "F/FB-111." Flying Safety, XXXIX (July 1983), 11-12.

Irelan, R. W. "Our Most Fantastic Flying Machine: The F-111." Nation's Business, LVI (March 1968), 48-50.

Jarolem, Stanley and Daniel Kahn. "TFX Trouble." New Republic, CLVIII (April 27, 1968), 18-21.

Johnston, Robert L. "The Navy Sanctions the F-111B." Ordnance, LI (January-February 1967), 340+.

Katzman, Jim. "Learning to Fly the 'Magic Show.'" Airman, XXVIII (April 1983), 24-30.

Kaye, Michael J. "F/FB-111." Flying Safety, XL (April 1984), 25-26.

Kempf, Fred. "F-111 Scope-Wizardry: Teaching New Dogs Old Tricks." TAC Attack, XV (April 1975), 22-24.

"Latest Proposals for the F-111A Fail to Meet Requirements of Navy." Aviation Week and Space Technology, LXXVI (June 18, 1962), 29-30.

Leonberger, Loren B. "Mighty Metal Men." Airman, XV (December 1971), 43-47.

Liggett, William R. "FB-111 Pilot Report." Air Force Magazine, LVI (February 1973), 30-37.

McDonald, Scott. "The F-111B: Can It Do the Navy's Job?" Armed Forces Management, XIII (December 1966), 36-41.

Miller, Jay. "F-111 Billion-Dollar Bonanza." Wings, IX (August 1979), 36-48.

"The Naval F-111 Starts Flight Tests." Interavia, XX (July 1965), 1126-1127.

"Navy's F-111B Down to a Wing and a Prayer." Business Week, (April 6, 1968), 89-91.

"New Bird in Town." Combat Crew, XXI (January 1970), 4-6.

O'Brien, F. Michael. "Wonder Bird of the Air Force." Airman, XV (March 1971), 38-39.

Parks, Al. "F-111 Attack Radar." USAF Fighter Weapons News Letter, no. 2 (June 1969), 2-6+.

____. "F-111 Bomb/Navigation System." USAF Fighter Weapons News Letter, no. 4 (December 1968), 34-37+.

Parks, Scott E. "Pilot Report: AFTI F-111." Air Force Magazine, LXXI (July 1988), 82-91.

Pearce, Ed. "Swing-Wing Geometry." Combat Crew, XXXV (August 1985), 4-7.

Peck, Vincent B. "A Navigator's View of the FB-111." Combat Crew, XXIII (January 1972), 16-18.

Pendry, Stephen H. "F/FB-111." Flying Safety, XLIII (March 1987), 20-21.

Pettijohn, Max L. "'Heads Up' and the F-111A." USAF Fighter Weapons News Letter, no. 3 (September 1965), 22-26.

Philpott, T. R. "F-111: The Wave of the Future." Air University Review, XXVI (May-June 1975), 74-79.

Plattner, C. M. "F-111 Expanding Variable-Wing Technology." Aviation Week and Space Technology, LXXXVIII (May 27, 1968), 80-81.

Prahl, Val. "F-111 Flight Test Report." Flight International, LXXXVIII (October 28, 1965), 727-729.

____. "Resume of F-111 Flight Test Programme." Aeronautical Journal, LXXII (January 1968), 34-42.

"Problem-Bird: F-111." Time, XC (September 1, 1967), 14-15.

Richardson, Michael E. "F-111 Armamemnt Systems." USAF Fighter Weapons New Letter, no. 4 (December 1969), 28-35.

Robinson, Clarence A., Jr. "USAF Commands Push Stretching 155 F-111s." Aviation Week and Space Technology, CXI (September 24, 1979), 16-19.

Schemmer, Benjamin F. "'I Like My Job': An F-111 Crew Shows Why." Armed Forces Journal, CIX (May 1972), 32-35.

Smith, Richard W. "The $7 Billion Contract That Changed the Rules." Fortune, LXVII (April-May 1963), 97-101, 182-188, 110-111, 191-194, 199-200.

Stanford, Neil. "F-111 on the Way as No. 1 Fighter-Bomber for Free World." U.S. Naval Institute Proceedings, XCI (June 1965), 168-169.

"State of the Art: F-111." Interceptor, IX (April 1967), 8-12.

Stoehrmann, Kenneth C. "New Roles for TAC's F-111s." Air Force Magazine, LXIII (February 1980), 46-51.

"Superplane or Megafolly?: F-111." Newsweek, LXIX (June 26, 1967), 66-67.

"The Swing-Wing Birds of Peace." Air Enthusiast, I (July 1971), 7+.

"Swing-Wings at Fort Worth." Flying Review International, XXI (February 1966), 327-330.

Tate, Grover C. "High, Low, Safe, and Slow." Aerospace Safety, XXI (December 1965), 10-11.

"Ten Million Dollar Gamble on a Plane--Now Its Paying Off: McNamara's TFX or F-111." U.S. News and World Report, LXI (August 15, 1966), 42-45.

"TFX Plane: What the Fight is About." U.S. News and World Report, LIV (March 25, 1963), 61-63.

Thomis, Wayne. "Whispering Death: The F-111 in Southeast Asia." Air Force Magazine, LVI (February 1973), 22-27.

Titus, Nathan T. "F/FB-111." Flying Safety, XLIV (January 1988), 20-21.

Turner, Marc C. "F-111 SRAM Airborne Mission Trainer." Combat Crew, XXXI (December 1981), 26-27.

"USAF Updates Nuclear-Capable F-111s." Aviation Week and Space Technology, CXIII (July 7, 1980), 48-49.

Warwick, Graham. "F-111: Strike Fighter for the 1990s?" Flight International, CXXI (January 9, 1982), 67-77.

Weiss, George. "The F-111: The Swing-Wing May Surprise You Yet." Armed Forces Journal, CVIII (July 1971), 22-25.

Wilson, G. C. "F-111 Weight, Drag Increases Navy Alarm." Aviation Week and Space Technology, LXXX (March 23, 1964), 16-17.

Winchester, James H. "The Facts About McNamara's Folly: TFX." Science Digest, LIV (July 1963), 14-18.

Winston, Donald C. "Senate Report Blasts Procurement of F-111." Aviation Week and Space Technology, XCIII (December 21, 1970), 19-20.

"With the Swing Wing." Aerospace Maintenance Safety, XXIII (June 1968), 1-5.

Witkin, Richard. "Why the Flak About the F-111?" New York Times Magazine, (April 2, 1967), 32-35+.

Witze, Claude. "F-111: Answer to Our Strategic Needs?" Air Force and Space Digest, XLIX (January 1966), 18-22.

____. "F-111A: The Men Who Fly It Like It." Air Force and Space Digest, L (December 1967), 45-47+.

____. "The F-111A's Real Struggle." Air Force and Space Digest, LI (July 1968), 14-15.

____. "Project Icarus and the F-111." Air Force and Space Digest, LIII (July 1970), 39-44.

Wright, Roger. "The F-111 in Europe." Aircraft Illustrated, VIII (August 1976), 330-333.

(Lockheed) F-117A Stealth
(see also B-2 Stealth)

Adam, John A. "How to Design an 'Invisible' Aircraft." IEEE Spectrum, XXV (April 1988), 26-31.

Ashley, Steven. "Revealed: Secret F-117A Stealth Fighter." Popular Science, CCXXXIV (January 1989), 92-93, 100.

Jones, J. Stealth Technology: The Art of Black Magic. Edited by Matt Thurber. Blue Ridge Summit, Pa.: TAB, 1989. 160p.

Morrocco, John D. "USAF Unveils Stealth Fighter; Black Weapons Probe Likely." Aviation Week and Space Technology, CXXIX (November 14, 1988), 28-29.

Nichols, Nick. "Our Most Secret Fighter." Popular Mechanics, CLXVI (January 1989), 66-69, 118.

"Top of the Top Guns." U.S. News and World Report, CV (November 28, 1988), 24-25.

(Boeing) FB Series

"The Boeing Shipboard Fighter." Aviation, XXIII (October 31, 1927), 1058-1059.

Bowers, Peter M. "Sea-Going Boeing: The First Shipboard Fighters." Airpower, VII (September 1977), 44-55.

FB-111 see F-111 Aardvark

FC see F11C Goshawk

(Grumman) FF-1

Baxter, L. Leroy. "The Pioneer Fighter Everyone Forgot: The Grumman FF-1." Air Classics, VIII (December 1971), 15-25.

Colerier, Luis R. "Grumman FF-1 Flies Again." American Aviation Historical Society Journal, XI (Fall 1966), 179-181.

Matt, Paul R. "The Grumman FF-1." Aero Album, III (1970), 32-37.

(McDonnell) FH Phantom

Andrews, Hal. "Naval Aircraft: FH Phantom." Naval Aviation News, (January-February 1989), 16-17.

Ginter, Steven J. McDonnell FH-1 Phantom. Naval Fighters Series, no. 3. Simi Valley, Calif.: Naval Fighters, 1981. 30p.

"Jets on a Carrier." *Flying,* XLIII (July 1948), 19-21.

"Jets Roost on a Flattop." *Popular Mechanics*, LXXXIX (June 1948), 108-111.

McLarren, Robert. "Phantoms Join the Atlantic Fleet." *Aviation Week*, XLVIII (June 21, 1948), 21-22.

___. "Thick Root Proved in Phantom." *Aviation Week*, XLVIII (June 21, 1948), 22-23.

"Mr. Mac's First Phantom." *Air International*, XXXIII (1987), 231+.

Perkins, Kenneth. "Design Development of the McDonnell Phantom." *Aviation,* XLV (November 1946), 40-48.

FJ Fury see F-86 Sabre

(Vought) FU-1 Corsair

McLaughlin, George F. "The Vought FU Fighter." *Aero Digest*, XI (November 1927), 540+.

"Vought Corsair Naval Fighter." *Aviation*, XXII (May 2, 1927), 934+.

(Grumman) GE-23

"The Strange Story of Grumman's First Fighter." *Flying Review International*, XXII (October 1966), 150-151.

(Hanriot) HD-1

Bruce, John M. *Hanriot HD-1.* Aircraft in Profile, no. 109. Windsor, Eng.: Profile Publications, 1967. 15p.

"The Hanriot HD-1." *Air Classics*, I (Summer 1964), 36-38; V (December 1969), 46-53; VI (February 1970), 56+.

"Hanriot HD-1." *Scale Models,* VIII (October 1977), 494-496.

(Curtiss) K-12

Trimble, Robert L. "Triplanes for the Navy." *Air Trails*, NS II (Winter 1977), 54-61.

(Martin) MBT/MT
(see also NBS)

Elliott, John M. "The Martin MBT/MT Story." *American Aviation Historical Society Journal*, XXXII (Summer 1987), 144-159.

(Curtiss) P-1C Hawk

Brounstein, Gary. "In Pursuit of a Dream: Bill Hodson and the Curtiss P-1C." Air Classics, XVIII (September 1982), 57-63.

Clements, N. V. "Recent Developments in the Curtiss Hawk." Slipstream, VIII (January 1927), 25-26.

"The Curtiss Pursuit Training Plane." Aviation, XXI (July 5, 1926), 24-25.

"The Development of the Curtiss Hawk." Aviation, XXII (January 3, 1927), 24-25.

"Hawks Available With Four Different Power Plants." US Air Services, XII (February 1927), 30-31.

Westburg, Peter. "Curtiss P-1C." RC Modeler, X (March-April 1980), 28-29, 25.

(Curtiss) P-6 Hawk
(see also P-1C Hawk)

Berlinger, Don. "Alchemy of the Curtiss P-6E Hawk." National Aeronautics, II (September-October 1974), 36-39.

Boyne, Walter J. "'P' for Pursuit." Wings, VI (March-April 1976), 8-21, 38-53.

"Curtiss Hawk Fighter." Aero Digest, XXIII (November 1933), 50-54.

"Curtiss P-6E Evaluation." American Aviation Historical Society Journal, XI (Spring 1966), 54-56.
Conducted at Wright Field, Ohio, April 19, 1932.

Hoyt, Ross G. "Flying the Early Birds: The Curtiss Hawk." Air Force Magazine, LIX (October 1976), 68-69.

Jarrett, Philip. "Mr. [Noel Pemberton] Billing's War Baby." Aeronautical Journal, LXXIX (June 1975), 261-266.

Jefferies, Walter M., Jr. "XP-6A." American Aviation Historical Society Journal, XIV (Fall 1969), 169-172.

Westburg, Peter. "Curtiss P-6E Hawk." Air Classics, XIV (March and May 1978), 18-25, 70, 18-24, 68.

(Boeing) P-12
(see also F4B)

Book/Document/Report Literature:

Aeronautical Staff. Boeing P-12/F4B. Aero Series, v. 5. Fallbrook, Calif.: Aero Publishers, 1966. 48p.

Bowers, Peter M. Boeing P-12E. Aircraft in Profile, no. 2. Windsor, Eng.: Profile Publications, 1965. 12p.

Wallick, S. L., Jr. and Peter M. Bowers. Flying the P-12. Seattle, Wash.: Red Baron Publications, 1982. 79p.

Periodical/Anthology Literature:

Bowers, Peter M. "Back in the Blue." Wings, IX (December 1979), 40-49.

Bradley, Mark. "The Boeing P-12E." Aerospace Historian, XXXII (December 1985), 239-245.

Cleveland, Carl M. "Oldest Boeing Still Flying." Air Classics, XIV (December 1978), 52-53.

Davisson, Budd. "The Junkman Cometh." Air Progress, XLIII (July 1981), 41+.

Ethell, Jeffrey L. "High-Altitude Pioneering in the P-12." Air Force Magazine, LXII (September 1979), 120+.

McElroy, Steve. "P-12 Comes Home." Air Force Magazine, LXIX (July 1986), 106-109.

Tallman, Frank. "Pilot Report: Boeing P-12, 1928 Fighter." Flying, LXXV (August 1964), 54-55+.

Westburg, Peter. "Boeing P-12E." Wings, XI (February-April 1981), 37, 30-31, 28-29.

(Boeing) P-26 Peashooter

Book/Document/Report Literature:

Bowers, Peter M. Boeing P-12A. Aircraft in Profile, no. 14. Windsor, Eng.: Profile Publications, 1967. 18p.

____. P-26 Variants. Minigraph, no. 8. Arlington, Tex.: Aerofax, Inc., 1984. 40p.

Maloney, Edward T. Boeing P-26 Peashooter. Aero Series, no. 22. Fallbrook, Calif.: Aero Publishers, 1973. 48p.

Periodical/Anthology Literature:

"Boeing's Fighter Finale: The Peashooter Chronicle." Air Enthusiast, XIV (1980), 1-12.

Bowers, Peter M. "Boeing P-26." Air Classics, IV (April 1968), 17-21.

____. "Evolution of the Boeing P-26 Peashooter." Air Classics, II (March-April 1965), 18-21.

Green, William. "The Peashooter." RAF Flying Review, XIV (May 1959), 30-33.

Hoyt, Ross G. "Flying the Early Birds: The P-26A." Air Force Magazine, LX (January 1977), 60-61.

____. "The P-26." Aerospace Historian, XXIII (June 1976), 62-64. Expanded in Robin Higham and Carol Williams, eds. Flying Combat Aircraft of the USAAF-USAF (Ames, Iowa: Iowa State University Press, 1978), pp. 169-174.

Mahurin, Walker. "Flying the Last P-26." Air Classics, IX (December 1973), 62-68.

"The Remarkable Peashooter." Flying Review International, XX (May 1965), 49-52.

Santick, Dan. "Peashooter." Model Airplane News, CV (December 1982), 62-66.

Sloan, James J. "Peashooter Closeup." American Aviation Historical Society Journal, XVII (Spring 1972), 1-7.

P-30 see PB-2

(Seversky) P-35/P-35A Guardsman and P-43 Lancer

Ackley, C. S. "The Spanish Seversky." American Aviation Historical Society Journal, XXVI (Spring 1981), 60-67.

Bowers, Peter M. "Seversky P-35." Air Classics, IV (February-April 1968), 26-39, 44-47.

"End of the Beginning." Air Enthusiast, X (1979), 8-21.

Holder, William G. "The P-35: First of a Breed." TAC Attack, XV (January 1975), 24-25.

Jensen, Ebbe. "Seversky EP-1 S/N 2134." Air Classics, XIII (December 1977), 41-43.

Lowe, Thomas E. and Earl Reinert. "The Last Lancer." American Aviation Historical Society Journal, XXIII (Summer 1978), 112-119.

Lustig, David. "The Last P-35." Air Classics, XI (November 1975), 44-53.

(Curtiss) P-36/Hawk 75/Mohawk

Book/Document/Report Literature:

Bowers, Peter M. Curtiss Hawk 75. Aircraft in Profile, no. 80. Windsor, Eng.: Profile Publications, 1966. 12p.

Periodical/Anthology Literature:

Beauchamp, Gerry. "Hawks for Hire." Wings, VIII (April 1978), 24-39, 59-63.

____. "Hawks of the Eastern Skies." Air Classics, XIII (September and November 1977), 20-31, 96, 74-82, 92; XIV (January 1978), 83-91.

Berlin, Don R. "Development of the Curtiss P-36A Pursuit." Aero Digest, XXXI (December 1937), 54-58.

Bowers, Peter M. "Hawk 75A: Evolution of the Curtiss P-36 Series." Air Classics, VI (August 1970), 6-23.

Boyne, Walter J. "A Veteran's Last Flight." RAF Flying Review, XVII (December 1961), 40-41.

"Curtiss P-36 Pursuit Monoplane." Aero Digest, XXIX (November 1936), 36-37.

Green, William. "It Fought on Both Sides: The Fantastic Story of the Curtiss Hawk Model 75." RAF Flying Review, XVI (February 1961), 19-20+.

____. "The Ubiquitous Hawk." Air International, I (November-December 1971), 307-314, 374-378.

Hoyt, Ross G. "Flying the Early Birds: The P-36." Air Force Magazine, LX (February 1977), 90-91.

Jane, David E. "Curtiss Hawk 75." Scale Models, XI (November 1980), 599-601.

Miller, Jay. "Mystery of the Missing Mohawk." American Aviation Historical Society Journal, XXVI (Spring 1981), 53-59.

"P-36A Hawk." TAC Attack, XXIV (June 1984), 16-17.

(Lockheed) P-38 Lightning

Book/Document/Report Literature:

Caidin, Martin. Fork-Tailed Devil: The P-38. New York: Ballantine Books, 1971. 369p.

Christy, Joe and Jeffrey L. Ethell. P-38 Lightnings at War. New York: Scribners, 1978. 144p.

Gurney, Gene. The P-38 Lightning. Famous Aircraft Series. New York: Arco, 1969. 60p.

Lockheed P-38 Lightning. Appleton, Wis. Aviat Publications, 1974. 78p.

Maloney, Ed ard T. Lockheed P-38 Lightning. Aero Series. Fallbrook, Calif.: Aero Publishers, 1968. 52p.

O'Leary, Michael, ed. Fighting Lightnings: The Complete Story of the Lockheed P-38 in Action During World War II. Air Combat Special. Canoga Park, Calif.: Challenge Publications, 1987. 100p.

Rice, Michael, ed. Pilot's Manual for the Lockheed P-38 Lightning. Glendale, Calif.: Aviation Book Company, 1972. 72p.
Commercial reprinting of AAF Manual 51-127-1 from 1945.

Shennan, Anthony. Lockheed P-38 Lightning: A History. Sun Valley, Calif.: JWC Publications, 1970. 40p.

Stafford, Gene B. The P-38 Lightning in Action. Aircraft, no. 38. Warren, Mich.: Squadron/Signal Publications, 1976. 50p.

Stanaway, John. Peter Three Eight: The Pilots' Story. Missoula, Mont.: Pictorial Histories Publishing Co., 1987. 140p.

Ward, Richard and R. Ernest McDowell. Lockheed P-38 Lightning. Aircam Aviation Series. New York: Arco, 1969. 48p.

Weber, LeRoy. Lockheed P-38J/M Lightning. Aircraft in Profile, no. 106. Windsor, Eng.: Profile Publications, 1967. 12p.

Periodical/Anthology Literature:

Anderson, Carroll B. "Ghost Sentai." Wings, VIII (February 1978), 54-61.

____. "Training Days in the P-38." Aerospace Historian, XXX (March 1983), 24-31.

Bodie, Warren M. "Man-Made Lightning: Lockheed P-38." Airpower, III (November 1973), 20-43.

____. "Sky Bolt." Wings, VI (April 1976), 8-21; Airpower, VI (September 1976), 20-33; VII (July 1977), 26-41, 48; VIII (March 1978), 24-44.

____. "Sky Bolt: The Lightning in Combat." Air Combat, V (January 1977), 40-64.

Boesen, Victor. "Evolution of the Lightning." Skyways, IX (April 1944), 34-41+.

Caidin, Martin. "The Lockheed P-38 Fighter." Air News, X (May 1946), 18-23+.

Carter, Dusty. "Vultee 'P-38.'" American Aviation Historical Society Journal, XXXIII (Spring 1988), 18-22.

"Combat Life With the Escorts." Air Force, XXVII (March 1944), 32-34.

"Development and Engineering Features of the Lockheed P-38 Fighter." Automotive and Aviation Industries, LXXXIX (November 1, 1943), 32-36.

"The Electrical System of the P-38." Aero Digest, XLIII (November 1943), 244-247.

Ethell, Jeffrey L. "Lightning Over Africa." *Aviation Quarterly*, V (Spring 1979), 88-104.

____. "Lightning Pilots." *American Aviation Historical Society Journal*, XVII (Fall 1972), 11-17.

____. "Night Lightning." *Air Classics*, XII (August 1976), 60-69. P-38M.

"Fork-Tailed Devil: Production of the P-38 Lockheed Lightning." *Flying*, XXXVII (August 1945), 26-28+.

Frey, Royal D. "P-38 Lightning." In: Robin Higham and Abigail T. Siddall, eds. *Flying Combat Aircraft of the USAAF-USAF*. Ames, Iowa, Iowa: State University Press, 1975. pp. 131-140.

Gaskill, Gordon. "Fifteen Feet Above Hell: A Death-Dealing Trip Over Libya in a Mysterious, Silent Super-Fighter." *American Magazine*, CXXXIII (February 1943), 14-15+.

Gray, William P. "P-38, Lockheed's Twin-Tailed Fighter, Lives Down Its Hoodoo." *Life*, XV (August 16, 1943), 51-54+.

Hegge, Robert D. "Der Gabelschwanz Teufel Flies Again: What It Takes to Restore a World War II Lightning." *Air Classics*, XI (April 1975), 38-41.

Hibbard, H. L. "Lockheed P-38 Lightning." *Aviation*, XLIII (August 1944), 123-147.

Hoelle, William J. "Lightnings in Africa." *Flying*, XXXII (May 1943), 24-25+.

Krims, Milton R. "Long Legs." *Air Force*, XXVIII (October 1945), 16-18.

"Lightning Progress." Aeronautics, V (August 1944), 46-49.

"Lightning Strikes the Axis: *AirTech* Presents the First Complete Story of the Lockheed P-38's Design and Structure." *AirTech*, II (June 1943), 21-39.

"A Lightning Test Flight." *Flying*, XXXIII (April 1945), 39-46.

"The Lockheed Lightning." *Flight*, XXXVI (November 4, 1943), 497-500.

"The Lockheed Lightning: A High Altitude Fighter." *Aero Digest*, XLII (June 1943), 202-203.

"Lockheed Lightning (P-38J)." *Flight*, XXXVII (June 22, 1944), 665-670.

"The Lockheed P-38." *Air Classics Annual*, III (1966), 34-45.

"The Lockheed P-38." *Aero Album*, (Summer 1969), 3-11.

"The Lockheed P-38 Fighter." Flying Review International, XXIV (April 1968), 224-226.

"The Lockheed P-38 Fighter." US Air Services, XXIX (May 1943), 16-17.

"Lockheed P-38 Has Wide Range of Uses." Aviation, XLII (January 1943), 101+.

"The Lockheed P-38 Lightning." Aero Digest, XXXII (May 1-15, 1945), 84-89+, 80-85+.

Miller, Jay. "Lefty's Lightning." Air Classics, XIII (April 1977), 32-43.
Confederate Air Force restoration.

Morrison, Samuel M. "The Lockheed P-38 Lightning." Aerospace Historian, XXX (September 1983), 150-162.

O'Leary, Michael. "The Restoration of 'Putt-Putt Marie.'" Air Classics, XXIV (May 1988), 50-55, 73-75.
P-38 of ace Charles MacDonald.

____. "Tribute to a Classic." Air Progress, XXXIX (December 1977), 20+.

"P-38 Lightning." TAC Attack, XXIV (May 1984), 16-17.

Pierce, Peter. "Riding the Lightning: Lockheed's P-38." Flying, XXXII (June 1943), 44-45+.

Roberts, Newell O. "Three Seconds to Fight: How Our Pilots Worked Their P-38s Against All Comers." Collier's, CXII (September 4, 1943), 11+.

Rust, Kenn C. "Lightnings at War." American Aviation Historical Society Journal, VIII (Summer 1963), 128-131.

Stone, Tim. "P-38: The Graceful Thing." Air Progress, XXXIX (July 1977), 80-81.

"Three Bullets on a Knife: Saga of the P-38." Air University Review, XVIII (January-February 1967), 46-59.

"Vega to Lightning: The Story of the P-38." Popular Science, CXLIV (April 1944), 96-97.

Von Hake, Robert A. "Building a Bolt of Lightning: The Lockheed P-38 Interceptor." Aviation, XL (June 1941), 50-51+.

(Bell) P-39/P-400 Airacobra

Book/Document/Report Literature:

Bell P-39 Airacobra. Appleton, Wis.: Aviat Publications, 1974. 48p.

Dial, Jay F. Bell P-39 Airacobra. Aircraft in Profile, no. 165. Windsor, Eng.: Profile Publications, 1967. 12p.

McDowell, R. Ernest. P-39 Airacobra in Action. Aircraft, no. 43. Carrollton, Tex.: Squadron/Signal Publications, 1980. 49p.

Pelletier, Alain. Bell P-39 Air Cobra. Paris, France: Quest-France, 1986. 160p.

Rice, Michael S., ed. Pilots Manual for the Bell P-39 Airacobra. Glendale, Calif.: Aviation Book Company, 1972. 72p.
Commercial reprinting of the official AAF manual.

Periodical/Anthology Literature:

"America's Airacobra." Popular Science, CXL (February 1942), 72D-72E.

Bell, Larry D. "Fighter Philosophy: How the Airacobra Was Designed." Aviation, XL (July 1941), 74-75, 90+.

"Bite of the Cobras: Bell P-39 Airacobra." Aviation Quarterly, II (Spring 1976), 97-123.

Bowers, Peter M. "Airborne Cobra." Airpower, VIII (November 1978), 20-37, 61; IX (January 1979), 10-30, 63.

____.____. Wings, IX (January 1979), 26-37, 60.

"The 'Calamitous Cobra.'" Air Enthusiast, I (January 1971), 134+.

"Cobras With Wings." Popular Mechanics, LXXX (August 1943), 1-5.

Davidson, Robert S. "Killer Cobra." Air Classics, VIII (December 1971), 10-13, 58-60, 64-65.

Furler, E. F., Jr. "Birth of the Cobra." Air Classics, XX (December 1984), 22-31.

____. "Klunker: The Combat History of the Bell P-400." Air Classics, XVI (March 1980), 34-36, 68.

Guerlick, Robert V. "One-Man Crew." In: Norman Carlisle, ed. The Air Forces Reader. Indianapolis, Ind.: Bobbs-Merrill, 1944. pp. 168-172.
Reprinted from the February 1944 issue of Air Force.

Hall, C. L. "Final Report of Development, Procurement, Performance, and Acceptance of XP-39E Airplanes: Air Corps Technical Report, no. 5022, 21 September 1943." American Aviation Historical Society Journal, XXVII (Spring 1982), 55-68.

Hudson, James J. "The P-39 in Europe." Aerospace Historian, XXIV (September 1977), 29-34.

Kent, Richard D. "P-39 Airacobra." In: Robin Higham and Abigail T. Siddall, eds. Flying Combat Aircraft of the USAAF-USAF. Ames, Iowa: Iowa State University Press, 1975. pp. 141-145.

Matthews, Barth J. "Cobra." American Aviation Historical Society Journal, VIII (Fall 1963), 185-190.

Miller, Edward E. "Design Analysis of the Bell Airacobra From Cannon to Tail." Aviation, XLII (May 1943), 126-155.

Owens, Rasher and James R. Hanson. "The Killer Cobra Strikes Again." Air Classics, VIII (May 1972), 26-29, 44.

Sparks, Bert. "Test Flying a Ball of Fire: Airacobra." Aviation, XL (July 1941), 84-85, 96+.

Trimble, Robert L. "Bell's Iron Dog." Air Classics, XVIII (December 1982), 28-31, 64-77; XIX (January and March 1983), 14-17, 76-78, 22-33, 44-47.

Woods, R. J. "Unveiling the Airacobra." Aviation, XXXIX (March 1940), 40-41+.

____. "Why a Rear Engine Installation?: Background Story of Driveshaft Design of Bell P-39 Airacobra." Aviation, XL (March-April 1941), 36-37+, 46-47+.

Woodson, C. L. "How Airacobras are Built." American Machinist, LXXXV (November 12, 1941), 1145-1160.

(Curtiss) P-40 Kittyhawk/Warhawk

Book/Document/Report Literature:

Christy, Joe and Jeffrey L. Ethell. P-40 Hawks at War. New York: Scribners, 1980. 128p.

Curtiss P-40 Warhawk. Appleton, Wis.: Aviat Publications, 1974. 36p.

Fighter Combat Comparison, No. 1: The Curtiss P-40C vs. the Mitsubishi A6M2 Model 21 Zero Sen. Teaneck, N.J.: Tacitus Publications, 1972. 24p.

McDowell, R. Ernest. Curtiss P-40 in Action. Aircraft, no. 26. Warren, Mich.: Squadron/Signal Publications, 1976. 50p.

____. The P-40 Kittyhawk. Famous Aircraft Series. New York: Arco, 1968. 64p.

Rice, Michael, ed. Pilots Manual for the Curtiss P-40 Warhawk. Glendale, Calif.: Aviation Book Company, 1972. 72p.
Commercial reprinting of AAF Manual 51-127-2 of 1945.

Scott, Robert L., Jr. Damned to Glory. New York: Scribners, 1944. 228p.

Vader, John. Pacific Hawk. Ballantine's Illustrated History of World War II. New York: Ballantine Books, 1970. 160p.

Wagner, Ray. Curtiss P-40. Aircraft in Profile, no. 35. Windsor, Eng.: Profile Publications, 1965. 12p.

____. Curtiss P-40F/N. Aircraft in Profile, no. 136. Windsor, Eng.: Profile Publications, 1967. 12p.

Periodical/Anthology Literature:

Beauchamp, Gerry. "Phantom Prototype." Wings, IV (December 1974), 16-23.

Bowers, Peter M. "Curtiss' Last Fighter: The Famed P-40." Air Classics, VII (December 1970), 6-18.

Bradley, Mark E. "The XP-40." Aerospace Historian, XXV (September 1978), 129-135.
Reprinted in Robin Higham and Carol Williams, eds. Flying Combat Aircraft of the USAAF-USAF (Manhattan, Kans.: Sunflower University Press, 1981), pp. 9-16.

"Curtiss-Wright's New Basic Combat Fighter." Aviation, XXXIX (May 1940), 52-53+.

Davis, Luther. "The 'B-40' Over Burma." In: Norman Carlisle, ed. The Air Forces Reader. Indianapolis, Ind.: Bobbs-Merrill, 1944. pp. 230-231.
Reprinted from the October 1943 issue of Air Force.

Der Tatevasion, Roosevelt. "Warpaths of the Warhawk." Skyways, III (August 1944), 45-46+.

Glover, B. A. "Assembling and Testing P-40s in Burma." Aviation, XLI (December 1942), 96-101+.

Hart, Eric H. "The Early Curtiss P-40s." American Aviation Historical Society Journal, XII (Winter 1967), 254-261.

____. "Hawk 81." American Aviation Historical Society Journal, XXV (Spring 1980), 19-25.
P-40 variant slated for export to France.

____. "A Study in Longevity: The Curtiss P-40 in World War II." American Aviation Historical Society Journal, XIV (Spring-Summer 1969), 26-29, 130-131.

"Hawk Monoplanes: The Second Generation." Air International, XII (February 1977), 75-81, 91.

Holloway, Bruce K. "The P-40." Aerospace Historian, XXV (Fall 1978), 136-140.
Reprinted in Robin Higham and Carol Williams, eds. Flying Combat Aircraft of the USAAF-USAF (Manhattan, Kans.: Sunflower University Press, 1981), pp. 17-22.

Marks, Donald M. "P-40 Kittyhawk." In: Robin Higham and Carol Williams, eds. Flying Combat Aircraft of the USAAF-USAF. Ames, Iowa: Iowa State University Press, 1978. pp. 175-180.

Mikesh, Robert C. "Bicentennial Warhawk." Wings, VI (October 1976), 32-39.

Milch, Robert J. "The P-40: Workhorse of the Skies." American History Illustrated, II (June 1967), 22-28, 30.

"The New Curtiss P-40." Aviation, XXXIX (August 1940), 60-61.

"The Old War Hatchet." RAF Flying Review, XVIII August 1963), 24-27+.

Sullivan, Edward P. "Curtiss and the Hawks." World War Enthusiast, 1939-1945, I (July-August 1974), 105-106, 110, 118.

"Tiger Warriors." Air Classics, IV (November 1967), 36-44.
AVG P-40s.

Tydon, Walter. "Pedigree of the P-40." Aircraft Age, II (June 1944), 6-9.

"The Ubiquitous Hawk." Air Enthusiast, I (May-June 1971), 307+, 374+; II (July 1971), 47+.

P-43 see P-35 Guardsman/P-43 Lancer

(Republic) P-47 Thunderbolt

Book/Document/Report Literature:

Davis, Larry. P-47 Thunderbolt in Action. Carrollton, Tex.: Squadron/Signal Publication, 1984. 50p.

Duvall, Geoffrey. Republic P-47 Thunderbolt Described. Newark, Del.: Air Enterprise Publications, 1974. 26p.

Freeman, Roger A. Republic P-47N Thunderbolt. Aircraft in Profile, no. 262. Windsor, Eng.: Profile Publications, 1974. 25p.

_____. Thunderbolt: A Documentary History of the Republic P-47. New York: Scribners, 1979. 152p.

Hess, William N. P-47 Thunderbolt at War. Garden City, New York: Doubleday, 1977. 160p.

McDowell, R. Ernest and Richard Ware. Republic P-47 Thunderbolt. Famous Aircraft Series. New York: Arco, 1968. 46p.

Morgan, Len. P-47 Thunderbolt. New York: Arco, 1963. 52p.

Rabbets, John B. Republic P-47D Thunderbolt. Oxford, Eng.: Aerodata International, 1978. 20p.

Republic P-47 Thunderbolt. Appleton, Wisc.: Aviat Publications, 1974. 50p.

Rice, Michael S., ed. Pilots Manual for the Republic P-47 Thunderbolt. Glendale, Calif.: Aviation Book Company, 1973. 72p.
Commercial reprinting of AAF Manual 50-5 of 1944.

Shacklady, E. G. The Republic P-47D Thunderbolt. Aircraft in Profile, no. 7. Windsor, Eng.: Profile Publications, 1965. 11p.

Stafford, Gene. P-47 Thunderbolt in Action. Aircraft, no. 18. Warren, Mich.: Squadron/Signal Publications, 1975. 49p.

United States. Army Air Forces. Training Division. Meet Peter 47. Washington, D.C., 1945. 20p.

____.____.____. Pilot Training Manual for the P-47N. AAF Manual 51-127-4. Washington, D.C., 1945. 72p.

Ward, Richard. Republic P-47 Thunderbolt. Aircam Aviation Series. New York: Arco, 1970. 46p.

Periodical/Anthology Literature:

Bodie, Warren M. "Legend of the P-47 Thunderbolt." Wings, IV (April 1974), 26-49.

____. "Wine From the Jug." Airpower, IV (May-June and August 1974), 8-25, 8-31, 26-33, 38-43.

Bradley, Mark E., with Helen Porter. "P-47 Thunderbolt: 'The Jug.'" In: Robin Higham and Carol Williams, eds. Flying Combat Aircraft of the USAAF-USAF. Manhattan, Kan.: Sunflower University Press, 1981. pp. 181-196.

Bugnon, M. A. "P-47 Thunderbolt." AirTech, II (December 1943), 21-35.

"CBI Thunderbolt." Air Trails, XXIII (February 1945), 32-34, 97.
P-47N.

Courtney, William B. "Thunder Bums: They Pilot the Big Thunderbolts." Collier's, CXIV (October 21, 1944), 18-19+.

Demme, Robert E. "Evolution of the Thunderbolt." Skyways, III (June 1944), 34-41+.

Dunn, William R. "P-47: The Beautiful Beast." Air Force Magazine, LVIII (September 1975), 91-94.

Farmer, James H. and Al Anderson. "Last of a Breed." Air Classics, VIII (November 1972), 24-37, 64-65.
P-47 "Miss Behave."

"Fighting Juggernaut." Air International, XIV (January, March, May 1978), 26-31, 133-139, 228-236.

"The Flying Organ." RAF Flying Review, XV (April 1960), 33+.

Frazer, Charles D. "Gallery of Fighters: P-47 Thunderbolt." Air Force, XXVII (February 1944), 57-59.

Friedman, William S. "It's the Thunderbolt." Popular Science, CXLI (December 1942), 108-113.

Fry, Garry L. "4th Fighter Group P-47 Markings." American Aviation Historical Society Journal, XVI (Spring 1971), 57-60.

Gask, R. C. "Thunderbolt." Flying, XXXII (February 1945), 34-38.

"Geneology of a Jug." Air Classics, II (May 1965), 18-20.

Gordon, Arthur. "'Blocking Back' in a P-47." Air Force, XXVII (January 1944), 44-48.

Hall, George F. "Thunder in the Skies." Air News, VI (February 1944), 28-29+.

Hendrix, William. "Thunder-Stricken." Airpower, VII (March 1977), 36-43.

Holder, William G. "Busted Jugs." TAC Attack, XIV (February 1974), 4-6.

Ivie, Tom. "Superbolt." Air Classics, XXI (October 1985), 14-21. P-47D-22.

Karten, Dave. "20 Year Old Jug." Airman, V (May 1961), 48+.

Larsen, Jim. "Big Stud." Air Progress, XLIV (June 1982), 58+.

Larson, George C. "Designing the P-47 Thunderbolt." Air Force Magazine, LXV (September 1982), 132-134+.

Lasker, H. E. "Thunderbolts to Smite the Axis: The Republic P-47 High-Altitude Fighter." Aviation, XLI (August 1942), 100-105+.

Masefield, Peter G. "First Analysis of the Thunderbolt." Flying, XXXI (August 1943), 47-48+.

Morgan, Len. "Sasha and the P-47." Wings, XII (December 1982), 16-27. Designer Alexander Kartveli.

O'Leary, Michael. "Tommy's Bolt: P-47D.Thunderbolt." Air Classics, X (May 1974), 48-51.

"P-47 Thunderbolt." TAC Attack, XXVIII (February 1988), 16-17.

Peeples, Lamar. "The Mighty 'N.'" Air Classics, XI (February 1975), 64-72.

Philips, Frederic M. "Back to the Jug--After 20 Years." Air Force and Space Digest, XLIV (June 1961), 118-120.

Powell, Hickman. "What It Takes to Be a Thunderbolt Ace." Popular Science, CXLIV (May 1944), 56a-56h.

"Republic P-47." Air Force, XXIX (December 1946), 25-29.

"Republic P-47 Thunderbolt Fighter." Aeronautics, V (November 1944), 28-33.

"The Saga of the Immortal Republic P-47 Thunderbolt." Air Classics, VII (July 1971), 18-38.

Sargent, Leslie K. "Existing P-47 Thunderbolts in the United States (Few Survive)." American Aviation Historical Society Journal, XXXII (Spring 1987), 48-57.

"Some Details of the P-47 Thunderbolt." Aero Digest, XXXII (January 1, 1945), 74-78+.

"Those Thunderbolts." American Aviation Historical Society Journal, VIII (Winter 1963), 282-285.

"Thunderbolts Over Britain." Flight, XXXVI (May 20, 1943), 521-524.

Turner, S. E. "Farewell to the Jug." Skyways, X (March 1951), 30-31,37.

Winchester, James H. "Republic's P-47: The Unbreakable Jug." Air Force, XL (December 1957), 100-102.

(North American) P-51 Mustang
(see also A-36, F-6, PA-48 Enforcer)

Book/Document/Report Literature:

Atkins, Richard. North American P-51B/C Mustang. Aircraft in Profile, no. 100. Windsor, Eng.: Profile Publications, 1969. 12p.

Coggan, Paul. Mustang Directory. Bishopmill, Scotland: Mustangs International, 1984. 90p.

____. Mustang Survivors. Osceola, Wis.: Motorbooks International, 1987. 216p.

Davis, Larry. P-51 Mustang in Action. Aircraft, no. 45. Carrollton, Tex.: Squadron/Signal Publications, 1981. 58p.

____. P-51 Mustang in Color. Fighting Colors Series. Carrollton, Tex.: Squadron/Signal Publications, 1983. 32p.

Ethell, Jeffrey L. Mustang: A Documentary History of the P-51. New York: Watts, 1981. 176p.

Fighter Combat Comparison, no. 5: The North American P-51D vs. the Focke-Wulf 190-A-8. Teaneck, N. J.: Tacitus Publications, 1973. 24p.

Freeman, Roger A. Mustang at War. Garden City, N.Y.: Doubleday, 1975. 160p.

Grant, William N. P-51 Mustang. New York: Prentice-Hall, 1983. 64p.

Grinsell, Robert. P-51 Mustang. New York: Crown, 1980. 48p.

Gruenhagen, Robert W. Mustang: Story of the P-51 Fighter. New York: Arco, 1980. 252p.

Hardy, Michael J. North American Mustang: The Story of the Perfect Pursuit Plane, P-51. Arco Aircraft Classics, no. 4. New York: Arco, 1979. 128p.

Hegg, William. P-51: Bomber Escort. Ballantine's Illustrated History of World War II. New York: Ballantine Books, 1971. 160p.

Hess, William N. Fighting Mustang: The Chronicle of the P-51. Garden City, N.Y.: Doubleday, 1970. 198p.

Holmes, Harry. North American P-51D Mustang. Oxford, Eng.: Aerodata International, 1978. 20p.

Mikesh, Robert C. "Excalibur III": The Story of a P-51 Mustang. Famous Aircraft of the National Air and Space Museum. Washington, D.C.: Smithsonian Institution Press, 1978. 76p.

Morgan, Len. The P-51 Mustang. Famous Aircraft Series. New York: Arco, 1963. 52p.

North American P-51 Mustang. Brooklands Aircraft Portfolio. Osceola, Wis.: Motorbooks International, 1987. 70p.

O'Leary, Michael. Mustang: A Living Legend. Osceola, Wis.: Motorbooks International, 1987. 128p.

____. Mustang: Chronicle of a Famous Fighter. Air Combat Special. Canoga Park, Calif.: Challenge Publications, 1986. 100p.

Rice, Michael S., ed. Pilots Manual for the North American P-51D Mustang. Glendale, Calif.: Aviation Book Company, 1973. 72p. Commercial reprinting of AAF Manual 51-127-5 from 1945.

Shacklady, E. G. North American P-51D Mustang. Aircraft in Profile, no. 8. Windsor, Eng.: Profile Publications, 1965. 13p.

Ward, Richard. North American P-51D Mustang. Aircam Aviation Series. New York: Arco, 1968. 48p.

____ and Ernest R. McDowell. North American P-51B/C Mustang in USAAF Service. Aircam Aviation Series. New York: Arco, 1969. 48p.

Periodical/Anthology Literature:

"Aerodynamic Weight and Servicing Refinements Featured in the North American P-51." Aviation, XLIV (July 1945), 197+.

Ashby, Edward H. "P-51: The Original Mustang." Airman, XII (May 1968), 38-40.

Blair, Clay, Jr. "Flying the Jet Stream." Skyways, X (September 1951), 12-13, 46.

Bowyer, Michael J. F. "Mustang Memories." Airfix Magazine, XII (October 1979), 80-82.

Bradley, Mark E. "The P-51 Over Berlin." Aerospace Historian, XXI (Fall 1974), 129-138.

Burkholder, Robert C. "New Thoughts on the Old Mustang." Aerospace Historian, XIX (June 1972), 66-68.

Caidin, Martin. "North American P-51." Air Force, XXIX (October 1946), 29-36.

Carson, L. K. ("Kit"). "Making of the Mustang." Wings, VI (May-June 1976), 8-21, 8-21.

Churchill, Edward. "The Mustang." Flying, XXXI (September 1944), 32-38.

Desmond, John. "Test Flight." Flying, CX (September 1983), 90-91.

DeVries, John A. "P-51 Mustang." In: Robin Higham and Carol Williams, eds. Flying Combat Aircraft of the USAAF-USAF. Ames, Iowa: Iowa State University Press, 1978. pp. 197-202.

Ethell, Jeffrey L. "The Foaling of the Mustang." Air Force Magazine, LXIV (June 1981), 82-85.

____. "'Man O' War.'" Air Classics, XIII (August 1977), 32-41.

____. "The Marvelous Mustang." Air Force Magazine, LXIV (September 1981), 144-146+.

____. "World War II Time Machine." Aviation Review, V (April 1981), 54-65.

Greenway, Sean. "Heritage of Valor." Air Classics, XVI (December 1980), 42-49.

Hillman, Bradford H. "Mustang Roundup: Saga of the P-51." Air Classics, VIII (June-July 1972), 16-25, 64-65, 52-63.

"The Immortal Mustang." RAF Flying Review, XV (April 1960), 26-27.

McCorkle, Charles M. "The Number One Fighter of World War II: Spitfire or Mustang?" Aerospace Historian, XX (December 1973), 170-177.
Both--for different reasons.

McLaren, David R. "ADC's Rarest Warhorse." Interceptor, XX (August 1978), 5-7.
F-51.

____. "Air Support in Korea--Mustang Style." Aerospace Historian, XXXVII (June 1986), 74-86.

____. "Mustangs in Combat: The Korean War." American Aviation Historical Society Journal, XXX (March 1985), 94-101.

____. "The North American P-51 Mustang." American Aviation Historical Society Journal, XXV (Summer-Fall 1980), 124-139, 302-337.

Martin, Patricia M. "Rebuilding a P-51 Mustang." Air Progress, XL (April 1978), 20+.

Matthews, Bird J. and Dustin W. Carter. "Racing Merlines: From Cleveland to Reno." American Aviation Historical Society Journal, XXXIII (Summer 1988), 126-138.

Mayor, David G. "'...and Not a Shot was Fired': The Story of a Near-Disasterous Flaw in the P-51 Mustang." Air Classics, XVIII (February 1982), 14-23.

"Milestone Mustangs." Airpower, V (November 1975), 12-19.

Miller, Jay. "Texas Mustangs." Air Classics, XIII (November 1977), 62-67.
Those of the Confederate Air Force.

Mizrahi, Joseph V., ed. "The P-51 Mustang Story." Airpower, XV (July 1985), 10+.

Moose, Wayne. "Fist Full of Mustang." Air Classics, XVIII (July 1982), 24-25, 77.

Morgan, Len. "Wild Mustang." Flying, CXV (October 1988), 64-73.

Mormillo, Frank B. "The Planes of Fame P-51A." Aircraft Illustrated, XV (February 1982), 64-66.

"The Mustang as Seen in Flight." Air Progress, V (May 1943), 54-57.

"Mustang--the Quarter-Century Fighter." Flying Review International, XXIII (April 1968), 175-177.

"Mustang--the Willing Warhorse." Air International, III (1983), 130+.

"The Mustang's Last Hurrah." Wings, XII (February 1982), 24-43.
P-51H.

Nelson, William R. "The North American P-51 Mustang." Aviation, XLIII (July 1944), 127-147.

"North American P-51 Has Novel Design Features." *Aviation*, XLII (June 1943), 234, 237+.

"The North American P-51 Mustang." *Aeronautics*, V (November 1944), 43-46.

"The North American P-51 Mustang." *AirTech*, I (May 1943), 19-23.

O'Keefe, Philip G. "The Saga of N13Y." *American Aviation Historical Society Journal*, XIX (Winter 1974), 287-296.

O'Leary, Michael. "Allison Mustang." *Air Classics*, XIX (June 1983), 46-51.
P-51A.

____. "'Jeanie.'" *Aeroplane Monthly*, IX (February 1981), 75-78.

____. "Military Mustangs: A Sword Through Freedom's Enemies." *Air Classics Quarterly Review*, I (Spring 1975), 1-116.

____. "NA-73K: First of a Famous Breed." *Air Classics*, X (July 1974), 10-20.

____. "P-51N:Last of the Mustangs." *Air Classics*, X (June 1974), 38-47.

____. "'Shangri-La.'" *Aeroplane Monthly*, X (February 1982), 74-75.

____. "Vintage Madera." *Air Progress*, XLIII (December 1981), 50+.

"The P-51 Mustang." *Aero Digest*, XXXI (January 1, 1944), 94-95.

"P-51 Mustang." *TAC Attack*, XXIV (April 1984), 16-17.

Powell, Hickman. "Fighting at 425 Miles an Hour: Meet the Riders of the Mustang." *Popular Science*, CXLV (August 1944), 108-114.

Rees, Edward. "A Tribute to Dutch Kindelberger: The Mustang, a Great Warhorse." *Aerospace Historian*, IX (October 1962), 197-206.

Reese, Fred. "P-51D Mustang." *RC Modeler*, XIII (June 1976), 24-31.

Rudd, Ralph H. "Mustangs by North American." *Aviation*, XLII (July 1943), 162-167+.

Schmued, Edgar. "Evolution of the Mustang." *Skyways*, III (January 1944), 22-26+.

Shores, Christopher F. "The Allison Mustangs." *Air Classics*, XVI (September-November 1980), 26-39, 60-74, 14-26, 69-72.

____. "Allison-Powered Mustang." *Air Enthusiast*, (Spring 1976), 190-206.

Taylor, Kelvin. "P-51 Mustang." *TAC Attack*, XXVII (June 1987), 16-17.

Tillman, Barrett. "Mustangs of Iwo Jima." Airpower, VII (January 1977), 30-43, 65.

Trimble, Robert L. "Military Markings of the World: ETO Mustangs." Air Combat, III (January 1975), 11-19.

"Wild Horses of the Sky: The P-51 Mustang Called the Outstanding Fighter of 1943." Popular Mechanics, LXXX (November 1943), 61-65.

(Bell) P-59 Airacomet

Book/Document/Report Literature:

Jensen, Edna. XP-59, XP-59A, YP-59A, P-59A, P-59B, and XP-59B, September 1940-May 1946. AMC Historical Study, no. 159. Wright-Patterson AFB, Ohio: USAF Air Materiel Command, 1947. 62p.

Periodical/Anthology Literature:

"Airacomet: A Jet Pioneer by Bell." Air International, XVIII (March 1980), 132-143.

"The Bell P-59 Airacomet." Flying Review International, XXIV (September 1968), 99-102.

Colby, Carroll B. "No Superman Needed: Flying the New Jet Plane." Popular Science, CXLIV (April 1944), 85-87+.

Craigie, Laurence G. "P-59." Society of Experimental Test Pilots Technical Review, XIV (Fall 1980), 147-165.

Fischer, C. M., "Flying the Airacomet." Flying, XXXV (November 1944), 29+.

Furler, E. F., Jr. "America's First Jet Fighter." Air Classics, XXI (February 1985), 28-39.

Green, William. "First Jet Fighter by Bell." Flying Review International, XXIV (September 1968), 99-102.

Haugland, Vern. "Jet Aircraft of World War II." Airpower Historian, IX (July 1962), 166-172.
Includes the P-59.

Holder, William G. "P-59: The Army Air Forces' First Jet." TAC Attack, XIV (May 1974), 4-6.

Kotcher, Ezra. "Our Jet-Propelled Fighter." Air Force, XXVII (March 1944), 6-8.
Expanded in Science Digest, XV (May 1944), 19-24, under the same title.

Neal, Ronald D. "The Bell XP-59A Airacomet: The United States' First Jet Aircraft." American Aviation Historical Society Journal, XI (Fall 1966), 155-178.

_____. "Mystery Jet." Air Classics, III (November 1966), 48-55.

"The P-59 Bell Airacomet Jet Fighter." Popular Science, CXLVI (May 1945), 96-97.

Pace, Steven J. "Airacomet." Wings, XVII (October 1987), 12-23.

Rogers, T. J. "The First American Jet." Flying, LX (March 1957), 44-46+.

Van Wyen, Adrian O. "They Flew the U.S. Navy's First Jets." Naval Aviation News, (March 1963), 6-13.

(Northrop) P-61 Black Widow

Book/Document/Report Literature:

Johnsen, Frederick A. Darkly Dangerous: The P-61 Black Widow Night Fighter. Glendale, Calif.: Aviation Book Company, 1981. 29p.

P-61 Black Widow. Appleton, Wis.: Aviat Publications, 1974. 72p.

Rice, Michael S., ed. Pilots Manual for the Northrop P-61 Black Widow. Glendale, Calif.: Aviation Book Company, 1973. 72p.
Commercial reprinting of AAF Manual 51-127-6 from 1945.

United States. Army Air Forces. Training Division. P-61 Series Airplanes: Descriptive Data. Washington, D.C., 1943. 62p.

Periodical/Anthology Literature:

"All in the Night's Work." Air Force, XXVII (December 1944), 8-10.

"Design Details of the Northrop P-61." Aero Digest, LI (November 1, 1945), 54-58.

Gallagher, James P. "The Not-So-Black Widow." Air Combat Quarterly Review, III (Winter 1976), 84-97.

Harald, Eric. "P-61 Black Widow." Skyways, III (November 1944), 18-20+.

Holder, William G. "Lady in the Dark." TAC Attack, XIV (June 1974), 18-20.

Littrell, Gaither. "The Black Widow Boys: Combat Flying the Big, Radar-Equipped Night Fighters." Flying, XXXVI (June 1945), 32-34+.

"The New P-61 Night Fighter." Aero Digest, XLVI (September 1, 1944), 76-77.

"Night Creature: Northrup P-61 Black Widow." Airpower, III (July 1973), 45-53.

Nolan, Robert R. "The Design and Production of the Northrop P-61 Fighter." Automotive and Aviation Industries, XCIII (October 15, 1945), 20-21+.

Paper, Garry R. and Ronald C. Harrison. "Dark Lady." Airpower, VI (November 1976), 8-25, 64; VII (January 1977), 10-29.

____.____. Wings, VI(October and December 1976), 10-31, 24-39. Two parts spread out over the two sister magazines.

Smith, C. C. "My Sweetheart is a Black Widow." Popular Science, CXLVII (July 1945), 116-119+.

Thompson, Warren. "The First F-15." Wings, XVIII (August 1988), 10-18.

____. "Northrop's Mean Lady." Aeroplane Monthly, VII (November 1980), 572-579.

____. "Terror in the Dark." Wings, XIV (October 1984), 10-75.

____. "Widow From Hawthorne." Air Enthusiast, (Fall 1977), 137-151.

(Bell) P-63 Kingcobra

Bell P-63 Kingcobra. Appleton, Wisc.: Aviat Publications, 1974. 52p.

Boesen, Victor. "Evolution of the Kingcobra." Skyways, IV (May 1945), 32-38.

Furler, E. H., Jr. "Kingcobra." Air Classics, XIX (November-December 1983), 14-21, 38-45.

"King Cobra." Air Trails, XXIII (October 1944), 38-39.

O'Leary, Michael. "Super Cobra." Air Classics, XIV (April 1978), 50-59.

"The P-63 Kingcobra Bell Fighter." Popular Science, CXLV (December 1944), 98-99.

Waag, Robert J. "Kingcobra: America's Great Giveaway Fighter." Airpower, III (September 1973), 16-23, 58-59.

Warren, Beatrice. "Last of the Kings." Aeroplane Monthly, III (July 1975), 336-340.

(Douglas) P-70
(see also A-20 Havoc)

Thompson, Warren. "Night Hunter." Wings, XVI (June 1986), 44+.

P-80 see F-80 Shooting Star

P-84 see F-84 Thunderjet

(North American/Piper) PA-48 Enforcer
(see also P-51 Mustang)

"Cavalier Treatment: A Turboprop Mustang for Close Air Support." Flight International, XCIV (November 21, 1968), 833-836.

Ethell, Jeffrey L. and Michael O'Leary. "Flying the Pucara." Air Progress, XLV (May 1983), 38-44.

Kolcum, Edward H. "Piper to Market Test PA-48 Enforcer." Aviation Week and Space Technology, CXVII (September 27, 1982), 59+.

____. "Piper to Modify Two P-51s as Enforcers." Aviation Week and Space Technology, CXV (September 1979), 28+.

Thomas, Barry K., Jr. "Mustangs Modified for Anti-Guerrilla Role." Aviation Week and Space Technology, LXXXVII (October 16, 1967), 67+.

(Boeing) PB-1

Bowers, Peter M. "Boeing PB-1." Air Classics, II (August-September 1965), 58-65.

____. "Boeing's First Fighters." Wings, VII (August 1977), 32-41.

Mikesh, Robert C. "The Boeing PB-1." American Aviation Historical Society Journal, IX (Spring 1964), 42-45.

(Consolidated) PB-2

"Consolidated PB-2, Formerly P-30." Flying Review International, XX (July 1965), 57+.

Dean, Jack. "Legion of the Lost." Wings, IV (October 1974), 36-45.

(Fokker) PW-5

Schrader, Richard K. "The Army's Forgotten Fokkers." Air Classics, XXIV (May 1988), 64-67.

(Curtiss) PW-8

"The Curtiss PW-8 Biplane." Flight, XVI (November 6, 1924), 705-706.

"The Curtiss PW-8 Pursuit Plane Described." Aviation, XVII (July 14, 1924), 232-233.

"The Ship That Crossed the Continent From Dawn to Dusk: The Curtiss PW-8 Pursuit Plane." USAir Services, IX (August 1924), 19-22.

(Boeing) PW-9

"Boeing Pursuit Ship." US Air Services, VIII (December 1923), 40-41.

Bowers, Peter M. "Early Boeing Biplane Fighters: PW-9." Air Classics, VII (March 1971), 40-49.

____. "The First Fighter: Boeing's PW-9." Wings, VII (August 1977), 42-55.

Cavanagh, Robert L. "Boeing PW-9 in the U.S. Army, 1925-1930." American Aviation Historical Society Journal, VII (Summer and Winter 1962), 100-114, 285-290+.

(Grumman) S-2 Tracker

Andrews, Hal. "Naval Aircraft: Tracker." Naval Aviation News, (July 1971), 20-21.

Green, William. "Tracker Ubiquitous." Flying Review International, XXVI (December 1970), 38-41+.

"Grumman's Sub-Killer." RAF Flying Review, XVII (March 1962), 22-24.

"Grumman's Tracker." RAF Flying Review, XII (August 1957), 12-13.

Hawkes, Robert. "S2F Design Sires Family of Economical Adaptations." Aviation Week and Space Technology, LXVI (July 1, 1957), 46-49.

"Marginally Qualified." Approach, XIX (December 1973), 6-7.

Redman, Rod. "Twilight for the Stoof: Grumman S2E Tracker." Air Classics, VIII (June 1972), 41-47.

Simmons, Jon. "Tracker Goodbye." Naval Aviation News, (January 1977), 8-13.

Weisbord, Marvin. "Nemesis of the Submarine: The S2F Tracker." Popular Mechanics, CIX (March 1958), 102-105+.

(Lockheed) S-3 Viking

Andrews, Hal. "Naval Aircraft: Viking." Naval Aviation News, (February 1975), 22-23.

Benjamin, Dick. "A Sea Rover for ASW." Naval Aviation News, (January 1972), 34-39.

Berent, Mark. "S-3 Viking: The U.S. Navy's Force Multiplier." Asian Defence Journal, no. 2 (February 1986), 44-46+.

Carius, Robert. "Viking Voice." Naval Aviation News, (October 1978), 13-15.

Curutta, Nick. "Keeping the Viking Watch." Wings of Gold, XII (Winter 1987), 18-21.

Godfrey, David W. H. "Carrier-Borne Sub Sinker." Aircraft, LIV (April 1975), 26-30, 32, 54.

____. "Fixer, Finder, Sinker: The S-3A Viking." Air International, VII (July 1974), 5-13.

Heppe, R. R., et al. "S-3A: A New Dimension in Airborne Sea Control." Journal of Aircraft, XI (September 1974), 577-583.

Hurlocker, Hal. "Lockheed's S-3A Viking." Aerophile, I (June 1979), 308-341.

"Introducing the S-3B Viking." Asian Defence Journal, no. 1 (January 1982), 66-67.

Joyce, Maurice S. "The S-3A Needs a Shot in the Arm, Not Major Surgery." U.S. Naval Institute Proceedings, CV (November 1979), 106-108.

Lawson, Robert L. "The Viking at Sea." U.S. Naval Institute Proceedings, CV (July 1979), 70-79.

"Lockheed Rolls Out an Anti-Sub Jet, the S-3A." Business Week, (November 6, 1971), 36-37.

"Lockheed S-3A Viking." Aircraft Engineering, XLVI (February 1974), 12-27.

"Lockheed S-3A Viking: A New Weapon Against the Submarine." Interavia, XXVII (January 1972), 59-62.

"Lockheed's Latest ASW Aircraft: The S-3A Viking." International Defence Review, V (February 1972), 30-32.

London, Michael P. "VSX: Sub Hunting From the Air." Space/Aeronautics, LI (April 1969), 42-51.

Lowndes, Jay C. "Studies Show Lift Coefficient Tripling." Aviation Week and Space Technology, CXIII (December 1, 1980), 119-123.

Mandeville, John. "Savage From Inglewood." Air Classics, X (September 1974), 20-29.

Mason, John F. "S-3A to Get Totally Integrated Avionics." Electronic Design, XVII (September 1969), 38-40.

Mouton, E. E. "The Right Aircraft at the Wrong Time." U.S. Naval Institute Proceedings, CVI (September 1980), 106-109.

Nevitt, R. D. "Reopen the S-3 Line." U.S. Naval Institute Proceedings, CXIII (October 1987), 136-137.

Newton, Bob. "Lockheed S-3A: The Navy's Viking Sub Hunter." Air Progress, XLI (October 1979), 72+.

Prina, L. Edgar. "A New ASW Plane." Navy, XII (September 1969), 10-12.

Rausa, Rosario. "The S-3A Viking." U.S. Naval Institute Proceedings, CII (October 1976), 131-134.

Richman, John P. "The Viking at Home in the Fleet." Approach, XXII (December 1976), 12-14.

Robinson, Clarence A., Jr. "S-3A Strengthens Carriers' ASW Role." Aviation Week and Space Technology, CV (November 22, 1976), 30-31+.

"The S-3A Joins the U.S. Fleet." International Defence Review, VII (June 1974), 322-323.

"S-3A Update." Naval Aviation News, (October 1978), 8-15.

Schultz, James B. "Lockheed Modifies S-3 Viking for Greater ASW Mission." Defense Electronics, XVI (April 1984), 91-92+.

Tuttle, Jerry O. "Viking Viewpoint." Naval Aviation News, (February 1983), 18-21.

"Vikings Sweep the Seas." Naval Aviation News, (February 1983), 12-17.

Voss, Philip D. "Harpooning the Viking." U.S. Naval Institute Proceedings, CVI (February 1980), 63-67.

Webb, George. "Direct Lift Control is Here, Finally." Approach, XXI (June 1976), 1-3.

Wilson, R. L. "So You're Going to Fly the Viking." Approach, XX (October 1974), 18-21.

(Curtiss) SBC Helldiver

Bowers, Peter M. "Helldiver: First in a Long-Lived Near Legendary Line That Never Quite Lived Up to Its Name." Wings, XII (April 1982), 10-27.

Miller, Jay. "Last of the Helldivers: Curtiss SB2C-5." Air Classics, XIII (July 1977), 60-67.

Roemer, Charles E. "SBC-4: The Navy's Last Combat Biplane." Naval Aviation Museum Foundation, IV (Fall 1984), 42-45.

"Uncle Sam's Last Combat Biplane." Air Enthusiast, no. 5 (November 1977-February 1978), 1-13.

Westburg, Peter. "Curtiss SBC-4." Model Builder, X (December 1980), 28-29.

(Douglas) SBD/A-24 Dauntless

Book/Document/Report Literature:

Brazelton, David. The Douglas SBD Dauntless. Aircraft in Profile, no. 196. Windsor, Eng.: Profile Publications, 1970. 15p.

Stern, Robert C. SBD Dauntless in Action. Aircraft, no. 64. Carrollton, Tex.: Squadron/Signal Publications, 1984. 90p.

Tillman, Barrett. The Dauntless Dive Bomber of World War II. Annapolis, Md.: Naval Institute Press, 1976. 232p.

Woods, Katie, Nancy Poage, and Margaret Scriven. A-24. ATSC Historical Study, no. 8. Wright Field, Ohio: USAAF Air Technical Service Command, 1944. 26p.

Periodical/Anthology Literature:

Andrews, Hal. "Naval Aircraft: Dauntless." Naval Aviation News, (August 1972), 24-25.

Arthur, Charles S. "Flying the Dauntless Dive Bomber." Aerospace Historian, XXIX (September 1982), 146-153.

Brown, Eric. "Doughty Dauntless." Air International, XVII (October 1979), 180-185, 196.

"The Dauntless." Flying, XXXIV (May 1944), 62-63.

"Dive Bomber." Air Classics, III (November 1966), 56+.

"Douglas Dauntless Design Sketches." Aero Digest, XLVI (August 1, 1944), 86-87.

Fischer, Vic and Jim Hall. "SBD Dauntless Flies Again." Sport Aviation, XXVI (December 1977), 31-33.

Francillon, Rene J. "Douglas Dauntless." Air Pictorial, XXVI (October 1964), 325-328.

Kimes, I. L. "The Scout Bomber." Flying, XXXII (February 1943), 85-88+.

Miller, Jay. "Nostaligic Sojourn of the SBD." Air Classics, XII (March 1976), 68-71.

Mizrahi, Joseph V. "The Saga of 'Speedy D.'" Air Classics, V (August 1969), 20-37.

Phelps, Donald W. "Diminutive Dauntless." Naval Aviation News, (November 1977), 30-31.

(Brewster) SB2A/A-34 Buccaneer

Andrews, Hal. "Naval Aircraft: Buccaneer." Naval Aviation News, (December 1979), 20-21.

Gemeinhardt, Fritz, David Lucabaugh, and Bob Martin. "The Brewster Blaster." American Aviation Historical Society Journal, XXXII (Spring 1987), 58-73.

"The Navy's New Brewster Dive Bomber: SB2A-1." Popular Science, CLXXXIX (September 1941), 50+.

(Curtiss) SB2C/A-25 Helldiver

Book/Document/Report Literature:

Andrews, Hal. The Curtiss SB2C-1 Helldiver. Aircraft in Profile, no. 124. Windsor, Eng.: Profile Publications, 1967. 12p.

Self, Mary and Margaret Scriven. A-25. ATSC Historical Study, no. 9. Wright Field, Ohio: USAAF Air Technical Service Command, 1944. 19p.

Stern, Bob. SB2C Helldiver in Action. Carrollton, Tex.: Squadron/Signal Publications, 1982. 50p.

Periodical/Anthology Literature:

Andrews, Hal. "Naval Aircraft: Helldiver." Naval Aviation News, (January 1974), 20-21.

Bowers, Peter M. "Helldiver." Wings, XII (April 1982), 10-27.

Brown, Eric. "Ill-Favoured Helldiver." Air International, XVII (December 1979), 281-288.

"Curtiss Helldiver, SB2C-4." Flying, XXXVII (September 1945), 51-54.

"Efficiency of Helldiver Production." Aero Digest, XLIX (June 1, 1945), 114+.

"The Helldiver." Flying, XXXIV (March 1944), 30-31.

"The Helldiver Meets South Pacific Battle Tests." Aero Digest, XLIV (January 15, 1944), 56-57.

Klemin, Alexander. "Dive Bomber." Scientific American, CLXV (July 1941), 35+.

Levine, William F. "Beauty of the Beast." Air Progress Aviation Review, IV (Summer 1980), 60-69.

McSurely, Andrew. "The Navy Releases Data on the Helldiver." Aviation News, I (January 3, 1944), 7-8.

Runyan, Joseph W. "I Ride the Beast: A Helldiver." Popular Science, CXLVI (February 1945), 131-133+.

Trimble, Robert L. "Big-Assed Bird." Air Classics, XVII (May-July 1981), 12-23, 14-26, 48-49, 14-26, 78.

Williams, Steve. "Evolution of the Helldiver." Skyways, III (October 1944), 34-39.

(Vought) SBU

Andrews, Hal. "Naval Aircraft: SBU." Naval Aviation News, (April 1974), 20-21.

(Vought) SB2U Vindicator

Andrews, Hal. "Naval Aircraft: Vindicator." Naval Aviation News, (March 1976), 20-21.

Bowers, Peter M. "Bomb Ten, Scout Five, Spot One." Wings, XVIII (February 1988), 3-23.

____. "Vought Vindicator." Air Classics, VI (February 1970), 14-35.

Weathers, Joseph H., Jr. "Super Baker, Two Uncle: Initial and Final Disposition." American Aviation Historical Society Journal, XIV (Spring 1969), 21-25.

(Grumman) TBF/TBM Avenger

Book/Document/Report Literature:

Francillon, Rene J. Grumman (Eastern) TBF/TBM Avenger. Aircraft in Profile, no. 214. Windsor, Eng.: Profile Publications, 1971. 15p.

Jackson, Berkeley R. and Thomas E. Doll. Grumman TBF/TBM Avenger. Aero Series, no. 21. Fallbrook, Calif.: Aero Publishers, 1984. 95p.

Scribner, Charles L. TBM/TBF Avenger in Action. Carrollton, Tex.: Squadron/Signal Publications, 1987. 49p.

Tillman, Barrett. Avenger at War. New York: Scribners, 1980. 128p.

Periodical/Anthology Literature:

Andrews, Hal. "Naval Aircraft: Avenger." Naval Aviation News, (July 1973), 20-21.

Brown, Eric. "Audacious Avenger." Air International, XIV (June 1978), 286-291, 308-309.

Cressman, Robert J. "Avengers." Naval Aviation News, (August 1979), 34-39.

"An Era Ends for a Veteran." Naval Aviation News, (June 1954), 12-13.

Friedman, William B. "The Avenger Scores Again." Popular Science, CXLIV (January 1944), 92-93.

"The Grumman Avenger." Naval Aviation News, (July 15, 1944), 40-47.

Miller, Jay. "Texas Turkey." Air Classics, XIII (December 1977), 44-49. Confederate Air Force TBF.

O'Leary, Michael. "Avenger." Air Classics, V (February 1969), 38-54.

____. "Pearl Harbor Avenger." Air Classics, IX (July 1973), 38-45.

"Portfolio of Design Features: The TBM-2." Aero Digest, XLVIII (March 15-April 1, 1945), 82-89, 88-93.

Smith, M. Lefevre. "The Avengers Strike." Skyways, III (July 1944), 19-20+.

Wickware, Francis S. "That's How Avengers are Born: The Team Behind One of the Navy's Deadliest Weapons." Saturday Evening Post, CCXV (September 5, 1942), 22-25+.

Wilcox, Donald E. "Torpedo Bomber." Flying, XXXII (February 1943), 97-99.

B. Bomber-Patrol Aircraft

(Rockwell International) B-1A/B

Book/Document/Report Literature:

Adams, Gordon. The B-1 Bomber: An Analysis of Its Strategic Utility, Constituency, and Economic Impact. New York: Council on Economic Priorities, 1976. 50p.

Andersen, Marlowe R. The Destiny of the B-1 Bomber. Research Paper. Maxwell AFB, Ala.: Air Command and Staff College, Air University, 1972. 66p.

Brown, Michael E. Flying Blind: Decision-Making in the U.S. Strategic Bomber Program. PhD. Dissertation. Ithaca, N.Y.: Cornell University, 1983.

Cantwell, Francis X. Bomber Defense: B-1B. Research Paper. Maxwell AFB, Ala.: Air Command and Staff College, Air University, 1986. 22p.

Holder, William G. B-1 Bomber. Aero Series, v. 32. Fallbrook, Calif.: Aero Publishers, 1986. 89p.

Kotz, Nick. Wild Blue Yonder: Money, Politics, and the B-1 Bomber. New York: Pantheon, 1988. 320p.

Kowalski, Harold W., Jr. The B-1 Bomber: Aircrew Selection. Research Paper. Maxwell AFB, Ala.: Air Command and Staff College, Air University, 1973. 78p.

Landen, John R. B-1B Lead-in Training. Research Paper. Maxwell AFB, Ala.: Air Command and Staff College, Air University, 1984. 42p.

Logan, Don and Jay Miller. Rockwell International B-1A/B. Minigraph, no. 24. Arlington, Tex.: Aerofax, Inc., 1983. 40p.

Lumpkin, Phillip R. The B-1 Program: How to Market a Manned Bomber. Research Paper. Maxwell AFB, Ala.: Air Command and Staff College, Air University, 1980. 54p.

Merkley, Jeffrey A. The B-1B Bomber and Options for Enhancements. Washington, D.C.: Congressional Budget Office, 1988. 94p.

Skantze, Lawrence A. B-1B Program in Perspective. Defense Issues, v. 2, no. 13. Washington, D.C.: GPO for the Department of Defense, 1987. 6p.

Spick, Mike. B-1B. Modern Fighting Aircraft, no. 11. New York: Arco, 1986. 64p.

United States. Congress. House. Committee on Armed Services. Subcommittees on Research and Development and on Procurement and Military Nuclear Systems. The B-1B Program Review: Report. 100th Cong., 1st sess. Washington, D.C.: GPO, 1987. 23p.

____. Government Accounting Office. Mission Analysis and Systems Acquisition Division. The B-1 Bomber Program: A New Start. Washington, D.C., 1983. 9p.

____.____. National Security and International Affairs Division. Logistics Support Cost for the B-1B Aircraft Can Be Reduced. Washington, D.C., 1984. 49p.

____.____.____. Supportability, Maintainability, and Readiness of the B-1B Bomber. Washington, D.C., 1987. 21p.

Vlach, William R. The B-1 and Congress. Research Report. Maxwell AFB, Ala.: Air War College, Air University, 1975. 74p.

Periodical/Anthology Literature:

Abgottspon, Peter. "Advances in the B-1B Programme." Armada International, X (April 1986), 56+.

"Air Force Nears SAC Upgrade Goal With B-1B Deployment." Aviation Week and Space Technology, CXXIV (June 2, 1986), 46-48+.

Allward, Maurice. "Rockwell International B-1 Strategic Bomber." Design Profile, no. 3. Air Pictorial, XXXVI (January 1974), 10-14.

Almy, David. "B-1." Flying, CXII (October 1985), 78-84+.

Anderton, David A. "B-1B: Out of the Shadows." Popular Mechanics, CLXII (November 1985), 88-90+.

"The B-1." Air Progress, XXXIX (October 1977), 66-67.

"B-1: America's New Strategic Bomber." Air Combat, III (January 1975), 46-56.

"B-1: Flight Testing America's Swing-Wing Bomber." Flight International, CX (July 24, 1976), 227-229.

"B-1:Tests Yield Penetrativity Data." Aviation Week and Space Technology, CXII (June 16, 1980), 187+.

"The B-1B: A Bomber in a Fix." Defence Update International, no. 93 (February 1989), 42-49.

"B-1B Enters USAF Service." Defence Update International, no. 62 (1985), 2-7.

"B-1B Manufacturing." Aviation Week and Space Technology, CXIX (August 1, 1983), 34-61.

"B-1B Readies for Service." Aviation Week and Space Technology, CXXIV (June 2, 1986), 46-48+.

"The B-1 Bomber." Air Classics, VIII (September 1977), 72-78.

"B-1 Bomber Crux of SAC Plans." Aviation Week and Space Technology, CIV (May 10, 1976), 39-45.

"B-1 Bomber Need Defended, Critics Hit." Aviation Week and Space Technology, CXIII (August 11, 1975), 53-55.

"The B-1 Bomber Up Close." Newsweek, XCVIII (October 12, 1981), 41-42.

"B-1, No, Cruise, Yes." Newsweek, XC (July 11, 1977), 14-17.

Berman, Bob. "The B-1 Bomber: Is It Worth $92 Billion?" Defense Monitor, IV (June 1975), 1-4.

Bezdek, Roger H. "B-1: A History." Bulletin of the Atomic Scientists, XL (November 1984), 10-13.

Blair, Edwin T. "First Flight." Airman, XVII (August 1973), 20-23.

"Building a Better Bomber." Aerospace Safety, XXX (July 1974), 10-12.

Burke, Kelly H. "'It's the Great Air Show of the World.'" Armed Forces Journal International, CXVIII (June 1981), 57-62.

Butchke, Mike. "The B-1: First Flight Plus One Year." Aerospace Safety,XXXII (March 1976), 10-14.

Butler, Robert K. "B-1B: High Tech for a Low Profile." Navigator, XXXIII (Spring 1986), 5-6.

Canan, James W. "The Magnificant B-1B." Air Force Magazine, LXVII (November 1984), 58-65.

"Carter and the B-1 Bomber." Newsweek, LXXXIX (June 20, 1977), 53-54.

"Carter Blocks Production of B-1." Aviation Week and Space Technology, CVII (July 4, 1977), 13-14.

"Carter's B-1 Shocker: Its Meaning." U.S. News and World Report, LXXXIII (July 11, 1977), 13-14.

"Carter's Big Decision." Time, CX (July 11, 1977), 8-12.

Center for Defense Information. "The B-1: Does the U.S. Need It?" Defense Monitor, II (January 22, 1973), 1-8.

Chapman, Stephen. "The Pentagon's Bad Penny." New Republic, CLXXXII (January 26, 1980), 15-17.

"Closeup: The B-1B." Defence Update International, no. 41 (1983), 24-40.

Cody, Richard N. "B-1: A View From the Aft Station." Combat Crew, XXVII (September 1976), 4-7.

Cole, T. H. "Fail Safe '88." Popular Mechanics, CLXV (January 1988), 54-57+.

"Controversy Over the B-1 Bomber Program." Congressional Digest, LV (December 1976), 289-314.

Correll, John T. "Here Comes the B-1B." Air Force Magazine, LXVIII (August 1985), 87-89.

Coyne, James P. "Bringing on the B-1B." Air Force Magazine, LXIX (June 1986), 63+.

Cutler, Katie. "The New Bomber--the B-1B." Airman, XXVI (October 1982), 2-7.

Darmstandler, Harry M. "The B-1 Bomber." Vital Speeches of the Day, XLI (September 15, 1975), 715-718.

____. "The B-1: Myth and Reality." Air Force Magazine, LVIII (October 1975), 60-61.

Daskal, Steven E. "LRCA: The Importance of Long-Range Combat Aircraft." National Defense, LXIX (March 1985), 49-51.

Elson, Benjamin M. "Boeing Facing B-1 Avinnics Challenge." Aviation Week and Space Technology, XCVIII (April 16-23, 1973), 44-47, 52-54.

"Experience, Computers Aid Transition." Aviation Week and Space Technology, CXXIV (June 2, 1986), 59-62.

Farrell, G. Francis. "The B-1: A Wise Investment." Combat Crew, XXII (December 1971), 4-7.

____. "B-1 Flyability." Combat Crew, XXXIII (December 1973), 4-11.

Fink, Donald E. "B-1 Tests Aim to Show Mission Capability." Aviation Week and Space Technology, C (June 3, 1974), 48-49, 51-52.

"First B-1 Bomber Prototype Rolls Out." Aviation Week and Space Technology, CI (November 4, 1974), 14-15.

Fleisher, Richard. "Economic Benefit, Ideology, and Senate Voting on the B-1." American Politics Quarterly, XIII (April 1985), 200-211.

"Flying the B-1B as a Crew." Combat Crew, XXXVII (June 1987), 4-6.

Ford, William A. "A Work of Wing and Wheel." Airman, XV (March 1971), 40-47.

Foxley-Norris, Christopher. "B-1: Loss of an Option." Army Quarterly and Defence Review, CVII (October 1977), 394-398.

Frisbee, John L. "The B-1 Blue Chip in the Deterrent Stack." Air Force Magazine, LIII (April 1970), 45-48.

Geddes, J. Philip. "The B-1 Strategic Bomber: A Costly Essential." International Defence Review, VIII (April 1975), 195-200.

____. "B-1B: The Bomber That Came Back." International Defence Review, XV (January 1982), 39-45.

____. "Progress on the B-1 Bomber." Interavia, XXVI (August 1971), 940-943.

____. "Rockwell International's B-1 Bomber at a Critical Point." Interavia, XXX (February 1975), 132-138.

Getlein, Frank. "Future of Aerial Warfare: The B-1 Program." Commonweal, CIII (June 18, 1976), 388-389.

Glines, Carroll V. "B-1: America's Most Studied Airplane." Air Line Pilot, XLIV (May 1975), 6-9.

____. "The B-1B: America's Newest Bomber." Aviation/Space, II (Fall 1984), 38, 52.

Graham, Vickie M. "Rolling Out the B-1B." Airman, XXVIII (November 1984), 41-43.

Greeley, Brendan M., Jr. "Crash of Third Operational B-1B Could Raise Mission, Funding Issues." Aviation Week and Space Technology, CXXIX (November 28, 1988), 28-29.

____. "Inquiry Finds Crash Was Inevitable After Bird Strike Caused Fire in B-1B." Aviation Week and Space Technology, CXXVIII (January 25, 1988), 20-21.

Greenway, Sean. "Camouflage for the B-1." Air Combat, VIII (May 1980), 34-36.

Gulcher, R. H. "B-1B: Designing for Supportability." Aerospace America, XXII (June 1984), 54-56.

Haystead, John. "Within Budget, Ahead of Schedule, Air Force Unveils First Production B-1B." Journal of Electronic Defense, VII (November 1984), 27-28.

Hieronymus, W. S. "B-1 Designed to New Standards." Aviation Week and Space Technology, XCV (July 26, 1971), 42-45.

Hirsch, Vincent T. "A Dubious Decision and a Disturbing Pattern." Sea Power, XX (August 1977), 3-4.
B-1 cancellation.

Hoeber, Francis P. "The B-1: A National Imperative." Strategic Review, IV (Summer 1976), 111-117.

Hubbell, John G. "Shall We Build This Superbomber, B-1?" Reader's Digest, CI (December 1972), 121-125.

"Inflation Hits the B-1 Bomber." Business Week, (September 21, 1974), 78-82.

Kaplan, Hy R. "New Muscle for America." Sergeants, XIX (October 1981), 28-31.

Katzaman, Jim. "The B-1B: On Target." Airman, XXIX (October 1985), 18-21.

Kenny, Michael A. "Flying the Fast Moving B-1B." Combat Crew, XXXVI (November 1986), 14-15.

____. "Pilot Report: B-1B." Air Force Magazine, LXIX (June 1986), 58-62.

Kitfield, James. "Logistics Fallout." Military Logistics Forum, III (May 1987), 32-36.

____. "Price of Concurrency." Military Logistics Forum, III (May 1987), 18-20+.

Klass, Philip J. "Air Force Examining Deficiencies in B-1B Electronic Warfare System." Aviation Week and Space Technology, CXXV (December 22, 1986), 20-21.

____. "ALQ-161 Flight Tests Set This Summer." Aviation Week and Space Technology, CXX (March 26, 1984), 62-63+.

Koeivar, Ben. "Our New B-1 Bomber: High, Low, Fast and Slow." Popular Science, CXCVII (November 1970), 86-87+.

Kotz, Nick. "The Chesapeake Bay Goose Hunt, the Beautiful Secretary, and Other Ways the Defense Lobby Got the B-1." Washington Monthly, XX (February 1988), 29-32+.

____. "Money, Politics, and the B-1 Bomber: Why the Military Program is Out of Control." Technology Review, XCI (April 1988), 30-40.

____. "The Mysterious Case of the B-1 Bomber." Common Cause Magazine, XIV (July-August 1988), 36-40.

Lake, Julian S. "The Beleaguered B-1B." Defence Science and Electronics, VI (April 1987), 31-32+.

Lambert, Mark. "B-1 Gets Airborne Again." Interavia, XXXVIII (May 1983), 455-458.

____. "B-1B Continues on to Target." Interavia, XXXIX (March 1984), 252-253.

Larson, George W., Jr. "First Impressions." Combat Crew, XXVI (January 1976), 4-7.

____. "Flying the B-1: A Pilot's View." Air Force Magazine, LIX (June 1976), 22-27.

Lenorovitz, Jeffrey M. "B-1 Proposed as Core Aircraft." Aviation Week and Space Technology, CXI (September 17, 1979), 14-15.

London, Michael P. "B-1: The Last Bomber?" Space/Aeronautics, LIII (April 1970), 26-33.

"A Look at the B-1." Navigator, XXIX (Spring 1982), 16-17.

Lowe, George E. "The Only Option?" U.S. Naval Institute Proceedings, XCVII (April 1971), 18-26.

Lynn, N. "B-1B: Born-Again Bomber." Flight International, CXXVIII (September 21, 1985), 39-42.

McCarthy, John F., Jr. "The Case for the B-1 Bomber." International Security, I (Fall 1976), 78-97.

McClelland, Lori A. "The B-1B Has Arrived." Defense Electronics, XVII (August 1985), 72-74+.

McCrossan, Eddie. "Prototype B-1." Airman, XIX (July 1975), 42-47.

McCullough, Bob. "Bomber Alert." Airman, XXXI (July 1987), 24-28.

McLucas, John L. "The Case for a Modern Strategic Bomber." AEI Defense Review, II (January 1978), 13-24.

"Mailed Fist for the Future: Rockwell International B-1 Strategic Bomber." Aircraft Illustrated, VIII (April 1975), 162-165.

"Managing New Technology--The B-1 Bomber: Special Report." Aviation Week and Space Technology, XCVI (June 26, 1972), 46-47+.

Mandros, Harry. "B-1B Rollout." Combat Crew, XXXIV (December 1984), 8-9.

Manning, Robert A. "The B-1: A Buy or a Bomb?" Encore, V (November 8, 1976), 14-16.

Marshall, E. "Bomber No. 1." Science, CCXXXIX (January 29, 1988), 452-455.

Martin, Abner B. "B-1: The Road to First Flight." Astronautics and Aeronautics, XIII (June 1975), 56-62.

____. "The B-1: Strategic Deterrence Into the 21st Century." Air University Review, XXVII (March-April 1976), 2-14.

Mason, Herbert M. "America's Deadly New Bomber, the B-1." Mechanics Illustrated, LXVI (October 1970), 74-76+.

Mathews, Mike. "B-1B Program Continues 2½ Months Ahead of Schedule." Defense Systems Review, II (September 1984), 22-24.

Milton, Theodore R. "Through Flaps and Flak." Air Force Magazine, LXXII (January 1989), 80-85.

Nelson, Douglas T. "Management Approaches to the B-1 Bomber Program." Ordnance, LVII (March-April 1973), 366-369.

Nordwall, Bruce D. "Eaton AIL Proposes B-1B Fixes as USAF Considers Alternatives." Aviation Week and Space Technology, CXXIX (October 10, 1988), 99-103.

Norman, C. "B-1's Problems Stir Debate on Bomber's Role." Science, CCXLI (September 9, 1988), 1282-1283.

North, David M. "B-1B Combines Brisk Low-Altitude Handling, More Capable Avionics." Aviation Week and Space Technology, CXXVII (September 14, 1987), 54-55+.

____. "Development Problems Delay Full B-1B Operational Capability." Aviation Week and Space Technology, CXXV (November 3, 1986), 34-35.

"Notes and Comment: The B-1." New Yorker, LXIV (March 7, 1988), 27-28.

Ognibene, P. J. "Hidden Costs of the B-1: Bombing Empty Holes." New Republic, CLXVIII (May 12, 1973), 18-21.

O'Mara, R. O. "The B-1 Strategic Bomber: A Necessary Weapons System." Royal Air Forces Quarterly, XVI (Autumn 1976), 236-242.

Ornstein, Norman J. and Shirley Elder. "The B-1 Bomber: Organizing at the Grass Roots." In: Eston T. White, ed. Studies in Defense. Washington, D.C.: National Defense University Press, 1983. pp. 43-70.

Paine, Christopher. "The Selling of the B-1 Bomber." Common Cause Magazine, VIII (October 1982), 16-21+.

"Palmsdale and the Bomber Connection." Air International, XXXI (August 1986), 65-73, 89.

Patton, W. G. "B-1: A New Approach to Building Planes." Iron Age, CCIX (June 15, 1972), 59-62.

Peck, Earle G. "Bomber One." Airpower, XVI (March 1986), 10+.

Pect, Hans. "An Electronic Umbrella: The B-1B Defensive Avionics System." Defense Systems Review and Military Communications, II (May 1984), 15-18.

Perini, Michael B. "Finally, the B-1B." Air Force Magazine, LXVI (January 1983), 58-63.

Priecko, John. "B-1B Aircrew Selection: New Crew, New Philosophy." Combat Crew, XXXVI (June 1986), 4-7.

____. "B-1B Next Stop: The Black Hills of South Dakota." Combat Crew, XV (October 1985), 24-26.

Raiklen, Hal. "The B-1 Bomber." National Defense, LXI (January-February 1977), 297-300.

Rice, Berkeley. "The B-1 Bomber: The Very Model of a Modern Major Misconcept." Saturday Review, LIV (December 11, 1971), 20-25, 59.

"Rockwell B-1 Cancelled." Flight International, CXII (July 9, 1977), 124-125.

Ropelewski, Robert R. "B-1 Design Keyed to Future Growth." Aviation Week and Space Technology, XCV (November 29, 1971), 34-37.

____. "B-1 Studied as Cruise Missile Launcher." Aviation Week and Space Technology, CXI (December 10, 1979), 48+.

____. "North American Gears to Produce B-1." Aviation Week and Space Technology, XCVI (June 26, 1972), 107+.

____. "Reagan Strategic Policy: Rockwell Begins B-1B Production Plans." Aviation Week and Space Technology, CXV (October 12, 1981), 26-28.

Rustenburg, John W. "B-1 Structural Test Program." Defense Systems Management Review, I (Winter 1977), 149-161.

"SAC Accepts First B-1B." Combat Crew, XXXV (September 1985), 6-9.

Schlesinger, James R. "The B-1 Bomber." Electronic Warfare, VI (November-December 1974), 13-14+.

Scholin, Allan R. "B-1." Air Progress, XXXVII (December 1974), 54-58.

Schultz, James B. "B-1B Avionics Readied." Defense Electronics, XV (January 1983), 44-45+.

Scott, William B. "B-1B Flight Tests Study Mission Needs." Aviation Week and Space Technology, CXVIII (May 2, 1983), 77-79.

Seamans, Robert C., Jr. "Review of the Past Brightens the Future: C-5 Problems Aid B-1 Development." Defense Management Journal, VIII (October 1972), 60-65.

Sherman, Ron. "The Future of the U.S. Manned Bomber." Military Technology, VI (July 1982), 121-122+.

Sherman, Stratford P. "How Rockwell Kept the B-1 Alive." Fortune, CIV (November 2, 1981), 106-108+.

Shifrin, Carole A. "Air Force Crews Cite Progress of B-1B in Initial Operations." Aviation Week and Space Technology, CXXVI (March 30, 1987), 21-23.

Simmons, N. "Carter Meets the B-1 Bomber." Astronautics and Aeronautics, XV (January 1977), 17-20.

Skantze, Lawrence A. "B-1B Program in Perspective." Program Manager, XVI (May-June 1987), 2-8.

Smith, Bruce A. "B-1A Testbed Crashes at Edwards Air Force Base." Aviation Week and Space Technology, CXXI (September 3, 1984), 47-48.

____. "B-1 B Program on Time Despite Component Shortages, Engine Problems." Aviation Week and Space Technology, CXXI (December 10, 1984), 17-19.

____. "Rockwell Completes First B-1B Fuselage." Aviation Week and Space Technology, CXX (January 23, 1984), 78-79.

____. "Rockwell Seeks to Use B-1 Expertise on Weapon System Modernization." Aviation Week and Space Technology, CXXVIII (January 25, 1988), 154-156.

Smith, J. S., III and F. A. Fiedler. "Review of Five Years of Flight Testing the B-1." Society of Experimental Test Pilots Technical Review, XIV (Fall 1979), 209-231.

Spick, Mike. "B-1B." In: William T. ("Bill") Gunston, ed. The Great Book of Modern Warplanes. New York: Crown, 1987. pp. 457-520.

Stambler, Irwin. "Engineering the B-1 Design." Interavia, XXVII (June 1972), 610-613.

Stefanides, E. J. "Giant Airborne Ball Screws Pivot B-1B Wing." Design News, XLIII (April 6, 1987), 126-127.

Stewart, James T. "Evolving Strategic Air Power and the B-1." Astronautics and Aeronautics, X (June 1972), 22-29.

Sturmthal, Emil. "B-1 Flight Test Program Report." Society of Experimental Test Pilots Technical Review, XIII (Spring 1976), 23-26.

Sund, Harald. "A Marred Takeoff for the B-1." Fortune, CX (October 1, 1984), 122-127.

Sweetman, Bill. "Blackjack: The Biggest Bomber by Far." International Defence Review, XVII (May 1984), 549-552.

____. "Survivable Bomber." International Defence Review, XX (August 1987), 1013-1014+.

Taylor, John W. R. "Rockwell International B-1." Air Force Magazine, LVIII (June 1979), 41-48.

Thompson, Warren. "Bombing in the B-1B." Airpower, XIX (January 1989), 26-31+.

Thurman, William E. "Doing It Right With the B-1B Program." Army Research, Development, and Acquisition, XXVII (May-June 1986), 7-8.

Tortolano, F. W. "B-1B Extends Air Force's Clout." Design News, XXXVIII (April 19, 1982), 76-78.

"Towards the B-1A." Flight International, XCVI (November 20, 1969), 817-818.

Ulsamer, Edgar. "B-1: Good Bargin for Deterrence." Air Force Magazine, LVII (December 1974), 38-40.

____. "The B-1: USAF's Coming Strategic Aircraft." Aerospace International, VI (March-April 1970), 37-41.

____. "USAF's Most Versatile Bomber." Air Force and Space Digest, LIII (April 1970), 37-44.

"USAF Keeps Tight Rein on B-1 Advanced Bomber Effort." Aviation Week and Space Technology, XCVI (June 26, 1972), 46-47.

"USAF, Rockwell Test Systems Needed to Expand B-1B's Flight Envelope." Aviation Week and Space Technology, CXXIX (November 28, 1988), 82-84.

Van Keuren, Terry. "B-1B: Flexible, Survivable Penetrator." Aerospace America, XXII (November 1984), 50-53.

Wade, Larry. "To Fly the New B-1B Bomber." Combat Crew, XXXIV (January 1984), 14-17.

Walsh, Bill. "Artful Dodger of the 1980s." Countermeasures, II (November 1976), 42-43+.

Welling, William. "Emerging Problems With the B-1B Program Prompt Thoughts on Weapon System Buying." Defence Science and Electronics, VI (August 1987), 21-24.

___. "The U.S. Air Force's B-1B." NATO's Sixteen Nations, XXIX (December 1984-January 1985), 80-82.

Witze, Claude. "The B-1: Still on the Hill." Air Force Magazine, LX (January 1977), 13-15.

Wood, Archie. "Modernizing the Strategic Bomber Force Without Really Trying: A Case Against the B-1." International Security, I (Fall 1976), 98-116.

Yonay, Ehud. "How the No. 2 Seller Sold the No. 1 Buyer $20 Billion Worth of California Product." California, X (October 1985), 94+.

Zumwalt, Elmo R., Jr. "An Assessment of the Bomber-Cruise Missile Controversy." International Security, II (Summer 1977), 47-58.

(Curtiss) B-2/R4C-1 Condor

"Curtiss Builds Largest All-Metal Bomber: Condor." Aviation, XXIII (December 12, 1927), 1404-1405.

"The Curtiss Condor." Flight, XIX (December 1, 1927), 819-820.

"Curtiss Condor." Flying Review International, XIX (February 1964), 60+.

"The Curtiss Condor Bomber." Aero Digest, XI (August 1927), 184-186.

"Curtiss Condor: Latest Night Bombardment Plane." US Air Services, XII (August 1927), 23-25.

Mathews, William R. "Curtiss Condor II." American Aviation Historical Society Journal, XII (Spring 1967), 3-27.

"Mighty War Machine Receives Plaudits of Army Flyers: Condor." Slipstream, VIII (September 1927), 24-25.

(Northrop) B-2 Stealth

Book/Document/Report Literature:

Jones, J. Stealth Technology: The Art of Black Magic. Edited by Matt Tharber. Blue Ridge Summit, Pa.: TAB, 1989. 160p.

Periodical/Anthology Literature:

Ashley, Steven and C. P. Gilmore. "Finally: Stealth." Popular Science, CCXXXIII (July 1988), 46-51, 94-95.

Biddle, W. "The Real Secret of Stealth." Discover, VII (February 1986), 20-31.

Braybrook, Roy M. "Observing the Unobservable; or, a Stealthy Look at Stealth." Pacific Defense Reporter, XIII (July 1986), 39-41.

Brown, Michael E. "B-2 or Not B-2?: Crisis and Choice in the U.S. Strategic Bomber Programme." Survival, XXX (July-August 1988), 351-366.

Browne, Malcolm W. "Will the Stealth Bomber Work?" New York Times Magazine, (July 17, 1988), 24-31.

Connolly, Ray. "'Invisible Plane' Has Many Goals." Electronics, LIII (September 11, 1980), 46-50.

Darrow, B. "Stealth: No Longer a Mystery." Design News, XLIII (April 6, 1987), 122-128.

Easterbrook, Gregg. "Sticker Shock: The Stealth is a Bomb." Newsweek, CXIII (January 23, 1989), 20-22.

Heppenheimer, T. A. "Stealth: First Glimpses of the Invisible Aircraft Now Under Construction." Popular Science, CCXXIX (September 1986), 74-79+.

Hicks, Donald A. "Stealth: Its Implications for the Future." Armed Forces Journal International, CXXIV (September 1986), 70-71.

Kaylor, Robert. "The Brave New World of Stealth Warfare." U.S. News and World Report, CV (November 28, 1988), 20-28.

Kittle, R. A. "The Plane That Isn't There." U.S. News and World Report, C (May 12, 1986), 43-44.

Kotz, Nick. "Shedding Light on Stealth." Harper's Magazine, CCLXXVII (August 1988), 54-55.

Lamar, Jacob V. "Will This Bird Fly?" Time, CXXXII (December 5, 1988), 20-21.

Mann, Paul. "New Defense Budget Requires Sweeping B-2 Cost Review." Aviation Week and Space Technology, CXXIX (July 18, 1988), 32-33.

Morganthau, Tom. "An $80 Billion Bust?" Newsweek, CXII (December 5, 1988), 18-20.

Morrocco, John D. "Air Force Unveils Design of Northrop B-2 Bomber." Aviation Week and Space Technology, CXXVIII (April 25, 1988), 16-18.

____. "Budget Constraints Force One-Year Delay in Peak Production of B-2." Aviation Week and Space Technology, CXXIX (December 19, 1988), 24-25.

____. "Management Problems Delay First Flight of Northrop B-2." Aviation Week and Space Technology, CXXVIII (January 11, 1988), 16-17.

Rabb, Charles. "Is the ATB Too Stealthy to Survive?" Defense Electronics, XVIII (July 1986), 66-70+.

Ramirez, A. "The Secret Bomber Bugging Northrop." Fortune, CXVII (March 14, 1988), 90-93+.

Schefter, John. "Invisible Bomber." Popular Science, CCXXII (February 1983), 60-63.

Schemmer, Benjamin F. "Financial Analysts Estimate Stealth Bomber Costs Within 4% of Each Other." Armed Forces Journal International, CXXV (October 1987), 26+.

Schultz, James B. "Air Force Presses for Maiden Flight of B-2 Bomber Before Year's End." Defense Electronics, XX (September 1988), 87+.

Scott, William B. "New Design, Production Tools Will Play Key Role in B-2 Cost." Aviation Week and Space Technology, CXXIX (December 5, 1988), 18-21.

_____, et al. "USAF, Northrop Unveil B-2 Next Generation Bomber." Aviation Week and Space Technology, CXXIX (November 28, 1988), 20-27.

Sears, W. R. "Flying Wing Could Stealthily Reappear." Aerospace America, XXV (July 1987), 16-19.

Smith, Bruce A. "Northrop Chief is Confident in B-2, Despite Tight Budgets." Aviation Week and Space Technology, CXXIX (December 5, 1988), 21-22.

Stine, G. Harry. "Stealth." Analog, CVII (March 1987), 120-123.

Sweetman, Bill. "B-2 Comes Out of the Black: IDR Predictions Come True." International Defence Review, XXI (July 1988), 787-788.

____. "The Bomber That Radar Cannot See." New Scientist, XCIII (March 4, 1982), 565-568.

____. "Stealth." International Defence Review, XVII (September 1984), 7-12.

____. "Stealth." Interavia, XL (November 1985), 1217-1219.

____. "Stealth in Service." Interavia, XLII (January 1987), 39-40.

____. "The Vanishing Air Force: Stealth Technology Goes Operational." International Defence Review, XVIII (August 1985), 1257-1259.

Thurman, William E. "USAF Defends Stealth Program." International Defence Review, XXI (May 1988), 577-578.

Varni, Gerard. "Air Force Reveals Ultrasecret B-2." International Combat Arms, VI (September 1988), 90-91.

____. "Stealth: America's 'Invisible' Air Force." International Combat Arms, V (January 1987), 50-57+.

____. "Stealth Bomber Taking Shape." International Combat Arms, V (September 1987), 27-31.

(Martin) B-10

Berliner, Don. "The Plane That Shook the World: Martin B-10." National Aeronautics, XI (July-August 1974), 24-27.

Boyne, Walter J. "Martin's Marvel." Airpower, II (June 1972), 51-67.

Greenspan, Michael M. "The First Modern Bomber." Navigator, XX (March 1973), 13-15.

Hall, Dick. "Martin B-10." Model Airplane News, CII (January 1981), 38-41, 76.

Marmo, Richard. "B-10: From Basketcase Bomber to Crown Jewel in the Air Museum's Collection." Airpower, VI (November 1976), 26-35.

"Martin B-10." Aeroplane Monthly, V (November 1977), 598-599.

"Martin Bomber for Export: Details of the B-10." Aviation, XXXIV (August 1935), 27-30.

"Martin's Bomber." Aviation, XXXIV (October 1935), 31-32.

"The Masterly Martin." Flying Review International, XX (August 1965), 49-51.

Olmstead, Merle C. "The Era of the Martin Bomber." American Aviation Historical Society Journal, VII (Fall 1962), 187-189+.

Trimble, Robert L. "Martin's Gallant Soldier." Air Classics, II (August-September 1965), 32-37.

____ and Mike Dario. "Magnificant Martin." Scale Modeller, X (January 1975), 22-29, 55-59, 92-95.

Virgin, Edward W. "The Martin Bomber, B-10." In: Robin Higham and Carol Williams, eds. Flying Combat Aircraft of the USAAF-USAF. Ames, Iowa: Iowa State University Press, 1978. pp. 27-32.

(Boeing) B-17/PB-1 Flying Fortress

Book/Document/Report Literature:

Birdsall, Stephen. B-17 Flying Fortress. Famous Aircraft Series. New York: Arco, 1965. 65p.

____. B-17 Flying Fortress in Action. Aircraft, no. 47. Warren, Mich.: Squadron/Signal Publications, 1973. 50p.

____. B-17 Flying Fortress in Color. Carrollton, Tex.: Squadron/ Signal Publications, 1987. 32p.

____. Hell's Angels. Canoga Park, Calif.: Grenadier Books, 1969. 48p.
Personalized names and markings painted on the Boeings.

Boeing Airplane Company. Familiarization Manual for Operation of Model B-17F Bombardment Airplane. Seattle, Wash., 1942.

Boeing B-17 Flying Fortress. Appleton, Wis.: Aviat Publications, 1974. 108p.

Bowers, Peter M. Boeing B-17, 1935-1985: 50th Anniversary. N.p.: Museum of Flight, 1985. 100p.

____. Fortress in the Sky. Grenada Hills, Calif.: Sentry Books, 1976. 275p.

Caidin, Martin. Everything But the Flak. New York: Duell, Sloan and Pearce, 1964. 204p.

____. Flying Forts. New York: Meredith Books, 1972. 516p.

Collison, Thomas. Flying Fortress: The Story of the Boeing Bomber. New York: Scribners, 1943. 168p.

Davis, Larry. B-17 Flying Fortress in Action. Aircraft, no. 106. Carrollton, Tex.: Squadron/Signal Publications, 1982. 58p.

Duerksen, Menno. "The Memphis Belle"--Home at Last. Memphis, Tenn.: Castle Books, 1988.

Ethell, Jeffrey L. F-17 Flying Fortress. Warbirds Illustrated, no. 41. New York: Sterling, 1987. 72p.

Freeman, Roger A. B-17 Flying Fortress. New York: Crown, 1983. 56p.

____. Boeing B-17G Flying Fortress. Aircraft in Profile, no. 205. Windsor, Eng.: Profile Publications, 1971. 20p.

Giering, Edward J. B-17 Bomber Crew Diary: A Story of the 8th Air Force in World War II. Manhattan, Kans.: Sunflower University Press, 1985. 142p.

Hardison, Priscilla and Anne Wormser. "The Suzy Q." New York: Literary, 1943. 170p.

Hess, William E. B-17 Flying Fortress. Ballantine's Illustrated History of World War II. New York: Ballantine Books, 1974. 159p.

Jablonski, Edward. Flying Fortress: The Illustrated Biography of the B-17 and the Men Who Flew Them. Garden City, N.Y.: Doubleday, 1965. 362p.

Johnsen, Frederick A. Winged Majesty: The Boeing B-17 Flying Fortress in War and Peace. Glendale, Calif.: Aviation Book Company, 1980. 41p.

Lloyd, Alwyn T. B-17 Flying Fortress in Detail and Scale: Pt. 1, Production Versions. Detail and Scale Series, v. 2. Fallbrook, Calif.: Aero Publishers, 1981. 72p.

____. B-17 Flying Fortress in Detail and Scale: Pt. 2, Derivatives. Detail and Scale Series, v. 11. Fallbrook, Calif.: Aero Publishers, 1986. 72p.

____. B-17 Flying Fortress in Detail and Scale: Pt. 3, More Derivatives. Detail and Scale Series, v. 20. Fallbrook, Calif.: Aero Publishers, 1986. 72p.

McDowell, R. Ernest. Boeing B-17B/H Flying Fortress. Aircam Aviation Series, no. 15. New York: Arco, 1970. 48p.

____. Flying Fortress: The Boeing B-17. Carrollton, Tex.: Squadron/Signal Publications, 1987. 80p.

O'Leary, Michael, ed. Flying Fortress: 50th Anniversary Pictorial History of America's Greatest Warplane. Air Combat Special. Canoga Park, Calif.: Challenge Publications, 1985. 64p.

Ray, James. The Inside Story of the Flying Fortress, B-17. Garden City, N.Y.: Doubleday, 1943. 30p.

Thompson, Charles D. Boeing B-17E/F Flying Fortress. Aircraft in Profile, no. 77. Windsor, Eng.: Profile Publications, 1966. 15p.

United States. Army Air Forces. 25 Missions: The Story of "The Memphis Belle." New York, 1943. 36p.

Willmott, W. P. B-17 Flying Fortress. War Planes in Colour, no. 4. New York: Prentice Hall, 1983. 69p.

White, William L. Queens Die Proudly. New York: Harcourt, 1943. 273p.

Periodical/Anthology Literature:

Adams, Nathan M. "Saga of a Valiant Lady." Reader's Digest, CXXVIII (March 1986), 92-100+.
"Shoo Shoo Baby."

Andrews, Paul M. "Seventeen Bits and Pieces--Boeing B-17Fs and Gs Assigned to the 8th US Army Air Force, August 1942-May 1945: 1st Air Division." American Aviation Historical Society Journal, XXIV (Fall 1979), 202-213.

____. "Seventeen Bits and Pieces--Boeing B-17Fs and Gs Assigned to the 8th US Army Air Force, August 1942-May 1945: 3rd Air Division." American Aviation Historical Society Journal, XXIV (Winter 1979), 303-312.

"B-17 Crewnames." Air Classics, V (December 1968), 14-20.

Balk, Alfred W. "Old Soldier of the Sky: The B-17 Flying Fortress." Coronet, XLVI (August 1959), 48-52.

Batchelor, John H. "The Boeing B-17F." Air Classics, XI (October 1975), 49-52.

"Battle Log of the Fortress." Western Flying, XXI (May 1943), 44-46.

Beall, Wellwood E. "The Boeing B-17 Flying Fortress." Aviation, XLIV (January 1945), 121-144.

____. "Design Analysis of the Boeing B-17 Flying Fortress." Aviation, XLIV (January 1945), 233-243.

Bergen, Florence. "The Story Behind the Fortress." Air Trails, XX (September 1943), 32-33+.

Blue, Allan G. "Fortress vs. Liberator." American Aviation Historical Society Journal, VIII (Summer 1963), 126-128.

"Boeing B-17." Air Force, XXX (April 1947), 19-23.

"The Boeing Flying Fortress, B-17G." Flying, XXXVI (January 1945), 35-38.

"A Bomber That Fights Anything: The Boeing Flying Fortress." Popular Mechanics, LXXX (October 1943), 82-87.

Bowers, Peter M. "The Boeing PB-1." Air Classics, II (August-September 1965), 58-66.

____. "The Forgotten Fortresses." Wings, IV (August 1974), 6-25.
Pre-World War II.

____. "Forgotten Fortress: From the Development X1B-17A Through the B-17D." Airpower, IV (September 1974), 8-31, 54-55.
A continuation of the previous citation.

____. "A Fortress is Forever." Wings, VII (March 1977), 24-50.

____. "The Saga of '5 Grand.'" Wings, VIII (June 1978), 9-19.

Bright, Charles D. "Navigating in the Big Leagues." Aerospace Historian, XXXV (December 1988), 259-265.
"Round Trip Ticket III."

Bristow, Dean. "First of the Big Sticks." Talon, IV (February 1959), 14-15.

Brooks, Mike. "Flying the B-17." Aircraft Illustrated, XII (December 1979), 592-599.

Bucey, B. K. "Good Tooling Simplifies Production of Flying Fortresses." Aviation, XLII (January 1943), 106-111.

Carr, Orrin I. "B-17: Flying Phoenix." Airpower, IV (November 1974), 72-78.

Carroll, James G. "'Goin' Home.'" Aerospace Historian, XXII (September 1975), 158-160.

Coker, William B. "America's Most Famous Bomber." Air University Review, XVII (July-August 1966), 80-86.

Davisson, Bud. "Time Trapping in 'Sentimental Journey.'" Air Classics, XVIII (August 1979), 34-53.

Doherty, Robert E. "Last Flight of the 'Mischief Maker II.'" Air Classics, XXV (March 1989), 14-19, 81-82.

Downie, Don. "Fortress Over the Atlantic." Airpower, XVIII (March 1988), 9-33.

Eaker, Ira C. "The Flying Fortress and the Liberator." Aerospace Historian, XXVI (June 1979), 66-68.

Easton, G. W. "New Methods Reduce Handwork on Boeing Fortress." American Machinist, LXXXIX (February 1, 1945), 115-120.

Edmundson, James V. "B-17 Flying Fortress." In: Robin Higham and Carol Williams, eds. Flying Combat Aircraft of the USAAF-USAF. Ames, Iowa: Iowa State University Press, 1978. pp. 33-44.

Ethell, Jeffrey L. "Our Still-Flying Fortress Turns 50." Popular Mechanics, CLXII (January 1985), 79-80+.

"Fabricating the Flying Fortress." American Machinist, LXXXIV (July 10, 1940), 498-499.

"Farewell to the B-17." Aeronautics, VII (July 1945), 32-35.

Farmer, James H. "Saga of the Civil Forts." American Aviation Historical Society Journal, XXII (Winter 1977), 292-302; XXIII (Fall 1978), 162-167.

____. "'Santa Maria' Diary." American Aviation Historical Society Journal, XIV (Fall 1969), 178-182.

"The Flying Fortress: A Combat Art Portfolio." Air Classics, VI (August 1970), 24-32.

Freeman, Roger A. "Forts in Films: Boeing B-17 Flying Fortress." Aircraft Illustrated, X (February 1977), 78-81.

Gallagher, William. "'Phyllis' Comes Home: A Flying Fortress Oyer France." In: Ben A. Williams, ed. Amateurs at War: The American Soldier in Action. Boston, Mass.: Houghton, Mifflin, 1943. pp. 449-454.

Glines, Carroll V. "The Fabulous Fortress." Air Force Magazine, LXVIII (July 1985), 118-121.

Graham, Vickie M. "To Save the 'Belle.'" Airman, XXXII (February 1988), 20-23.
Restoration of "The Memphis Belle."

Gribble, Roger. "Restoration of 'The Memphis Belle.'" National Guard, XLII (February 1988), 37-38+.

Hartley, Harold. "Fortress in the Sky." Collier's, CVII (April 19, 1941), 18-19+.

"'Hell's Angels.'" Air Classics, V (April 1969), 36-54.

Hersey, John. "Portrait of a B-17." Air Force, XLII (December 1959), 116-118.

Holder, William G. "The Return of 'Shoo Shoo Baby.'" Air University Review, XXIV (January-February 1973), 22-32.

____. "That Tough Old Bird." Air Classics, VIII (July 1972), 42-51.
"Shoo Shoo Baby."

____. "You've Come a Long Way 'Shoo Shoo Baby.'" Air Classics, IX (October 1973), 16-21, 78.

Iverson, Caroline. "'Suzy-Q' Comes Back From a Year of Action." Life, XIV (January 18, 1943), 82-84+.

Johansen, H. O. "Requiem for the B-17." Popular Science, CLXXVI (January 1960), 76-80.

Johnsen, Frederick A. "The Flying Fortress Turns 50." Air Classics, XXI (November 1985), 14-19.

Kaplan, Hy R. "'The Memphis Belle.'" Sergeants, XX (March 1982), 16+.

Keyt, H. R. "In Memory of a Queen: The B-17." American Mercury, LXXXIV (February 1957), 78-82.

Landry, Frank A. "We Fly in the Mighty Fortress." Air Progress Aviation Review, III (Winter 1979), 26-31.

Levitt, Saul. "'Rosie's Riveters.'" In: Yank, Editors of The Best From Yank, the Army Weekly. New York: E. P. Dutton, 1945. pp. 85-91.

Marmo, Richard. "Year of the Sky Queen." Airpower, VI (January 1976), 18-30.

Masefield, Peter G. "Fort vs. Lib." Flying, XXXIV (February 1944), 49+.

Mikesh, Robert C. "The Boeing PB-1." American Aviation Historical Society Journal, IX (Spring 1964), 42-45.

____. "Japan's Little Fleet of Big American Bombers: Japanese-Built B-17s in World War II." Air Force and Space Digest, LII (August 1969), 81-85.

Minnick, Mike. "The Queen's Last Battle." Wings, XI (February 1981), 20-23.

Oldfield, Barney. "The 'Near Miss' Who Didn't Miss Much." Aerospace Historian, XIV (Spring 1967), 10-14.

Parke, Robert B. "The B-17." Flying, C (March 1977), 72-79.

Reed, Boardman C. "Flying the Fortress Into Combat." American Aviation Historical Society Journal, XXXII (Spring 1987), 27-33.

Ryker, H. E. "Fortress by Vega." Aviation, XLII (July 1943), 124-125+.

Schreier, Konrad, Jr. "The Boeing B-17E." Air Classics, VIII (February 1972), 22-29.

Schweinfurth, Richard. "African Combat Diary." Air Classics, XVI (December 1980), 14-26; XVII (January 1981), 14-17, 76.

Smith, Gene. "B-17 Ferry Flight." Air Progress, XL (March 1978), 24+.

Spenser, Jay P. "The Narrow Margin." Airpower, XIV (January 1984), 10-23.

Stoddard, Robin. "Flying the B-17 'Sally B.'" Air Force Magazine, LXIV (September 1981), 98-99+.

Streetly, Martin. "Aircraft of 100 Group." Scale Models, XI (May 1980), 260-262.

"The 'Suzy-Q.'" Skyways, II (March 1943), 24-29.

"Task Force: A Big Bomber Learns Its Job." Life, XII (April 6, 1942), 62-69.

Thompson, Scott. "The Story of Flying Fortress 44-83525." Air Classics, XVIII (May 1982), 30-39, 74-79.

Thorp, Walter N. "Baby Reborn." Air Classics, XXV (March 1989), 34-45. "Shoo Shoo Baby."

Umstead, S. M. "Testing the World's Biggest Bomber." Popular Mechanics, LXXXVI (November 1941), 28-31+.
Abridged in Science Digest, X (December 1941), 26-28.

"Uncle Sam's Flying Fortress." Popular Mechanics, LXX (September 1938), 329-336.

Van Pelt, Charles B. "An Airplane You Can Trust." American History Illustrated, (October 1969), 22-31.

Wolfe, C. S. "Peacetime Combat for the Flying Forts." Pegasus, XXVI (November 1957), 6-9.

(Douglas) B-18 Bolo

"A Classic Revisited." Wings, VII (June 1977), 28-35.

Close, Winton R. "The B-18: A Reminiscence." In: Robin Higham and Carol Williams, eds. Flying Combat Aircraft of the USAAF-USAF. Manhattan, Kans.: Sunflower University Press, 1981. pp. 110-113.
Reprinted in Aerospace Historian, XXIX (June 1982), 90-92.

Goodall, Geoff. "Bolos Down Under." Air Classics, XVII (October 1981), 20-23.

Johnsen, Frederick A. "The Great Bolo Boondoggle." Wings, III (August 1973), 8-19.

Trimble, Robert L. "Bolo: America's Forgotten Warrior." Air Classics, XVI (May 1980), 22-33, 62-67.

(Douglas) B-23 Dragon

Bowers, Peter M. "Reluctant Dragon." Airpower, X (November 1980), 28-36, 38-41.

"Daly's Dragon." Aeroplane Monthly, VIII (January 1980), 19-20.

O'Leary, Michael. "A Dragon's Last Flight." Air Classics, XXII (April 1986), 34-43.

(Consolidated) B-24 Liberator
(see also PB4Y)

Book/Document/Report Literature:

The B-24 Liberator. Appleton, Wisc.: Aviat Publications, 1977. 146p.

Birdsall, Stephen. The B-24 Liberator. Famous Aircraft Series. New York: Arco, 1968. 64p.

____. The B-24 Liberator in Action. Aircraft, no. 21. Warren, Mich.: Squadron/Signal Publications, 1975. 50p.

____. Log of the Liberators: An Illustrated History of the B-24. Garden City, N.Y.: Doubleday, 1973. 340p.

Blue, Allan G. The B-24 Liberator: A Pictorial History. New York: Scribners, 1977. 223p.

Bowman, Martin. B-24 Liberator, 1939-1945. Chicago, Ill.: Rand McNally, 1980. 124p.

Freeman, Roger A. Consolidated B-24J Liberator. Aircraft in Profile, no. 19. Windsor, Eng.: Profile Publications, 1965. 12p.

Hallstead, William F. Ghost Plane of Blackwater. New York: Harcourt, Brace, Jovanovich, 1974. 156p.
B-24 lost over South Carolina in 1944.

McClendon, Dennis E. "'The Lady Be Good': Mystery Bomber of World War II. New York: John Day, 1962. 192p.

Munday, E. A. USAAF Heavy Bomb Group Markings and Camouflage, 1941-1945: Consolidated B-24 Liberator. Berkshire, Eng.: Osprey Publications, 1972.

Sheehan, Susan. A Missing Plane. New York: G. P. Putnam, 1986. 200p.
B-24 # 42-41081 lost over New Guinea.

Periodical/Anthology Literature:

"B-24 Nose Modifications." Air Classics, VII (September 1971), 8-16.

"Battle Log of the Liberators." Popular Mechanics, LXXX (September 1943), 28-32.

Blue, Allan G. "LB-30s in Latin America: A Forgotten Chapter in the History of the Liberators in World War II." American Aviation Historical Society Journal, XV (Spring 1970), 27-29.

Boone, Andrew R. "Liberator, Our Consolidated B-24 Bomber, Carries a Promise of Deliverance to the World." Popular Science, CXLII (May 1943), 86-91+.

Carigan, William. "The B-24 Liberator: A Man's Airplane." Aerospace Historian, XXXV (Spring 1988), 11-25.

Chaplain, Donald G. "Return of a Gallant Lady." Airman, XIII (October 1969), 24-29.

"The Consolidated B-24." Aeronautics, VI (February 1945), 43-46.

"The Consolidated Liberator." The Aeroplane, LII (May 12, 1944), 530-534.

"The Consolidated-Vultee B-24 Liberator." US Air Services, XXVIII (June 1943), 30+.

"The Desert Gives Up Its Secret: The 17-Year Mystery of 'The Lady Be Good.'" Life, XLVIII (March 7, 1960), 20-27.

"Evolution of the Liberator." Skyways, II (November 1943), 34-38+.

Famme, J. H. "Design Analysis of the Consolidated B-24 Liberator." Aviation, XLVI (July 1945), 121-143.

____. "Design Details of the Liberator, B-24." Automotive and Aviation Industries, LXXXIX (December 15, 1943), 28-32.

Fleet, Reuben H. "Building the B-24." Aviation, XLI (January 1942), 66-69+.

"Flying the Consolidated Liberator." Air Enthusiast, II (August 1971), 87+.

Forbes, Esther H. "Final Fixings for Liberators." US Air Services, XXIX (June 1943), 30+.

Fritsche, Carl H. "B-24 Liberator." In: Robin Higham and Carol Williams, eds. Flying Combat Aircraft of the USAAF-USAF. Ames, Iowa: Iowa State University Press, 1978. pp. 45-58.

Granholm, Jackson. "Those People Could Shoot." World War II, III (Winter 1988), 26-33.
B-24 gunners.

Gwinn, J. M., Jr. "B-24s in Quantity Production." Aviation, XLI (July 1942), 98-103, 186-190.

Hanna, William. "Ordeal of 'The Lady Be Good.'" American History Illustrated, XVI (November 1981), 8-15, 49.

Hendrix, Lin. "Requiem for a Heavy Weight." Wings, VIII (February 1978), 20-37, 52.

Henney, Bill. "Log of a Liberator." Skyways, III (September 1944), 34+.
The "Ginny."

Holder, William G. "Epitaph to the 'Lady': 30 Years After." Air University Review, XXIV (March-April 1973), 41-51.
"The Lady Be Good."

____ and Clifford Glassmeyer. "B-24, the Liberator." Aviation Quarterly, V (Fall 1979), 288-304.

"In Focus--The B-24D: A Pictorial." Military Journal, I (January-February 1977), 13-16.

"Two Missions: Lieutenants W.V. Core and M. K. Harms Describe the Last Flight of Their B-24 Liberator, 'The Chamber Maid.'" New Yorker, XX (January 6, 1945), 45-52.

Johnsen, Frederick A. "A Lady With a Hundred Faces." Air Classics, VII (September 1971), 8-15.

____. "Liberator Logbook." Wings, XV (February 1985), 24-33.

Koch, R. W. "Mystery of the Saline Valley Liberator." Air Classics, XV (June 1979), 31-35.

LaRoue, Jim. "A B-24 Liberator Comes Home." Michigan History, LXXII (March-April 1988), 24-31.

"The Liberator Family." Flying, XXXII (October 1945), 50-54+.

Little, W. E. "Bridge-Busting Liberator." Air Progress Aviation Review, V (June 1981), 8-21.

Marmo, Richard. "Those Garish, Gaudy Liberators." Air Classics, IX (February 1973), 24-31, 48.

Merrick, Ken and Pat Lloyd. "Tribute to the Liberator." Scale Models, VIII (February 1977), 72-75.

Morgan, Frederick W. "'The Lady Be Good.'" Airman, XIV (August 1970), 39-40.

Newman, G. S. "Liberators by Consolidated-Vultee." Aviation, XLII (July 1943), 146-151+.

Oughton, James D. "Liberator: Introduction to War." Aircraft Illustrated, VI (October 1973), 394-399.

Paule, J. M. "The Secret of the Sahara." RAF Flying Review, XV (January 1960), 13-15.
"The Lady Be Good."

Reynolds, George A. "Final Approach." Air Classics, XXI (October 1985), 22-26.
By a 458th Bomb Group (H) B-24.

Rust, Kenn C. "The Big Fighters." RAF Flying Review, XVI (June 1961), 30-31+.

Sheehan, Susan. "A Missing Plane." New Yorker, LXII (May 12-26, 1986), 46-48+, 43-44+, 39-40+.
Later published as the book cited above.

Smith, Ernest L. "Night Flight: Training in the B-24." Aerospace Historian, XXX (September 1983), 163-164.

Taylor, Henry A. "Flying the Consolidated Liberator." Air International, II (February 1972), 87-90.

Terry, Thursel Q. "The Last Flight of the 'Cactus Kid.'" Air Force Magazine, LIV (April 1971), 76-80.

"The Twin-Tailed Time Bomb." *Air Classics*, V (April 1969), 16-26.

"Versatile Bomber." *Flying*, XXX (July 1943), 28-29.

(North American) *B-25/PBJ Mitchell*

Book/Document/Report Literature:

McDowell, R. Ernest. *B-25 Mitchell in Action*. Aircraft, no. 34. Warren, Mich.: Squadron/Signal Publications, 1978. 49p.

Mizrahi, Joseph V. *The North American B-25 Mitchell*. North Hollywood, Calif.: Challenge Publications, 1965. 48p.

North American B-25 Mitchell. Appleton, Wis.: Aviat Publications, 1972. 84p.

Scutts, Jerry. *B-25 Mitchell at War*. Osceola, Wis.: Motorbooks International, 1985. 143p.

United States. Army Air Forces. Training Division. *Pilot Training Manual for the Mitchell Bomber, B-25.* AAF Manual 51-126-4. Washington, D.C., 1944.

Wagner, Ray. *North American B-25A/G Mitchell*. Aircraft in Profile, no. 59. Windsor, Eng.: Profile Publications, 1966. 12p.

Periodical/Anthology Literature:

Andrews, Hal. "Naval Aircraft: PBJ Mitchell." *Naval Aviation News*, (March-April 1986), 20-21.

Avery, Norman L. "North American's B-25 Transports." *American Aviation Historical Society Journal*, XXII (Summer 1977), 136-141.

____. "The Whisky Epxress." *American Aviation Historical Society Journal*, XVIII (Fall 1973), 192-193.

"B-25 Flying Cannon." *Air Classics*, III (July 1967), 24-31.

"The B-25 Mitchell." *Aero Digest*, XLV (May 1, 1944), 76-78.

Beavers, Jim. "AAF's Flying Artillery: The 75mm Baker Two Five." *Air Force Magazine*, LIX (April 1976), 65-69.

"Beginning of the 'N.'" *Flying Safety*, X (November 1954), 16-18.

Bry, A. J. "Flying Cannon." *Air News*, VI (February 1944), 46-47.

Calvert, Denis J. "The Mitchells of 'Hanover Street.'" *Air Classics*, XV (June 1979), 36-47.
Those appearing in the motion picture of that name.

Davisson, Bud. "Baker Two Five." Air Progress, XLVII (June 1985), 70+.

____. "The Remarkable Mitchell." Air Progress, XLIV (March 1982), 40+.

"Flying Cannoneers." Popular Mechanics, LXXXI (March 1944), 34-37+.

Gault, Owen. "Anatomy of a Killer: The Mitchell." Air Classics, VIII (December 1972), 32-45, 56-59.

Hansen, Charles J. "Design Analysis of the North American B-25 Mitchell." Aviation, XLIV (March 1945), 119-142.

____. "Evolution of the Mitchell." Skyways, III (May 1944), 19-24+.

Houston, N. S. "North American B-25C and D Bomber." Automotive and Aviation Industries, XC (February 1-15, 1944), 26-31+, 34-37+.

Janczarek, Ted. "North American's Classic Bomber." In: Michael O'Leary, ed. Air Progress Pilot Reports. Canoga Park, Calif.: Challenge Publications, 1979. pp. 12-16.

Keeler, H. L. "The B-25 Mitchell." AirTech, III (November 1943), 21-33.

Lardner, John. "Seven Deadly Young Americans: On Board a B-25 That Smashed a Jap Base at Lae." Saturday Evening Post, CCXV (November 14, 1942), 19+.

Larkins, William T. "U.S. Civil B-25s." American Aviation Historical Society Journal, XXIII (Spring 1978), 29-35.

Larsen, Jim. "'Executive Sweet.'" Air Classics, XIV (August 1978), 50-55.

McDonald, Chuck. "The Invisible Task Force." American Aviation Historical Society Journal, XXVII (Winter 1982), 258-261.
B-25 restoration.

Matzinger, Keith R. "B-25 Mitchell." In: Robin Higham and Carol Williams, eds. Flying Combat Aircraft of the USAAF-USAF. Ames, Iowa: Iowa State University Press, 1978. pp. 59-64.

"The North American Aircraft, B-25." Air Classics, IV (April 1968), 28-44.

"The North American B-25 Mitchell." Air Pictorial, XV (August 1953), 238-239.

"The North American Mitchell, B-25H." Flying, XXXV (December 1944), 59-63.

Picinich, R. C. "Aboard a Mitchell Bomber." Aero Digest, XLII (May 1943), 114-115.

Redman, Rod. "B-25 Checkout." Air Progress, XXXIX (September 1977), 31-33.

"75mm Cannon Gives Mitchell Formidable Firing Power." Air News, I (December 6, 1943), 10-11.

Thomas, Arthur J. "B-25 Mitchell." In: Robin Higham and Abigail T. Siddall, eds. Flying Combat Aircraft of the USAAF-USAF. Ames, Iowa: Iowa State University Press, 1975. pp. 3-10.

Trimble, Robert L. "Deadly Mitchells." Air Progress Aviation Review, VI (Summer 1982), 24-48, 76.

Wright, Roger. "Mitchell Flight." Aircraft Illustrated, XV (January 1982), 20-21.

(Martin) B-26 Marauder
(see also A-26 Invader)

Book/Document/Report Literature:

B-26 Marauder. Appleton, Wis.: Aviat Publications, 1977. 130p.

Birdsall, Steven. B-26 Marauder in Action. Aircraft, no. 50. Carrollton, Tex.: Squadron/Signal Publications, 1981. 50p.

Fox, Anne and Eugene Porter. B-26, March 1939-May 1944. ATSC Historical Study, no. 11. Wright Field, Ohio: USAAF Air Technical Service Command, 1944. 51p.

Francis, Devon E. Flak Bait: The Story of the Men Who Flew the Martin Marauders. New York: Duell, Sloan and Pearce, 1948. 331p.

Freeman, Roger A. B-26 Marauder at War. New York: Scribners, 1978. 192p.

Havener, Jack K. The Martin B-26 Marauder. Blue Ridge Summit, Pa.: TAB, 1988. 272p.

Moore, Carl H. World War II: Flying the B-26 Marauder Over Europe. Blue Ridge Summit, Pa.: TAB, 1980. 176p.

United States. Army Air Forces. 12th Air Force. Getting Better Results With Marauder Bombers: Where the Bombs Fell in 434 Rail Bridge Attacks. N.p.: 941st Engineer Battalion, 1945. 29p.

Wagner, Ray. Martin B-26B/C Marauder. Aircraft in Profile, no. 112. Windsor, Eng.: Profile Publications, 1967. 12p.

Periodical/Anthology Literature:

"B-26 Surprise Package." Flight Magazine, XLVIII (October 1959), 32+.

Besecker, Roger F. "The Civil B-26 Marauder." American Aviation Historical Society Journal, XXV (Fall 1980), 228-233.

Campbell, C. "Iron in Her Soul: The B-26." Flying, LI (November 1952), 18-19.

Conley, Douglas C. "B-26 Marauder." In: Robin Higham and Abigail T. Siddall, eds. Flying Combat Aircraft of the USAAF-USAF. Ames, Iowa: Iowa State University Press, 1975. pp. 11-17.

Duval, G. R. "Martin B-26." Scale Models, IV (May 1973), 331-336.

"Elegant Widow Maker." RAF Flying Review, XVIII (December 1963), 24-27.

Engel, Leonard. "The Truth About the B-26." Air Trails, XXII (July 1944), 22-23+.

"Flak Bait." Air Classics, VI (June 1970), 22-34.

Friedman, William S. "The Army Asked for a Miracle, the Answer Was the B-26." Popular Science, CXLII (March 1943), 70-75.

Frisbee, John L. "Marauders at Midway." Air Force Magazine, LXIX (April 1986), 140+.

Giant, Robert M. "They Said It Was Too Hot to Fly: The B-26 Marauder." Popular Mechanics, LXXXI (May 1944), 61-64+.

Havener, Jack K. "Ingot to Scrap Pile: Full Cycle of B-26-B-50 Marauder." Aerospace Historian, XVI (Summer 1969), 41-43+.

"The Martin B-26." Air Classics Annual, III (1966), 66+.

"The Martin B-26 Marauder Medium Bomber." Aero Digest, XLVI (July 15, 1944), 79-82.

"Martin Marauder." RAF Flying Review, XVIII (September 1963), 24-27.

Mendenhall, Charles. "Anatomy of a Marauder." Air Classics, XVII (May 1981), 62-69, 77-82.

Miller, R. F. "For Loss of a Bolt." Flying Safety, XI (June 1955), 20-22.

Mitchell, Kent A. "The Tandem Landing Gear Martin XB-26H." American Aviation Historical Society Journal, XXIII (Spring 1988), 23-25.

Morgan, Len. "The Last B-26." Flying, CXII (November 1985), 26-34.

Mundey, Eric. "The B-26 Marauder." Military Journal, II (Winter 1978-1979), 26-30.

"The Murderous Marauder." Wings, XI (October 1981), 25-31.

O'Leary, Michael. "Widow Maker." Air Progress Aviation Review, V (Fall 1981), 38-47.

Smith, Harry. "'Hell's Belle II' is Still No. 1." Aerospace Historian, XXXIII (Fall 1986), 199-202.

Smith, M. LeFeure. "Evolution of the Marauder." Skyways, III (November 1944), 18-22+.

Spenser, Jay P. "Flak Bait: Biography of an Intrepid Marauder." Airpower, VIII (September 1978), 36-57.

Stuart, John. "Marauding Veterans: Back From Combat Came the Oldest B-26s, Bearing the Highest Praise." Flying, XXXIII (August, 1943), 90+.

Tannehill, Victor C. "Mediterranean Marauder Markings." American Aviation Historical Society Journal, XXIV (Winter 1979), 55-67.

Ward, Roger and Phelps F. Mace. "The Martin Marauder, B-26." AirTech, VI (January 1945), 17-27.

Wehman, Dwight. "Five Years, Five Thousand Hours." Air Progress, XLVI (January 1984), 61-65.
B-26 restoration.

"The Widow Maker." Air Classics, I (Fall 1964), 16-22.

(Boeing) B-29 Superfortress

Book/Document/Report Literature:

Berger, Carl. B-29: The Superfortress. Ballantine's Illustrated History of World War II. New York: Ballantine Books, 1970. 160p.

Birdsall, Steven. B-29 Superfortress in Action. Aircraft, no. 31. Warren, Mich.: Squadron/Signal Publications, 1977. 48p.

Chant, Christopher F. B-29 Superfortress. Super Profile Series. Osceola, Wisc.: Motorbooks International, 1988. 56p.

Clark, Freddie W. Problems Associated With Deploying the B-29 to the China-Burma-India Theater. Research Paper. Maxwell AFB, Ala.: Air Command and Staff College, Air University, 1982. 47p.

Collison, Thomas. Superfortress is Born: The Story of the Boeing B-29. New York: Duell, Sloan and Pearce, 1945. 218p.

Gurney, Gene. B-29 Story: The Plane That Won the War. Greenwich, Conn.: Fawcett Publications, 1962. 112p.

____. Journey of the Giants: The Story of the B-29 Superfort, the Weapon That Won the War in the Pacific. New York: Coward-McCann, 1961. 280p.

Holley, Irving B. B-29, 1933-1944. ATSC Historical Study, no. 98. Wright Field, Ohio: USAAF Air Technical Service Command, 1945. 52p.

Johnsen, Frederick A. The B-29 Book. Mesa, Ariz.: Champlin Fighter Museum Press, 1978. 28p.

LeMay, Curtis E. and Bill Yenne. Superfortress: The B-29 and American Air Power. New York: McGraw-Hill, 1988.

Lloyd, Alwyn T. B-29 Superfortress in Detail and Scale: Pt. 1, Production Version. Detail and Scale Series, v. 10. Fallbrook, Calif.: Aero Publishers, 1983. 72p.

____. B-29 Superfortress in Detail and Scale: Pt. 2, Derivatives. Detail and Scale Series, V. 25. Fallbrook, Calif.: Aero Publishers, 1986. 72p.

McDonough, Roger. History of the Supply, Maintenance, and Training For the B-29, 1939-1945. AMC Historical Study, no. 141. Wright Field, Ohio: USAAF Air Materiel Command, 1946. 111p.

Mayborn, Mitchell. Boeing B-29 Superfortress. Aircraft in Profile, no. 101. Windsor, Eng.: Profile Publications, 1967. 12p.

Midlam, Donald S. Flight of the "Lucky Lady." Portland, Ore.: Bindfords, 1954. 208p.
Story of one 504th Bomb Group (VH) B-29.

Morrison, Wilbur H. Hellbirds: The Story of the B-29 in Combat. New York: Duell, Sloan and Pearce, 1960. 181p.

Pimlott, John. B-29 Superfortress. New York: Prentice-Hall, 1983. 64p.

Sinclair, William B. The Big Brothers: The Story of the B-29s. San Antonio, Tex.: Naylor, 1972. 132p.

Periodical/Anthology Literature:

Altman, O. L. and C. G. Goor. "Acturial Analysis of the Operating Life of B-29 Aircraft Engines." Journal of the American Statistical Association, XLI (1946), 190-203.

Anderton, David A. "The Biggest Year of My Life." Air Force Magazine, LX (September 1977), 138+.

"Armament Secrets of the Superfortress." Aero Digest, XLVII (October 1, 1944), 55+.

"The B-29 Superfortress." Aero Digest, XLVI (July 1, 1944), 70-75+.

Bachmann, Lawrence P. "The B-29--Modified by Battle." Air Force, XXVIII (April 1945), 6-8+.

"Battleship of the Air: The Superfortress, B-29." Scientific American, CLXXI (September 1944), 121-122+.

Bliven, Bruce. "The B-29." New Republic, CXII (January 1, 1945), 19-21. Abridged in Reader's Digest, XLVI (March 1945), 80-82.

Boesen, Victor. "B-29 Superfortress." Skyways, III (September 1944), 30-33+.

"A Bomber is Born." Flying, XXXIII (October 1943), 108-110.

Bowers, Peter M. "Forgotten Fortress B-29: Building and Testing the Forerunner B-299." Wings, IV (August 1974), 6-25.

____. "Super Bomber: The Boeing B-29." Wings, III (October 1973), 10-39.

Boyle, James M. "This Dreadnought Can Fly." Aerospace Historian, XIV (Summer 1967), 85-92.

____. "How the Superfortresses Paced the Attack Against Japan." Air Force and Space Digest, XLVII (December 1964), 63-69.

Burger, Knox. "Wing Talk: The 'American Maid's' Third Trip to Nagoya." Collier's, CXV (June 9, 1945), 8+.

Calkins, Selby W. "The B-29 Brings in a New Era in Bombing." Popular Science, CXLV (November 1944), 94-99+.

Close, Winston R. "B-29s in the CBI." Aerospace Historian, XXX (March 1983), 6-14.

"Construction Details of the Boeing B-29 Superfortress." Automotive and Aviation Industries, XCI (September 1, 1944), 36-37.

Davis, Luther. "The B-29 and You." Air Force, XXVII (October 1944), 16-18.

"The Deadly B-29, Star of a Mission." New York Times Magazine, (June 24, 1945), 6-7.

Dean, Jack. "By Bombing Alone." Wings, XVIII (June 1988), 10-21.

"Design Details of the Boeing B-29." Aero Digest, L (August 15, 1945), 61-66; LI (October 1, 1945), 51-55.

Hailey, Foster. "This is the Superfortress." In: Norman Carlisle, ed. The Air Forces Reader. Indianapolis, Ind.: Bobbs-Merrill, 1944. pp. 188-190.

Hansell, Haywood S., Jr. "B-29 Superfortress." In: Robin Higham and Abigail T. Siddall, eds. Flying Combat Aircraft of the USAAF-USAF. Ames, Iowa: Iowa State University Press, 1975. pp. 19-28.

Hokanson, Henry T. "The Armament of the Superfortress." Electrical Review, XLVIII (March 1945), 7-12.

____. "How the B-29 Concentrates Its Firepower." Machine Design, XVII (January 1945), 113-116.

Jenstrom, Eino E. "The Last Mission of the 'Eddie Allen." Air Force Magazine, LXIII (August 1980), 94-99.

Johnson, Robert E. "Why the Boeing B-29 Bomber and Why the Wright R-3350 Engine?" American Aviation Historical Society Journal, XXXV (Fall 1988), 174-189.

Karant, Max. "The B-29, Boeing Superfortress." Flying, XXXV (August 1944), 48+.

Keenan, Richard M. "The Aircraft That Won the War: The Last of the Superfortresses." Aerospace Historian, XVII (Spring 1970), 20-27.

____. "The Not-So-Lucky 'Lucky Irish.'" Aero Album, IV (1971), 23-25.

Krimes, Milton R. "From Kansas to Tokyo." Air Force, XXVIII (June 1945), 18-21, 26.

Kucera, J. M. "'Fertile Myrtle' Flies Again." Aeroplane Monthly, VI (November 1978), 579-581.

Lansburgh, L. M. "B-29, Take One." Flying, CVI (March 1980), 22-23.

"Let's Look Inside a B-29." Popular Science, CXLVI (March 1945), 86-88.

Martinez, Marty. "The 'Gray Ghost.'" Combat Crew, XXIII (October 1973), 16-18.

Moore, Robert. "Rebirth of a Veteran." Airman, XVII (December 1973), 17-18.
The "Gray Ghost."

"New Planes Stress Long Range to Carry the War to the Jap Homeland." Aviation News, II (July 31, 1944), 17+.

Phillips, Ed. "Boeing's Battle of Wichita." Airpower, XI (July 1981), 36-41, 44.

Phillips, Ren. "10 Big Years." Boeing Magazine, XXII (October 1952), 6-9.

____. "A Warrior Comes Home." Boeing Magazine, XXIV (December 1954), 10-11.

Powell, Hickman. "The Boeing B-29 Superfortress." Aviation, XLIII (July 1944), 110-111+.

Price, William. "Saga of the Super Bomber." Science Digest, XV (April 1944), 54-58.

Putt, Donald L. "The B-29 Superfortress." Air Force, XXVII (July 1944), 4-8.

Rust, Kenn C. "Bomber Markings of the 20th Air Force." American Aviation Historical Society Journal, VII (Fall-Winter 1962), 165-179+, 257-263.

"Some B-29 Design Features." Automotive and Aviation Industries, XCI (July 15, 1944), 32-34.

"Systematic Production Cycles Speed Superfortresses to Combat." American Machinist, LXXXVIII (December 7, 1944), 83-90.

Walker, Wayne T. "The B-29 Superforts." World War II Magazine, III (April 1974), 6-19.

West, David. "Homebound With the U.S. Air Force." Aeronautics, XXXIII (July 1950), 23-33.

Whittaker, Wayne. "Meet the B-29." Popular Mechanics, LXXXII (August 1944), 8-13.

Wible, John T. "Boeing B-29 Serial Numbers." American Aviation Historical Society Journal, VII (Fall 1962), 180+.

(Consolidated) B-32 Dominator

"Consolidated-Vultee B-32: Design and Construction Details." Aviation, XLIV (September 1945), 168-169.

Harding, Steven and James Long. Dominator: The Story of the Consolidated B-32 Bomber. Missoula, Mont.: Pictorial Histories Publishing Co., 1984. 64p.

Johnsen, Frederick A. "Last and Unluckiest of the Hemisphere Bombers: The Consolidated B-32 Dominator." Wings, IV (February 1974), 8-17.

Pertinent Facts About the B-32." Automotive and Aviation Industries, XCIII (August 1, 1945), 24-25.

Ruliffson, Willard S., with Robin Higham. "B-32 Dominator." In: Robin Higham and Carol Williams, eds. Flying Combat Aircraft of the USAAF-USAF. Ames, Iowa: Iowa State University Press, 1978. pp. 75-81.

"A Tougher Liberator." Flying, XXXIII (November 1943), 45+.

Whittaker, Wayne. "Here's the B-32, Our Newest Superbomber." Popular Mechanics, LXXXIV (September 1945), 12-15+.

Y'Blood, William T. "The Second String." American Aviation Historical Society Journal, XIII (Summer 1968), 80-92.

(Lockheed) B-34/B-37 Lexington
(see also PV-1 and C-60)

"The Lockheed B-34 Lexington." Flight, XXXVIII (May 16, 1946), 492-493.

(Convair) B-36 Peacemaker

Book/Document/Report Literature:

Green, Murray. Stuart Symington and the B-36. PhD. Dissertation. Washington, D.C.: American University, 1960.

Jacobsen, Meyers K. B-36 Peacemaker in Action. Aircraft, no. 42. Carrollton, Tx.: Squadron/Signal Publications, 1980. 49p.

Johnsen, Frederick A. Thundering Peacemaker: The B-36 Story in Words and Pictures. Tacoma, Wash.: Bomber Books, 1978. 29p.

Johnson, Elliott L. The B-36 Controversy: A History of Air Force-Navy Rivalry Over Strategic Bombardment. MS Thesis. Madison, Wis.: University of Wisconsin at Madison, 1961.

Loegering, J. B. The B-36:Design, Procurement, and Materiel Development Through 1950. Offutt AFB, Neb.: History Office, Strategic Air Command, 1951. 60p.

Scott, Jessie C. Conclusion of the B-36 Airplane Phase-Out, July 1957-May 1959. AFLC Historical Study, no. 448. Kelly AFB, Tex.: History Office, San Antonio Air Materiel Area, 1960. 34p.

Stringer, D. L. Roles, Missions, and Politics: The U.S. Air Force and the B-36, 1945-1950. Research Paper. Maxwell AFB, Ala.: Air Command and Staff College, Air University, 1987. 39p.

United States. Air Force. Air Materiel Command. B-36, April 1941-May 1948. AMC Historical Study, no. 188. Wright-Patterson AFB, Ohio: USAF Air Materiel Command, 1949. 266p.

____. Congress. House. Committee on Armed Services. Investigation of B-36 Program: Hearings. 81st Cong., 1st sess. Washington, D.C.: GPO, 1949. 664p.

____.____.____.____.______:Report. 81st Cong., 1st sess. Washington, D.C.: GPO, 1949. 34p.

Periodical/Anthology Literature:

Argyle, Christopher J. "Convair's Ten-Engined Bomber." Aeroplane Monthly, II (July 1974), 744-750.

"The B-36." Life, XXVI (June 6, 1949), 46-47.

"The B-36: A Procurement Study." Aviation Week, LI (September 26, 1949), 83-86.

"The B-36: Biggest Bomber." New York Times Magazine, (May 15, 1949),8+.

"B-36 Investigation." Air Force, XXXII (October 1949), 7-8.

Baer, H. S. "FISCON: USAF's New Role for Old Bombers." American Aviation, XVII (August 31, 1953), 15-16.

Baldwin, Hanson W. "Postscript on the Revolt of the Admirals: The B-36 Bomber." Time, LXV (February 21, 1955), 28-29.

Broadhurst, Edwin B. "Veteran of the Dry-Run War." Air Force, XXXIX (April 1956), 82-89.

"Building Big Bombers." Interavia, VII (December 1952), 686-689.

"Design Features of Convair's B-36." Aviation Age, XVIII (August 1952), 36-37.

Dodson, J. T. "B-36: Global Bomber." Flying, XLV (July 1949), 16-17+.

Duffy, K. E. "The Convair GRB-36D." Scale Models, XII (September 1981), 452-455.

Erickson, B. A. "It's Easy to Fly the B-36." Flying, XLVIII (April 1951), 13-15+.

"Exposing the 'Milk Wagon' Myth." Air Force, XXXII (March 1949), 11-13, 47-48.

Gaston, James. "Salute to the Peacemaker." Airman, XV (September 1971), 19-20.

Goldsworthy, Harry E. "The B-36 Peacemaker." Aerospace Historian, XXX (December 1983), 261-267.

Griswold, W. S. "Remember the B-36?" Popular Science, CLXXIX (September 1961), 98-101.

Hinckley, H. W. "The B-36 Story." Aviation Age, XVIII (August-September 1952), 33-35, 34-35.

Hotz, Robert. "New B-36 Performance Revealed at Probe." Aviation Week, LI (August 22, 1949), 13-15.

____. "New B-36 to Give USAF Greater Range." Aviation Week, XLIX (October 18, 1948), 12-14.

____. "What the B-36 is Costing the USAF." Aviation Week, LI (September 12, 1949), 37-38.

____. "Why the B-36 Was Made USAF Top Bomber." Aviation Week, LI (August 15, 1949), 13-16.

Jacobsen, Meyers K. "Design and Development of the SB-36." American Aviation Historical Society Journal, XV (Winter 1970), 224-235; XVI (Summer 1971), 124-130.

____. "Double Trouble." Aerospace Historian, XXIII (December 1976), 207-213.

____. "Names and the B-36." American Aviation Historical Society Journal, XVI (Fall 1971), 186-187.

____. "Peacemaker." Airpower, IV (November 1974), 8-27, 50-55.

____. "The Red-Tailed Bandits of the 7th Bomb Wing." American Aviation Historical Society Journal, XXIV (Spring 1979), 19-40.

Karten, Dave. "The Bomber That Never Bombed." Aerospace Historian, XIII (Winter 1966), 161-163.

____. "Requiem for the B-36." Flying, LXIV (May 1959), 34-35+.

____. "Twilight for the Peacemaker." Airman, I (November 1957), 30-31.

Kay, Tom. "Hell Comes to a B-36." Skyways, X (March 1951), 8-9, 41.

Koby, Victor. "SAC's Sunday Punch." Canadian Aviation, XXIX (March 1956), 26-27, 49-51.

Long, H. A. "B-36 is a Tanker." Freeman, I (March 26, 1951), 393-396. Inadequacies.

Loosbrock, John F. "Inside the B-36." Popular Science, CLVI (June 1950), 144-146.

Lorge, R. J. "Flying a Really Big Plane, the B-36." Science Digest, XXXII (July 1952), 32-35.

McLarren, Robert. "The Convair B-36." Aero Digest, LXVIII (January 1954), 30-32+.

Miller, Jay and Roger Cripliver. "B-36: The Ponderous Peacemaker." Aviation Quarterly, IV (Fall 1978), 332-370.

Muller, William W. "The B-36 Peacemaker." Navigator, XXIX (Winter 1982), 9-10.

Naugle, R. G. "The Big, Bad B-36." Flying, LXIX (July 1961), 38-39+.

"Never a Shot in Anger." Combat Crew, IX (April 1959), 8-9.

Ott, Richard L. "B-36 Remembered." Combat Crew, XXXV (March 1985), 4-5.

Philips, Frederic M. "B-36 Deterrence: Yesterday's and Tomorrow's." Air Force, XXXIX (May 1959), 87-91.

Rector, W. R. "B-36 Operation at 40° Below." Combat Crew, VI (November-December 1955), 4-7, 24-28.

Reed, B. G. "Building the World's Biggest Bombers: Convair's Fort Worth Plant." Automotive Industries, CVI (April 1, 1952), 38-41+.

Saxon, H. "B-36: Fortissimo or Fiasco?" Aviation Week, LI (July 18, 1949), 30+.

Sebolt, R. C. "Temperature Testing the B-36." Aero Digest, LXIII (July 1951), 20-21.

"30 Hours Airborne." Flight, LXI (February 8, 1952), 153-154.

Timmins, Edward A. "I Flew 36 Hours in a B-36." Popular Mechanics, XCIV (September 1950), 98-102+.

Trimble, Robert L. "Aluminum Overcast." Air Classics, XX (November 1984), 14-21.

____. "Big Stick From Texas." Air Progress Aviation Review, III (Fall 1979), 46-61.

Tyson, Newby G. "10,000 Hours in the B-36." Combat Crew, VII (September 1956), 22-23.

Vandenberg, Hoyt S. "B-36: Superplane or Sitting Duck?" U.S. News and World Report, XXVII (July 1, 1949), 38-40.

"What the High-Flying B-36 Taught Convair." Aviation Week, LVI (August 11, 1952), 21-22.

"The XB-36 Super Bomber." Popular Science, CXLVIII (May 1946), 96-97.

B-37 see B-34/B-37 Lexington

(North American) B-45 Tornado

"At the Controls of the B-45." Flight, LXVII (April 22, 1955), 523-525.

"B-45 Tornado." Joint Services Recognition Journal, VIII (December 1953), 271-272.

Bodie, Warren H. "Breaching the Walls of Fortress Europe: B-45A Tornado." Airpower, III (September 1973), 36-55.

Gunston, William T. ("Bill"). "North American Tornado." Aeroplane Monthly, X (December 1982), 632-637.

"North American XB-45: First of the Four-Jet Bombers." Air Force, XXX (June 1947), 44-46.

Sorrells, A. R. "Salute to an Old Warrior." Skyline, XVI (Fall 1958), 16-19.

(Boeing) B-47 Stratojet

Book/Document/Report Literature:

Bagwell, Margaret C. B-47 Production Program, 1949-1955. AMC Historical Study, no. 257. Wright-Patterson AFB, Ohio: USAF Air Materiel Command, 1954. 72p.

____. XB-47, B-47, November 1943-February 1950. AMC Historical Study, no. 201. Wright-Patterson AFB, Ohio: USAF Air Materiel Command, 1950. 128p.

Bassett, Cecil A. Phase Out of the B-47: A Mistake? Research Paper. Maxwell AFB, Ala.: Air Command and Staff College, Air University, 1966. 48p.

Davis, Paul E. The Role of SAC and the B-47 in Limited War. Research Paper. Maxwell AFB, Ala.: Air Command and Staff College, Air University, 1965. 39p.

Lloyd, Alwyn T. B-47 Stratojet in Detail and Scale. Detail and Scale Series, v. 18. Fallbrook, Calif.: Aero Publishers, 1986. 72p.

Loegering, J. R. The B-47: Design, Procurement, and Materiel Development, 1943-1951. Offutt AFB, Neb.: History Office, Strategic Air Command, 1952. 96p.

____. The B-47: Modification, Armament, Equipment, and Training, July-December 1951. Offutt AFB, Neb.: History Office, Strategic Air Command, 1952. 45p.

Moulton, Harland B. The SAC Conversion Program to B-47 Wings Through June 1953. Offutt AFB, Neb.: History Office, Strategic Air Command, 1953. 121p.

Ott, Franklin D. The B-47 Aircraft in Limited or Counterinsurgency Warfare. Research Paper. Maxwell AFB, Ala.: Air Command and Staff College, Air University, 1965. 61p.

Peacock, Lindsay. Boeing B-47 Stratojet. Osceola, Wisc.: Motorbooks International, 1987. 200p.

United States. Air Force. Strategic Air Command. Select B-47 Operational Exercises, January-June 1953. Offutt AFB, Neb.: History Office, Strategic Air Command, 1954. 61p.

Weeks, Allen R. B-47 Pilot Instrument Proficiency and Training in the Strategic Air Command. Research Paper. Maxwell AFB, Ala.: Air Command and Staff College, Air University, 1962. 59p.

Periodical/Anthology Literature:

Adams, H. L. "Structural Testing the Stratojet." Aero Digest, LXII (June 1951), 20-21+.

Alexander, John V. and Tom Meyer. "B-47." Aerospace Historian, XXVIII (March 1981), 32-37.

Arthur, Dennis S. "Flying the B-47 is a Three-Man Job." Combat Crew, XI (February 1961), 8-10.

"Assembly Line Magic." Boeing Magazine, XXV (December 1955), 6-7.

Atkinson, John E. "Back in the Race." Boeing Magazine, XXIV (August 1954), 10-11, 14.

"B-47 Swings Air War Balance to Offense." Life, XXIX (December 4, 1950), 157-160.

Bartes, Darrell. "Practical Picture Book." Boeing Magazine, XXVII (May 1957), 10-11.

____. "Ten Years of Swept-Wings." Boeing Magazine, XXVIII (January-February 1958), 10-11.

"The Biography of a Bomber." Air Force, XXXIII (July 1950), 28-31.

"Birth of a Bomber." Aviation Age, XIX (June 1953), 149-150.

"Boeing B-47 Stratojet." Aero Digest, LXI (November 1950), 22-25+.

"Boeing B-47 Stratojet Opens Transonic Era in Bomber Design." Aero Digest, LXI (November 1950), 22-25+.

"Boeing B-47E in the Air." Flight, LXVI (August 20, 1954), 239-243.

Boghosian, S. Samuel. "B-47 Goes Home." Air Classics, XXIII (January 1987), 24-27.

Bollinger, L. E. "More Thrust for the B-47." Flying, LVI (May 1955), 45+.

Butterwick, R. G. "The Big Wind." Honeywell Flight Line, VI (February 1955), 1-3.

Campbell, Ernest E. "B-47 Refueling Know-How." Combat Crew, V (April 1955), 26-29.

Davey, Chris. "Boeing B-47 Stratojet." Aviation News, IX (September 12, 1950), 7-10.

"Development of a Jet Bomber." Aviation Age, XIX (June 1953), 151-157.

Evans, Richard E. "Operational Angles on the B-47." Flight, LXXI (January 25, 1957), 115-118.

Findley, E. N. "Wichita and Points West." US Air Services, XXXVIII (September 1953), 14-15, 18-22.

"Flying the Atom Bomber." RAF Flying Review, VIII (August 1953), 29-31.

Foote, J. R. "Jet Streamers." Boeing Magazine, XXVII (July 1957), 8-9.

Fornasaro, John B. "The Big Difference is Performance." Boeing Magazine, XXIII (June 1953), 5-7, 14.

____. "A Test Pilot Reports on the B-47." Flying Safety, IX (April 1953), 6-8.

Fraser, John. "I Fly the World's Fastest Bomber: The Boeing B-47 Stratojet." Popular Science, CLV (November 1949), 139-143.

Gamiem, T. E. "Growth of a Bomber." Boeing Magazine, XXVI (August 1956), 6-7.

Gray, Jay O. and Kenneth G. McGraw. "Strategic Sleigh Ride." Combat Crew, VII (November 1956), 24-29.

Green, William. "Stratojet." RAF Flying Review, X (March 1955), 30-33.

Gunston, William T. ("Bill"). "Looking Back at the B-47B." Aircraft Illustrated, VI (September 1973), 358-359.

Hamilton, W. H. "It's in the Book." Flying Safety, XII (April 1956), 12-13.

Higgins, H. C. "B-47 Takeoff Technique." Combat Crew, VII (April 1957), 19-21.

"Last of the B-47 Line." Boeing Magazine, XXVI (November 1956), 14-15.

Lennox, William M. "Stratojet." Airpower, IV (January 1974), 32-45.

McCoy, M. N. W. "Breaking in the Stratojet." Air Force, XXXIX (April 1956), 90-96.

McDaniel, Dennis K. "The B-47 From the Back Seat, SAC From the Bottom." Aerospace Historian, XXXIV (March 1987), 39-45.

McNay, Curtis E. "B-47 Flight Control Problems." Combat Crew, XI (March 1961), 12-13.

McSurely, Alexander. "Air Force Pushes Building of B-47 Jet Bomber." Aviation Week, LIV (April 16, 1951), 13-14.

Miller, A. "Eyeball in the Milk Bottle: Modification of B-47 Medium Bomber." Industrial Quality Control, XVII (August 1960), 8-9.

Mizrahi, Joseph V., ed. "B-47 Stratojet." Wings, XI (August 1981), 10+.

Morrison, Joseph C. "30 in 8." Combat Crew, XV (August 1964), 16-18.

"Not for the Records." Boeing Magazine, XXIII (June 1953), 3-4.

Peck, Earl G. "The B-47 Stratojet." Aerospace Historian, XXII (Summer 1975), 61-64.
Reprinted in Robin Higham and Carol Williams, eds., Flying Combat Aircraft of the USAAF-USAF (Ames, Iowa: Iowa State University Press, 1978), pp. 82-88.

Phillips, Charles. "B-47s Abroad." Boeing Magazine, XXIII (November 1953), 10-11.

"Photo Review." Air Pictorial, XIV (July 1952), 209-211.

Pitts, T. C. "Building the Stratojet." Aviation Age, XIX (February 1953), 24-26.

Rechenschieber, S. R. "The Mighty B-47." Flying, XLV (August 1949), 15-17+.

Robbins, R. M. and W. H. Cook. "Boeing Builds Swept-Wing Bombers: Flight Characteristics of the Boeing B-47 Stratojet." Interavia, VIII (January 1953), 32-34.
Reprinted in Aviation Week, LIV (April 30, 1951), 25-68.

Ross, Edwin C. "The Sad, Sad Saga of 'Snafu 72.'" Combat Crew, XV (October 1964), 22-25.

Schraeder, Richard K. "Stratojet." Air Classics, XXIV (February 1988), 30-40, 78-79.

Simmons, William M. "Stratojet Checkout." Air Training, V (September 1955), 19-24.

Soth, Martin W. "No Dodo Yet." Airman, XII (December 1968), 44-46.
WB-47 configuration.

"A Successful Jet Bomber." Aeronautics, XXIX (January 1954), 42-43.

"Testing the Stratojet, XB-47." Aero Digest, LVII (August 1948), 26-27.

Weber, Kenneth L. "Flying the Forty-Seven." MATS Flyer, X (August 1963), 6-8.

Wecker, Jack. "The Blue-Plate Passenger." Boeing Magazine, XXIII (August 1953), 6-7.

____. "Shaking Down the Stratojet." Boeing Magazine, XXV (January 1955), 3-5.

____. "Target for Today." Boeing Magazine, XXIII (April 1953), 3-5.

Woodruff, Walter W. "Navigator Heritage: The B-47." Navigator, XXXIII (Fall 1986), 17-18.

Y'Blood, William T. "B-47 Odds and Ends." American Aviation Historical Society Journal, XIII (Fall 1968), 204-206.

Young, Barnett B. "No Boom for Error." Combat Crew, VIII (April 1958), 12-15.

Zedler, Donald L. "Stratojet Field Check." Combat Crew, VI (April 1956), 4-9.

(Boeing) B-50 Superfortress

"Boeing B-50D Superfortress." Aviation Age, XVI (August 1951), 28-29.

Bourgeois, Harold. "'Lucky Lady's' Flight." Aviation Quarterly, VI (Summer 1980), 180-191.

Bowers, Peter M. "Boeing B-50: Late Model B-29 Version." Air Classics, IV (August 1968), 8-17.

Castle Air Museum Foundation. "Return of the B-50." Air Classics, XVII (June 1981), 38-39.

Eastman, James N., Jr. "Flight of the 'Lucky Lady II.'" Aerospace Historian, XVI (Winter 1969), 9-11+.

Fouquet, Douglas M. "Last of the Superforts." Air Force, XXXVIII (August 1955), 46, 49-52.

"Globe Hop Sets B-50's New Role." Aviation Week, L (March 14, 1949), 14-15.
"Lucky Lady II."

"Mission to Somewhere." Boeing Magazine, XXII (May 1952), 11-13.

Silsbee, Nathaneil F. "To Any Spot on Earth: Flight of the 'Lucky Lady II.'" Aero Digest, LVIII (April 1949), 19-21+.

"Twilight of a Giant Warrior." Airman, IX (June 1965), 16-18.

(Boeing) B-52 Stratofortress

Book/Document/Report Literature:

Arnold, Francis W., et al. Strategic Air Command B-52 Navigator/Bombardier Training. Research Paper. Maxwell AFB, Ala.: Air Command and Staff College, Air University, 1975. 79p.

Bagwell, Margaret C. XB-52, November 1945-November 1948. AMC Historical Study, no. 189. Wright-Patterson AFB, Ohio: USAF Air Materiel Command, 1949. 34p.

Barker, Joseph K. B-52 Combat Crew Effectiveness. MBA Thesis. Chicago, Ill.: University of Chicago, 1965. 33p.

Bates, Albert R. An Analysis of B-52 Combat Crew Flying Training. Research Paper. Maxwell AFB, Ala.: Air Command and Staff College, Air University, 1965. 51p.

Beck, Douglass D., et al. B-52 Operations in Southeast Asia. Research Paper. Maxwell AFB, Ala.: Air Command and Staff College, Air University, 1969. 182p.

Birkett, Robert C. Increased B-52 Alert Capability, Research Report. Maxwell AFB, Ala.: Air War College, Air University, 1973. 50p.

Blanton, John W. The Role of the B-52 in Limited War, Research Paper. Maxwell AFB, Ala.: Air Command and Staff College, Air University, 1966. 42p.

Bowers, Peter M. Boeing B-52A/H Stratofortress, Aircraft in Profile, no. 245. Windsor, Eng.: Profile Publications, 1974. 24p.

Boyne, Walter J. Boeing B-52: A Documentary History. Washington, D.C.: Smithsonian Institution Press, 1982. 158p.

Cook, D. G., C. H. Horne, and W. W. Manning. B-52 Maritime Operations: The Anti-Surface Warfare Mission. Research Report. Maxwell AFB, Ala.: Air War College, Air University, 1987. 58p.

Daugherty, Robert G. B-52 Terrain Avoidance Training, Research Paper. Maxwell AFB, Ala.: Air Command and Staff College, Air University, 1966. 60p.

Davis, Robert P. and John G. Walther, Jr. A Historical Perspective of the Required Characteristics for Future Strategic Bombers. Research Paper. Maxwell AFB, Ala.: Air Command and Staff College, Air University, 1982. 60p.
Stratofortress case history.

Dilworth, Billy G. The Problems of Measuring B-52 Effectiveness in Limited War. Research Paper. Maxwell AFB, Ala.: Air Command and Staff College, Air University, 1972. 54p.

Drendel, Lou. B-52 Stratofortress in Action, Warren, Mich.: Squadron/Signal Publications, 1975. 50p.

Eickhoff, Bruce E. New Training Realism for B-52 Crews. Research Paper. Maxwell AFB, Ala.: Air Command and Staff College, Air University, 1981. 16p.

Ethell, Jeffrey L. B-52 Stratofortress, New York: Scribners, 1981. 128p.

Funk, Dale R. Study of the Use of B-52 Bombers in the War in Vietnam, Research Report. Maxwell AFB, Ala.: Air War College, Air University, 1967. 89p.

Gaddie, Richard D. Is There a Maritime Role for the B-52? Research Paper. Maxwell AFB, Ala.: Air Command and Staff College, Air University, 1984. 33p.

Gooch, Stanford R., et al. The B-52D in Europe, Research Paper. Maxwell AFB, Ala.: Air Command and Staff College, Air University, 1980. 39p.

Green, Cameron K. An Analysis of Combat Readiness of B-52 Units Tasked for Contingency Commitments. Research Paper. Maxwell AFB, Ala.: Air Command and Staff College, Air University, 1984. 39p.

Greene, Warren E. B-52, 1945-1953. AMC Historical Study, no. 271. Wright-Patterson AFB, Ohio: USAF Air Materiel Command, 1956. 175p.

Harford, James D. Assignments for the B-52 Officer Crewmember. Research Paper. Maxwell AFB, Ala.: Air Command and Staff College, Air University, 1981. 91p.

Hastie, Robert T. The B-52 and the Nixon Doctrine. Research Paper. Maxwell AFB, Ala.: Air Command and Staff College, Air University, 1973. 52p.

Hinch, Nicholas J. The B-52H as a Contingency Bomber Replacement for the B-52D.. Research Paper. Maxwell AFB, Ala.: Air Command and Staff College, Air University, 1982. 36p.

Holder, William G. Boeing B-52 Stratofortress. Aero Series, v. 34. Fallbrook, Calif.: Aero Publishers, 1975. 104p.
A 112-page 2nd ed. was published by the same firm in 1988.

Honeywell, David R. The Role of the B-52 in Force Projection in the 1980s and 1990s. Research Paper. Maxwell AFB, Ala.: Air Command and Staff College, Air University, 1984. 23p.

Johnson, Vaughn A. B-52 Navigator's Guide. Research Paper. Maxwell AFB, Ala.: Air Command and Staff College, Air University, 1980. 72p.

Kappmeier, Jay A. An Analysis of the SAC B-52 Integral Crew System. Research Paper. Maxwell AFB, Ala.: Air Command and Staff College, Air University, 1975. 72p.

Kormondy, Robert C. Increased Non-Nuclear Capabilities for the B-52 Aircraft. Research Report. Maxwell AFB, Ala.: Air War College, Air University, 1968. 105p.

Linnee, Roger E. SRAM and SCAD: New Teeth for an Old Bird. Research Paper. Maxwell AFB, Ala.: Air Command and Staff College, Air University, 1973. 84p.

Lloyd, Alwyn T. B-52 Stratofortress in Detail and Scale. Detail and Scale Series, v. 27. Blue Ridge Summit, Pa.: TAB, 1988. 64p.

Long, Francis J. Do Strategic Bombers Have a Role in Conflicts Short of Central War? Research Report. Maxwell AFB, Ala.: Air War College, Air University, 1966. 72p.

Lother, Joseph S., Jr. Obsolescence of the Strategic Bomber. Research Paper. Maxwell AFB, Ala.: Air Command and Staff College, Air University, 1967. 42p.

McCandless, William G. Retention of the B-52D. Research Paper. Maxwell AFB, Ala.: Air Command and Staff College, Air University, 1972. 68p.

Mandeles, Mark D. The Air Force's Management of R & D: Redundancy in the B-52 and B-70 Development Programs. PhD. Dissertation. Bloomington, Ind.: Indiana University, 1985.

Martin, Willie Rae. Progress of B-52 Modernization at San Antonio Air Materiel Area, July 1959-June 1961. AFLC Historical Study, no. 460. Kelly AFB, Tex.: History Office, San Antonio Air Materiel Area, 1963. 76p.

Micklas, Paul J., Jr. Guide for the New B-52 Co-Pilot. Research Paper. Maxwell AFB, Ala.: Air Command and Staff College, Air University, 1980. 35p.

Mihura, Max L. The Non-Nuclear Role of the B-52 in Southeast Asia. Research Paper. Maxwell AFB, Ala.: Air Command and Staff College, Air University, 1967. 45p.

Miller, Jay. B-52 Stratofortress. Minigraph, no. 6. Arlington, Tex.: Aerofax, Inc., 1986. 48p.

Nelson, Jack A. Evolution of the B-52 Weapon System--Past, Present, and Future. Wichita, Kans.: Boeing Military Airplane Company, 1980. 20p.

Orr, Loren E. Quick Strike Plan for B-52H Aircraft. Research Paper. Maxwell AFB, Ala.: Air Command and Staff College, Air University, 1965. 41p.

Owens, Max G. B-52: Mothballs or What? Research Paper. Maxwell AFB, Ala.: Air Command and Staff College, Air University, 1967. 52p.

Patchin, Kenneth L. and James N. Eastman. B-52 Management, 1952-1961. AFLC Historical Study, no. 369. Tinker AFB, Okla.: History Office, Oklahoma City Air Materiel Area, 1961.

Peters, Karl M. Use of the B-52 in an Antisubmarine Role. Research Paper. Maxwell AFB, Ala.: Air Command and Staff College, Air University, 1985. 26p.

Rambo, Larry D. The Future of Simulation on the B-52 Crew Training Program. Research Paper. Maxwell AFB, Ala.: Air Command and Staff College, Air University, 1973. 64p.

Rousseau, Lamont E. The Manned Strategic Bomber in Limited War: In Vietnam--an Analysis. Research Paper. Maxwell AFB, Ala.: Air Command and Staff College, Air University, 1967. 64p.

Sanders, James D. Utilization of the B-52 in Vietnam: A Case Study for B-52 Operations in Limited War. Research Paper. Maxwell AFB, Ala.: Air Command and Staff College, Air University, 1966. 60p.

Shaw, James A. and Vincent F. Speziale. Conventional and Contingency Use of the B-52: A Reference Guide. Research Paper. Maxwell AFB, Ala.: Air Command and Staff College, Air University, 1980. 55p.

Sweeney, Llewellyn R. What If? No B-1.... Research Paper. Maxwell AFB, Ala.: Air Command and Staff College, Air University, 1975. 80p.

Szymczak, Michael A. B-52 Combat Readiness: How Long Between Flights? Research Paper. Maxwell AFB, Ala.: Air Command and Staff College, Air University, 1986. 39p.

Temple, Robert J. Do Strategic Bombers Have a Role During Non-Nuclear Conflicts? Research Report. Maxwell AFB, Ala.: Air War College, Air University, 1966. 64p.

Thornton, Richard W. Thesis for Change?: The B-52 in Limited War. Research Paper. Maxwell AFB, Ala.: Air Command and Staff College, Air University, 1967. 51p.

Periodical/Anthology Literature:

"Air Force Decides on Boeing B-52 for Intercontinental Bomber." U.S. Air Services, XXVII (November 1952), 8-11.

Aitken, Douglas B. "Guam--12 O'Clock, 200 Miles." Combat Crew, XXIX (November 1979), 10-13.

Alexandrakos, Peter. "Giant Thrust II." Combat Crew, XXVIII (November 1978), 4-7.

Arnold, Dave. B-52 Stratofortress." Internatinnal Combat Arms, V (May 1987), 26-29+.

Astrella, Ralph J. "Giant Strike and the Blue Steel Winner." Combat Crew, XXI (July 1970), 4-7.

"B-52 Heavyweights in SAC." Defence Update International, no. 75 (September 1986), 1-10.

"The B-52 in the Torture Chamber." Interavia, X (November 1955), 854-857.

"B-52: Lifetime Extension Effort Pushed." Aviation Week and Space Technology,CIV (May 10, 1976), 140-142.

"B-52 Production." Flight, LXVII (March 4, 1955), 278-279.

"B-52: 20th Anniversary in SAC." Aerospace Historian, XXII (December 1975), 198-200.

Bartes, Darrell. "Have Skill, Will Travel." Boeing Magazine, XXVIII (March-April 1958), 10-11.

____. "More Sock for SAC." Boeing Magazine, XXIX (September 1958), 3-5.

____. "Old Faithful of the Space Age." Space World, III (July 1962), 38-39+.

____. "Pattern for Precision." Boeing Magazine, XXVII (April 1957), 12-13.

"Behemoth Joins the Air Force." Aviation Week, LXIII (September 5, 1955), 22-23.

Benjamin, Ronald G. "Pacer Plank." Combat Crew, XXVII (September 1977), 22-23.

"Boeing B-52 Stratofortress." Air Progress, XXIV (December 1969), 50-53.

"Boeing B-52 Stratofortress." Joint Services Recognition Journal, IX (December 1954), 312-313.

Boyne, Walter J. "The B-52 Story." Air University Review, XXXIV (November-December 1983), 60-67.

____. "B-52: The Once and Future Emperor of Airpower." Airpower, XII (March 1982), 12-49.

____. "New Mexico Nightmare: An F-100 Actually Shoots Down a Fully-Manned B-52." RAF Flying Review, XVII (June 1962), 15-17.

Brandon, Harold E. "The B-52." Aerospace Safety, XXII (December 1966), 22-23.

Breemer, Jan S. and Todd Hoover. "SAC Goes to Sea With Harpoon." National Defense, LXXI (February 1987), 41-45.

Brownlow, Cecil. "B-52s Prove Tactical Value During Siege of Khe Sanh." Aviation Week and Space Technology, LXXXVIII (May 13, 1968), 26-30.

Buck, Fred. "Egress Update: The B-52." Combat Crew, XXIX (January 1979), 12-14.

Calkins, Ken. "Stratoforts on the Line." Boeing Magazine, XXIV (December 1954), 8-9.

____. "A Work of Art." Boeing Magazine, XXIV (July 1954), 3-5, 14.

Calvert, Denis J. "B-52 Deployments to the U. K." Aircraft Illustrated, XIII (October 1980), 462-465.

Chipman, Donald S. and David Lay. "Sea Power and the B-52 Stratofortress." Air University Review, XXXVII (January-February 1986), 45-50.

Chubbuck, Robert M. "The Big Stick: SAC and the B-52." Air Classics, VIII (September 1972), 50-55.

Clark, J. J. "Platform in the Sky." Boeing Magazine, XXVIII (May-June 1958), 6-7.

Coble, Donald W. "Last of the Tail-End Charlies." Airman, VII (July 1963), 17-21.
B-52 rear gunners.

Correll, John T. "Deterrence Today." Air Force Magazine, LXVII (February 1984), 40-45.

Cushman, Robert H. "Air Force Will Switch Drives in B-52." Aviation Week and Space Technology, LXVI (June 24, 1957), 81+.

Day, Bonner. "The B-52: Growing More Vital With Age." Air Force Magazine, LXII (February 1979), 32-37.

"Delivery System: B-52 Chronology." Flight, LXXVII (April 29, 1960), 596-597.

Dettmer, H. William. "All This Harassement." Combat Crew, XXV (September 1975), 16-18.

Drachlis, David B. "New Day For an Old Warrior." Airman, XXIV (January 1980), 15-19.

Eckstein, P. M. "We Take a Ride in the B-52." Mechanics Illustrated, LXXIV (November 1978), 48-49+.

Elliott, Robert J. "Unslave the B-52G at Low Level." Navigator, XX (January 1973), 25-27.

Elsdon, Ronald C. "B-52G: Mastering the Magnificent Monster—a Pilot's Report." Air Force Magazine, LVII (November 1974), 46-51.

Elson, Benjamin M. "Bomb/Nav Changes Key to B-52 Update." Aviation Week and Space Technology, CIX (August 14, 1978), 60-65. Reprinted in Combat Crew, XXVIII (December 1978), 16-20.

Engle, Georgia M. "B-52D Tailgunner: Last of a Dying Breed." Aerospace Historian, XXX (December 1983), 268-270.

Eubank, William E., Jr. "SAC's Newest Recruit--the Stratofort." Air Force, XXXIX (April 1956), 99-103.

"'52s Shrink a World." Life, XLII (January 28, 1957), 21-27.

"First Flight: YB-52." Boeing Magazine, XXII (April 1952), 3-5.

Fisher, Chuck. "Introduction to the 'H.'" Combat Crew, XI (April 1961), 4-6.

Flickinger, Don, R. L. Linforth, and G. M. Townsend. "Crew Effectiveness in the B-52 Strategic Bomber." Journal of Aviation Medicine, XXVI (January 1955), 2-19.

Gillert, Doug. "A Fix in Time." Airman, XXXII (February 1988), 29-33.

____. "Those Seagoing B-52s." Airman, XXXIII (January 1989), 29-33.

Goodell, James H. "There've Been Some Changes Made." Combat Crew, IX (May 1959), 24-27.

Griswold, Wesley S. "Mightiest Bomber Fires 1,000-Mile Missile." Popular Science, CLXXIII (September 1958), 88-92.

Gunston, William T. ("Bill"). "Boeing B-52." Flight, LXIX (June 8, 1956), 732-733; LXXIX (November 15, 1957), 771-778.

_____. "Stratofortress." Aeroplane Monthly, IX (August 1981), 416-423.

Hale, Reid. "From Beehive to B-52." Boeing Magazine, XXVII (February 1957), 12-13.

Hanna, Sidney B., Jr. "Grandpa's Thoughts on the B-52." Combat Crew, XXVIII (October 1978), 16-17+.

Hartman, Richard. "B-52 Update." Military Electronics/Countermeasures, IV (March 1978), 18-20+.

Harvey, David. "Flying With SAC's 20-Year Old Deterrent." Defense and Foreign Affairs, IX (June 1981), 9-11+.

Hays, Joseph A. "The Air Force's 'Long-Rifle.'" Air Training, IV (May 1955), 8-17.

Hillman, D. M. "The Birth of a Bomber." Air Force, XXXIX (April 1956), 104-105.

_____. "A True Test." Combat Crew, VII (October 1957), 4-8. Winter operations.

Holder, William G. "Still Going Strong: The B-52 in Its Third Decade." Air University Review, XXVI (September-October 1975), 48-62.

Howard, R. T. "Bail Out—B-52." Combat Crew, XX (February 1969), 4-9.

Huddleston, James R. "B-52." Flying Safety, XL (May 1984), 6-7; XLI (April 1985), 6-7.

Hudlow, Richard G. "New Horizons in the C thru G." Combat Crew, XVI (July 1965), 4-6.

Hurn, William P. "The 'D' Will Never Die." Combat Crew, XXV (March 1975), 4-7.

"Is the B-52 the Last of the Line?" Business Week, (October 4, 1952), 84+.

"It's Hard to See a B-52." Combat Crew, XXVII (November 1977), 20-21.

Jackson, Peter H. "Winning Team." Combat Crew, XXIII (June 1973), 16-17.

Jaszka, Paul R. "B-52 Stratofortresses Prepare for New Mission." Defense Electronics, XX (September 1988), 71-72+.

Johnson, Vaughn A. "Automated Offset: An Adventure in Advanced Bombing." Combat Crew, XXV (August 1975), 4-5+.

Johnston, A. M. ("Tex"). "Test Flying the B-52." Boeing Magazine, XXIII (November 1953), 3-5.

____. "Flying the B-52." Combat Crew, V (June 1955), 4-7.

____. "The Plan Behind the Plane." Flying Safety, XI (January 1955), 16-17.

Kachkley, Paul D. "SAC's New Stratofortress." Airman, V (September 1961), 28-29.
B-52H.

Katzaman, Jim. "Sea Spy in the Sky." Airman, XXVII (July 1983), 41-48.

____. "Twilight of the Tall-Tail Gunners." Airman, XXVI (December 1982), 41-47.

"Keeping the Boeing B-52 Operational Until the End of the Century." Interavia, XXXIII (December 1978), 1181-1184.

King, Kenneth G., Jr. and Kenneth W. Baisden. "'Busy Figure II." Combat Crew, XVIII (October 1967), 20-23.

Kipp, Robert M. "Counterinsurgency From 30,000 Feet: The B-52 in Vietnam." Air University Review, XIX (January-February 1968), 10-18.

Kozicharow, Eugene. "USAF Asks B-52 Upgrade to Carry Cruise Missile." Aviation Week and Space Technology, CXXI (September 3, 1984), 270-271.

Kritt, Robert R. "B-52 Arc Light Operations." In: Carl Berger, ed. The United States Air Force in Southeast Asia, 1961-1973. Washington, D.C.: GPO for the Office of Air Force History, 1977. pp. 149-168.

Lee, B. "Air Force Decides to Buy Boeing XB-52." Aviation Week, LIV (March 12, 1951), 14-16.

Lert, Peter. "The B-52." Air Progress, XXXVI (March 1975), 62-66.

Liggett, William R. "Long-Range Combat Aircraft and Rapid Deployment Forces." Air University Review, XXXIII (July-August 1982), 72-81.

Lloyd, Alwyn T. "B-52: 30 Years Young." Combat Crew, XXXVII (April 1987), 22-26.

"'Long Rifle.'" Boeing Magazine, XXIV (March-April 1954), 8-9.

Lubkin, Yale J. "Embarrassment of Riches--the B-52G." Defense Science and Electronics, VII (January 1988), 15-16+.

McMahon, Bradford. "Mighty Machine's Surprise Mission." Vietnam, I (Spring 1988), 34-41.
"Arc Light" tactical support.

McPherson, Ray. "High 'G' Performance." Boeing Magazine, XXIX (January 1959), 10-11.

McQueen, Barry A. "Triangle-K Returns." Aerospace Historian, XXIX (December 1982), 269-273.

McSurely, Alexander. "Veil Lifted on Boeing B-52 Details." Aviation Week, LVII (August 18, 1952), 12-17.

McVeigh, George M., Jr. and Joel M. Graybeal. "Facelift Gives B-52 New Lease on Life." Aerospace America, XXII (November 1984), 62-65.

Manning, Stephen O., III and Len Brady. "Twenty Years on the Line." Airman, XX (January 1976), 2-7.

Mansfield, Robert. "Test to Destruction." Boeing Magazine, XXV (October 1955), 3-6.

Meyer, Charles B. "Boeing B-52Fs on Guam." American Aviation Historical Society Journal, XI (Fall 1966), 186-187.

Milberg, Ray. "They're Not Getting Older--They're Getting Better!" Combat Crew, XXXIV (November 1974), 12-15.

Miller, Barry. "USAF Studies B-52G/H Retrofit Plan." Aviation Week and Space Technology, CIV (February 23, 1976), 56-57.

Mitchell, Linda. "B-52 Gunners Defend the Skies." Profile, XXX (December 1986), 18-20.

Monaco, Gregg A. "Offensive Avionics System: Something New for the B-52." Navigator, XXX (Summer 1983), 25-27.

Morrison, Joseph G. "CAT on the B-52." Combat Crew, XIV (January 1964), 6-7.

Morrocco, John D. "Air Force Approves SAC Plan to Modify B-52G Bombers for Conventional Role." Aviation Week and Space Technology, CXXIX (July 4, 1988), 47-48.

Naugle, R. G. "Air Force Buys a Plane." Flying, L (April 1952), 34+.

Newton, George L., III. "Busy Prairie." Combat Crew, XXX (December 1980), 4-7.

Newton, Bob. "B-52 Stratofortress." Air Progress, XLII (April 1980), 62-65.

Nicol, James M. "B-52." Flying Safety, XLII (April 1986), 1-3.

Nordick, Casper D. "Shhh--Don't Tell Anyone, Especially B-52D/G/H Pilots." Combat Crew, XXXIII (February 1983), 22-23.

Peisher, Robert J. "What's Holding It Together: B-52." Combat Crew, XXII (October 1971), 4-7.

Peterson, Mark A. and Peter B. Mapes. "Planning to Conserve." Combat Crew, XXXI (January 1981), 18-19.

Phillips, Ren. "Behind the '52." Boeing Magazine, XXII (July 1952), 3-5, 14.

"Pilot With a Briefcase." Boeing Magazine, XXVII (March 1957), 8-9.

Plattner, C. M. "B-52 Equipped With SRAM Evaluation." Aviation Week and Space Technology, LXXXIX (December 9, 1968), 53-54.

____. "Tactical Raids by B-52s Stun Viet Cong." Aviation Week and Space Technology, LXXXIII (November 29, 1965), 16-21.

Preble, Cecilia. "There's a First Time For Everything." Flying Safety, XXXIX (May 1983), 6-7.

Prieve, Donald G. "A Must For Professional Crew." Combat Crew, VII (August 1956), 24-29.

Pugh, Craig. "Target: Nevada." Airman, XXVII (February 1983), 2-11.

Riedinger, Tom. "Fireball on Wings." Boeing Magazine, XXV (November 1955), 6-7.

Robinson, Clarence A., Jr. "B-52Ds Designated for Support of NATO." Aviation Week and Space Technology, CXI (October 15, 1979), 14-16.

Robinson, Edward L. "B-52 Bomber-Tanker Technique." Combat Crew, VI (May 1956), 9-11.

Rogers, Coleman. "B-52 Role Facing Change." Military Electronics/Countermeasures, V (February 1979), 96-97+.

Ropelewski, Robert R. "U.S. B-52 Fleet Being Upgraded." Aviation Week and Space Technology, CXII (June 16, 1980), 192-193+.

"SAC Shows B-52 Electronics." Electronics, XXX (January 1, 1957), 10+.

Schmidt, Stephen R. "B-52 In-Flight Emergency." Combat Crew, XXVII (September 1976), 16-17.

Schopen, Gerhard L. "B-52 Power Steering." Combat Crew, XIII (June 1963), 16-18.

"Seattle Giant." Flight, LXVIII (October 14, 1955), 614-615.

Shoultz, Leroy L. "All That Glitters." Combat Crew, XXI (April 1970), 26-27.

Showalter, N. D., Jr. "A Major Change." Combat Crew, XII (October 1961), 4-7.

Spriggs, Fred. "With a Giant Thrust." Airman, XXIII (January 1979), 11-15.

Starch, Stephen. "Formation Flying for B-52 Pilots." Combat Crew, XI (June 1969), 4-7.

Stein, Kenneth J. "SAC Integrates B-52 Weapons Training." Aviation Week and Space Technology, CXX (May 14, 1983), 81-84+.

Storm, Allen. "Night Eyes on the B-52." Air Force Magazine, LVII (October 1974), 36-40.

"Stratofortress." Flight, LXII (December 5, 1952), 702-704.

"Stratofortress." Interavia, X (July 1952), 576-577.

"Stratofortress." Scale Models, XIII (April-May 1982), 180-181, 244-245.

Stricklin, Charles. "Alert and Waiting." Airman, XVIII (July 1974), 12-16.

Sweeney, Robert. "B-52 Gives Air Force More Authority, Headaches." Aviation Week, LXIV (March 19, 1956), 26-30.

"Test for a B-52 Crew." Life, XLIV (June 9, 1958), 51-54.

Tilson, Joseph F. "Low Level and Big Aircraft." Flying Safety, XLIV (May 1988), 4-7.

"USAF Demonstrates B-52 Capabilities." Aviation Week and Space Technology, LXV (December 3, 1956), 31-32.

"The USAF's Big Stick." RAF Flying Review, XVIII (October 1962), 30-33+.

Waddell, Milton H. "B-52." Flying Safety, XLIII (April 1987), 2-4; XLIV (February 1988), 2-4.

Wanstall, Brian. "Stand-off to Strike and Survive." Interavia, XXXIX (July 1984), 673-674.

"Weapon Advances Raises B-52 Capability." Aviation Week and Space Technology, CIV (May 10, 1976), 132-135.

Welch, John F. "The B-52 and You." Combat Crew, VIII (February 1958), 26-29.

Wheeler, Frederic. "The Sound of Trombones." Combat Crew, XXXIII (August 1983), 18-19.

"Where Do All the Wheels Go on a B-52?" American Aviation, XVIII (September 27, 1954), 42-43.

Witkin, Richard. "Most Formidable Plane, the B-52." New York Times Magazine, (February 5, 1955), 10-11.

Wolke, Gerald G. "Maintaining the Stratofort." Combat Crew, VI (June 1956), 22-27; VII (July 1956), 24-28.

Wolkomir, Richard. "The Once and Future Bomber." Smithsonian Air and Space, III (February-March 1989), 68-79.

Worden, William L. "Old Original Goes Away." Boeing Magazine, XXIX (October 1959), 8-9.

Zaiman, Robert. "The X-15's Winged Booster." BeeHive, XXXVII (Summer 1962), 8-12.

Zeybel, Henry. "Old, But Tough." National Defense, LXVIII (July-August 1984), 64-67.

(Martin) B-57 Canberra

Book/Document/Report Literature:

Alling, Frederick. History of the B-57 Airplane, July 1953-January 1958. AMC Historical Study, no. 293. Wright-Patterson AFB, Ohio: USAF Air Materiel Command, 1958. 55p.

Anderton, David A. Martin B-57 Night Intruders and General Dynamics RB-57F. Aircraft in Profile, no. 247. Windsor, Eng.: Profile Publications, 1973. 25p.

Mesko, Jim. B-57 Canberra in Action. Carrollton, Tex.: Squadron/Signal Publications, 1986. 50p.

Mikesh, Robert C. B-57 Canberra at War, 1964-1972. New York: Scribners, 1980. 160p.

Pfau, Richard A. and William H. Greenhalgh. The B-57G--Tropic Moon III, 1967-1972. Washington, D.C.: Office of Air Force History, 1978. 127p.

Schultz, Helen. B-57, 1950-1953. AMC Historical Study, no. 258. Wright-Patterson AFB, Ohio: USAF Air Materiel Command, 1954. 36p.

Periodical/Anthology Literature:

"America Gets the Canberra." Aeronautics, XXIV (April 1951), 40-42.

Archer, Robert D. "Oldie But Goodie: The Canberra Jet Bomber." Scale Modeler, IX (December 1974), 56-62, 96.

Beaumont, Roland. "Canberras For America." Aircraft Illustrated, XIV (March 1981), 109-113.

Beggerly, Douglas J. "And There I Was." Aerospace Safety, XVI (November 1960), 22-23.

Bradford, M. V. N. "Canberra: Her Bridge and Navigation Equipment." Navigation, X (Summer 1963), 136-142.

Brown, Terry A. "30 Years of Canberras: Overseas Orders and Production." Airfix Magazine, XXI (February 1980), 288-290.

Blevins, Dallas R. "High Flight to Pakistan." Navigator, XXVII (Summer 1980), 27-30.

"Building the B-57: Two Approaches." American Aviation, XVII (March 29, 1954), 28-29.

"The Canberra in Detail." Flight, LXXIII (March 14, 1958), 335-339.

Carrow, Roy J., Jr. "B-57 Canberra." In: Robin Higham and Carol Williams, eds. Flying Combat Aircraft of the USAAF-USAF. Ames, Iowa: Iowa State University Press, 1978. pp. 89-100.

Chandler, Frank R. "The Unsung Canberras." Airman, XI (December 1967), 12-15.

Christy, Douglas. "The Magpies Always Score." Airman, XIV (June 1970), 44-45.

Gething, Mike. "Canberra Connections." Aircraft Illustrated, IX (October 1976), 407-413.

Green, William. "Design Report: Canberra." Skyways, X (August 1951), 10-12.

Keeley, Robert F. "The Only Way to Fly." Aerospace Safety, XXIII (April 1967), 22-25.

Lee, Irvin H. "Canberras Over Vietnam." Airman, XII (April 1968), 19+.

"Long, Lean and Lethal." Flying Safety, XI (June 1955), 10-11.

Lorenzo, Peter, Jr. "The Canberra Goes to Southeast Asia." Navigator, XV (Spring 1968), 44-45.

McLarren, Robert. "Canberra." Aero Digest, LXII (April 1951), 17-18+.

McSurely, Alexander. "B-57 Shows Off." Aviation Week, LIX (August 31, 1953), 17-18.

"Martin Canberra." Aeronautics, XXIX (November 1953), 54, 57.

"Martin Canberra Developments." Flight, LXV (January 15, 1954), 60-61.

Mikesh, Robert C. "Buy British, Fly American." Airpower, VII (September 1977), 26-43; Wings, VII (October 1977), 32-47, 55.

____. "Farewell to the '57." Wings, XII (August 1982), 28-29.

____. "Flying the B-57." Aircraft Illustrated, XIII (July 1980), 317-321.

"The Name of the Game." Flying Safety, XV (June 1959), 20-22.

Nielsen, Dennis E. "An Old Breed of Cat." Navigator, XV (Summer 1968), 27-28.

Simmons, W. M. "Night Intruder." Air Training, IV (July 1955), 25-30.

Sprotta, R. H. "Martin Light Bomber Hangs on Garden Gate." Machinist, LXII (July 1956), 154-161.

Stone, Irving. "Martin Boosts Canberra Performance." Aviation Week, LX (January 4, 1954), 28-30+.

____. "Martin Speeds Plans for B-57 Production." Aviation Week, LV (July 23, 1951), 19-20.

"Swift Blackbird." Flying Safety, XI (June 1955), 30-33.

Trimble, Robert L. "The Iron Cranberry." Air Classics, XIII (June 1977), 78-85.

Walters, R. C. "Canberra to Night Intruder: Problems in Conversion of British-Designed Bomber to U.S. Type." Aero Digest, LXIX (August 1954), 74+.

Witze, Claude. "Martin B-57 Proves Versatile Performer." Aviation Week, LXIII (November 28, 1955), 13-15.

(Convair) B-58 Hustler

Book/Document/Report Literature:

Durham, John M. B-58 Hustler: The System Gone Awry. Research Paper. Maxwell AFB, Ala.: Air Command and Staff College, Air University, 1982. 34p.

General Dynamics Corporation. Convair Division. The Convair B-58 Airplane: Pilot Indoctrination. Long Beach, Calif., 1956. 56p.

Miller, Jay. Convair B-58. Aerograph, no. 4. Arlington, Tex.: Aerofax, Inc., 1986. 136p.

Murray, Loren P., Jr. B-58 Program: Testbed for Weapon System Management. Research Report. Maxwell AFB, Ala.: Air War College, Air University, 1961. 89p.

Robinson, Douglas M. B-58 Hustler. Famous Aircraft Series. New York: Arco, 1967. 63p.

Scott, Jessie C. The B-58 and the San Antonio Air Materiel Area Mission, July 1959-March 1961. AFLC Historical Study, no. 457. Kelly AFB, Tex.: History Office, San Antonio Air Materiel Area, 1962. 99p.

Thomas, Richard D. History of the Development of the B-58 Bomber, November 1965-June 1970. Wright-Patterson AFB, Ohio: History Office, Aeronautical Systems Division, Air Force Systems Command, n. d. 403p.

Yon, Kenneth W. A Study of B-58 Low-Level Training. Research Paper. Maxwell AFB, Ala.: Air Command and Staff College, Air University, 1965. 63p.

Periodical/Anthology Literature:

Allen, Francis J. "Delta Queen: The Story of the B-58 Hustler." Air Classics, XXII (August 1986), 46-61.

"Almost the Plane of the Future: B-58 Hustler." Business Week, (July 27, 1957), 75-76.

Anderton, David A. "B-58 Hustler Packs a Big Punch in a Small Frame." Aviation Week and Space Technology, LXV (December 17, 1956), 26-29.

Barrett, John O. "Tokyo to London--8 hrs., 35 min." Navigator, XII (Winter-Spring 1965), 56-58.

Bentz, A. W. "B-58 Owes Performance to Materials, Design Breakthroughs." American Aviation, XXI (July 29, 1957), 30-31.

Bright, G. L. "Hyperhoming Hustler." Navigator, V (Winter 1958), 13-15.

"The Convair B-58 Hustler Multiple Weapons System." American Aviation, XX (January 14, 1957), 28-29.

"Convair B-58 Hustler Supersonic Combat Aircraft." Interavia, XV (July 1960), 825-827.

Cross, Roy. "Convair's B-58 Hustler." Air Pictorial, XIX (February 1957), 54-55.

Drake, Francis V. "Bomb Carrier Nobody Can Stop: The B-58 Hustler." Reader's Digest, LXXXIV (April 1964), 153-164.

Dugard, G. A. "One of Those Days." Combat Crew, XX (November 1969), 4-7.

Erickson, D. A. "Development of the B-58:The Mach 2 Environment." Navigation, VIII (Summer 1961), 112-115.

____. "Flight Characteristics of the B-58 Mach 2 Bomber." Royal Aeronautical Society Journal, LXVI (November 1962), 665-671.

Esemwein, A. C. "Weapon System Management and the B-58 Hustler." Sperryscope, XIV (June 1957), 6-9.

Farnsworth, D. B. "Celestial Navigation for the B-58." Combat Crew, XV (May 1965), 16-18.

Fusca, J. A. "First Details of B-58 Bomb/Nav System." Aviation Week and Space Technology, LXXI (August 3, 1959), 76-79.

Griswold, Wesley S. "The World's Fastest Bomber: B-58." Popular Science, CLXXIX (August 1961), 71-74+.

Gunston, William T. ("Bill"). "Thoughts on the Hustler." Flight, LXX (September 21, 1956), 507-509.

Haddaway, G. E. "More Than Speed Alone." Flight Magazine, XLVI (August 1957), 32-33.

Hall, R. Cargill. "The B-58 Hustler." Air University Review, XXXIII (November-December 1981), 44-56.

____. "To Acquire Strategic Bombers: The Case of the B-58 Hustler." Air University Review, XXXII (September-October 1980), 2-20.

Hamilton, C. L. "Hustler for Peace: The Convair B-58." Flying, LXI (September 1957), 30+.

Harper, Jennifer. "Supersonic Hustler." Air Force Magazine, LXV (August 1982), 62-63.

Hinnant, Robert E. "B-58 Hustler." In: Robin Higham and Abigail T. Siddall, eds. Flying Combat Aircraft of the USAAF-USAF. Ames, Iowa: Iowa State University Press, 1975. pp. 29-36.

Hirsch, Henry R. "The B-58." Air University Review, XV (September-October 1964), 51-60.

"How Hustler Handles Its Payload." Aviation Week and Space Technology, LXV (December 31, 1956), 30-31.

Johnson, James K. "Bombs Away From 60,000 Feet." Air Force, XLIV (August 1961), 48-52.

_____. "The Hustling Hustler." Combat Crew, X (June 1960), 10-14.

Kardong, Abe. "A Date With the Hustler." Combat Crew, XVIII (March 1967), 18-21.

Karten, Dave. "Hustler." Airman, V (October 1961), 7-11.

Lewis, Cecil. "Irvine Details B-58 Design Advances." Aviation Week and Space Technology, LXVII (July 15, 1957), 28-30.

McHenry, Francis T. "Weather Factors in Support of B-58 Operations." Navigation, VIII (Winter 1961-1962), 305-311.

Polhemus, William L. "The B-58 as a Competitor in SAC's Bombing-Navigation Contest." Navigation, VIII (Spring 1961), 40-47.

____. "Navigating the B-58." Navigator, VI (August 1960), 10-19.

____. "Navigating the Supersonic B-58." Navigation, VII (Summer-Autumn 1960), 127-139.

Powers, Jack A., et al. "Mating Men and Machine." USAF Instructors Journal, III (July 1965), 52-56.

Schanche, D. "Missile With Men in It: The B-58." Life, XLII (April 15, 1957), 193-194+.

Smith, Don. "B-58 Wing Operations." Air University Review, XV (September-October 1964), 61-73.

____. "Greased Lightning." Air Force and Space Digest, XLVII (February 1964), 83-87.

Snow, John L. "Last Flight." Combat Crew, XXI (April 1970), 9-13.

Solomon, Howard N. "First Impressions of a Hustler Driver." Combat Crew, XIV (November 1963), 25-27.

"Styled for Supersonic Speed: Convair B-58 Hustler." Business Week, (December 8, 1956), 78+.

"Supersonic Delta Wing Bomber." Shell Aviation News, no. 225 (March 1957), 12-13.

Test, James. "The Hustler's Record." Aerospace Historian, XX (June 1973), 71-75.

"Test Force." Airman, III (December 1959), 26-27.

Weeks, John M., Jr. "Recipe and Receipt." Navigator, XII (Winter-Spring 1965), 59-60.

Wendell, Roy M. "Cooling the Hustler." BeeHive, XXXIII (Spring 1958), 24-25.

Woodruff, Walter W. "The B-58 Hustler." Navigator, XXX (Spring 1983), 13-15.

"The World's Fastest Bomber." U.S. News and World Report, XLIII (July 19, 1957), 14-15.

(Douglas) B-66 Destroyer

Alling, Frederick. History of the B/RB-66 Weapon System, 1952-1959. AMC Historical Study, no. 300. Wright-Patterson AFB, Ohio: USAF Air Materiel Command, 1960. 37p.

Boyer, Frank O. "From the Ground Up." Flying Safety, XII (August 1956), 6-11.

"The Douglas B-66 Destroyer." Air Progress, XXIV (October 1968), 46-48.

Francillon, Rene J. and Mick Roth. Douglas B-66 Destroyer. Minigraph, no. 19. Arlington, Tex.: Aerofax, Inc., 1988. 48p.

Schrader, Richard K. "The USAF's Destroyer." Air Classics, XXIV (April 1988), 20-22, 78.

(Caproni) Ca. 46

"An American-Built Caproni Bombing Plane." World's Work, XXXVII (November 1918), 117-120.

"The Caproni." Aerial Age Weekly, V (April 16, 1917), 146-147.

(DeHavilland) DH-4/02B-1

Book/Document/Report Literature:

Bowers, Peter M. American DH-4. Aircraft in Profile, no. 97. Windsor, Eng.: Profile Publications, 1967. 12p.

Boyne, Walter J. DeHavilland DH-4: From Flying Coffin to Living Legend. Famous Aircraft of the National Air and Space Museum, no. 7. Washington, D.C.: Smithsonian Institution Press, 1984. 120p.

Bruce, John M. DeHavilland DH-4. Aircraft in Profile, no. 26. Windsor, Eng.: Profile Publications, 1965. 12p.

Emmons, Roger M. Colors and Markings on DH-4 Aircraft Used by the U.S. Marine Day Wing in World War I. Baltimore, Md.: 1st Marine Aviation Force Veterans Association, 1970. 4p.

Gerhardt, W. F. and Robert Anderson. Economy Test of the DH-4. U.S. Air Service Information Circular, v. 5, no. 461. Washington, D.C.: GPO, 1924. 4p.

United States. Army Air Service. Performance Test of DH-4 With 400-hp Liberty 12 Engine, Equipped as Two-Seater Observation Airplane. U.S. Air Service Information Circular, v. 3, no. 287. Washington, D.C.: GPO, 1921. 7p.

Periodical/Anthology Literature:

Andrews, Hal. "Naval Aircraft: DeHavilland." Naval Aviation News, (October 1976), 20-21.

Bowers, Peter M. "Damnable DH-4." Wings, X (December 1980), 37-45.

____. "The Diabolical DH-4." Wings, VIII (April 1978), 44-58.

Crean, Michelle. "DH-4M-489." World War I Aero, no. 114 (April 1987), 3-7, 12.

Freidhoff, James W. "Navy and Marine Corps DH-4 Aircraft." American Aviation Historical Society Journal, X (Fall 1965), 202-203.

Gann, Ernest K. "The Golden Age of Air Transport: The DeHavilland DH-4." Flying, XC (April 1972), 22-24.

Gemeinhardt, Walter. "DH-4s in Marine Corps Service." American Aviation Historical Society Journal, VII (Winter 1962), 243-253.

Hart, P. G. "Observations From a DH-4." Cross and Cockade Journal, II (Winter 1961), 330-334.

"Passing of the DeHavillands." Aviation, XVIII (June 29, 1925), 730+.

Rust, Kenn C. "The DH-4B and the DH-4M." Aero Album, IV (1971), 10-18.

Trimble, Robert L. "Flaming Coffins." Air Trails, NS III (Summer-Fall 1978), 44-55, 21-23.

Whiteley, John F. "The Old DH: A Memorable Airplane." Aerospace Historian, XXVI (December 1979), 245-248.
Reprinted in Robin Higham and Carol Williams, eds., Flying Combat Aircraft of the USAAF-USAF (Manhattan, Kans.: Sunflower University Press, 1981), pp. 87-90.

(N.A.F.) F-5L

Diehl, W. S. F-5L Seaplane Comparative Performance With Direct-Drive and Geared Engines. Technical Note, no. 116. Washington, D.C.: NACA, 1922. 12p.

____. F-5L Seaplane: Performance Characteristics. Technical Note, no. 118. Washington, D.C.: NACA, 1922. 8p.

United States. Navy Department. The F-5L Flying Boat Handbook. Washington, D.C.: GPO, 1918. 43p.

"U.S. Navy F-5L Flying Boat." Flight, XI (July 31-August 7, 1919), 1024-1026, 1058-1062.

Williams, S. T. "F-5L Flying Boat." Automotive Industries, XL (April 10, 1919), 809-811.

(Thomas/N.A.F.) HS

Andrews, Hal. "Naval Aircraft: Thomas HS." Naval Aviation News, (November-December 1988), 28-29.

"The Navy HS-1L and 2L Flying Boats." Aerial Age, IX (April 7, 1919), 200-202.

(Keystone/Huff-Daland) LB

Brockman, William. "Jungle's Victims." Airman, XIV (February 1970), 29-31.
A recovered LB-7.

Donnelly, Jan R. "Last of the Keystones." Air Force Magazine, LXVI (July 1984), 100-102.
LB-7.

Hansell, Haywood S., Jr. "The Keystone Bombers: Unhonored and Unloved." Air Force Magazine, LX (September 1977), 130-132+.

"The Huff-Daland LB-1 Bomber." Aviation, XIX (November 2, 1925), 632+.

"Huff-Daland Bomber, XLB-1." Aviation, XXI (August 9, 1926), 252-254.

"The Keystone All-Metal Bomber." Aviation, XXIII (September 12, 1927), 596-597.

"The Keystone Panther Bomber." Aero Digest, XIII (July 1928), 102-103.

"The Keystone Pirate." Aviation, XXIV (March 26, 1928), 776, 792.

"Keystone Pirate Bomber." Aero Digest, XII (April 1928), 588-589.

(Barling) NBL-1

"The Barling Bomber." Aeronautical Digest, III (October 1923), 249-284.

"The Barling Bomber: An American Six-Engined Giant." Flight, XV (December 12, 1923), 749-751.

"The Barling Bomber is Ready for Trial Flights." Aviation, XV (July 9, 1923), 38-40.

"The Barling Bomber: Largest American Airplane." Aviation, XIII (October 23, 1922), 562+.

"Barling Bomber Successfully Flight-Tested." Aviation, XV (September 1, 1923), 281-282.

Haskins, George W. "The Barling Bomber of the Army Air Service." US Air Services, VII (December 1922), 24+.

Lincoln, Marshall. "The Barling." Air Classics, II (February 1965), 28-32.

Ludd, Neddy. "Mr. Barling's Bomber." Air Classics, XIX (July 1983), 16-32.

Tilford, Earl H., Jr. "The Short, Unhappy Life of the Barling Bomber." Air Force Magazine, LXI (February 1978), 68-70.

(Martin) NBS Martin Bomber
(see also MBT/MT)

"Bombing of Warships Proves Air Power: Martin Bombers of Army Air Service Sink Destroyer G-102 and Cruiser Frankfurt." Aviation and Aircraft Journal, XI (July 25, 1921), 96-98.

Hicks, Frederick C. Aircraft Versus Dreadnoughts. Washington, D.C.: GPO, 1921. 14p.

Olmstead, Merle C. "First of the Heavies: The Martin Bomber." Aerospace Historian, XIX (September 1973), 150-154.

"Ostfriesland Sunk by 2000-lb. Aircraft Bombs: Martin Bombers of Army Air Service Sink the 27,000-ton Battleship in 14 Minutes." Aviation and Aircraft Journal, XI (August 1, 1921), 128-132.

United States. Army Air Service. Official Performance Test of Martin Bomber NBS-1, Equipped With Two 400-hp Liberty 12 Engines. U.S. Air Service Information Circular, v. 3, no. 290. Washington, D.C.: GPO 1921. 10p.

Whitely, John F. "Airways 1923." Aerospace Historian, XX (Summer 1973), 88-92.
Transcontinental air mail flight of an NBS-1.

Youngman, David L. "The Day They Sank the Ostfriesland." Airman, XIV (July 1970), 23-25; XXII (December 1978), 6-8.

(Navy/Curtiss) NC

Book/Document/Report Literature:

Curtiss Aeroplane and Motor Corp., Department of Education. Flight Across the Atlantic. Buffalo, N.Y., 1919. 88p.

Smith, Richard K. First Across: The U.S. Navy's Transatlantic Flight of 1919. Annapolis, Md.: Naval Institute Press, 1973. 279p.

Steirman, Hy and Glenn D. Kittler. Triumph: The Incredible Saga of the First Transatlantic Flight. New York: Harper & Row, 1961. 199p.

Westervelt, George C., Holden C. Richardson, and Albert C. Reed. The Triumph of the NCs. New York: Doubleday, Page, 1920. 308p.

Periodical/Anthology Literature:

"Aboard the NC-9 Enroute to Plymouth." Spur, XXIV (August 1, 1919), 34-35.

Andrews, Hal. "Naval Aircraft: The NCs." Naval Aviation News, (July-August 1986), 20-21.

Bellow, Steve. "NC-4: They Called Her a 'Lame Duck.'" All Hands, no. 747 (April 1979), 10-13.

Cagle, Malcolm W. "Nancy Flew the Atlantic First." Air Line Pilot, XLVIII (June 1979), 15-19, 48.

"The Construction of the Navy-Curtiss Flying Boats." Society of Automotive Engineers Journal, IV (June 1916), 439-442.

"Development of the NC Seaplanes." Aviation, VI (June 1, 1919), 468-474.

"Famous Flights: The NC-4 Trip." Popular Aviation, I (December 1927), 27-31.

Harrison, Kirby. "Flight Into History." Naval Aviation News, (July-August 1986), 4-9.

Hunsaker, Jerome C. "Details of Tests of the American Naval Curtiss." Aeronautics, XVII (August 28, 1919), 216-220.

____. "How American Ingenuity Designed the NC Boats." Automotive Industries, XLI (July 10-24, 1919), 68-72, 120-123, 172-176.

Johnston, James. "The NC-4: History Preserved." U.S. Naval Institute Proceedings, XCV (May 1969), 94-105.

Knott, Richard C. "Journey to Plymouth." Naval Aviation News, (November 1976), 22-27.

Martin, Glenn L. "Lessons of the Transatlantic Flight." U.S. Air Services, II (November 1919), 12-13.

"NC Boats Were Built to Fight Submarines." Aircraft Journal, IV (May 24, 1919), 8-9, 22.

"The NC-1 Naval Flying Boat." Aerial Age, IX (March 24, 1919), 110-111.

"The NC-4 Transatlantic Flight." Air Classics, VI (August 1970), 42-52.

"NC-4's Great Flight to the Azores." Aircraft Journal, IV (May 24, 1919), 3-4, 14-15.

Reed, Albert C. "NC-4: The Flight." American Aviation Historical Society Journal, XXIV (Summer 1979), 144-146.

Trimble, Robert L. "Nancy Boats vs. the Atlantic." Air Progress Aviation Review, IV (January 1980), 12-23, 96.

"The U.S. Navy Flying Boat NC-1." Flight, XI (May 15, 1919), 637-639.

"The U.S. NC-4 Makes First Trans-Atlantic Flight." Air Power, IV (June 1919), 47-54, 74-77.

Van Vleet, Clarke. "Flight of Ye Goode U.S. Ship NC-4." Aviation Quarterly, II (Winter, Summer-Fall 1976), 64-89, 273-299, 464-474.

Westervelt, George C. "The Design and Construction of the NC Flying Boats." U.S. Naval Institute Proceedings, XLV (December 1919), 1529-1581.
Reprinted in American Aviation Historical Society Journal, XXVI (Winter 1981), 279-303.

____. "The Flight of the NC Boats." World's Work, XXXVIII (August 1919), 424-438.

Wilbur, Theodore. "The First Flight Across the Atlantic." Naval Aviation News, (May 1969), 7-36.

____. "The Glory of Being First." True Magazine, LVII (October 1975), 35+.

____. "Paint and Pluck--and the NC-4." U.S. Naval Institute Proceedings, XCV (May 1969), 146-149.

Woodhouse, Henry. "The Trans-Atlantic Flight." Flying, VIII (June 1919), 445-455.

(Handley-Page) 0/400

Bagley, John. "The Handley-Page Agreement of 1918." American Aviation Historical Society Journal, XXI (Spring 1976), 38-45.

Barnes, C. H. "Handley-Page's 'Bloody Paralyser.'" Air Enthusiast, V (August 1973), 74-81.

David, Evan J. "Christening for the First American-Made Handley Page." Flying, VII (August 1918), 610-615.

"First American-Built Handley-Page Plane Launched." Automotive Industries, XXXIX (July 11, 1918), 46, 85.

"The Handley-Page 0/400." Aviation Week and Space Technology, LXXXV (April 8, 1963), 37-38.

"The Handley-Page Type 0/400 Bomber." Aerial Age, VIII (February 10, 1919), 1082-1083, 1095.

"Triumph of America's Handley-Page Bomber." Air Service Journal, III (July 11, 1918), 41-45.

Underwood, John. "Anglo-American Blockbuster: Full Details on America's First Heavy Bomber." Air Classics, XII (October 1976), 20-31, 94-95.

____. "Postwar Use of the Handley-Page in the U.S." American Aviation Historical Society Journal, XXI (Spring 1976), 46-49.

(Lockheed) P2V Neptune

Book/Document/Report Literature:

Anderson, Holmes G. Lockheed P2V Neptune. Aircraft in Profile, no. 204. Windsor, Eng.: Profile Publications, 1973. 24p.

Sullivan, Jim. P2V Neptune in Action. Aircraft, no. 68. Carrollton, Tex.: Squadron/Signal Publications, 1985. 49p.

Periodical/Anthology Literature:

Andrews, Hal. "Naval Aircraft: Neptune." Naval Aviation News, (October 1977), 20-21.

Beltz, Stan. "The Snorkel Sniper." Skyways, IX (July 1950), 10-11.

Brown, Chuck. "Sweet Sixteen and Still Going Strong." All Hands, no. 544 (May 1962), 16-17.

Davies, Thomas D. and H. L. Hansen. "Design Turned the Trick: Lockheed P2V 'Truculent Turtle.'" Aviation, XLVI (January 1947), 70-71.

____ and Edward P. Stafford. "The Story of the Historic Flight." Navy, VIII (December 1965), 23-24+.
"Truculent Turtle" 1946 Australia-Bermuda flight, via Washington, D.C.

Hansen, Chuck. "Nuclear Neptunes." American Aviation Historical Society Journal, XXIV (Winter 1979), 262-268.

Kirkland, Thomas. "Like Walking a Tightrope: P2V Radar Patrol Plane." Life, LII (June 22, 1962), 62-63.

Kroger, Walter B. "Lockheed Neptune Record Flight Reveals Efficiency of New Design." Aviation News, VI (October 7, 1946), 7-8.

"The Lockheed P2V Neptune." Aero Digest, LXVI (April 1953), 28-32+.

"Lockheed P2V Neptune." Air Pictorial, XIV (February 1952), 55-56.

"Lockheed P2V-6 Neptune." Air Pictorial, XV (January 1953), 29-30.

"The Long-Lived Neptune." RAF Flying Review, XVI (April 1961), 26-27.

Mangion, Godfrey. "Neptune." Scale Models, X (November 1979), 751+.

Owen, Kenneth. "The Neptune in Service." Flight, LXIV (October 30, 1953), 589-594.

Pace, Steve. "Neptune." Airpower, XIX (March 1989), 10-47.

Reinertsen, Stephen P. "Search for the Neptune's Soul." Naval Aviation News, (September-October 1985), 20-25.

Trimble, Robert L. "Old Man of the Sea." Air Classics, XVII (December 1981), 14-26, 74-78; XVIII (January-March 1982), 24-29, 66-82, 24-29, 56-69, 16-29, 59-66.

"'Truculent Turtle.'" Air Classics, IV (August 1968), 26-39.

"'Truculent Turtle.'" Naval Aviation News, (September 1976), 24-28.

Wassall, J. B. "How the P2V Meets Patrol Objectives." Aviation Week, LI (July 25, 1949), 20-22+.

(Lockheed) P-3 Orion

Book/Document/Report Literature:

Golder, Thomas V. P-3 Pilot Errors: A Conceptual Approach. Advanced Research Paper. Newport, R. I.: Center for Advanced Research, Naval War College, 1978. 78p.

United States. Government Accounting Office. The P-3 Aircraft Overhaul Program Can Be Improved. Washington, D.C.: GPO, 1987. 29p.

Periodical/Anthology Literature:

Ackerman, Mark T. "Nobody's Helpless, Lumbering Grape." U.S. Naval Institute Proceedings, CXIV (January 1988), 101-104.

Andrews, Hal. "Naval Aircraft: Orion." Naval Aviation News, (October 1970), 20-21.

Beasley, Jay. "Landing the P-3 (or Whatever)." Approach, XIX (February 1974), 28-29.

Beebe, Jeff. "Mr. P-3." Naval Aviation News, (September 1981), 26-27.

Booda, Larry L. "P-3C Orion Flies Navy's New ASW Defense System." Undersea Technology, IX (November 1968), 64-65, 67-69.

Burgess, Rick. "P-3 Orion: 20 Years Young." Naval Aviation News, (August 1982), 14-17.

Butler, Charles T. and John Christiansen. "Three-Engine Ferry Considerations." Approach, XXIII (July 1977), 18-23.

Christmann, Timothy J. "P-3C Orion: Durable Plane, Demanding Mission." Naval Aviation News, (January-February 1986), 18-21.

Connolly, Robert. "Updates Add to P-3C Capabilities." Electronics, LI (June 8, 1978), 90+.

Dornheim, Michael A. "Modified P-3 Gives Customs Long-Range AEW Capability." Aviation Week and Space Technology, CXXIX (July 4, 1988), 34-36.

____ and Brendan M. Greeley, Jr. "Navy Chooses ASW Aircraft for 21st Century Forces." Aviation Week and Space Technology, CXXIX (October 24, 1988), 16-19.

Geddes, J. Philip. " ASW From the Air: The P-3C Orion in Operation." International Defence Review, V (February 1972), 27-29.

Godfrey, David W. H. "Orion: ASW Star Performer." Air International, IX (September 1975), 118-123, 143-147.

Hull, E. W. "A New Dimension in ASW." Ordnance, LIII (January-February 1969), 404-407.

Humphrey, Jim. "P-3 Crew Co-ordination During In-Flight Emergencies." Approach, XIX (October 1973), 22-24.

Jacobs, Gene. "The Lockheed AEW and C Aircraft." Asian Defence Journal, no. 12 (December 1984), 62+.

Jenkins, A. K. "Your Mission: SAR." Approach, XXI (September 1975), 16-17.
SAR = Search and Rescue.

King, Chet. "Searching For Big Fish: P-3 Orion." All Hands, no. 807 (May 1984), 16-17.

Layton, P. B. "Keeping the Orion a Hunter." U.S. Naval Institute Proceedings, CXIII (October 1987), 159-161.

Lefschultz, L. "The Flight of the Orion." Far Eastern Economic Review, CXXXIV (December 18, 1986), 24-25.

"Lockheed Completes Assembly of New P-3C Prototype." Design News, XXXIX (February 21, 1983), 24+.

"Low, Low Approach." Approach, XXI (August 1975), 16-17.

Luehring, Davidson. "The Never-Never Sea." U.S. Naval Institute Proceedings, XCV (August 1969), 142-146.

McClelland, Lori A. "Versatile P-3C Orion Meets Growing ASW Challenge." Defense Electronics, XVII (April 1985), 132+.

Marshall, William G. "The Airborne Destroyer." U.S. Naval Institute Proceedings, C (March 1974), 26-34.

Mihalak, Edward. "Where is Kedalion?" U.S. Naval Institute Proceedings, CIV (January 1978), 109-112.

Nordwall, Bruce D. "Navy Solicits Proposals for P-3C ASW Successor." Aviation Week and Space Technology, CXXVII (September 28, 1987), 30-32.

"Options for the Orion." Officer, LXII (January 1986), 15-18.

"Orion--ASW Answer." Flying Review International, XIX (March 1964), 9-14.

"Orion is Replacing Neptune in Patrol Squadrons." All Hands, no. 562 (November 1963), 562+.

"Orion Update." Flight International, XCIV (October 31, 1968), 708-709.

"Orion vs. the Submarine." Interavia, XVIII (September 1963), 1392-1396.

"P-3 Pilot Training." Approach, XIX (November 1973), 20-22.

"P-3C Decision Impact Weighed." Aviation Week and Space Technology, CXV (December 21, 1981), 16-17.

"P-3C Update." Naval Aviation News, (August 1982), 26-27.

Porth, Jacquelyn S. "Submarine Hunter." Defense and Foreign Affairs, X (July 1982), 10-13+.

Robinson, Clarence A., Jr. "Computer Enhances P-3C Effectiveness." Aviation Week and Space Technology, CV (November 15, 1976), 34-36.

Ropelewski, Robert R. "Lockheed Plans to Update P-3C Orion." Aviation Week and Space Technology, CXIII (December 8, 1980), 41+.

Sherman, Kenneth B. "Orion the Hunted." U.S. Naval Institute Proceedings, CXII (October 1986), 90-92.

Simonsen, Erik. "Countering the Soviet Sub Threat." Combat Arms, IV (July 1986), 24-28.

Stein, Kenneth J. "Orion Avionics Systems Being Updated." Aviation Week and Space Technology, CXVII (September 13, 1982), 101+.

Stenzoski, J. J. "P-3 Update." Naval Aviation News, (September-October 1987), 11-12.

Sullivan, Arthur. "Putting the P-3 Reserve to Work." U.S. Naval Institute Proceedings, CXII (October 1986), 93-94.

Waller, Edward C. "Orion: Meeting the Global Threat." Wings of Gold, XII (Winter 1987), 14-17.

Welch, Ken. "Orion the Hunter." Flying, CXIV (October 1987), 102-104.

Westlake, William R. "The P-3 Orion: A Politico-Military Instrument." U.S. Naval Institute Proceedings, XC (November 1970), 99-100.

Wilmoth, George. "Lockheed's Antisubmarine Aircraft: Watching the Threat." Defense Systems Review, III (June 1985), 18-20+.

Wilson, Tug. "P-3C Orion Accomodates System Changes." Military Electronics/Countermeasures, IX (March 1983), 39-42+.

(Martin) P4M Mercator

Trimble, Robert L. "Forgotten Warrior." Air Classics, XV (May 1979), 56-61.

(Martin) P5M Marlin

Andrews, Hal. "Naval Aircraft: Marlin." Naval Aviation News, (September 1978), 24-25.

Fey, W. G. "The Flying Boat Becomes a Lady." Pegasus, XXIV XXIV (March 1955), 8-12.

"Martin P5M-1 Marlin." Aero Digest, LXV (July 1952), 24-32.

"Misadventures of a P5M." Naval Aviation News, (June 1957), 32-33.

"Navy Sub Stalker." Skyways, XI (July 1952), 44-45.

"The P5M-2." Joint Services Recognition Journal, X (August 1955), 206-207.

"Seaborne Striking Force." Marine Corps Gazette, XLI (July 1957), 32-33.

Stone, Irving. "How Martin Tailored P5M to ASW Job." Aviation Week, LXII (January 17, 1955), 28-45.

"Swan Song of the Marlin." Ryan Reporter, XXVIII (May-June 1967), 29-33.

Trimble, Robert L. "Pig Boat." Air Classics, XVI (April 1980), 30-37, 60-69.

PB-1 see B-17 Flying Fortress

PB-2 see PBY Catalina

(Martin) PBM Mariner

Andrews, Hal. "Naval Aircraft: Mariner." Naval Aviation News, (April 1977), 20-21.

Gunston, William T. ("Bill"). "Martin's PBM Mariner." Aeroplane Monthly, IV (May 1976), 228-233.

Kelly, Frank. "Martin PBM Mariner Bomber." Model Airplane News, XCVI (April 1978), 29-31, 68.

Quinn, D. C. "Confessions of a PBM Pilot." Air Classics, XII (May 1976), 72-79.
Reprinted in Michael O'Leary, ed., Air Progress Pilot Reports (Canoga Park, Calif.: Challenge Publications, 1979), pp. 28-33.

Shanline, Rex. "The Dependable PBM Martin Mariner." Aerospace Historian, XXXII (March 1985), 33-40.

Smith, Bob. PBM Mariner in Action. Aircraft, no. 72. Carrollton, Tex.: Squadron/Signal Publications, 1986. 58p.

(Consolidated) PBY/OA-10 Catalina

Book/Document/Report Literature:

Cassangnares, Everett. Consolidated PBY Catalina. Aircraft in Profile, no. 183. Windsor, Eng.: Profile Publications, 1967. 15p.

Creed, Roscoe. PBY: The Catalina Flying Boat. Annapolis, Md.: Naval Institute Press, 1986. 351p.

Crocker, Mel. Black Cats and Dumbos: World War II's Fighting PBYs. Blue Ridge Summit, Pa.: TAB, 1987. 288p.

Hendrie, Andrew. Flying Cats: An Account of the Catalina. Annapolis, Md.: Naval Institute Press, 1988. 256p.

Knott, Richard C. Black Cat Raiders of World War II. Annapolis, Md.: Nautical and Aviation Publishing Company of America, 1981. 198p.

Scarborough, W. E. PBY Catalina in Action. Aircraft, no. 62. Carrollton, Tex.: Squadron/Signal Publications, 1983. 50p.

Thorburn, Lois and Donald. No Tumult, No Shouting: The Story of the PBY. New York: Holt, Rinehart and Winston, 1945. 148p.

Wagner, Ray. The Story of the PBY. Aero Biographies, no. 1. San Diego, Calif.: Flight Classics, 1973. 22p.

Periodical/Anthology Literature:

Andrews, Hal. "Naval Aircraft: Catalina." Naval Aviation News, (June 1972), 20-21.

Bachmann, L. P. "Dumbo, I Love You." Air Force, XXVII (June 1944), 18-20.

Bishop, Jim. "Dumbo: The Magnificent Navy PBY." Saturday Evening Post, CCXVII (August 5, 1944), 12-13+.

Bovelt, Allan. "Catalinas in the South Pacific." American Aviation Historical Society Journal, XVI (Winter 1971), 273-278.

Chappell, Bud. "PBY-2 Catalina." RC Modeler, XIII (July 1976), 24-29.

"Consolidated Catalina." Aircraft Illustrated, VII (September 1974), 373-376.

Cross, David W. J. "Consolidated Catalina PBY-6A L866." Air Classics, XII (November 1976), 41-43.

Dean, Jack. "Legion of the Lost: Consolidated's PB-2." Wings, IV (October 1974), 36-45.

"Dumbo." Flying, XXXV (October 1944), 52-53.

"Here Come the Cats." Popular Mechanics, LXXXIX (February 1943), 19-27.

Marshall, James L. "Old Lady With a Punch." Collier's, CXII (July 24, 1943), 18-19+.

Mundorff, George T. "The Catalina Patrol Bomber." Aerospace Historian, XXIII (December 1976), 217-222.

"Naval Aircraft Material Center's PBN-1." Aero Digest, XLIV (January 15, 1944), 88-94.

Oliver, Robert. "The Cats Have Nine Lives." Skyways, II (September 1943), 34-36+.

Poulos, George F. "Pegasus Flew Low, Slow, and Long During World War II." Naval Aviation Museum Foundation, IV (Fall 1984), 30-36, 45-48.

Rawlings, Charles A. "I Heard the Black Cats Singing: The Navy's PBY Pilots." Saturday Evening Post, CCXVII (December 2, 1944), 12-13+.

Reynolds, George A. "Flying, Fighting, and Floating Angels." Aerospace Historian, XVIII (December 1971), 200-201.

Richards, Dave. "You'll Never Get to Heaven in a PBY." Air Classics, X (July 1974), 24-27.

Rust, Kenn C. "Army Cats." Air Classics, VIII (November 1971), 11-16.

Scarborough, William E. "The Consolidated PBY: Catalina to Canso." American Aviation Historical Society Journal, XVI (Spring-Fall 1971), 27-38, 113-123, 200-203.

____. "P-Boat Flying." Airpower, V (May 1975), 8-21.

"Track of the Cat." Air Classics, V (October 1968), 62+.

Van Vleet, Clarke. "Mad Jack Cram and the 'Blue Goose.'" Naval Aviation News, (May 1982), 22-25.
Personal pilot/plane of Gen. Roy Geiger on Guadalcanal.

(Consolidated) PB2Y Coronado

"The Consolidated XPB2Y-1: The Navy's Newest Big Boat." Aviation, XXXVII (January 1938), 30-31.

DeLorenzo, Frank L. "The Coronado." Naval Aviation Museum Foundation, IX (Fall 1987), 18-26.

Gwinn, J. M., Jr. "How Coronados are Born." Air Trails, XXI (January 1944), 28-30.

Klemin, Alexander. "The World's Largest Naval Patrol Bomber." Scientific American, CLVIII (March 1938), 162-163.

Mohan, Charles Q. "Flying Barge." Skyways, III (August 1944), 50-52+.

"Some Design Features of the Coronado." Aero Digest, XLVI (July 1, 1944), 82-85.

Stout, George. "Details of the Coronado PB2Y-3 Design." Automotive and Aviation Industries, XCIII (July 15, 1945), 18-22+.

White, I. "Bombers Into Transports: Conversion of Coronado Flying Boats From Patrol Bombers to Cargo and Passenger Transports." Air Transportation, II (April 1944), 49-50.

(Consolidated) PB4Y-1 Liberator/PB4Y-2 Privateer
(see also B-24 Liberator)

Book/Document/Report Literature:

Consolidated-Vultee Aircraft Corporation. Flight Manual: PB4Y-2 Bombardment Airplane. San Diego, Calif., 1945. 87p.

Johnsen, Frederick A. Bombers in Blue: PB4Y-2 Privateers and PB4Y-1 Liberators. Glendale, Calif.: Aviation Book Company, 1979. 28p.

Periodical/Anthology Literature:

Andrews, Hal. "Naval Aircraft: PB4Y-1 Liberator." Naval Aviation News, (September-October 1987), 16-17.

____. "Naval Aircraft: PB4Y-2 Privateer." Naval Aviation News, (November 1979), 20-21.

Boone, Andrew R. "The Liberator is Reborn as the Navy's Privateer." Popular Science, CXLVI (April 1945), 96+.

"Convair PB4Y-2 Privateer." Air Pictorial, XV (June 1953), 179-180.

Hartman, Gary P. "'Old Blue's' Last Hurrah." Air Classics, XIX (September 1983), 58-63.
A restored PB4Y-2.

Mitchell, Mitch. "Charlie-30." Air Classics, XXIV (June 1988), 54-59.
The PB4Y-2.

"The Privateer PB4Y-2." Aero Digest, XLVIII (February 15, 1945), 86+.

Scutts, Jerry. "Privateer: In Detail." Scale Models, XIII (August 1982), 414-419.

Trimble, Robert L. "Still Going Strong." Air Classics, XV (October 1979), 49, 98.
A restored PB4Y-2.

(Douglas) PD-1

Andrews, Hal. "Naval Aircraft: Douglas PD-1." Naval Aviation News, (July 1975), 20-21.

(General Aviation) PJ Flying Lifeboat Antares

Boone, Andrew R. "Flying Lifeboats Make Thrilling Rescues at Sea." Popular Science, CXXXI (December 1937), 56-57.

Hall, Norman B. "Life Boats in the Sky." National Aeronautic Magazine, X (August 1932), 16-17.

Kalinowski, Ludwig B. "General Aviation's Antares, New Flying Boat for U.S. Coast Guard Embodies Many New Features of Design." Aviation Engineering, VII (August 1932), 10-12.

Lewis, H. Latane, II. "Flying Lifeboats of the Coast Guard." US Air Services, XVII (August 1932), 38-40.

Nalty, Bernard C. "The Coast Guard's Flying Lifeboat." Air Classics, XV (October 1979), 14-23.

Reisinger, Harold G. "The Flying Lifeboat of the Coast Guard." U.S. Naval Institute Proceedings, LIX (January 1933), 81-88.

Thaden, H. V. "The Coast Guard's Flying Lifeboat Antares." Coast Guard Magazine, (July 1932), 3-34.

PM see PN

(Martin/N.A.F.) PN

Book/Document/Report Literature:

Messimer, Dwight R. No Margin for Error: The U.S. Navy's Transpacific Flight of 1925. Annapolis, Md.: Naval Institue Press, 1981. 176p.

Periodical/Anthology Literature:

Andrews, Hal. "Naval Aircraft: PM." Naval Aviation News, (November 1976), 20-21.

Cocklin, Henry S. "The PN-12 Navy Patrol Plane." Aero Digest, XIII (July 1928), 44-46, 174.

Coleman, Charles H. "Given Up as Lost, Navy Hawaiian Flight Plane [PN-9] Found After Nine Days." US Air Services, X (October 1925), 14-18.

____. "The 2,000 Mile Flight of Navy Planes From San Francisco to Honolulu." US Air Services, X (September 1925), 13-17.

Murphy, William B. "They Sailed to the Flying Boat's Future." U.S. Naval Institute Proceedings, XCI (January 1965), 82-91.

"The Navy PN-12 Seaplane." Aviation, XXIV (May 14, 1928), 1366, 1404.

"PN-9 No. 1 Found Off Hawaiian Islands." Aviation, XIX (September 21, 1925), 350-352.

Rodgers, John. "Rodgers Tells His Story of the Flight." US Air Services, X (October 1925), 19-21.

"The Seaplane That Sailed to Glory." Air Classics, II (May 1966), 14-17.

(Lockheed) PV-1 Ventura
(see also B-34/B-37 Lexington)

Andrews, Hal. "Naval Aircraft: Ventura." Naval Aviation News, (August 1974), 20-21.

"A Big League Trainer: The Combat-Tested Venturas Start Where Advanced Trainers Stop." Flying, XXXIII (December 1943), 72-73+.

Bowers, Peter M. "Ventura: The Borrowed Bomber." Wings, V (February 1975), 40-53.

Greenway, Sean. "Sea-Search Ventura." Air Classics, XIV (June 1978), 34-35.

"Integral Wing Tanks Extend PV-1 Range." Aviation Age, XXIV (October 1955), 184-185.

Rawlings, Neal. "The Hudson-Ventura Story." Air Classics, VII (September 1971), 33-39.

Scrivner, Charles L. and William E. Scarborough. PV-1 Ventura in Action. Carrollton, Tex.: Squadron/Signal Publications, 1982. 49p.

"Versatile Ventura, Navy Patrol Ship, the PV-1." Popular Science, CXLVI (June 1945), 116-117.

(Lockheed) PV-2 Harpoon.

Andrews, Hal. "Naval Aircraft: Harpoon." Naval Aviation News, (August 1974), 20-21.

Scrivner, Charles L. "Harpoon: On Patrol in a Lockheed PV-2." Airpower, III (November 1973), 50-60.

(Vought) VE-7

Andrews, Hal. "Naval Aircraft: VE-7." Naval Aviation News, (December 1970), 20-21.

"VE-7 Airplane." Aerial Age, XII (October 4, 1920), 110-11.

(Vought) VE-10

"The Vought Flying Boat, Model VE-10." Aviation, VII (November 1919), 301-303.

"Vought Model VE-10 Makes Its Appearance." Aircraft Journal, V (September 13, 1919), 7, 18.

C. Reconnaissance and Observation, Electronic Warfare, and Utility Aircraft (Recce/EW/Utility Aircraft)

(Grumman) E-1 Tracker

Andrews, Hal. "Naval Aircraft: Tracker." Naval Aviation News, (July 1971), 20-21.

"One of These Days." Approach, XXI (June 1976), 6-7.

(Grumman) E-2 Hawkeye

"Airborne Control of Viet Strikes." Flight International, XC (August 11, 1966), 237-238.

"Airborne Navonics." Countermeasures, II (February 1976), 43-44+.

Andrews, Hal. "Naval Aircraft: Hawkeye." Naval Aviation News, (September 1970), 20-21.

Booda, Larry. "E-2A Compatible With Carrier Operations." Aviation Week and Space Technology, LXXVIII (March 4, 1963), 68-69.

Boyle, Dan. "Is AEW Getting More Affordable?" International Defence Review, XIX (September 1986), 1241-1244.

Brownlow, Cecil. "E-2A Controls Navy North Viet Strikes: The Hawkeye Twin-Turboprop Early Warning Aircraft." Aviation Week and Space Technology, LXXXV (August 1, 1966), 57+.

DeSantis, S. "Hawkeye: Terror of the Florida Drug Smugglers." Popular Mechanics, CLIX (February 1983), 79-81+.

Fenn, Richard W. "The E-2A Hawkeye." U.S. Naval Institute Proceedings, XC (November 1964), 156-159.

Geddes, J. Philip. "Airborne Early Warning for the U.S. Navy." International Defence Review, VIII (October 1975), 679-682.

Geisenheyner, Stefan. "Early Warning and Control." Asian Defence Journal, no. 1 (January 1981), 28-33.

Gray, Jeanne. "Patrolling the Skies." Naval Aviation News, (June 1982), 42-43.

Guarino, Thomas A. and Charles F. Muller, Jr. "The Eyes of the Fleet." U.S. Naval Institute Proceedings, CII (October 1976), 117-120.

"Hawkeye: A New Dimension in Tactical Warfare." Air International, XII (January 1977), 7-13.

Haystead, John. "Yesterday, Today and Tomorrow: The Navy's E-2C Hawkeye Grows With the Threat." Journal of Electronic Defense, VIII (October 1985), 57-58+.

Lerner, E. J. "Giant Surveillance Radar Squeezed Into Wing." Aerospace America, XXIII (July 1985), 29-32.

McClelland, Lori A. and David M. Russell. "AEW Aircraft Market Soars." Defense Electronics, XVI (September 1984), 124-126.

Magnet, M. "Grumman's Comeback." Fortune, CVI (September 20, 1982), 62-64+.

Moore, Bob. "E-2C Hawkeye: Sentinel in the Sky." Naval Aviation News, (November 1974), 8-15.

Mulquin, James J. "E-2A Early Warning System Operations in the Western Pacific." U.S. Naval Institute Proceedings, XCIII (July 1967), 132-134.

North, David M. "E-2C Cements Early Warning Role With Other Pilots." Aviation Week and Space Technology, CVI (January 31, 1977), 108-109.

Page, Richard S. "Eyes and Ears for the Tactical Air Force." Aerospace International, XIII (January-February 1977), 22-24+.

Philip, John S. "Early Warning and Control." Aerospace International, XI (September-October 1975), 6-10.

Place, Cam. "The Saga of the Humble Hummer." Approach, XXII (December 1976), 6-8.

Richeson, Jim. "Sharpening the Eyes of the Fleet." Naval Aviation News, (November-December 1987), 14-17.

Rosenblatt, A. "Navy Gets a New Airborne Watchdog." Electronics, XLII (July 9, 1969), 133-134+.

Russell, J. B. "The E-2C Hawkeye: A Most Versatile Bird." U.S. Naval Institute Proceedings, CIII (June 1977), 104-105.

Welling, William. "E-2C Airborne Early-Warning Aircraft." International Combat Arms, VI (March 1988), 40-43.

(Boeing) E-3 Sentry

Book/Document/Report Literature:

Allen, John K. AWACS Diplomacy. Research Report. Maxwell AFB, Ala.: Air War College, Air University, 1985. 23p.

Allen, Melvin A. A New E-3A Employment Concept for Air Defense: What is Needed for the 1980s? Research Report. Maxwell AFB, Ala.: Air War College, Air University, 1979. 75p.

Koiner, Wylie J. A User's Guide for the Airborne Warning and Control System. Research Report. Maxwell AFB, Ala.: Air War College, Air University, 1986. 29p.

Lloyd, Alwyn T. Boeing 707 and AWACS in Detail and Scale. Detail and Scale Series, v. 23. Fallbrook, Calif.: Aero Publishers, 1986. 72p.

Miller, Jack C., II. Evolution of the AWACS. Research Paper. Maxwell AFB, Ala.: Air Command and Staff College, Air University, 1986. 72p.

O'Neill, Thomas P. Utilization of the E-3A in Iceland. Research Paper. Maxwell AFB, Ala.: Air Command and Staff College, Air University, 1982. 21p.

Sun, Jack K., et al. AWACS Radar Program: "The Eyes of the Eagle." Baltimore, Md.: Westinghouse, 1985. 23p.

West, Walter P. Considerations for the Deployment of the E-3 AWACS. Research Paper. Maxwell AFB, Ala.: Air Command and Staff College, Air University, 1986. 43p.

Periodical/Anthology Literature:

Alder, Konrad. "Airborne Surveillance and Early Warning System." Armada International, X (August 1986), 6+.

Beach, John B. "First AWACS Update." Defense Electronics, XII (July 1980), 21-22+.

Beal, Clifford F. "AWACS:Finally Cracking the World Market?" Journal of Defense and Diplomacy, V (June 1987), 64-67.

Burton, John, III. "Under His Wings Shalt Thou Trust." Defense and Foreign Affairs, IX (April 1981), 9+.

Chacko, George K. and Bruce W. Stratvert. "Systems Approach to Multinational Acquisition: NATO AWACS." Concepts, IV (Autumn 1981), 90-106.

Christy, John. "Angels at 30,000 Feet." Combat Arms, IV (March 1986), 38-42.

Church, G. J. "Flying Into Trouble." Time, CXVII (May 4, 1981), 14-16.

Clark, Graham. "AWACS Teams Train With a Will." Military Technology, XI (February 1987), 122-125+.

Coleman, Herbert J. "AWACS Tested in NATO Role." Aviation Week and Space Technology, CII (April 14, 1975), 12-13.

Correll, John R. "With the Eye of an Eagle." Airman, XXI (March 1977), 33-37.

Cutler, Katie. "NATO's Own Air Force." Airman, XXVII (June 1983), 17-25.

Davis, Richard T. "Can AWACS Survive an Attack?" Electronic Warfare, IX (May-June 1977), 43-46+.

Day, Bonner. "AWACS in Operation." Air Force Magazine, LXII (June 1979), 52-56.

Elson, Benjamin M. "Advanced Version of AWACS Expected." Aviation Week and Space Technology, CXIII (November 10, 1980), 69+.

___. "Air Force Gets First Operational E-3A." Aviation Week and Space Technology, CVI (March 28, 1977), 19-21.

___. "E-3A Expands USAF Defense Capabilities." Aviation Week and Space Technology, CVI (April 18, 1977), 49+.

___. "TAC Readies for Deployment of E-3A." Aviation Week and Space Technology, CVIII (February 6, 1978), 105-112.

English, R. D. and D. I. Bolaf. "Defense Against Bomber Attack: AWACS." Scientific American, CCXXIX (August 1973), 11-19.

Ethell, Jeffrey L. "We Fly the AWACS." Air Progress, XLII (September 1980), 74+.

___. "Fistful of Dollars: Flying With the Airborne Sentry." Air Progress, Aviation Review, IV (Summer 1980), 26-33, 78.

Fuchs, Ron. "Tough, But Worth It." Airman, XXIX (April 1985), 14-15.

___. "Hunting for Bear." Airman, XXIX (April 1985), 8-13.

Geddes, J. Philip. "AWACS Takes Off." Interavia, XXVII (April 1972), 392-393.

Glasser, Frederick J. "E-3A: An Untapped Maritime Support Resource." U.S. Naval Institute Proceedings, CV (August 1979), 108-111.

Gordon, Ray. "E-3." Flying Safety, XLII (April 1986), 18-19.

Hart, John W. "It's Called AWACS--An 'Eye in the Sky.'" Airman, XV (June 1971), 15-17.

Hughes, David. "USAF Will Develop Major Radar Upgrade for Its E-3 AWACS Fleet." Aviation Week and Space Technology, CXXX (January 23, 1989), 45-49.

Jung, Kineson. "AWACS--Needed Now." Tactical Air Warfare Center Quarterly Report, II (June 1970), 14-17.

Kessler, M. J. "Reconnoitering the Electromagnetic Spectrum." Defense Electronics, XVII (September 1985), 107-108+.

Klass, Philip J. "AWACS Future Hinges on Feasability Tests." Aviation Week and Space Technology, LXXXII (February 22, 1965), 80-81+.

____. "AWACS Meets Heavy Traffic Challenge." Aviation Week and Space Technology, CII (May 5, 1973), 18-19.

____. "AWACS Seen Gaining Long-Term Impact." Aviation Week and Space Technology, LXXXVIII (March 18, 1968), 285-288.

____. "GAO Study Expands AWACS Data." Aviation Week and Space Technology, CIII (September 15, 1975), 47-51.

____. "GAO Study Hits AWACS Vulnerability to ECM." Aviation Week and Space Technology, CII (March 1, 1975), 45+.

Meikel, Arthur P., III. "E-3A." Flying Safety, XL (May 1984), 20-21.

Meurer, Fred. "Building a Better Bubble." Air Force Magazine, LVIII (April 1975), 32-37.

Mitchelmore, Gary. "Big World of the Sentry." Air Force Magazine, LXVII (June 1984), 70-74+.

____. "A Long, Deep Look." Airman, XXV (May 1981), 28-32.

"Multistage Improvements Proposed for AWACS." Aviation Week and Space Technology, CXXIII (December 9, 1985), 102-103.

North, Jim and Mike King. "Eye in the Sky." Air Force Magazine, LXI (July 1978), 86-90.

Pearce, Harry A. "AWACS to Bridge the Technology Gap." Air University Review, XXIII (May-June 1972), 55-61.

Petersen, Stefan. "The Guardians: The Multinational E-3A Component of NATO's Airborne Early Warning Force." Defence Update International, no. 64. (1985), 24-29.

Phillip, John S. "Early Warning and Control." Aerospace International, XI (September-October 1975), 6-10+.

Rimmek, Klaus. "Early Warning in NATO: NATO's Multinational Air Force." NATO's Sixteen Nations, XXVIII (December 1983-January 1984), 48-49+.

Riolo, Dre , "You've Come a Long Way, Nav." Aerospace Safety, XXXV (September 1979), 2-5.

Robins, Yves and Siegfried Boy. "AWACS: Portrait of a Sentry." NATO's Fifteen Nations, XXIV (June-July 1979), 89-90+.

Robinson, Clarence A., Jr. "Air Defense Role for AWACS Stressed." Aviation Week and Space Technology, C (March 11, 1974), 19-21.

Rutter, George W. "E-3A." Signal, XXXIII (October 1978), 9-12+.

Schumack, Thomas A. "AWACS: Where the Action Is." Navigator, XXV (Spring 1978), 12-14.

Sheffield, Dick. "A Mighty Tall Story." Airman, XXII (October 1978), 44-48.

Skantze, Lawrence A. "How the E-3A Gives the Big Battle Picture." Air Force Magazine, LX (July 1977), 58-62.

Stein, Kenneth J. "AWACS Capabilities Tested in Europe." Aviation Week and Space Technology, CVIII (January 30, 1978), 60-71.

Tagnesi, John P. "Flying the E-3A." TAC Attack, XVII (November 1977), 4-7.

Thompson, Geoffrey R. "Flying Computer Programmers." Aerospace Safety, XXXIV (March 1978), 5-7.

Ulsamer, Edgar. "AWACS: Mobile 'Eye in the Sky' for Airborne Surveillance." Aerospace International, VI (September-October 1970), 18+.

____. "AWACS: USAF's Most Successful Development Program." Air Force Magazine, LVII (July 1974), 69-76.

Walsh, Bill. "An Eagle in the Air." Countermeasures, II (July 1976), 30+.

Warzinski, Victor. "E-3A Flight Training on the Ground." Flying Safety, XXXVII (April 1981), 20-21.

Widell, Carl A. "AWACS--a What?" Air Defense Magazine, (October-December 1977), 30-34.

Witze, Claude. "AWACS: Fast, Mobile and Versatile." Air Force and Space Digest, LIII (June 1970), 41-45.

____. "Versatile AWACS: Its Expanding Tactical Potential." Aerospace International, VI (September-October 1970), 19+.

Wright, Dwayne P. "The Day of the E-3A Tactical Test." Air University Review, XXIX (September-October 1978), 58-73.

(Boeing) E-4 Advanced Airborne National Command Post

Book/Document/Report Literature:

Pederson, Arvid P. Advanced Airborne Command Post Aircraft: A Case Study. Research Paper. Maxwell AFB, Ala.: Air Command and Staff College, Air University, 1982. 41p.

Periodical/Anthology Literature:

Babyak, Blythe. "Cruising to Armageddon: Presidential Plane." Washington Monthly, X (December 1978), 32-36.

Hirst, Mike. "E-4B: The Supreme Airborne Commander." Flight International, CXIII (June 3, 1978), 1702-1704.

Liebman, Marc. "Pacer Link--Updating the Worldwide Airborne Command Post." Defense Electronics, XX (September 1988), 95-96+.

"Modified E-4 Project Calls for Six Aircraft." Aviation Week and Space Technology, CIV (May 10, 1976), 59-61.

O'Lone, Richard G. "Boeing Sees Military Role for 747." Aviation Week and Space Technology, C (April 8, 1974), 48-50.

Phillips, Susanne L. "Salvation From Above." Airman, XXII (February 1978), 27-31.

Rudd, Douglas L. "Navigating the Great White Bird." Navigator, XXXIII (Summer 1986), 5-6.

Sams, John B., Jr. "E-4A: A Pilot's Report." Combat Crew, XXVIII (October 1978), 4-7.

Shockley, W. L. "The E-4 Airborne Command Post." Countermeasures, II (July 1976), 34-36.

Stein, Kenneth J. "E-4B Boosts SAC's Communications." Aviation Week and Space Technology, CXII (June 16, 1980), 77+.

Ulsamer, Edgar. "Boeing's Heavyweight Contender." Air Force Magazine, LV (January 1972), 68-69.

Van Riper, Frederick. "The President's Doomsday Plane." Saturday Evening Post, CCL (September 1978), 52-54.

(Grumman) EA-6 Prowler

(see also A-6 Intruder)

Book/Document/Report Literature:

Francillon, René J. Grumman EA-6A/B. Minigraph, no. 7. Arlington, Tex.: Aerofax, Inc., 1984. 40p.

Periodical/Anthology Literature:

Evans, Julius E. "Jamming." Naval Aviation News, (July-August 1988), 24-27.

Gershanoff, Hal. "The EA-6B Prowler: Five Generations and Still Going Strong." Journal of Electronic Defense, VI (October 1983), 51-52+.

Godfrey, David W. H. "All-Weather Tacticians." Air International, X (January 1976), 7-14.

Henderson, Breck W. "EA-6B Flight." Aviation Week and Space Technology, CXXIX (October 24, 1988), 48-57, 61.

"Improvements in Works on EA-6B Aircraft." Aviation Week and Space Technology, CXV (September 21, 1981), 88-90.

Jackson, Grady L. "Maintaining the Edge: The Navy's EA-6B AVCAP Program." Journal of Electronic Defense, X (April 1987), 41-42+.

Krech, Kenneth. "The HARM'ed Prowler: A Potential Platform for Lethal or Non-Lethal ECM." Journal of Electronic Defense, VI (October 1983), 85-86+.
HARM = High-Speed Anti-Radiation Missile.

____. "Harming the Prowler." Defense Electronics, XVIII (October 1986), 165-166+.

Robinson, Clarence A., Jr. "EA-6B Demonstrates ECM Capabilities." Aviation Week and Space Technology, CV (December 11, 1976), 38-39+.

Shohat, Murray. "The EA-6B Weapon Systems Trainer." Military Electronics/Countermeasures, V (April 1979), 50+.

Watkins, Robert G., Jr. "A Simulated Approach to EA-6B Aircraft Training." U.S. Naval Institute Proceedings, CV (April 1979), 117-118.

(Boeing) E-6A TACAMO

Christmann, Timothy J. "New TACAMO Aircraft: E-6A to Replace EC-130." Naval Aviation News, (November-December 1986), 24-25.

"Military Report: E-6A." Air Progress, XLIX (May 1987), 68-69.

(Martin) EB-57 Canberra
(see also B-57 Canberra)

"ADC's High Altitude Faker." Interceptor, XI (February 1969), 16-17.

Grant, James R. "The EB-57: Like Most Antiques, Its Value Increases With Age." Navigator, XXVII (Summer 1980), 12-13.

Hutcherson, Norman. "Canberras to Europe." Interceptor, XVIII (December 1976), 8-9.

(Douglas) EB-66 Destroyer
(see also B-66 Destroyer)

"EB-66 Destroyer." TAC Attack, XXVIII (March 1988), 16-17.

Moore, Courtland C. EB-66 Out-Country Electronic Reconnaissance, 1965-1967. Research Report. Maxwell AFB, Ala.: Air War College, Air University, 1968. 79p.

(I.A.F.) EC-18 Aria

Greeley, Brendan M., Jr. "USAF Readies Range Instrumentation Aircraft for First Flight." Aviation Week and Space Technology, CXXII (February 25, 1985), 23-24.

Meyer, Donald K. "They Call Me ARIA: Advanced Range Instrumentated Aircraft." Navigator, XXXI (Summer 1984), 22-24.

(Douglas) EC-24

Hunter, F. S. "Douglas Aims DC-8F at New Markets." Airlift, XXV (June 1961), 25-27.

(Lockheed) EC-121 Constellation
(see also C-121 Constellation)

Andrews, Hal. "Naval Aircraft: Constellation." Naval Aviation News, (February 1972), 20-21.

Cole, Dave. "The Connies of College Eye." Airman, XVII (March 1973), 4-7.

Harvey, Frank L. "Warroom in the Sky." Popular Science, CLXXVII (November 1960), 94-97+.

Herranen, Harry. "Airborne Ohaha." Skyline, XIX (February 1961), 40-47.

Meredith, Charles. "Airborne Command Post." Airman, IV (December 1960), 14-17.

Witze, Claude. "Can an Airborne Alert Prevent a Space-Age Pearl Harbor?" Air Force, XLIII (March 1960), 33-37.

(Lockheed) EC-130 Hercules
(see also C-130 Hercules)

Allsup, Dan. "When Compass Called." Airman, XXXII (June 1988), 28-31.

Apap, Antonio. "TACAMO: A New Airborne Weapons System and Warfare Specialty." U.S. Naval Institute Proceedings, XCIX (September 1973), 113-115.

"Lockheed Surveillance Concept Gains." Aviation Week and Space Technology, CXIV (May 25, 1981), 24-25.

Nordwall, Bruce D. "New Airborne Command Center to Support Forward Forces." Aviation Week and Space Technology, CXXIX (July 18, 1988), 63-65.

Pratt, Laurence R. "The Herk's Newest Mission: EC-130H Compass Call." Navigator, XXXI (Winter 1984), 25-26.

Wetmore, Warren C. "EC-130 Fleet Nearing Full Strength." Aviation Week and Space Technology, CVI (February 28, 1977), 38-39.

(Boeing) EC-135 Looking Glass

(see also C/KC-135 Stratofreighter/Stratotanker)

Book/Document/Report Literature:

Stooke, W. N., Jr. Retrofit of SAC EC-135C and RC-135 Aircraft With CFM-56 Engines. Research Paper. Maxwell AFB, Ala.: Air Command and Staff College, Air University, 1985. 42p.

Periodical/Anthology Literature:

Cutler, Katie. "Inside the Looking Glass." Airman, XXV (September 1982), 7-12.

Denmark, James H. "The Jimmy Durante Air Force." Airman, XIV (January 1970), 59-61.

Gillert, Doug. "Through the Looking Glass." Airman, XXXI (October 1987), 13-15.

"Looking Glass Capabilities Improved: SAC Airborne Command Post Aircraft." Aviation Week and Space Technology, CIV (May 10, 1976), 52-54.

Patton, William E., Jr. "You and the Airborne Battle Staff." Combat Crew, XVII (June 1966), 13-14.

Scholin, Allan R. "Airborne 'CP' of SAC Now in Fifth Year of Continuous Operation." Air Force and Space Digest International, I (April 1965), 45-46.

Schuermann, Robert S. "Operation Looking Glass." Combat Crew, XVII (June 1966), 12-13.

Stein, Kenneth J. "EC-135 Looking Glass Provides New Capabilities." Aviation Week and Space Technology, CXII (June 16, 1980), 85-87.

"TAC's One and Only." Airman, IX (May 1965), 4-5.
EC-135K.

Ulsamer, Edgar. "Nuclear-Proof Flying Command Post." Air Force Magazine, LVI (January 1973), 64-66.

EC-137 see E-3 Sentry

(General Dynamics) EF-111 Raven
(see also F-111 Aardvark)

"Blinding the Enemy." Journal of Electronic Defense, VI (April 1983), 57-58.

Brown, David A. "EF-111A Deployment Provides Radar Countermeasures Ability." Aviation Week and Space Technology, CXXIII (August 5, 1983), 44-45.

Dorio, Mark G. "The EF-111 Raven." Navigator, XXXII (Summer 1985), 26-27.

"EF-111 Raven." TAC Attack, XXVI (September 1986), 16-17.

"EF-111A Raven: Airborne EW Specialists in Europe." Defence Update International, no. 66 (1985), 24-32.

Fuchs, Ron. "Jamming With the Jammers." Airman, XXIX (July 1985), 37-40.

Herd, Doug. "Electronic Warfare Officer: An Electronic Warfare Scenario." Defense Science and Electronics, II (November 1985), 25-27+.

"Jammer's Delight." Air International, XXXI (1986), 281+.

Klass, Philip J. "EF-111A Exceeds Goals of Defense Department." Aviation Week and Space Technology, CXI (October 20, 1979), 68-71.

"Modern War Wizards." Defence Update International, no. 78 (December 1986), 45-51.

Pugh, Craig. "The EF-111A: Fox on the Prowl." Airman, XXVI (August 1982), 16-21.

Saltzmann, A. Robert. "EF-111: Newest Electronic Warrior." Grumman Aerospace Horizons, XIV (Summer 1978), 2-9.

____. "Tactical Support Jamming." Journal of Electronic Defense, VI (December 1983), 29-30+.

Scire, David H. "EF-111A." Navigator, XXVIII (Winter 1981), 18-20.

Stein, Kenneth J. "EF-111A Geared for Electronic Thicket." Aviation Week and Space Technology, CVIII (February 6, 1978), 93-95.

___. "EF-111A to Begin Full Flight Tests." Aviation Week and Space Technology, CVI (June 13, 1977), 65+.

Thompson, Warren. "Jammers." Wings, XVII (June 1987), 46+.

Ulsamer, Edgar. "EF-111A: The U.S. Air Force's Supersonic Electronic Shield." Air Force Magazine, LX (August 1977), 54-59.

___. "New Era in Electronic Warfare." Air Force Magazine, LXIV (September 1981), 116-118.

F-1 see C-8

F-2 see C-45

F-3 see A-20 Havoc

F-4 see P-38 Lightning

(Lockheed) F-5 Lightning
(see also P-38 Lightning)

Urbank, Walter J. "Photo Joe." Air Classics, XX (July 1984), 16-25.

(North American) F-6 Mustang
(see also P-51 Mustang)

"Flying Undertaker." Air Classics, V (December 1968), 6-14.

Ivie, Tom. "Recon Ace Supreme." Air Classics, XXI (March 1985), 14-23.

(Consolidated) F-7 Liberator
(see also B-24 Liberator)

United States. Army Air Forces. Air Technical Service Command. F-7, February 1943-July 1945. ATSC Historical Study, no. 54. Wright Field, Ohio: USAAF Air Technical Service Command, 1945. 19p.

(DeHavilland) F-8 Mosquito

Bishop, Edward. Mosquito. Ballantine's Illustrated History of World War II. New York: Ballantine Books, 1971. 160p.

___. The Wooden Wonder: The Story of the DeHavilland Mosquito. London, Eng.: Parrish, 1959. 168p.

Hardy, Michael J. The DeHavilland Mosquito. New York: Arco, 1977. 128p.

Holliday, Joseph. Mosquito: The Wooden Wonder Aircraft of World War II. Garden City, N.Y.: Doubleday, 1970. 236p.

"Mosquito." Air Classics, V (October 1968), 54-62.

Moyes, Philip J. B. DeHavilland Mosquito I-IV. Aircraft in Profile, no. 52. Windsor, Eng.: Profile Publications, 1966. 12p.

Sharp, Cecil M. and Michael J. F. Bowyer. Mosquito at War. New, rev. ed. London, Eng.: Faber and Faber, 1971. 498p.

F-9 see B-17 Flying Fortress

F-10 see B-25/PBJ Mitchell

(Boeing) F-13 Superfortress
(see also B-29 Superfortress)

Morse, David B. "Eye in the Sky: The Boeing F-13." American Aviation Historical Society Journal, XXVI (Summer 1981), 150-168.

F-14 see F-80 Shooting Star

F-15 see P-61 Black Widow

(Grumman) HU-16/SA-16/JR2F Albatross

Book/Document/Report Literature:

Ginter, Steven J. Grumman HU-16 Albatross. Naval Fighters Series, no. 11. Simi Valley, Calif.: Naval Fighters, 1984. 74p.

Periodical/Anthology Literature:

"Albatross Modified for ASW." National Defense Transportation Journal, XVII (January-February 1961), 4+.

Andrews, Hal. "Naval Aircraft: Albatross." Naval Aviation News, (December 1981), 20-21.

"Grumman Albatross." Aero Digest, LXVII (November 1953), 28-32.

"Grumman Albatross." Skyways, XI (February 1952), 42-43.

"Grumman SA-16B Albatross: Ubiquity and Versatility." RAF Flying Review, XVIII (November 1962), 24-27.

Koch, R. W. "The CIA's Death Valley Albatross." Air Classics, XV (April 1979), 68-73, 98.

Kovitt, Bern. "Rebirth of the Albatross." Grumman Aerospace Horizons, XVIII (Summer 1982), 16-20.

Lunde, Erik. "Watchdog of the North." NATO's Fifteen Nations, VII (December 1962-January 1963), 66-68.

Mosley, W. F. "I'll Take the Albatross." Flying, LIX (November 1956), 41+.

Powers, Robert J. "Flying the SA-16A Albatross." Aerospace Historian, XXXII (December 1985), 252-260.

Steven, F. J. "Flying Boat Pilotage." Shell Aviation News, nos. 382-384 (Spring-Fall 1970), 2-7, 22-24, 2-8.

Stone, Irving. "Albatross: Multi-Mission Amphibian." Aviation Week, LXII (July 17, 1950), 22+.

(Dassault-Breguet) HU-25A Falcon

Andrews, Hal. "Naval Aircraft: Falcon." Naval Aviation News, (July 1981), 24-25.

"ATF-3-6 Engine Problem Delays HU-25A." Aviation Week and Space Technology, CXIII (September 15, 1980), 86-87.

"Final HU-25A Delivery Due This Month." Aviation Week and Space Technology, CXIX (December 12, 1983), 26-27.

Geddes, J. Philip. "Choosing Civil Maritime Patrol Aircraft: It Isn't All That Simple." Interavia, XXXV (February 1980), 147-149.

Grangier, Marc. "Guardian Bizjet Watches Waves." Interavia, XXXVI (September 1981), 901-902.

"HU-25A Delay Spurs Contingency Plan." Aviation Week and Space Technology, CXII (February 25, 1980), 50+.

McClellan, J. Mac. "The Art of Falconry." Flying, CXV (May 1988), 28-39.

North, David M. "HU-25A Increases Coast Guard Mission Capabilities." Aviation Week and Space Technology, CXVII (July 26, 1982), 112-113+.

____. "Other Nations Monitor HU-25A Effort." Aviation Week and Space Technology, CX (March 19, 1979), 62-65.

Thorsen, Ho ard B. "The Coast Guard's New Medium-Range Surveillance Aircraft." U.S. Naval Institute Proceedings, CVII (April 1981), 112-114.

"The U.S. Coast Guard's Falcon 20G Sets New Standards for MRS Aircraft." Interavia, XXXII (June 1977), 607-609.

"Uncle Sam's Gallic Guardian." Air International, XX (April 1981), 173-179, 189.

(Grumman) J2F Duck

Buchanan, James B. "Eminent Duck." Naval Aviation News, (December 1980), 44-45.

Shea, Tim. "Duck on the Wing." Air Progress, XLV (March 1983), 4-5.

Tallman, Frank. "The Duck That Won 'Murphy's War.'" Flying, XCI (July 1972), 58-63.

Trimble, Robert L. "The Duck That Roared." Air Classics, XIV (December 1978), 32-41, 75-76.

(Fairchild) J2K

Stone, Tim. "A Fairchild Joins the Coast Guard." Air Progress, XXXIX (January 1977), 61+.

(Grumman) J4F/OA-14 Widgeon

Bowers, Peter M. "The Grumman Widgeon." The AOPA Pilot, XIX (December 1976), 65-67.

"Grumman J4F-2 Widgeon." Aero Digest, XLVI (September 15, 1944), 98-102.

(Grumman) JRF/OA-9/OA-13 Goose

Bowers, Peter M. "The Grumman Goose." The AOPA Pilot, XXI (May 1980), 121-123.

"Buying Off the Drawing Board: Grumman G-21 Amphibian." Scientific American, CLVIII (February 1938), 103-104.

Lert, Peter and Jim Larsen. "Wet and Wild." Air Progress, XLIV (December 1982), 70+.

(Taylorcraft) L-2

Davisson, Bud. "The Littlest Warbird." Air Progress, XXXVII (December 1975), 67-70.

____. "Warbug." Air Progress, XLVIII (March 1986), 78-82.

Smith, Ken. "1937 Model A Taylorcraft." Plane and Pilot, XVI (October 1980), 64-68.

"Taylorcraft's Liaison Planes: L-2 Series." Aero Digest, XLVIII (February 15-March 1, 1945), 71-75, 76-79.

Walker, David A. "Taylorcraft." Aviation Quarterly, VI (Fall 1980), 309-317.

(Piper) L-4/NE Cub

Book/Document/Report Literature:

Triggs, James M. The Piper Cub Story. New York: Sports Car Press, 1963. 108p.

Periodical/Anthology Literature:

Bech, F. M. "Grasshoppers Prove Worth in War." Aero Digest, XLV (April 1, 1944), 75+.

"Cubs For Combat." Skyways, X (February 1951), 18-19, 38.

"The Dauntless Piper Cub at Cape Gloucester." Marine Corps Gazette, LXIX (March 1985), 43-45.

Davisson, Bud. "The Classic Cub." Air Progress, XLIX (February 1987), 45-51+.

____. "Perspective on a Legend." Air Progress, XLIV (April 1982), 74+.

Francis, Devon. "The Piper Cub Goes to War." American Aviation Historical Society Journal, XXI (Spring 1976), 72-79.

Gault, Owen. "Cubs at War." Air Classics, IX (March 1973), 18-27.

Glines, Carroll V. "50 Years of the Cub." The Air Line Pilot, XLIX (September 1980), 14-20.

Golbey, Seth B. "Charisma." The AOPA Pilot, XXIX (July 1986), 46-53.

Horne, Thomas A. "Selling the Dream." The AOPA Pilot, XXIX (July 1986), 54-59.

Hutton, C. I. "Cubs in Combat." U.S. Army Aviation Digest, III (June 1957), 31-38.

Munson, Russell. "Bear Country." Flying, CXIV (October 1987), 64-69.

____. "Cub." Flying, CI (December 1977), 62-67, 103.

Ragle, G. L. "Dragonflies Over Korea." Combat Forces Journal, I (November 1950), 32-33.

Stitt, N. S. P. "In Italy." Field Artillery Journal, XXXIV (May 1944), 280-281.

Stubblefield, Bernard. "Grasshoppers Go to War." Aviation, XLIII (October 23, 1944), 114-115+.

(Stinson) L-5 Sentinal

Davisson, Bud. "Sentinal." Air Progress, XLVIII (November 1986), 54+.

___. "The Littlest Warbird." Air Progress, XXXVII (December 1975), 66-70.

"Guinea Shortlines." Air Classics, II (May 1966), 9-12.

Parnell, Neville M. "Antipodean Guardian." Aeroplane Monthly, IX (June 1981), 337-338.

Smith, Gene. "Mr. Johnson's Marvelous Machine." Air Progress, XLV (April 1983), 68-74.

"Stinson L-5 Sentinal." Air Pictorial, XV (March 1953), 82-83.

"Test Flights Show Army's Sentinal to Be Rugged Flying Jeep." Steel, CXI (November 23, 1942), 52-54+.

United States. Army Air Forces. Training Command. Pilot Training Manual for the L-5. AAF Manual 51-128-1. Washington, D.C., 1945.

L-9 see L-5 Sentinal

(Convair) L-13

"Convair L-13A." Air Pictorial, XV (January 1953), 18-19.

"Versatile Grasshopper: L-13 Liaison Plane." Flying, XLIII (August 1948), 21-23.

(Aeronca) L-16

"The New Grasshopper: L-16." Field Artillery Journal, XXXVII (November-December 1947), 355-356.

(Piper) L-18 SuperCub

Twombly, Mark R. "Ursa Major." The AOPA Pilot, XXIX (July 1986), 68-72.

L-19 see O-1 Bird Dog

(Piper) L-21 SuperCub

Drake, Charlie. "The L-21 Serves Again." Translog, IX (March 1978), 15-16.

(Piper) L-23 Seminole

Apple, W. V. "L-23 Operations Hints." U.S. Army Aviation Digest, I (November 1955), 5-7.

Hill, E. D. "L-23 Modifications." U.S. Army Aviation Digest, II (March 1956), 9-11.

L-26 see U-4 Commander

L-27 see U-3 Blue Bird

L-38 see U-10 Super Courier

(Curtiss) O-1 Falcon
(see also A-3 Falcon)

"The Curtiss Falcon Observation Plane." Aviation, XI (July 19, 1926), 92-94.

"Falcon Observation Plane Undergoing Service Tests." US Air Services, XI (July 1926), 40-42.

Westberg, Peter. "The Unforgettable Curtiss Falcon." Air Classics, XI (November 1975), 66-81.

(Cessna) O-1/L-19 Bird Dog

Book/Document/Report Literature:

Adcock, Al. O-1 Bird Dog in Action. Carrollton, Tex.: Squadron/Signal Publications, 1988. 50p.

Krull, Larry D. VNAF O-1 Operations, 1962-1968. Research Paper. Maxwell AFB, Ala.: Air Command and Staff College, Air University, 1969. 56p.

Periodical/Anthology Literature:

Bourne, Harold O. "Can the Bird Dog Fly IFR?" U.S. Army Aviation Digest, VI (October 1960), 4-7.

Browning, Joseph. "Bird Dog Over the Jungle." National Guardsman, XXI (March 1967), 10-12.

Broyles, Hugh L. "L-19 High Altitude Operations." U.S. Army Aviation Digest, VII (April 1961), 18-19.

Burnham, Frank. "Bird Dog." Air Progress, XL (January 1978), 52-55,78.

____. "Cessna Bird Dog." Plane and Pilot, XII (May 1976), 20-24.

"Cessna L-19A." Air Pictorial, XV (January 1953), 19-20.

"Cessna L-19A Bird Dog." Flight, LXV (March 26, 1954), 364-365.

Coleman, Herbert J. "Turbine Aids L-19 Performance." Aviation Week and Space Technology, XCV (September 13, 1971), 58-60.

Deken, George T. "Adaptable Birds for Counterinsurgency." Tactical Air Warfare Center Quarterly Report, II (September 1970), 12-21.

Dittmer, Karl K. "Hunt With a Bird Dog." Aerospace Safety, XXII (June 1966), 14-15.

"Forward Air Controller: The Vietnam L-19." Air Classics, VII (September 1971), 56-61.

Hoadley, Ed. "Cessna L-19A: Reconnaissance-Observation Airplane." Flying, XLVIII (February 1951), 30-31+.

Johnson, Darel S. "The Gallant Bird Dog--Opps--O-1." Army, XVII (January 1967), 49-52.

Kent, G. S. "Land Where You Want to Land." U.S. Army Aviation Digest, II (November 1956), 10-16.

Landrum, Malcolm. "Ground Loops Following Power Approaches." U.S. Army Aviation Digest, V (April 1959), 18-19.

McCloskey, Charles C. "The Deadliest Enemy--Boredom." U.S. Army Aviation Digest, XV (October 1969), 22-23.

Marti, John H. "O-1 Orientation." U.S. Army Aviation Digest, XIII (July 1967), 8-9.

"Powered Lift Put to Proof." Aeronautics, XXXII (April 1955), 42-44.

Roth, H. F. "Bird Dogs and Guerrillas." Ordnance, LI (March-April 1967), 483-484.

"Veteran L-19 Retires." Air Pictorial, XX (November 1958), 400-401.

White, F. Lloyd. "Overlooking the Obvious--and the Not So Obvious." U.S. Army Aviation Digest, XVI (September 1970), 32-33.

(Douglas) O-2

Chapman, Frank. "Ugly Duckling." Air Classics, XVII (July 1981), 28-29.

"The Douglas O-2 Observation Airplane." Slipstream, VI (July 1925), 18-19.

(Cessna) O-2 Super Skymaster

Blair, Edwin T. "The Catbird Seat." Airman, XIV (November 1970), 38-41.

"Cessna Twin to Follow Bird Dog." Flight International, XCI (January 19, 1967), 104-105.

Correll, John T. "Smokey the Deuce." Airman, XIX (August 1975), 4-8.

Golden, Gordon N. "O-2." Flying Safety, XXXIX (July 1983), 14-15.

Harris, Hap. "Welcome Big Brother O-2." Airman, XI (October 1967), 24-25.

Harrison, Bob. "New Bird for FACs." Aerospace Safety, XXIII (August 1967), 18-19.

Hoey, Brian. "With a Little Help From Our Friends." Airman, XXVIII (November 1984), 9-16.

Kline, Timothy. "O-2A Super Skymaster." In: Robin Higham and Abigail T. Siddall, eds. Flying Combat Aircraft of the USAAF-USAF. Ames, Iowa: Iowa State University Press, 1975. pp. 119-124.

Newton, Bob. "O-2 Forward Air Controller." Air Progress, XLII (March 1980), 74+.

Nicholson, Robert W., Jr. "The Duck Lives." Air Force Magazine, LXVIII (July 1985), 128-129.

Sweetman, Bill. "Robertson Converts the O-2 for New Roles." Interavia, XXXVIII (November 1983), 1216-1217.

O2B see DH-4

O2C see O-1/OC Falcon

(Vought) O2U/O3U Corsair

Andrews, Hal. "Naval Aircraft: Corsair." Naval Aviation News, (November 1973), 20-21.

____. "Naval Aircraft: O3U/SU Corsair." Naval Aviation News, (November 1977), 22-23.

Gault, Owen. "Chance Milton Vought and the First Corsair." Air Classics, VIII (January 1972), 44-55, 62-63, 66.

"The New Corsair." Aviation, XXV (August 4, 1928), 404, 434.

"A New Vought Corsair With a New Engine." US Air Services, XVII (May 1932), 44-46.

Trimble, Robert L. "The First Corsairs: Vought's Battling Biplanes." Air Classics, XV (September 1979), 14-27.

"The Vought Corsair." Aero Digest, X (May 1927), 416-417.

"The Vought Corsair." Aviation, XXIII (December 26, 1927), 1518-1519.

"Vought Corsair Model O2U-2." Aero Digest, XIII (November 1928), 926-928.

"The Vought Corsair Naval Airplane." Aviation, XXII (May 2, 1927), 934-940.

"Vought Model V-80 Corsair." Aviation, XXXII (July 1933), 225-226.

"Vought V-100 Corsair." Aero Digest, XXVI (March 1935), 40-41.

(Douglas) O-5 World Cruiser

Book/Document/Report Literature:

First Around the World. Santa Monica, Calif.: Douglas Aircraft Corp., 1974.

McKay, Ernest A. World to Conquer: The Epic Story of the First Around-the-World Flight. New York: Arco, 1981. 201p.

Thomas, Lowell. The First World Flight: Being the Personal Narratives of Lowell Smith, Erik Nelson, Leigh Wade, Leslie Arnold, Henry Ogden, and John Harding. Boston, Mass.: Houghton, Mifflin, 1925. 328p.

Periodical/Anthology Literature:

"All Aboard for the World Tour in U.S. Army Planes." Literary Digest, LXXXII (March 15, 1924), 81-85.

Bowers, Peter M. "Around the World in 175 Days: Douglas World Cruisers." Wings, IV (December 1974), 8-15, 48-53.

"A Brief History of the Round-the-World Flight." Aviation, XVII (September 22, 1924), 1014-1021.

Brown, Robert J., Jr. "America Girdles the Globe in the Air." Current History, XXI (November 1924), 189-196.

Christy, Joe. "The First Round-the-World Flight." Air Force Magazine, LVII (March 1974), 52-59.

Craven, Harold L. "Around-the-World in 175 Days." Translog, VI (September 1975), 2-8.

Glines, Carroll V. and Marty Martinez. "First Flight Around the World." Air Progress, XXXV (October 1974), 26-29, 64-65, 69, 72-73.

Harvey, Alva. "'Seattle' Has Crashed in Alaska." Air Force Magazine, LVII (September 1974), 104-107.

McDarment, Corley P. "Around-the-World by Air." Scientific American, CXXX (October 1924), 230-232+.

Reed, Thomas R. "Isolation Vanquished." US Air Services, IX (November 1924), 33-38.

Wade, Leigh. "We Flew Round-the-World--First." Reader's Digest, CVII (September 1975), 166-170+.

____. "The World Flight: An All-American Venture." Air Force Magazine, LVII (March 1974), 60-61+.

0-9 see 0-2

0-10 see 0A-1/2

0-11 see 0-1

0-12 see 0-1

0-13 see 0-1

0-16 see 0-1

0-18 see 0-1

0-22 see 0-2

(Douglas) 0-25
(see also 0-2)

Westburg, Peter. "Douglas Observation Monoplanes: The 0-25." Air Classics, XIII (May 1977), 82-101, 114-116.

0-26 see 0-1

0-28 see 02U/03U Corsair

0-29 see 0-2

(Douglas) 0-31
(see also 0-43/0-46)

Westburg, Peter. "The Douglas Observation Monoplanes: The YO-31A/0-31B." Air Classics, XII (November 1976), 18-25, 68-77.

0-32 see 0-2

0-34 see 0-2

(Douglas) O-38
(see also O-2/O-25)

Miller, Ed Mack. "Operation O-38: A Salvage Adventure in Alaska's Deep Freeze." Air Force and Space Digest, LII (February 1969), 56-60.

"The Saga of 33-324." Air Classics, XI (September 1975), 36-47.

Westburg, Peter W. "Douglas Observation Biplanes: The O-38." Air Classics, XIII (June 1977), 94-111.

O-39 see O-1

(Douglas) O-43
(see also O-31/O-46)

Westburg, Peter W. "The Douglas Observation Monoplanes: The O-43A." Air Classics, XIII (January 1977), 16-26, 68-74.

___ and Peter M. Bowers. "The Parasole of Santa Monica." Wings, IV (April 1974), 57-67.

O-45 see B-10

(Douglas) O-46
(see also O-31/O-43)

Westburg, Peter W. "The Douglas Observation Monoplanes: XO-46A/O-46A." Air Classics, XIII (March 1977), 20-31, 83-91.

(North American) O-47

Gault, Owen, "The Flying Milk Bottle: North American's O-47." Air Classics, VII (September 1971), 40-45.

Glassbrook, Rich. "Flying the North American O-47 and the Curtiss-Wright O-52." Aerospace Historian, XXV (March 1978), 5-11.

"O-Bird." Air Classics, II (May 1966), 46-51.

Smith, Gene. "O-Bird Resurrection." Air Classics, XVIII (November 1982), 34-45.

Trimble, Robert L. "Spy in the Sky." Air Classics, XV (September 1979), 42-49, 82-86.

O-48 see O-46

(Curtiss-Wright) O-52 Owl

Glassbrook, Rich. "Flying the North American O-47 and the Curtiss-Wright O-52." Aerospace Historian, XXV (March 1978), 5-11.

O-53 see A-20 Havoc

O-56 see B-34/B-37 Lexington

O-57 see L-2

O-59 see L-4

O-62 see L-5

(Loening) OA-1/2 and OL Amphibian

Book/Document/Report Literature:

Loening, Grover C. Amphibian: The Story of the Loening Biplane. Greenwich, Conn.: New York Graphic Society, 1973. 196p.

Periodical/Anthology Literature:

Andrews, Hal. "Naval Aircraft: Loening Amphibian." Naval Aviation News, (February 1976), 20-21.

Bowers, Peter M. "Loening's Flying Shoehorns." Wings, IV (February 1974), 48-67.

Groves, Patricia T. "The Loening Monoplanes." American Aviation Historical Society Journal, XXIX (Winter 1984), 242-255.

"The Loening Amphibian." Flight, XVII (May 21, 1925), 302-304.

"The Loening Amphibian: America's Most Unusual Airplane." Slipstream, VIII (January 1927), 14-19.

"Loening Firm Scores Triumph." Slipstream, VI (February 1925), 14-17.

"The Loening OA-1." Aeroplane Monthly, VI (July 1978), 382-383.

"The Loening OL-8 Amphibian." Aviation, XXIII (October 24, 1927), 1002-1008.

Nalty, Bernard C. and Truman R. Strobridge. "The OL-5 and the Beginnings of Coast Guard Aviation." American Aviation Historical Society Journal, XIX (Fall 1974), 200-203.

(Curtiss) OC Falcon

Andrews, Hal. "Naval Aircraft: OC Falcon." Naval Aviation News, (July 1981), 24-25.

(Berliner-Joyce) OJ-2

Andrews, Hal. "Naval Aircraft: OJ-2." Naval Aviation News, (August 1978), 24-25.

(Vought-Sikorsky) OS2U Kingfisher

Andrews, Hal. "Naval Aircraft: Kingfisher." Naval Aviation News, (February 1977), 20-21.

Bowers, Peter M. "Kingfisher." Air Classics, II (May 1966), 34-43.

Doll, Thomas E. and Berkeley R. Jackson. Vought-Sikorsky OS2U Kingfisher. Aircraft in Profile, no. 251. Windsor, Eng.: Profile Publications, 1974. 25p.

"A Page From History: OS2U, the Kingfisher." Approach, XVII (March 1972), 14-17.

Schoeni, Arthur L. "The Mayday Kingfisher." Air Classics, XIII (July 1977), 16-23.

___. "Mayday! Mayday!" Aeroplane Monthly, VII (November 1979), 570-576.

Schrader, Richard. "Kingfisher." Air Classics, XXII (September 1986), 28-31.

Semmes, Raphael. "The Kingfisher Odyssey." Aerospace Historian, XXVII (December 1980), 231-238.

U.S.S. North Carolina Battleship Commission. Sitting Duck: The U.S.S. North Carolina's Kingfisher. Wilmington, N.C., 1971. 16p.

(Grumman) OV-1 Mohawk

Appleton, D. O. "Denying the Guerrilla His Best Protection: Grumman's Mohawk." Data, XIII (October 1968), 36+.

Bailly-Cowell, G. M. "A Flying Watchdog for Front Line Field Commanders." NATO's Fifteen Nations, XIII (August-September 1968), 66-69.

Barkley, James R. "Airborne Surveillance." U.S. Army Aviation Digest, XI (December 1965), 26-30.

___. "AO-1A Mohawk Logistical Evaluation." U.S. Army Aviation Digest, VIII (July 1962), 41-43.

Berdux, Sylvester C., Jr. "The Army's Special Electronic Missions Aircraft." Grumman Aerospace Horizons, XVII (Spring 1981), 3-7.

Bernstein, Donald J. "A Challenge: Mohawk Infrared Mission." U.S. Army Aviation Digest, XIV (October 1968), 18-23.

Blacker, Michael F. "The 'Hawk' Performs in Europe." U.S. Army Aviation Digest, XXII (August 1976), 6-7.

Boone, James M. "'Hawk' in the Sky." U.S. Army Aviation Digest, XVI (September 1970), 2-7.

Clark, Theron O. "Mohawk Evasive Maneuvers: The Rest of the Story." U.S. Army Aviation Digest, XXXI (June 1985), 14-15+.

"Combat Surveillance Mohawk." RAF Flying Review, XVIII (June 1963), 34-35.

Crawford, James C. "A Hole in the Airmobility Concept?--Find It With the Mohawk!" U.S. Army Aviation Digest, XIII (October 1967), 2-3.

Donnell, Ralph. "How to Get More Out of the Mohawk." U.S. Army Aviation Digest, IX (June 1963), 20-23.

Durwen, Simon. "New Wine in Old Bottles." Asian Defence Journal, no. 7 (July 1983), 55+.

Freeman, Bobby H. "Additional Duty for the Mohawk." U.S. Army Aviation Digest, XI (October 1965), 6-7.

Gagen, Pat. "The OV-1D Penetration Mission." Military Intelligence, VIII (April-June 1982), 32-34.

George, Jonathan D. "Mohawk." Infantry, LIX (March-April 1969), 37-39.

"Grumman Mohawk--The Army's Flying Scout." Interavia, XV (November 1960), 1402-1405.

"Grumman's Mohawk--An Eye in Troubled Skies." Flying Review International, XXII (February 1967), 341-342+.

Hiestand, Harry H. "Military Intelligence and Enemy Air Defense." U.S. Army Aviation Digest, XX (September 1974), 2-3+.

"Home is the Hunter." Army, XXXI (January 1981), 24-25.

Johnston, Donald P. "Time to Punch Out." U.S. Army Aviation Digest, XXIII (March 1977), 32-33.

Kolcum, Edward H. "Grumman to Export Updated Mohawks." Aviation Week and Space Technology, CXV (September 14, 1981), 130+.

Kontos, Ted. "'Ol' Yeller': The Saga of an OV-1 Operating in the Republic of Vietnam." U.S. Army Aviation Digest, X (May 1964), 2-6.

Lambert, Mark. "Grumman AO-1AF Mohawk in the Air." Flight, LXXX (July 6, 1961), 10-12.

_____. "Through the Paces." U.S. Army Aviation Digest, X (January 1964), 8-11.

"Mohawk AO-1AF." U.S. Army Aviation Digest, VI (September 1960), 1-5.

"The Mohawk Surveillance System." Interavia, XXI (March 1966), 338-339.

Nielsen, E. L. "Mohawk Outlook." U.S. Army Aviation Digest, XIII (April 1967), 14-15.

Paine, Lauren, Jr. "Replacing the OV-1 Mohawk With the A-10." Military Intelligence, XX (July-September 1983), 26-27+.

Plattner, C. M. "Mohawk Helps Confirm Army Air Concept." Aviation Week and Space Technology, LXXXIV (February 28, 1966), 70-72+.

Roberts, Donald A. "Aboard the Mohawk in Vietnam." U.S. Army Aviation Digest, XIV (October 1968), 28-31.

Rux, William A. "Battle Basics." U.S. Army Aviation Digest, XXXIV (October 1988), 27-29.

"A STOL Reconnaissance Aircraft for Export: The Grumman Mohawk." Interavia, XVI (July 1961), 990-992.

Thorpe, Gerald T. "Speaking of the OV-1." U.S. Army Aviation Digest, XI (June 1965), 23-25.

Tsakis, H. U. "Grumman Mohawk Features Short-Field Performance." Space/Aeronautics, XXXIII (February 1960), 65-66+.

Ulery, Vincent. "The Mohawk." Army Information Digest, XXI (January 1966), 18-21.

Weaver, William C. "Mohawks in Hostile Airspace." Military Intelligence, VII (October-December 1981), 49-52.

Wolfe, Richard P. "The Mohawk Meets New Challenges." U.S. Army Aviation Digest, XXI (August 1975), 14-19.

(North American) OV-10 Bronco

Book/Document/Report Literature:

Reed, Lawrence L. The OV-10A: Can It Perform the Airborne FAC Mission? Research Paper. Maxwell AFB, Ala.: Air Command and Staff College, Air University, 1968. 129p.

Periodical/Anthology Literature:

Andrews, Hal. "Naval Aircraft: Bronco." Naval Aviation News, (October 1972), 20-21.

Bartos, Henry S. "The OV-10: The U.S. Air Force's Battling Bronco." Air Force Magazine, LVI (April 1973), 36-40.

"Bronco, the Willing Workhorse." Air Enthusiast, I (February 1971), 231+.

Butz, J. S., Jr. "Better Eyes and More Muscle for the FACs." Air Force and Space Digest, LI (December 1968), 39-43.

Carmen, Jonathan. "OV-10: Versatile Battlefield Aircraft." Army, XVI (November 1966), 16+.

Golden, Gordon N. "OV-10." Flying Safety, XXXIX (July 1983), 15-16.

Jacobi, Leverett H. "Tough Tests for Ten-A." Skyline, XXXIV (April 1966), 12-17.

Larsen, Jim. "Naval Bronco." Air Classics Quarterly Review, VI (Fall 1979), 46-52.

Morton, A. O. "The COIN Aircraft." U.S. Naval Institute Proceedings, XCI (February 1965), 138-143.

Mulquin, James J. "The Case for Carrier-Based COIN Aircraft." U.S. Naval Institute Proceedings, XCI (August 1965), 111-114.

Nance, Keith J. and Roger B. Dunn. "It's Pave Nail--and That's a FAC!" USAF Tactical Air Warfare Center Review, III (August 1972), 10-13.

"The North American OV-10A." Interavia, XX (September 1965), 1413-1416.

"OV-10 Bronco." TAC Attack, XXVI (April 1986), 16-17.

"OV-10 Combat Workhorse." Naval Aviation News, (November-December 1983), 18-21.

"OV-10A Trials Progress." Flying Review International, XXI (April 1966), 448-449.

Peregoy, Nelson G. "Surveillance at Its Best." U.S. Army Aviation Digest, XX (August 1974), 20-23.

Rees, William H. "Picture Day." Interceptor, XVIII (September 1976), 5-7.

"Reforging NATO: Rockwell International OV-10 European Role Seen as Command, Control." Aviation Week and Space Technology, CXV (November 23, 1981), 64-65.

Richmond, James A. "OV-10 Bronco." In: Robin Higham and Abigail T. Siddall, eds. Flying Combat Aircraft of the USAAF-USAF, Ames, Iowa: Iowa State University Press, 1975. pp. 125-130.

Ropelewski, Robert R. "Marine Corps Considers OV-10D Buy." Aviation Week and Space Technology, CXVI (May 31, 1982), 63-66.

"Rounding Up the Broncos: The OV-10A Joins the USAF and the U.S. Marines." Flying Review International, XXIII (June 1968), 311-313.

Thomas, B. K., Jr. "Forward Controllers to Train in OV-10A." Aviation Week and Space Technology, LXXXVIII (January 1, 1968), 71-73.

Tomlinson, Frank J. "The New Breed of Bird." Aerospace Safety, XXII (October 1966), 1-3.

Verdi, J. M. "Missions of OV-10A Limited." Marine Corps Gazette, LII (November 1968), 104-105.

Wickersham, F. G., III. "The OV-10 as a Helicopter Killer." Marine Corps Gazette, LXIV (May 1980), 26+.

Williams, Dent. "Potent Package." Skyline, XXIII (February 1965), 27-39.

(North American) RA-5 Vigilante
(see also A-5 Vigilante)

Carroll, Daniel M. "RVAH: The Heavy is Dead, But the Recce is Alive and Flying." Approach, XXII (March 1977), 1-2.

"The Lean and Hungry Eye." Skyline, XXIV (March 1966), 20-29.

Mersky, Peter B. "Vigilante Fadeout." U.S. Naval Institute Proceedings, CVI (June 1980), 63-69.

"Panoramic Cameras for the Vigilante." Interavia, XX (September 1965), 1410-1411.

"Vigilante: The Eyes of the Fleet." Air International, IX (September 1975), 215-222, 258.

(McDonnell-Douglas) RA-18 Hornet
(see also F/A-18 Hornet)

Bauer, William D. and James H. Rutiffson. The RA-18: Expanded Roles for the Hornet. Advanced Research Report. Newport, R.I.: Center for Advanced Research, Naval War College, 1980. 76p.

Clausen, P. "Navy to Test F-18 Recon Modification." Aviation Week and Space Technology, CXVII (July 19, 1982), 125+.

Pruitt, Paul. "The Recon Hornet." The Hook, X (Spring 1982), 38-39.

RB-17 see B-17 Flying Fortress

RB-29 see F-13/B-29 Superfortress

RB-36 see B-36 Peacemaker

(Boeing) RB-47 Stratojet
(see also B-47 Stratojet)

Heiman, George G., Jr. "High, Fast and Lonesome." Boeing Magazine, XXV (August 1955), 10-11.

"Nikita and the RB-47." Time, LXXVI (July 23, 1960), 21-22.

Norman, L. B. "Mystery of the RB-47." Newsweek, LVI (July 23, 1960), 36-37.

"Russians Stir Up a Phony Spy Crisis: Shooting Down of an RB-47 Over the Barents Sea." Life, XLIX (July 25, 1960), 30-31.

White, William L. The Little Toy Dog: The Story of Two RB-47 Flyers. New York: E. P. Dutton, 1962. 304p.

(Martin) RB-57 Canberra
(see also B-57 Canberra)

Bennett, George F. A Study of the Use of the RB-57 Aircraft in Southeast Asia, 1965-1967. Research Report. Maxwell AFB, Ala.: Air War College, Air University, 1968. 73p.

Blevins, Dallas R. "High Flight to Pakistan." Navigator, XXVII (Summer 1980), 27-30.

Bulban, Erwin J. "F-111, B-58 Techniques Used in RB-57F." Aviation Week and Space Technology, LXXXI (July 20, 1964), 51, 56-57.

____. "RB-57F Built for Varied Special Missions." Aviation Week and Space Technology, LXXXI (July 13, 1964), 22-24.

Kinney, Harry and Robert L. Trimble. "Flying the Stratosphere." Air Classics, IX (October 1973), 26-35.

"Swift Blackbird." Flying Safety, XI (June 1955), 30-33.

Watson, Ted M. "20 Years With the Same Lady." Interceptor, XX (June 1978), 5-6.

(Douglas) RB-66 Destroyer
(see also B-66 Destroyer)

Anderton, David A. "Training Pays Off in RB-66 Wing's Score: Royal Flush Five, Aerial Reconnaissance Exercise." Aviation Week and Space Technology, LXXII (June 20, 1966), 146-147.

Fricker, John. "USAF Reconnaissance Re-equipment." The Aeroplane, XCII (March 29, 1957), 441-444.

"High Pilot Jet." Flight, LXXI (March 29, 1957), 411-414.

Leonberger, Loren B. "War of the Wizards." Airman, XIV (November "H1970), 17-19. ' Flight, LXXI (March 29, 1957), 411-414.

"Soviet Fighters Down USAF RB-66 Over East Germany." Air Force and Space Digest, XLVII (April 1964), 23-24.

(Beech) RC-12 King Air
(see also C-12 King Air)

"RC-12D to Fill Communications Role." Aviation Week and Space Technology, CXVIII (February 7, 1983), 79-80.

RC-45 see C-45

(Lockheed) RC-121 Warning Star
(see also C/EC-121 Constellation

Bateman, Cecil E. "Navigating the Warning Star." Navigator, XVIII (Spring 1963), 4-5.

Light, Roger B. "End of an Era." Combat Crew, XXV (August 1975), 15-18.

Mohney, Russell E. "The RC-121D Warning Star." In: Robin Higham and Carol Williams, eds. Flying Combat Aircraft of the USAAF-USAF. Manhattan, Kans.: Sunflower University Press, 1981. pp. 69-72.

"Radar Patrol." Interavia, XI (February 1956), 120-121.

(Lockheed) RC-130 Hercules
(see also C/EC-130 Hercules)

Leonberger, Loren. "Airborne Flashlight." Airman, XV (May 1971), 12-14.

(Boeing) RC-135
(see also C/EC-135)

Bell, David G. Replacement Aircraft for the RC-135. Research Paper. Maxwell AFB, Ala.: Air Command and Staff College, Air University 1985. 35p.

Chamberlin, Alton E. Supporting the One-of-a-King Weapon System, Research Report. Maxwell AFB, Ala.: Air War College, Air University, 1971. 22p.

Falvey, Michael J. "The RC-135: The Air Force as Advertised." Navigator, XXX (Spring 1983), 5-7.

Murphy, John J. "RC-135 Reconnaissance Enhancement Program." Combat Crew, XXVII (July 1977), 2-3.

(McDonnell-Douglas) RF-4 Phantom II
(see also F-4 Phantom II)

Book/Document/Report Literature:

Curtis, Edward H. A Study of the RF-4C Weapon System in Out-Country Operations in Southeast Asia, 1965-1967, Research Report. Maxwell AFB, Ala.: Air War College, Air University, 1968. 102p.

Ripley, Anthony J. Additional Capabilities of the RF-4C. Research Paper. Maxwell AFB, Ala.: Air Command and Staff College, Air University, 1973. 41p.

Schanzenbach, Walter. Peacetime Uses of the RF-4C. Research Paper. Maxwell AFB, Ala.: Air Command and Staff College, Air University, 1975. 42p.

Periodical/Anthology Literature:

"ANG's New Camera on Wings: The RF-4C." Air Reservist, XXIII (May-June 1971), 10-11.

Biles, Graham C. "RF-4C Navigator." Navigator, XVI (Fall 1968), 14-16.

Hart, Jim. "'Exposed' to Practical Experience." Air Reservist, XXVII (November 1975), 8-9.

Katzaman, Jim. "The Tactical Recce Force." Airman, XXVII (September 1983), 14-20.

Masse, Richard F. "RF-4C Radar Predictions." Navigator, XXVII (Winter 1980), 12-14.

Morrocco, John D. "USAF Demonstrates Electro-Optical Recon System in RF-4C." Aviation Week and Space Technology, CXXVI (May 18, 1987), 123+.

Oldfield, Barney. "Forerunners of the Phantom." NATO's Fifteen Nations, XV (August-September 1970), 70-78.

"Peace Jack: An Enigma Exposed." Air International, XXIX (July 1985), 18-23.

Petersen, Stefan. "Photo Phantoms Around the World." Defence Update International, no. 74 (August 1986), 50-54+.

____. "Total RECCE: USAFE's Photo Phantoms on the Continent--European Front Line Tactical Reconnaissance." Defence Update International, no. 71 (May 1986), 24-29.

"RF-4 Photo Phantoms." Aerophile, II (April 1981), 3-39.

Stambolis, Mike and Paul W. Post. "Tactical Electronic Reconnaissance Operational Employment." Journal of Electronic Defense, V (October 1982), 115-116+.

(Northrop) RF-5 Tigereye
(see also F-5 Freedom Fighter)

Fink, Donald E. "F-5E Nose Modified for Reconnaissance." Aviation Week and Space Technology, CIX (July 31, 1978), 47-49.

Gibb, Ronald M. "Design and Development of the RF-5E Aircraft." Journal of Aircraft, XX (May 1983), 418-424.

"Northrop RF-5E Tigereye." Air Progress, XLV (March 1983), 26-27.

Warwick, Graham. "Northrop RF-5E Recce Tiger." Flight International, CXVIII (November 1, 1980), 1699-1701.

RF-51 see F-6/P-51 Mustang

RF-61 see P-61 Black Widow

RF-80 see F-80 Shooting Star

(Republic) RF-84 Thunderflash
(see also F-84 Thunderjet)

"Thunderflash." RAF Flying Review, XII (January 1957), 8-9.

"USAF Authorizes RF-84F as Thunderflash." American Helicopter, XXXIV (March 1954), 4, 14.

RF-86 see F-86 Sabre Jet

(McDonnell) RF-101 Voodoo
(see also F-101 Voodoo)

Anderton, David A. "Ground, Air Planning Accelerate RF-101C Scramble Rate." Aviation Week and Space Technology, LXXII (June 13, 1960), 54-55+.

Hill, Donn P. The RF-101 Weapon System in Out-Country Reconnaissance Operations in Southeast Asia, 1965-1967. Research Report. Maxwell AFB, Ala.: Air War College, Air University, 1968. 142p.

Stevens, Paul D. "RF-101s in Southeast Asia." American Aviation Historical Society Journal, XIV (Winter 1969), 282-288.

(Lockheed) RF-104 Starfighter
(see also F-104 Starfighter)

Oldfield, Barney. "Family Reunion--Starfighter Style." NATO's Fifteen Nations, XIV (October-November 1969), 59-61.

RF-105 see F-105 Thunderchief

RF-111 see F-111/EF-111

(Beech) RU-21
(see also T-44 King Air)

Miller, Edward C. "The Flying Porcupine." Air Combat, III (January 1975), 57-59, 74.

(Curtiss) SC Seahawk

Andrews, Hal. "Naval Aircraft: Seahawk." Naval Aviation News, (August 1975), 20-21.

(Curtiss) SOC Seagull

Andrews, Hal. "Naval Aircraft: Seagull." Naval Aviation News, (June 1975), 20 21; (June 1977), 20-21.

Gunston, William T. ("Bill"). "Last of the Seagulls." Aeroplane Monthly, IV (December 1976), 620-625.

Larkins, William T. Curtiss SOC Seagull. Aircraft in Profile, no. 194. Windsor, Eng.: Profile Publications, 1972. 24p.

Stinson, Patrick. "Eyes of the Battle Fleet." U.S. Naval Institute Proceedings, CXII (April 1986), 87-89.

Walter, A. Dean. "A Cross-Country Flight in an SOC-3." Naval Aviation Museum Foundation, V (Fall 1985), 70-72.

(Lockheed) SR-71 Blackbird
(see also YF-12)

Book/Document/Report Literature:

Crickmore, Paul F. Lockheed SR-71 Blackbird. Osceola, Wis.: Motorbooks International, 1986. 200p.

Drendel, Lou. SR-71 Blackbird in Action. Aircraft, no. 55. Carrollton, Tex.: Squadron/Signal Publications, 1982. 49p.

Graham, Richard H. Employment of the SR-71 in the European Theater. Research Report. Maxwell AFB, Ala.: Air War College, Air University, 1982. 57p.

Miller, Jay. Lockheed SR-71 (A-12/YF-12/D-21). Minigraph, no. 1. Rev. ed. Arlington, Tex.: Aerofax, Inc., 1985. 40p.

Periodical/Anthology Literature:

"The Blackbird." Air International, XXIV (January 1983), 188+.

Blair, Edwin T. "Seeing the Earth Supersonically." Airman, XVII (February 1973), 38-41.

Boone, Earle M. "Blackbird SR-71." Aerospace Safety, XXX (November 1974), 1-3.

Brown, Buddy C. "The Remarkable SR-71." Combat Crew, XX (June 1969), 8-10.

Brown, Kevin V. "America's Supersecret Spy Plane." Popular Mechanics, CXXIX (June 1968), 59-62+.

"A Day Out With the Blackbirds." Aircraft Illustrated, X (January 1977), 40-42.

Dean, Jack. "Back to the Future." Wings, XVI (October 1986), 22+.

"Detailed Preparations Mark SR-71 Flight." Aviation Week and Space Technology, CIV (May 10, 1976), 99-100.

Fink, Donald E. "Simulator Displays SR-71 Characteristics." Aviation Week and Space Technology, CIV (May 10, 1976), 93-97.

"Flying With the Big Black Bird." Airman, XIV (October 1970), 4-8.

George, James A. "From the Starting Gate." Airman, XXII (September 1978), 24-25.

Graham, Vickie M. "Mach 3 Plus and Climbing." Airman, XXIX (March 1985), 21-28.

Gunston, William T. ("Bill"). "Burbank's Blackbirds." Aeroplane Monthly, II (March 1974), 532-539.

Halloran, Patrick J. "Happiness is Mach 3 Plus." Tactical Air Reconnaissance Digest, III (August 1969), 4-7.

____. "SR-71 at Mach 3.0 Cruise." Air Force Magazine, LIV (September 1971), 62-64+.

Hartz, Jim. "Alone, Unarmed, Untouchable: The Amazing SR-71." Reader's Digest, CXVIII (January 1981), 132-136.

Hewitt, Albert G., Jr. "Computer Flight Planning--Mach 3 Plus." Combat Crew, XXI (April 1970), 20-21.

Johnson, Clarence L. ("Kelly"). "The Blackbird Story." Popular Mechanics, CLVIII (July 1982), 70-73+.

Kardong, G. Abe. "High, Hot, and Headin' Out--More About the SR-71." Air Force Magazine, LIV (December 1971), 44-49.

Miller, Jay. "Burbank's Blackbirds." Air International, XXVIII (February 1985), 59-64, 84-87.

Morgan, George T. "SR-71: The Fabulous Blackbird." Navigator, XXIV (Spring 1977), 5-7.

O'Lone, Richard G. "First Operational SR-71 Delivered to SAC." Aviation Week and Space Technology, LXXXIV (January 17, 1966), 33-34.

Pugh, Craig. "Surviving in High Places." Airman, XXVI (September 1982), 42-48.

Rich, B. R. "Designing and Building the SR-71 Blackbird." Design News, XXXVIII (April 19, 1982), 82-84+.

___. "Sixteen Years of Mach 3 Flight." Interavia, XXXV (July 1980), 636-637.

Ropelewski, Robert R. "SR-71 Impressive in High-Speed Regime." Aviation Week and Space Technology, CXIV (May 18, 1981), 46-51+.

Smith, Gene. "Partridge of War." Air Progress, XLV (April 1983), 46+.

"SR-71." Combat Crew, XXVII (October 1976), 4-7.

"SR-71 Imposes Burden on Maintenance Units." Aviation Week and Space Technology, CXIV (May 18, 1981), 105-108.

Stampf, Frank W. "SR-71: Mach 3 Plus." Navigator, XXIX (Spring 1982), 4-8.

Witze, Claude. "Yankee Ingenuity in England." Air Force Magazine, LVII (October 1974), 29-32.

(DeHavilland-Canada) U-1 Otter

Dickens, Raymond E. "Homeliness--a Thing Called Otter." U.S. Army Aviation Digest, XII (May 1966), 10-12.

Fowler, Bob. "Getting the Most From Otter." U.S. Army Aviation Digest, V (March 1959), 5-11.

Hemphill, Ernie. "U.S. Army Show Features DHC." Canadian Aviation, XXXIII (June 1960), 43+.

Raines, Samuel C. "The Longest Mission." *U.S. Army Aviation Digest*, XVIII (February 1972), 6-9+.

Stevens, James H. "DH Canada Otter." *Flight*, LXI (April 11, 1952), 428-429, 433.

(Lockheed) U-2/TR-1

Book/Document/Report Literature:

Beschless, Michael R. *Mayday: Eisenhower, Khruschev, and the U-2 Affair--The Untold Story of the Greatest U.S.-U.S.S.R. Spy Scandal.* New York: Harper & Row, 1986. 417p.

Bonsi, D. J. *Ejection Situations in the U-2/TR-1: An Analysis of Emotions and Events*. Research Paper. Maxwell AFB, Ala.: Air Command and Staff College, Air University, 1988. 30p.

Cook, Fred J. *The U-2 Incident*. New York: Watts, 1973. 64p.

Davis, Larry. *U-2 Spyplane in Action*. Carrollton, Tex.: Squadron/ Signal Publications, 1988. 50p.

Gann, Ernest K. *The Black Watch: The True Inside Story of America's Most Elite and Secret Air Corps.* New York: Random House, 1989. 288p.

Miller, Jay. *Lockheed U-2.* Aerograph, no. 3. Arlington, Tex.: Aero-Fax, Inc., 1985. 120p.

Slizeski, Robert S. *The U-2 Incident: A Case Study in Planning and Coordination.* Research Paper. Maxwell AFB, Ala.: Air Command and Staff College, Air University, 1962. 55p.

Smith, Dale C. *Simulation Requirements for U-2/TR-1 Pilot Training*. Research Paper. Maxwell AFB, Ala.: Air Command and Staff College, Air University, 1982. 18p.

Wise, David and Thomas B. Ross. *The U-2 Affair.* New York: Random House, 1962. 269p.

Periodical/Anthology Literature:

"Alconbury's TR-1s: Ready for Reconnaissance Breakthrough." *Air International*, XXVII (April 1984), 206-208.

Bergman, Jules. "The Amazing Saga of the U-2." *Reader's Digest*, LXXVII (October 1960), 75-79.

Beschless, Michael R. "The Spy Flight That Killed Ike's Dream." *U.S. News and World Report*, C (March 24, 1986), 36-41+.

Fink, Donald E. "Role of U-2 High-Altitude Surveillance Aircraft to be Expanded." *Aviation Week and Space Technology*, CXII (June 16, 1980), 200-202+.

"The Flight of the U-2." Newsweek, LV (May 16, 1960), 27-29.

"The Flight to Sverdlovsk." Time, LXXV (May 16, 1960), 15-18.

"Flying High--and Hairy." Time, CXXX (October 19, 1987), 60-61.

"Flying Spies." Flying Review International, XXI (February 1966), 331-332+.

Ford, William A. "Walking a Tightrope." Air Force Magazine, LXIV (June 1981), 86-89.

Gabriel, Charles A. "Tactical Reconnaissance for the 1980s: A Close-up Look at the TR-1 and QSB." Signal, XXXIV (October 1979), 9-11.

Gann, Ernest K. "Dragon Lady 13 Miles Up." Flying, CVI (June 1980), 50-52+.

Gaskin, Robert. "Flying the U-2." Air Force Magazine, LX (April 1977), 66-71.

Green, William. "U-2: Facts and Fiction." RAF Flying Review, XV (August 1960), 18-19, 22.

Hane, Raymond, III. "Lockheed's TR-1: The Latest in Tactical Reconnaissance." Pacific Defense Reporter, X (March 1984), 53+.

Hotz, Robert. "Lockheed U-2 Over Sverdlovsk." Aviation Week and Space Technology, LXXII (May 16, 1960), 20-21.

Hubbard, Todd P. "Making the Right Decisions Before You Fly." Flying Safety, XL (November 1984), 3-4.

____. "'Solum Volamus': Alone We Fly." Combat Crew, XXXIV (August 1984), 4-5.

Humphries, Orin L. "High Flight: Flying the U-2, Testing the TR-1." Wings, XIII (June 1983), 10-31; Airpower, XIII (July 1983), 8-29.

Jarvinsen, Phillip L. "Everything You've Ever Wanted to Know About a U-2 Navigator--But Didn't Know Where or How to Ask." Navigator, XXV (Spring 1978), 5-9.

Lawson, Bob. "Et Tu, U-2?: Lockheed's U-2 and CVA-66." The Hook, X (Summer 1982), 9-12.
Landing on an aircraft carrier.

Leavitt, Lloyd R., Jr. "It's Not a 'Thing.'" Combat Crew, XI (November 1960), 20-22.

Miller, Jay. "U-2R/TR-1: Lockheed's Black Ladies." Air International, XXVII (October 1984), 185-190, 203-208.

Nathan, James A. "A Fragile Detente: The U-2 Incident Reexamined." Military Affairs, XXXIX (October 1975), 97-104.

Park, Bob. "Another Way to Spell Recce." Combat Crew, XXIX (August 1979), 18-19.

Perry, Glenn, II. "Learning to Land the U-2." Air Force Magazine, LIX (January 1976), 42-45.

Randle, Kevin D. "What Really Happened to Gary Powers and His U-2?" Air Classics, XV (April 1979), 64-67.

Smith, Bruce A. "First Production TR-1 Rollout Nears." Aviation Week and Space Technology, CXV (July 13, 1981), 45-48.

"Spying From on High." Time, CXV (April 7, 1980), 76-77.

Sturm, Ted R. "HASP and the Lonely Ones." Airman, V (February 1961), 25-27.

____. "The Lonely Ones." Space World, III (July 1962), 42-44.

Sweetman, Bill. "New Life for the Dragon Lady." Flight International, CXXIII (May 7, 1983), 1219-1222.

Townsend, Roy. "And You Thought They Weren't Building the U-2 Any More?" Air Classics, X (April 1974), 59-63.

"TR-1s Provide High Altitude Reconnaissance and Surveillance." Aviation Week and Space Technology, CXXIII (August 5, 1983), 59-61.

"U-2 Developmental and Operational Chronology." Congressional Record, CVI (June 7, 1960), 11977-11978.

Weir, K. W. "The U-2 Story." Society of Experimental Test Pilots Technical Review, XIV (Spring 1979), 186-192.

Wise, David and Thomas B. Ross. "The Secret History of the U-2." Look, XXVI (May 8 and 22, 1962), 128+, 101-102+.

(Cessna) U-3 Blue Bird/Administrator

"The Blue Canoe." MATS Flyer, VI (July 1959), 14-15.

Bunyan, F. J. "Check Pilot Report on the Cessna 310." Flying, LVI (June 1955), 26-27+.

Colby, Doug. "Snub Nose '56." Plane and Pilot, XXII (May 1986), 32-35, 61-63.

Collins, Richard L. "Cessna 310, Plane With a Past." Flying, LXXXIV (June 1969), 44-51+.

____. "Pilot Report: Cessna 310." Flying, XC (June 1972), 32-37.

Freed, C. D. "The U-3 and Thee." Combat Crew, XI (July 1960), 16-17.

Gidlof, Lars F. "The Early Cessna 310." Plane and Pilot, XII (June 1976), 58-61.

Landsberg, Bruce. "First in the Class." Flying, CVII (September 1980), 74-76+.

Van Gorder, F. C. "The Blue Canoe and You." Combat Crew, XIII (February 1963), 16-17.

(Aero) U-4 Commander

Kunc, Richard A. "Ride in a Warbird." Air Progress, XXXIX (August 1977), 49-50, 84.

Lert, Peter. "Built for Comfort, Not Speed." Air Progress, XLIII (April 1981), 34-40.

(DeHavilland-Canada) U-6 Beaver

Angstadt, I. F. "Aerial Pickup Truck: DeHavilland Beaver." Aero Digest, LIX (August 1949), 21-22+.

"Beavers Bought." Aeronautics, XXV (December 1951), 52-53.

Brooks, Alvin L. "Beavers and Bumps." U.S. Army Aviation Digest, VI (July 1960), 16-19.

Carlyle, John R. "They're Still Flying Airplanes." Combat Crew, XII (January 1962), 26-28.

Cooksley, Peter G. "Combat Colours: Beaver." Airfix Magazine, XXII (July 1981), 512-513.

Monroe, David M. "The Beaver." U.S. Army Aviation Digest, X (April 1964), 24-27.

Paul, J. F. Christopher. "The DeHavilland Beaver." Air Pictorial, XXII (December 1960), 416-419.

"Pilot's Report: DeHavilland Beaver." Skyways, IX (June 1950), 20-21.

Siuru, William D. "Leave It to Beaver." Airpower, XI (September 1981), 44-51.

U-7 see L-21

(Piper) U-8 Seminole
(see also L-28 Seminole)

Barthel, William H. "U-8 Pilots--Are You Confused?" U.S. Army Aviation Digest, XIII (July 1967), 60-61+.

Collins, Richard L. "Light Twin From Piper." Flying, CIII (October 1978), 36-40, 92.

Field, Hugh. "Seminole in the Air." Flight International, CXV (January 13, 1979), 108-112.

Marsh, Alton K. "Seminole Aimed at Transitioning Pilots." Aviation Week and Space Technology, CIX (July 31, 1978), 37-41.

"280 Feet Too Low." U.S. Army Aviation Digest, XVIII (July 1972), 36-42.

(Helio) U-10 Super Courier

Gilbert, John. "Pilot Report: The Helio Stallion." Flying, LXXIX (July 1966), 102-108.

Wilkinson, Stephen. "Pilot Report: The Helio Stallion." Flying, LXXXVIII (May 1971), 50-55+.

(Piper) U-11 Aztec

Collins, Richard L. "The Iron Horse: Piper Aztec." Flying, CVI (January 1980), 46-49+.

Davis, Art. "Aztec." The AOPA Pilot, XXII (December 1979), 80-87.

Durosko, Ted. "Pilot Report: Piper Aztec." Flying, LXVI (May 1960), 50-51+.

Gilbert, John. "Pilot Report: Aztec." Flying, LXXXIII (September 1968), 52-57.

Shattuck, Dennis. "Piper's Durable Dowager." Aero, XIII (February 1980), 30-37.

U-16 see HU-16/SA-16 Albatross

(Cessna) U-17 Skywagon

Brink, Randall. "Gusto Machine." Air Progress, XLIII (January 1981), 10-17.

Colby, Douglas. "Cessna's King Kong Taildragger." Plane and Pilot, XIV (June 1978), 32-35, 66.

Illing, R. G. "Pilot Report: Cessna 185." Flying, LXVIII (June 1961), 34-35+.

Lert, Peter. "Cessna 185." Air Progress, XLV (May 1983), 8+.

U-21 see T-44 King Air

(Cessna) U-26 Stationaire

Collins, Richard L. "Pilot Report: Eight-Seat Express." Flying, CVI (March 1980), 38-43.

Cox, Bill, "Cessna 206 Stationaire." Plane and Pilot, X (July 1974), 54-58.

Lert, Peter. "Stationaries." Air Progress, XLIII (October 1983), 46+.

McClellan, J. Mac. "Pilot Report: Utility Fielder." Flying, CVIII (May 1981), 54-58+.

Wilkinson, Stephen. "Pilot Report: Cessna Stationaire." Flying, LXXXIX (November 1971), 58-61+.

(Cessna) U-27 Caravan

Aarons, Richard N. "Cessna's Caravans." Business and Commercial Aviation, LXIII (July 1988), 48-55.

"The Cessna Caravan." Air Progress, XLV (March 1985), 42-43.

Collins, Richard L. "Singular Sensation." Flying, CX (April 1983), 34-38.

Cook, LeRoy. "Cessna's Caravan I: A Super Sky-Hulk." Private Pilot, XX (August 1985), 44-49.

Grangier, Marc. "Cessna Caravan I: Rethinking the Utility Aircraft." Interavia, XXXVIII (December 1983), 1321-1325.

North, David M. "Single-Engine Turboprop Displays Suitability for Utility Missions." Aviation Week and Space Technology, CXXII (March 25, 1985), 69-71+.

Walters, Brian. "U-27A: A Special Mission Aircraft With a Difference." Asian Defence Journal, no. 3 (March 1988), 86-87.

UCs see Transport Aircraft

(Vought) UO/FU

Andrews, Hal. "Naval Aircraft: UO/FU." Naval Aviation News, (May 1979), 20-21.

(DeHavilland Canada) UV-18 Twin Otter

Brown, David A. "Twin Otter Adds Ease of Piloting to STOL Capabilities." Aviation Week and Space Technology, LXXXIV (February 14, 1966), 104-108+.

"DHC Introduces the Turboprop Twin Otter." Interavia, XIX (November 1964), 1663-1664.

Gerard, Doug. "Twin Otter: Surveillance in the Last Frontier." U.S. Army Aviation Digest, XXVII (January 1981), 14-26.

Sutton, Glenn H. "A New Bird in the ADCOM Nest." Interceptor, XIX (November 1977), 20-23.

Weeghman, Richard B. "From Canada: DeHavilland Twin Otter." Flying, LXXXIII (August 1968), 38+.

(Pilatus-Porter) UV-20 Porter

Fink, Donald E. "Stretched Twin-Engined Porter Designed." Aviation Week and Space Technology, LXXXVII (October 9, 1967), 77-78+.

Fricker, John. "Pilatus PC-7." Air International, XXI (September 1981), 110-117.

(Skytrader) UV-23 Scout

Scott, William B. "STOL Scout Begins Flight Tests With Astazou 16 Powerplants." Aviation Week and Space Technology, CXXX (February 13, 1989), 64-65.

(Boeing) WB-29
(see also B-29 Superfortress)

"Flying Through a Typhoon." Boeing Magazine, XXII (November 1952), 12-13.

"Graveyard Duty." Air Training, IV (December 1954), 28-29.

(Boeing) WB-47
(see also B-47 Stratojet)

Lee, Lawrence B. "WB-47 Unsymmetrical Takeoff Thrust." MATS Flyer, XI (November 1964), 20-23.

Soth, Martin W. "No Dodo Yet." Airman, XII (December 1968), 44-46.

(Boeing) WB-50
(see also B-50 Superfortress)

Fuller, John F. "A Lesson From History: The WB-50D." Friends Bulletin, IX (Summer-Fall 1986), 18-21.

WB-66 see B-66 Destroyer

(Lockheed) WC-121
(see also C-121 Constellation)

Clarke, Richard. "Typhoon History." Air Classics, X (December 1974), 38-45.

(Lockheed) WC-130
(see also C-130 Hercules)

Roll, Richard. "The Air Weather Reconnaissance Squadron Navigation System." Navigator, XXIII (Summer 1976), 20-21.

(Boeing) WC-135
(see also C/KC-135 Stratolifter/Stratotanker)

King, Jerry D. and Douglas J. Greenheck. "Ever Heard of the WC-135B?" Navigator, XXXI (Winter 1984), 20-21.

D. Transport-Glider Aircraft

(Douglas) C-1
(see also O-2)

"Douglas C-1 Transport Plane." Aviation, XVIII (June 1, 1925), 606-607.

(Fokker) C-2/RA Trimotor

"The New Fokker Army Transport Plane." Aviation, XXIII (August 29, 1927), 473-474.

(Grumman) C-2 Greyhound

Appleton, D. O. "The Fleet Gets Its COD." Data, XIII (February 1968), 29+.

Brown, David A. "Flight Tests of C-2A are Geared to 1966 Fleet Service." Aviation Week and Space Technology, LXXXI (December 7, 1964), 20-22.

Carpenter, A. E., Jr. "C-2A Matches Corps Needs." Marine Corps Gazette, LIV (June 1970), 48-49.

Grumman Aerospace Horizons, Staff of. "Special Delivery to the Carrier." Grumman Aerospace Horizons, XVI (Spring 1980), 2-7.

Marquez, Toby. "The Navy's New Greyhound Cargo Plane Lands on the Kitty Hawk Off Vietnam." Navy, X (February 1967), 36-37.

(Ford) C-4/JR/RR Trimotor

Book/Document/Report Literature:

Ford Motor Company. Airplane Division. Ford Trimotor Instruction Manual. Edited by Dan R. Post. Arcadia, Calif.: Post-Era Books, 1977. 114p. Reprint of the 1929 service/operating handbook.

The Ford Trimotor. Appleton, Wisc.: Aviat Publications, 1974. 120p.

Ingells, Douglas J. Tin Goose: The Fabulous Ford Trimotor. Fallbrook, Calif.: Aero Publishers, 1968. 96p.

Larkins, William T. Ford Trimotor. Aircraft in Profile, no. 156. Windsor, Eng.: Profile Publications, 1966. 16p.

Weiss, David A. The Saga of the Tin Goose. New York: Crown, 1971. 278p.

Periodical/Anthology Literature:

Andrews, Hal. "Naval Aircraft: Ford Trimotor." Naval Aviation News, (October 1974), 20-21.

Baker, John L. "Trimotor." The AOPA Pilot, XXX (November 1987), 48-66.

Gann, Ernest K. "Tin Goose." Flying, XCV (August 1974), 92-93+.

Morgan, Len. "Flying the Ford Trimotor." Flying, CIV (January 1979), 50-54+.

Towle, Tom and John G. Lee. "Who Designed the Ford Trimotor?" American Aviation Historical Society Journal, XV (Fall 1970), 181-193.

(Grumman) C-4 Gulfstream I

Beller, W. "Design Report: Grumman's Gulfstream Turboprop." American Aviation, XXI (April 7, 1958), 22+.

Bulban, Erwin J. "Gulfstream Stresses Fuselage Strength." Aviation Week and Space Technology, LXIX (September 22, 1958), 82-84+.

Knight, Fred J. The Story of the Grumman Gulfstream. Worthington, Eng., 1979. 182p.

Stanfield, Robert I. "Gulfstream is Easy to Handle, Rugged." Aviation Week and Space Technology, LXXI (October 5, 1959), 52-54+.

C-5 see C-2/RA Fokker Trimotor

(Lockheed) C-5 Galaxy

Book/Document/Report Literature:

Brannum, Edward S. C-5 Training: Time for a Change. Research Paper. Maxwell AFB, Ala.: Air Command and Staff College, Air University, 1984. 55p.

Butts, Ray F. C-5A Aircraft: Panacea or Waste? Research Paper. Maxwell AFB, Ala.: Air Command and Staff College, Air University, 1967. 48p.

DiFerdinando, Anthony F. The Role of the C-5A in Combat Airlift. Research Paper. Maxwell AFB, Ala.: Air Command and Staff College, Air University, 1972. 73p.

Fritzsch, Ralph S. The C-5A and the Army Deployment Mission: The Relationship Between Technology, Economy, and Strategy. MS Thesis. Columbus, Ohio: The Ohio State University, 1967. 72p.

Garrard, Wilfred C. The Lockheed C-5A: A Case Study in Aircraft Design. New York: American Institute of Aeronautics and Astronautics, n.d. 28p.

Lockheed-Georgia Company. Public Information Office. C-5B Galaxy. Marietta, Ga., 1987. 2p.

Rice, Berkeley. The C-5A Scandal: An Inside Story of the Military-Industrial Complex. Boston, Mass.: Houghton, Mifflin, 1971. 238p.

United States. Congress. Joint Economic Committee. Subcommittee on Priorities and Economies in Government. Economics of Defense Procurement--the C-5A and Strategic Mobility: Hearings. 94th Cong., 2nd sess. Washington, D.C.: GPO, 1981. 614p.

____.____.____. Subcommittee on International Trade, Finance, and Security Economics. C-5A and Air Force Defense Profits Policy: Hearings. 98th Cong., 1st sess. Washington, D.C.: GPO, 1984. 72p.

Periodical/Anthology Literature:

Albertson, James J. "The C-5A." Military Review, XLVIII (May 1968), 3-10.

Allen, Ken. "Make Way for the Giant." Airman, XII (March 1968), 4-8.

____. "Men to Make the Giant Great." Airman, XIII (March 1969), 62-64.

____. "Training the King's Men." Airman, XIV (August 1970), 4-9.

Alliso, H. Bard. "C-5B: A Giant Reborn." Aerospace America, XXIII (July 1985), 62-69.

Allsup, Dan. "The Galaxy's New Look." Airman, XXIX (March 1984), 16-19.

Anderson, Dick. "A Week With the Big One." Airman, XVII (November 1973), 2-7.

Ashlock, James R. "Lockheed's C-5A Builds on C-141 Advances." Aviation Week and Space Technology, LXXXII (June 28, 1965), 50-52+.

Barnett, Ned. "The Largest Galaxy in the Sky." Air Classics, IX (June 1973), 60-65, 76.

"The Big Bird is Here." Aerospace Maintenance Safety, XXIV (October 1969), 1-5.

Braunch, Bill. "Questions and Answers About the C-5A." MAC Flyer, XIV (July 1967), 4-7.

Brown, David A. "C-5A Exhibits Agility Despite Size." Aviation Week and Space Technology, XCII (June 1, 1970), 40-42+.

____. "C-5 Offers Future Development Lessons." Aviation Week and Space Technology, C (January 7, 1974), 49-52.

____. "Israel Airlift Flights Underscore C-5 Rapid Deployment Capability." Aviation Week and Space Technology, XCIX (December 10, 1973), 16-19.

Brown, Garnett C., Jr. "Just Another Airplane?" MAC Flyer, XVII (April 1970), 4-8.

Brown, Harold. "Revolution in Airlift." Air Force and Space Digest, LI (June 1968), 114-118.

Budd, Phillip R. "28 Tires, 4 Fires, and Me." MAC Flyer, XVIII (September 1971), 12-15.

Butz, J. S., Jr. "C-5A: Even More Than Meets the Eye." Air Force and Space Digest, XLVIII (December 1965), 33+.

"C-5A." Flying Review International, XXI (June 1966), 607+; XX (April 1965), 13+.

"C-5A Symposium." Aviation Week and Space Technology, LXXXVII (November 20, 1967), 164-169+.

"C-5 Jumbo is Here." Aerospace Safety, XXVI (June 1970), 2-5.

Collins, Richard L. "Latter-Day Load Star." Flying, CXIV (March 1987), 17-18.

"Fat Albert." Combat Crew, XXV (August 1975), 28-29.

"Flight Testing the C-5A Galaxy." Interavia, XXIII (September 1968), 1094-1095.

Flemming, Donald J. "C-5: In and Out in 5 Minutes." Air Force Magazine, LVI (November 1973), 38-42.

Fricker, John. "A New Galaxy in the Sky." Flying Review International, XXIV (September 1968), 74-79.

"The Giants of Georgia." Air International, XXVI (February 1984), 61-68, 87-90.

Gilbert, Kimberly H. "The C-5A." Navigator, XXIX (Summer 1982), 5-7.

Goodrum, John. "The Biggest Airplane Ever Built: C-5A Galaxy." Popular Science, CXCIV (March 1969), 98-100.

Griffiths, David R. "Tests Find C-5 Capable of Off-Runway Operations." Aviation Week and Space Technology, CXIV (January 5, 1981), 54-55.

Harrison, Neil. "Lockheed C-5A: The Giant From Georgia." Flight International, LXXXIX (February 10, 1966), 223-228.

Kane, Pamela A. "C-5A Galaxy: Giant in the Sky." National Guard, XL (May 1986), 26-30.

Kaplan, Fred. "The Flying Lazarus: How the Pentagon Brought the C-5 Transport Plane Back From the Dead." Washington Monthly, XIV (February 1983), 48-52+.

Katzaman, Jim. "The Range Riders." Airman, XXII (November 1978), 39-43.

Kolcum, Edward H. "C-5B Production Planned for Cost, Time Efficiency." Aviation Week and Space Technology, CXX (May 28, 1984), 138+.

Lambert, Mark. "C-5A Galaxy Moves Towards First Flight." Flight International, XCIII (March 14, 1968), 390-391.

Lee, James M. "C-5: The First 15 Years." MAC Flyer, XXX (June 1983), 3-5.

"Lockheed C-5A: A New Transport System for the U.S. Air Force." Interavia, XXII (August 1967), 1262-1267.

"Lockheed-Georgia C-5A Galaxy." Aircraft Engineering, XL (June 1968), 17-21.

"Lockheed Readies C-5A for First Flight." Aviation Week and Space Technology, LXXXIX (July 1, 1968), 24-27.

Lucas, Hal. "C-5A May Revamp Strategic Airlift." American Aviation, XXIX (November 1965), 29-31.

Mayhew, David L. "The Sky Giant." Airman, XV (July 1971), 48-55.

Mecklin, John. "The Biggest, Cheapest Lift Ever." Fortune, LXXII (November 1965), 179-180+.

Miller, E. Mack. "The Monster is Well Behaved." Air Force and Space Digest, LI (October 1968), 54-55+.

Moore, John A. "C-5A: MAC's Biggest and Newest." Navigator, XXI (Fall 1974), 16-18.

Munson, Russell. "C-5A: Nimble Giant." Flying, XCVI (April 1980), 118-124.

O'Leary, Michael. "Fat Albert and the Alconbury Aggressors." Air Classics, XII (December 1976), 32-41, 76.

Perini, Michael B. "Airlift for Far and Near." Air Force Magazine, LXVI (October 1984), 44-51.

Powell, Craig. "The Air Force's C-5 Transport: A Program Under Pressure." Armed Forces Management, XIII (August 1967), 36-38+.

____. "The Saga of Fat Albert." Congressional Record, CXX (May 7, 1974), H3668-H3671.

Poyer, Joe. "Gargantuan Galaxy." International Combat Arms, V (September 1987), 74-77.

Prochoroff, Alan. "Frontier of Galaxies." Citizen Airman, XXXVIII (June 1986), 4-7.

Rice, Berkeley. "C-5A: Postmortem on a Mess." Look, XXXV (May 18, 1971), 80-81.

Schlitz, William P. "The C-5 Team at Dover." Air Force Magazine, LXVI (August 1983), 62-67.

Seiderman, Paul. "The C-5: Extending It Into the 21st Century." National Defense, LXV (December 1980), 34-37+.

Shister, Neil. "Our Piece of the Pie in the Sky." Atlanta, XXII (October 1982), 70+.

Sibbald, David M. H. "MAC and Me and the C-5 Make Three." MAC Flyer, XXVI (March 1979), 9-12.

Stambler, Irwin. "The C-5A Transport." Space/Aeronautics, XLIII (March 1969), 40-45.

Thomas, Benjamin K., Jr. "C-5A Meets Milestone With 1hr, 34 min. First Flight." Aviation Week and Space Technology, LXXXIX (July 8, 1968), 22-25.

Tortolano, F. W. "Delivering the Goods." Design News, XXXVIII (April 19, 1982), 65-70+.

Ulsamer, Edgar E. "How the World's Largest Aircraft Revolutionizes Strategy and Technology." Air Force and Space Digest, LI (April 1968), 64-77.

____. "Lessons of the C-5 Program." Air Force and Space Digest, LII (September 1969), 51-55.

____. "Such a Nimble Giant." Air Force and Space Digest, LII (April 1969), 79-80.

Young, J.W. "Flight Testing the Galaxy." MAC Flyer, XV (July 1968), 4-8.

C-6 see RS

(Beech) C-6 King Air
(see also T-44 King Air)

Baer, John. "Beech King Air E90." Plane and Pilot, XI (April 1975), 58-63.

Brechner, Beryl. "Royal Ride." Flying, CVI (February 1980), 58-61+.

Collins, Richard L. "King Air C90: Trusty Turboprop." Flying, CI (August 1977), 62-65.

Lert, Peter. "King Air F90." Air Progress, XLIV (April 1982), 10-23.

McClellan, J. Mac. "Long Live the King." Flying, CIV (June 1982), 24-27+.

Trammell, Archie. "Pilot Report: King Air." Flying, LXXXI (November 1967), 42-46.

C-7 see C-2/RA Fokker Trimotor

(DeHavilland-Canada) C-7 Caribou

Beach, Dwight E. "The Caribou in Southeast Asia." Army, XIII (December 1962), 36-39.

Booda, Larry. "STOL Caribou Calls for Special Handling." Aviation Week and Space Technology, LXXIV (January 23, 1961), 56-57+.

Butz, J. S., Jr. "Army Attempts Major STOL Advance." Aviation Week and Space Technology, LXXI (November 30, 1959), 59+.

"Caribou Debut." U.S. Army Aviation Digest, V (November 1959), 1-4.

"The DeHavilland Caribou." *Aviation Week and Space Technology,* LXXIII (January 23, 1961), 56-69.

Grimshaw, J. E. "The Caribou: Flexible Logistical Capability for the U.S. Army." *Interavia*, XVII (January 1962), 76-78.

Herron, Randall A. "Going Home on a Caribou." *Airman*, XXXI (September 1987), 20+.

Hulme, J. M. "Down, But Not Out." *Airman,* XIII (October 1969), 52-53.

Koster, Ed. "A Vietnam Vet Joins the Air Force." *Airman*, X (December 1966), 8-9.

Martinez, Ed. "Ferrying Caribous." *Airman,* XII (November 1968), 48-51.

Newman, Denis. "Background of the Caribou." *Flight,* LXXIX (April 6, 1961), 436-438.

O'Leary, Michael. "Charge of the Caribou." *Air Combat,* XVI (October 1988), 34-41.

Stephenson, Charles A., III. "What am I?" *U.S. Army Aviation Digest,* VIII (March 1962), 26-28.

Verrico, Anthony. "Flying the DHC-4 Caribou." *Interavia,* XV (February 1960), 181-182.

Zeiman, Robert. "Caribou." *BeeHive,* XXXVI (Summer 1961), 29-35.

(Fairchild) C-8/UC-96

Ferguson, William P. "Canada's Fairchild: The Story of the Unique Super 71 Transport." *Air Classics,* XIII (September 1977), 68-73.

C-9 see C-4/JR/RR Ford Trimotor

(McDonnell-Douglas) C-9 Nightingale/Skytrain II

Anderson, Thomas G. "The Flying Hospital." *Air Reservist,* XXII (February 1970), 12-13.

Andrews, Hal. "Naval Aircraft: Skytrain II." *Naval Aviation News,* (February 1974), 20-21.

Bayer, Harold. "A Contractor's View of System Support: C-9 Case Study." *Defense Management Journal*, XIII (July 1977), 22-28.

Brown, Gainett C., Jr. "Notes on the Nightingale." *MAC Flyer,* XVI (February 1969), 4-7.

"The C-9: Aircraft and Mission." *MAC Flyer*, XV (June 1968), 4-6.

Gann, Harry. "She Ain't Got a Tailhook, But That's No Sin--C-9B." *The Hook,* VIII (Winter 1979), 47-48.

Hart, J. W. "The First C-9A Joins the Air Force." *Airman,* XII (June 1968), 23-24.

Haskins, Gale L. "Routine Nightingale." *MAC Flyer,* XVII (October 1970), 9-11.

Stone, Irving. "Aeromedical Airlift Joins the Jet Age." *Air Force and Space Digest,* LI (March 1968), 102-105.

Willworth, James. "Nightingale." *Airman,* XIII (May 1969), 30-33.

(Curtiss) C-10 Robin

Hattendorf, Dan. "Story of a Robin." *International Antique Airplane Digest,* XXIV (April-June 1976), 43-46.

"Texas Lady." *Aviation Quarterly,* V (Fall 1979), 204-209.

(McDonnell-Douglas) C/KC-10 Extender

Book/Document/Report Literature:

Bailey, Charles E. *Strategic Mobility and the KC-10: A Pilot's Perspective.* Research Paper. Maxwell AFB, Ala.: Air Command and Staff College, Air University, 1983. 21p.

Foley, Jack O. *The KC-10: Concept, Capabilities, and Future.* Research Paper. Maxwell AFB, Ala.: Air Command and Staff College, Air University, 1984. 47p.

Godson, John. *The Rise and Fall of the DC-10.* New York: David McKay, 1975. 351p.

Helms, Ramsey H., Jr. *Employment of the KC-10A.* Research Paper. Maxwell AFB, Ala.: Air Command and Staff College, Air University, 1979. 53p.

Holubik, T. E. *History of the KC-10A Aircraft Acquisition.* Research Paper. Maxwell AFB, Ala.: Air Command and Staff College, Air University, 1988. 82p.

Kuehner, Gregory A. *Air Refueling: The KC-10 Connection.* Research Report. Washington, D.C.: Industrial College of the Armed Forces, 1983. 57p.

Neill, Curtis N., III. *KC-10: A Study in Commercial Derivative Aircraft Acquisition.* Research Paper. Maxwell AFB, Ala.: Air Command and Staff College, Air University, 1980. 147p.

Schmitt, Michael W. *KC-10 Flight Test Management: A Study of the Predelivery Testing of a Commercial Derivative.* Research Paper. Maxwell AFB, Ala.: Air Command and Staff College, Air University, 1982. 44p.

Tapleshay, Thomas N. KC-10A: The SIOP Tanker. Research Paper. Maxwell AFB, Ala.: Air Command and Staff College, Air University, 1982. 39p.

Periodical/Anthology Literature:

Bailey, Charles E. "The Maturing of the KC-10." Air Force Magazine, LXVI (May 1983), 52-55.

Batezel, Anthony L. "Just Plane Partners." Airman, XXV (May 1981), 2-7.

Brunz, Wayne. "The KC-10 Extender." Combat Crew, XXX (August 1980), 12-13.

Castledine, Steven. "Air Force Deploys KC-10s." Aviation Week and Space Technology, CXIV (June 15, 1981), 64+.

"Exercising With the Extender." Air International, XXXII (1987), 111+.

Fink, Donald E. "DC-10 Tanker/Cargo Program Starting." Aviation Week and Space Technology, CVIII (January 16, 1978), 41-43.

Fullerkamp, Bernie. "Double-Duty Tankers." U.S. Naval Institute Proceedings, CXIV (February 1988), 103-106.

Gillert, Doug. "Ready to Respond." Airman, XXX (February 1986), 9-16.

Jacobs, Gene. "The KC-10A Extender Long-Range Airlifter." Asian Defence Journal, no. 7 (july 1985), 68+.

Jones, Brad. "Refueling With the KC-10." Combat Crew, XXXIII (August 1983), 4-6.

Mordoff, Keith F. "SAC Attaining Full Operational Use of KC-10A Extender." Aviation Week and Space Technology, CXXII (May 20, 1985), 50-51+.

Peacock, Lindsay. "The KC-10A: USAF's Newest Range-Extender." Air International, XXIII (December 1982), 265-270.

Roberts, Thomas. "KC-10A Extender." MAC Flyer, XXVIII (February 1981), 16-19.

Ropelewski, Robert R. "KC-10 Readied for Delivery to USAF." Aviation Week and Space Technology, CXIV (March 2, 1981), 43-44.

Siuru, William D., Jr. "New KC-10 Air Force Tanker." National Defense, LXIII (January-February 1979), 32-33.

Smith, W. S. "KC-10 Test Flight Program." Society of Experimental Test Pilots Technical Review, XVI (Spring 1981), 136-143.

Vairo, Joseph M. "SAC's Newest Tanker." Combat Crew, XXVIII (July 1978), 16-18.

(Handley-Page/BAe) C-10 Jetstream

Lee, G. H. C. "Aerodynamic Design Philosophy of the Handley-Page Jetstream." Aircraft Engineering, XXXIX (September 1967), 10-14+.

Trammell, Archie. "Progress Report: Handley-Page Jetstream." Flying, LXXXIV (February 1969), 50-53.

Whitaker, Richard. "High Performance Jetstream 31." Flight International, CXX (October 10, 1981), 1087-1093.

(Consolidated) C-11 Fleetster

Bowers, Peter M. "The Consolidated Fleetster." The AOPA Pilot, XXII (April 1979), 71-73.

(Lockheed) C-12 Vega

Strasser, Emil. "Aviation Scrapbook: Lockheed Vega." Air Classics, XIV (July 1978), 56-59.

Tallman, Frank. "Lockheed Vega: A Pilot's Report." Flying, LXXVII (September 1965), 74-79.

(Beech) C-12 Huron

Book/Document/Report Literature:

Gilson, R. L. A Preliminary Analysis of C-12 Aircraft Usage by the Naval Air Logistics System. Thesis. Monterey, Calif.: Naval Postgraduate School, 1984. 100p.

Watte, Joseph C., et al. Airworthiness and Flight Characteristics Evaluation C-12A Aircraft: Final Report. Edwards AFB, Calif.: Army Aviation Engineering Flight Activity, 1976. 203p.

Periodical/Anthology Literature:

Anderson, David A. "Flight Inspection of Army Navigation Aids." U.S. Army Aviation Digest, XXXV (January 1989), 22-23.

Barger, Millard. "Flying High Through Turbulence: The Story of the Beech C-12." Armed Forces Journal International, CXXII (January 1985), 94+.

Bridges, Alastair G. "King of the Air." Flying Safety, XLII (September 1986), 6-7; XLIII (February 1987), 23-24.

Clouse, Gale E., Jr. "Operational Support Aircraft." MAC Flyer, XXXI (January 1984), 16-19.

Forsyth, Robert F. "A New Member of the Team." U.S. Army Aviation Digest, XXII (July 1976), 24-27.

Hinton, Al. "Check Plane." U.S. Army Aviation Digest, XXVI (August 1980), 22-23.

McClellan, J. Mac. "B200 Super K." Flying, CVIII (July 1981), 47-51.

"Military Report: C-12F." Air Progress, XLIX (January 1987), 68-71.

Treff, Tom. "Propwash Around the C-12 Only Hogwash." Armed Forces Journal International, CXXII (January 1985), 95+.

C-17 see C-12 Vega

(McDonnell-Douglas) C-17

Book/Document/Report Literature:

Hooten, B. R. C-17: We Need It Yesterday. Student Essay. Carlisle Barracks, Pa.: Army War College, 1985. 37p.

Johnson, Charles L., II. Acquisition of the C-17 Aircraft: An Historical Account. Research Paper. Maxwell AFB, Ala.: Air Command and Staff College, Air University, 1986. 521p.

Thomas, R. William. Improving Strategic Mobility: The C-17 Program and Alternatives. Washington, D.C.: Congressional Budget Office, 1986. 54p.

Periodical/Anthology Literature:

"C-17 Will Fill Long-Haul Airlift Gap." Flight International, CXXVII (April 27, 1985), 9-11.

Cooper, Thomas E. "Filling the Airlift Gap: The C-17 is Still the Best Solution." Logistics Forum, II (April 1986), 51+.

DeMeis, Richard. "Designing an Airlifter: McDonnell-Douglas' C-17." Aerospace America, XXIV (January 1986), 66-68+.

Denny, Jeffrey. "The C-17's Uncertain Future." Military Logistics Forum, I (January-February 1985), 58-61+.

Grimes, Vincent. "C-17: The Airlift Answer?" International Combat Arms, VI (September 1988), 60-67.

Kozicharow, Eugene. "DOD Seeks C-17 Full-Scale Development." Aviation Week and Space Technology, CXX (March 19, 1984), 19-20.

McDonnell-Douglas Corporation. "C-17: The Direct Delivery Airlifter." Airlift Operations Review, IV (January-March 1982), 12-16.

Mohr, Henry. "Deterrence and the C-17." Journal of Defense and Diplomacy, V (February 1987), 50-55.

Pace, Steve. "Pumping Iron at Mach 7." Airpower, XVIII (March 1988), 34-42.

Pauley, J. D. "Reliable and Maintainable: C-17 Design Features." Airlift, IX (Spring 1987), 4-7.

Rhodes, Jeffrey P. "The First C-17." Air Force Magazine, LXXI (August 1988), 54-61.

Turk, Paul. "C-17 Cleared for Takeoff." Interavia, XL (May 1985), 511-512.

(Boeing) C-18 Monomail

"Boeing's Model 200 Monomail." Aviation, XXIX (October 1930), 232-240.

Bowers, Peter M. "Airmail Special Delivery: The Story of Boeing's Monomail." Wings, II (April 1972), 21+.

(Northrop) C-19 Alpha

Wilson, Kenneth D. "Jack Northrop's Metal Miracle." Model Aviation, VI (October 1980), 55-60, 122.

(Boeing) C-19
(see also E-4)

Gilcrist, Peter. Boeing 747. London, Eng.: Ian Allan, 1985. 80p.

Marsh, Alton K. "USAF to Keep 747s Out of Inventory." Aviation Week and Space Technology, CXVI (June 21, 1982), 23-24.
Employed by Civil Reserve Air Fleet and Air National Guard.

Seo, Hiroshi. Boeing 747. New York: Jane's, 1984. 84p.

(Grumman) C-20 Gulfstream III

Almy, David W., Richard N. Aarons, and John W. Olcott. "B/CA Analysis: Gulfstream III." Business and Commercial Aviation, LII (January 1983), 77-83.

Clouse, Gale E., Jr. "C-20A Gulfstream III: The 89th MAW Gets New Aircraft." MAC Flyer, XXX (September 1983), 16-19.

"Grumman Readies New Gulfstream." Aviation Week and Space Technology, CVIII (April 10, 1976), 17-18.

Kozicharow, Eugene. "Gulfstream IIIs Meet Military Requirements." Aviation Week and Space Technology, CXXI (October 1, 1984), 63-64.

Mead, Lawrence M. The Grumman Aerospace/Gulfstream American Gulfstream III: A Case Study in Aircraft Design. New York: American Institute of Aeronautics and Astronautics, 1980. 30p.

Proctor, Paul. "Air Force Exploits Speed, Range of Gulfstream C-20 in VIP Role." Aviation Week and Space Technology, CXXVI (June 8, 1987), 74-75.

___. "The Mission is the Challenge." Professional Pilot, XVIII (August 1984), 66-71.

(Douglas) C-21 Dolphin

Bowers, Peter M. "Douglas Dolphin." Airpower, XII (November 1982), 10-27.

(Lear) C-21 Lear Jet

Bridges, Alastair C. "C-21." Flying Safety, XLII (April 1986), 7-8.

Collins, Richard L. "Lear Jet Legacy." Flying, CIX (December 1982), 62-67.

Hirst, Mike and John Marsden. "Longhorns Into Service." Flight International, CXV (February 10, 1979), 401-495.

Howard, Byron L. "Flying the C-21: Much Power Plus Great Avionics Equals One Efficient People-Mover." MAC Flyer, XXXII (January 1985), 16-19.

Slocum, Jerard A. "Reporting for Duty." Flying, CXII (January 1985), 42-46+.

United States. Government Accounting Office. National Security and International Affairs Division. ANG's Need for C-21 Aircraft. Washington, D.C., 1986. 20p.

"Welcome the C-21A." Airman, XXVIII (June 1984), 18-19.

C-22 see C-11 Fleetster

(Boeing) C-22

Gunston, William T. ("Bill"). "Boeing 727." Flight, LXXXIII (May 9, 1963), 675-688.

Morgan, Len. "Flying the Three-Holer." Flying, CI (October 1977), 68-73, 106.

___ and Terry Morgan. Boeing 727 Scrapbook. Fallbrook, Calif.: Aero Publishers, 1978. 104p.

(Shorts) C-23 Sherpa

Feazel, Michael. "European Distribution System Uses C-23As to Optimize Supplies." Aviation Week and Space Technology, CXXIII (August 19, 1985), 111-112.

Howard, Byron L. "The C-23A Sherpa." MAC Flyer, XXXI (June 1984), 12+.

Nolan, David S. "Long and Short of the Sherpa." Airman, XXX (September 1986), 20-21.

"Sherpa: Prime Mover for the USAF." Air International, XXVII (1984), 225+.

(Fairchild) C-26 Metro III

Fairchild Aircraft Corporation. Metro III: Built to Be the Best Turboprop in the World. San Antonio, Tex., 1981. 16p.

"Fairchild Boosting Production Rate of Metro III to Meet Rising Demand." Aviation Week and Space Technology, CXXII (May 20, 1985), 20+.

(Bellanca) C-27 Airbus

Bowers, Peter M. "The Bellanca Airbus." The AOPA Pilot, XXIV (February 1981), 99-101.

Poupitch, Vernet. "Damn the Torpedoes!" Flying Safety, XLIV (August 1988), 12-14.

(BAe) C-29

Parrish, Robert L. "Operator Survey: BAe 125-800." Business and Commercial Aviation, LVII (December 1985), 52-56.

(Curtiss) C-30/R4C Condor II
(see also B-2 Condor)

Gann, Ernest K. "Condor." Flying, XC (June 1972), 20-21+.

Matthews, William R. "Curtiss Condor II." American Aviation Historical Society Journal, XII (Spring 1967), 3-27.
Abridged in Flying Review International, XXII (May 1967), 579-582.

(Douglas) C-32/C-33/R2D-1

Justin, Joseph W. "The Douglas DC-2: A Catalogue--Usage and Disposition." American Aviation Historical Society Journal, XV (Winter 1970), 284-288; XVI (Summer 1971), 105-110.

Trimble, Robert L. "Silver-Winged Thoroughbreed." Air Classics, XXI (May 1985), 28-39, 60-61, 82.

(Lockheed) UC-36/UC-37 Electra

Daniels, C. M. "Electra: Twin-Tailed Lockheed." Airpower, X (November 1980), 8-11, 14-23.

Emmert, Thomas M. and William T. Larkins. "Lockheed's Model 10 Electra." American Aviation Historical Society Journal, XXIII (Summer 1978), 141-153.

UC-38/UC-39 see C-32/C-33 and C-47/C-48

(Lockheed) UC-40 Electra Junior

Emmert, Thomas M. and William T. Larkins. "Electra Junior: Lockheed's Model 12." American Aviation Historical Society Journal, XIX (Winter 1974), 307-319.

Purnell, Neville M. "Long-Lived Lockheed." Aeroplane Monthly, IV (December 1976), 656-658.

C-41/C-42 see C-47/C-48

(Beech) UC-43/GB Staggerwing

Cook, LeRoy. "Classic Encounter With a Staggerwing Beech." Private Pilot, XX (August 1985), 30-43.

Cox, Bill. "Beech Staggerwing." Plane and Pilot, XI (September 1975), 26-29, 41.

Holcomb, Mal. "Eagle and the Turtle." Wings, X (February 1980), 10-17.

Lemen, W. T. "Staggerwing." In: Michael O'Leary, ed. Air Progress Pilot Reports. Canoga Park, Calif.: Challenge Publications, 1979. pp. 58-61, 98-99.

Phillips, Ed. "Staggerwing." Wings, IX (October 1979), 47-53.

Smith, Robert T. "Beech Staggerwing." Air Classics, II (January-February 1965), 24-28+.

___. "Staggerwing." Aviation Quarterly, I (Spring 1974), 32-65.

___. Staggerwing: The Story of a Classic Beech Biplane. Medina, Pa.: Maney, 1967. 249p.

(Beech) C-45/JRB/SNB Expeditor

Bowers, Peter M. "The Beech 18." The AOPA Pilot, XXVI (February 1983), 83-86.

Cantrell, Christopher. "Old Snort." Air Progress, XXXVIII (December 1976), 50+.

"Exit the Expeditor." Airman, VIII (February 1964), 10-11.

Hearde, Basil. "Lady of Distinction." Air Classics, II (July-August 1965), 58+.

Papineau, Ned. "Beechcraft's Legendary Twin: Model 18." Plane and Pilot, XV (February 1979), 14-19.

Smith, Ken. "Looking Back at the Beechcraft 18." Plane and Pilot, XII (May 1976), 30-33.

(Curtiss) C-46/R5D Commando

Book/Document/Report Literature:

Bagwell, Margaret C., et al. C-46, January 1940-August 1945. ATSC Historical Study, no. 51. Wright Field, Ohio: USAAF Air Technical Service Command, 1945. 85p.

Eastwood, A. B. Curtiss-Wright C-46 Commando. 2 vols. Cowley, Eng.: LAAS International, 1971.

Hump Pilots Association. Flying the C-46 Type Aircraft, Popular Bluffs, Mo., 1976. 15p.

United States. Army Air Forces. Training Command. Pilot Training Manual for the Commando C-46. AAF Manual 51-129-1. Washington, D.C., 1945.

Periodical/Anthology Literature:

"Cargo Commando." Air Force, XXXII (December 1949), 44-45.

"The Curtiss C-46 Commando." AirTech, III (February 1945), 16-25+.

"Curtiss Commando." Air Pictorial, XV (October 1953), 308-309.

"The Curtiss Commando." Automotive and Aviation Industries, LXXXVII (November 1, 1942), 24-25.

Foster, John, Jr. "Design Analysis of the Curtiss Commando." Aviation, LXII (August 1943), 130-153.

Gibson, T. J. "The Broken Nose Commando." American Aviation Historical Society Journal, XXII (Spring 1977), 39-47.

Jansen, P. N. "The Curtiss Commando." Machinist, XLIX (July 1943), 134-143.

Klemin, Alexander. "Troopship of the Air." Scientific American, CLXVII (November 1942), 230-231.

Miller, P. W. "C-46 Plus Jet Pod." Skyways, XII (February 1953), 18-20.

Morgan, Len. "The Whale's Tale." Flying, CXIV (December 1987), 64-70.

Trimble, Robert L. "Big-Assed Bird." Air Classics, XVII (July 1981), 14-26, 75.

Wiley, John. "The Cantankerous C-46 Commando." Professional Pilot, XXI (September 1987), 90-95.

Woznek, William S. "C-46 Commando." In: Robin Higham and Abigail T. Siddall, eds. Flying Combat Aircraft of the USAAF-USAF, Ames, Iowa: Iowa State University Press, 1975. pp. 37-42.

(Douglas) C-47/R4D Skytrain/Dakota

Book/Document/Report Literature:

Borge, Jacques and Nicholas Viasnoff. The Dakota: The DC-3 Story. New York: Frederick Warne, 1982. 192p.

Bowers, Peter M. DC-3: 50 Years of Legendary Flight. Fallbrook, Calif.: Aero Publishers, 1986. 259p.

Glines, Carroll V. and Wendell F. Mosley. The Grand Old Lady: The Story of the DC-3. Cleveland, Ohio: Pennington Press, 1960. 250p. Reprinted by the Philadelphia, Pa., firm of Lippincott in 1966.

Grandidge, J. M. G. The Douglas DC-3 and Its Predecessors. Toubridge, Eng.: Air Britain, Ltd., 1984. 780p.

Hooks, Michael J. Douglas DC-3 Dakota. Super Profile Series. Newberry Park, Calif.: Interbook, 1985. 56p.

Ingells, Douglas J. The Plane That Changed the World: A Biography of the DC-3. Fallbrook, Calif.: Aero Publishers, 1973. 250p.

Morgon, Len. Great Aircraft: The Douglas DC-3. Dallas, Tex.: Aviation Books, 1964. 95p.

____. The Douglas DC-3. Famous Aircraft Series. New York: Arco, 1964. 56p.

Pearcy, Arthur. The DC-3 Dakota, Ballantine's Illustrated History of World War II. New York: Ballantine Books, 1975. 160p.

____. DC-3 Survivors. Bourne End, Eng.: Aston Publications, 1987. 215p.

____. The Douglas DC-3. Aircraft in Profile, no. 96. Windsor, Eng.: Profile Publications, 1966. 16p.

____. Douglas DC-3 Survivors, V. 2. Bourne End, Eng.: Aston Publications, 1988. 256p.

____. Douglas Dakota Mks. I-IV. Aircraft in Profile, no. 220. Windsor, Eng.: Profile Publications, 1971. 24p.

____. Douglas R4D Variants. Aircraft in Profile, no. 249. Windsor, Eng.: Profile Publications, 1972. 24p.

United States. Army Air Forces. Training Command. Pilot Training Manual for the Skytrain, C-47. AAF Manual 51-129-2. Washington, D.C., 1945.

____. Department of the Air Force. C-47 Standardization/Evaluation Manual: Flying. AFM 60-47. Washington, D.C., 1961. 28p.

Periodical/Anthology Literature:

"Age Cannot Wither Her, Nor Custom Stale Her Infinite Variety." Flight, (November 1, 1962), 706-709.

"Ageless--The Douglas DC-3." American Aviation Historical Society Journal, X (Spring 1965), 58-61.

Anderson, Tom and Randy Trott. "A Plane in Olympus." Airman, XVII (October 1973), 15-16.

Baxter, Raymond. "A Touch of the Gregory Pecks." Aeroplane Monthly, IV (April 1976), 190-193.

Block, T. E. "Pilot Report: Dakota." Flying, CIV (March 1979), 34-38.

Bowers, Peter M. "DC-3 in Uniform." Airpower, XI (January 1981), 34-50; Wings, XI (February and June 1981), 9-19, 36-49.

Brodgen, Stanley. "The DC-3: 50 Years and Still a Trojan." Airlift, (December 1985), 42+.

Brooks, Peter W. "The DC-3 is 25." Flight, LXXVIII (December 23, 1960), 981-985+.

Churchill, Jan. "Resolute R4D." Naval Aviation News, (November-December 1983), 10-14.

Cole, James L., Jr. "Lamplighters and Gypsies." Aerospace Historian, XX (Spring 1973), 30-35.

____. "An Old Airplane in a New War." Aerospace Historian, XVIII (June 1971), 71-74.

Collins, Helen M. "They Fly...and Fly....and Fly." Naval Aviation News, (May 1977), 32-36.

Courtney, William B. "Douglas Nets the Earth." Collier's, XCVIII (July 4, 1936), 20+.

"Donald Douglas and the Indestructible DC-3." FAA Aviation News, XI (August 1972), 12-13.

Douglas, Donald W. "The Super DC-3." Aviation Week, L (January 31, 1949), 14-15.

Dykes, Marion A., Jr. "C-47 Worldwide." Navigator, XV (Summer 1968), 36-37.

Eaton, William W. "The Gooneybird's Last Stand." Airman, VII (June 1963), 26-29.

Elliott, John. "Marines and the Gooney Bird." Naval Aviation News, (November-December 1983), 15-16.

Eskow, Dennis. "Sentimental Journey." Popular Mechanics, CLXIII (May 1986), 110-112, 210.

"The Faithful Gooney." Combat Crew, VIII (May 1958), 26-29.

Fleming, Mike. "Last Flight of the Ellsworth 'Goon.'" Aerospace Historian, XXXII (December 1985), 246-248.

Frazer, Charles D. "Workhorse of the AAF." Air Force, XXVII (September 1944), 14-16.

Gann, Harry K. "Mark of Distinction." Naval Aviation News, (June 1978), 32-33.

Gann, Mike. "Legend of the DC-3." Aviation/Space, (Winter 1985), 39-40.

Glines, Carroll V. "The Grand Old Gooney Bird (DC-3 or C-47)." Air Force Magazine, LXVIII (December 1985), 94-100.

____. "Never Say Farewell to the Gooney Bird." Aerospace Historian, XIII (Autumn 1966), 121-125.

____. "Tribute to the 'Goon.'" Airman, X (March 1966), 18-21.

Greenway, Sean. "Donald Douglas and the Gooney Bird." Air Classics, XVII (June 1981), 28-37.

Harrison, Neil. "New Safety For Old." Flight, LXXXII (August 9, 1962), 185+.

Higham, Robin and James L. Cole, Jr. "C-47 Dakota and AC-47." In: Robin Higham and Abigail T. Siddall, eds. Flying Combat Aircraft of the USAAF-USAF. Ames, Iowa: Iowa State University Press, 1975. pp. 43-52.

Holden, Henry M. "Douglas DC-3: Wings of the World." American Aviation Historical Society Journal, XXX (Winter 1985), 242-259; XXXI (Spring 1986), 26-43.

Howe, Paul H. "Queen of the Transports." Flying,XXXVIII (April 1946), 27-28+.

Johnson, Robert E. "Who Really Gave Us the DC-3?" American Aviation Historical Society Journal, XXIX (Fall 1984), 202-213.

Jones, John P. "Making a Legend: The DC-3." *Flight Operations*, LXIV (December 1975), 42-44; XLV (February 1976), 34-36, 46.

Kahn, Ely J., Jr. "Profiles: The Three." *New Yorker*, XXXVI (September 10, 1960), 49-50+.

Kilpatrick, Bill. "Glory Days of a Grand Old Lady: the DC-3." *Popular Mechanics*, CXLIX (June 1978), 89-93+.

Kropf, Gene. "Peripatetic Gooney Bird." *Naval Aviation News*, (May 1977), 36-37.

"Looney Over the Gooney." *Air Classics*, I (Fall 1964), 44-48.

McCann, John A. "The Ugly Duckling of Air Power." *Air Power Historian*, V (January 1958), 54-61.

Martin, Jimmie D. "50 Years of Aviation Fame." *Flying Safety*, XLII (January 1986), 14-16.

Messex, Curtis L. "C-47 Short Field Landings." *Aerospace Safety*, XXII (February 1966), 10-12.

Miller, Ronald and David Sawers. "The Story of the DC-3: A Digression." In: their *The Technical Development of Modern Aviation*. New York: Praeger, 1968. pp. 98-123.

Morgan, Len. "Fond Memories of the Famed DC-3." *Popular Mechanics*, CXLIX (June 1978), 160-161.

____. "Happy Birthday DC-3." *Flying*, CXII (December 1985), 50-57.

Moore, F. L. "Defense Puts Spotlight on DC-3." *Aviation Week*, LIII (December 25, 1950), 39-40.

Munson, Russell. "DC-3." *Flying*, CIII (July 1978), 68-73.

Nistal, Gerald E. "The Douglas C-47." *AirTech*, V (September 1944), 15-22.

Oakey, Michael. "50 Years of the DC-3." *Flight International*, CXXVII (December 21, 1985), 40-44.

O'Doherty, John K. "Never Laugh at a Gooney Bird." *Airman*, IV (November 1960), 8-10.

Parfit, Michael. "Homage to the DC-3." *Smithsonian*, XIX (September 1988), 144-169.

Pearcy, Arthur. "DC-3: 40 Glorious Years." *Aeroplane Monthly*, III (December 1975), 509-615; IV (January 1976), 39-45.

____. "DC-3 Survivors." *Exxon Air World*, XXXIX (Fall 1987), 18-19.

____."When the Army Flew Goonies." *Air Classics*, XXI (December 1985), 53-67.

Richter, Bill. "A Golden Anniversary for the DC-3." OAG Frequent Flyer, (December 1985), 52-60.

Satterlee, H. Y. "The Conversion of the C-47 Skytrain for Airborne Operations." Aero Digest, XLVII (October 1, 1944), 74-76.

Schnepf, Ed. "Golden Hours: Douglas DC-3/C-47 Approaches a Half Century of Existence." Air Progress, XLV (December 1983), 5-10+.

"Short Stories DC-3 Style." Naval Aviation News, (May 1977), 38-39.

Smith, Edgar A. "Stuka Chaser: The True Story of a World War II C-47 Transport." Aerospace Historian, XXXIII (Winter 1986), 238-241.

Smith, Gene. "Kilroy and Friends." Air Classics, XXII (January 1986), 14-23, 59.

Stocker, Joseph S. "The Saga of the DC-3." In: Richard Harmet, ed. Highlights From These Exciting Times. Chicago, Ill.: Popular Mechanics Press, 1959. pp. 87-91.

Sweet, Harold. "Whistling Willie, the Flying Sieve." In: Norman Carlisle, ed. The Air Forces Reader. Indianapolis, Ind.: Bobbs-Merrill, 1944. pp. 226-229.
The story of a DC-3 in the CBI Theater as reprinted from the April 1943 issue of Douglas Airview.

Volney, Joseph and Harry S. Baer, Jr. "The Impact of the Immortal DC-3." Flying, LIX (October 1956), 47-49.

Waldo, Richard K. and Herbert J. Guth. "DC-3: The Plane That Won't Wear Out." Harper's Magazine, CCIX (August 1954), 75-82.

Weeks, C, S. "'Will O' the Wisp.'" Aircraft, XX (May 1958), 29-33, 76-80.

Welling, William. "Evolution of a Legend." Airline Executive, VIII (December 1985), 19-23.

Winchester, James H. "Winged Work Horse." Collier's, CXXVI (December 30, 1950), 20-21+.

Wright, Roger. "Birth of a Legend: History of the DC-3." Air Extra, no. 48 (June-August 1985), 4-9.

C-48 to C-53 see C-47/R4D Skytrain/Dakota

(Douglas) C-54 Skymaster

Book/Document/Report Literature:

The Douglas DC-4. Hutton, Eng.: Air Britain, Ltd., 1969. 104p.

United States. Air Force. Air Transport Command. Reconditioning of the C-54 Fleet, 1 January-30 November 1947. Scott AFB, Ill.: History Office, Air Transport Command, 1947. 88p.

___. ___. Military Air Transport Service. C-54/R5D Operating Instructions Manual. MM-55-2. Washington, D.C., 1955.

___. Army Air Forces. Air Transport Command. Pilot Training Manual for the Skymaster, C-54. AAF Manual 51-129-3. Washington, D.C., 1945.

Periodical/Anthology Literature:

Bossen, Victor. "Evolution of the C-54." Skyways, IV (January 1945), 34-40.

Burton, E. G. "Design of the DC-4/C-54 Airplane." Automotive and Aviation Industries, XCI (November 1, 1944), 88+.

Clithero, Robert G. "C-54 Postscripts." Combat Crew, XXII (October 1971), 28-29.

"The Douglas C-54: The Nation's Largest Army Transport." Automotive and Aviation Industries, LXXXVI (April 1, 1942), 38-39.

Gann, Ernest K. "Purebred." Flying, XCV (July 1974), 18-19+.

Grant, Robert B. "The Four." Air Classics, XXII (March 1986), 28-37, 74-75.

Ohlinger, John F. "C-54 Skymaster." In: Robin Higham and Abigail T. Siddall, eds. Flying Combat Aircraft of the USAAF-USAF. Ames, Iowa: Iowa State University Press, 1975. pp. 53-60.

"Roosevelt's C-54." Western Flying, XV (July 1945), 48-49.

Taylor, Henry A. "The Skymaster Story." Air Enthusiast, XV (April-July 1981), 34-48, 78-80.

C-55 see C-46 Commando

(Lockheed) C-56/R50 Lodestar

"Lodestar as a Glider Towplane." Aero Digest, XLV (April 15, 1944), 79+.

C-57 see C-56/R50 Lodestar

C-58 see B-18 Bolo

C-59 see C-56/R50 Lodestar

C-60 see C-56/R50 Lodestar

(Fairchild) UC-61/GK-1 Forwarder

Ames, Bob. "Fairchild 24." Air Classics, II (September-October 1965), 10-13+.

Davisson, Budd. "The Clean Machine: We Fly a Super Fairchild F-24W." Air Progress, XLVII (August 1985), 14-15+.

Smith, Gene. "Fairchild 24: A Gentleman's Airplane." Air Progress, XXXVIII (September 1976), 58-62, 84.

(Waco) C-62

Cody, Francis M. and Margaret C. Bagwell. C-62, August 1941-January 1944. ATSC Historical Study, no. 52. Wright Field, Ohio: USAAF Air Technical Service Command, 1945. 20p.

C-63 see A-29 Hudson

(Noorduyn) C-64 Norseman

Halford, Robert G. "Saga of the Norseman." Canadian Aviation Historical Society Journal, XVII (Fall 1979), 67-75.

Larson, George C. "Noorduyn Norseman: Prosaic Makes Perfect." Flying, XCVIII (June 1976), 71-78.

"Noorduyn Norseman Serves AAF." Aviation, XLII (June 1943), 241+.

O'Leary, Michael. "Norseman Recovery." Air Classics, XXIV (April 1988), 14-19.

C-66 see C-56/R5O Lodestar

C-67 see B-23 Dragon

C-68 see C-47/R4D Skytrain/Dakota

(Lockheed) C-69/R70 Constellation
(see also C-121 Super Constellation)

Book/Document/Report Literature:

Anderson, Holmes. Lockheed Constellation. Aircraft in Profile, no. 120. Windsor, Eng.: Profile Publications, 1971. 24p.

Hardy, Michael J. The Lockheed Constellation. New York: Arco, 1973. 128p.

Johnson, Clarence M. ("Kelly"). Development of the Lockheed Constellation. Burbank, Calif.: Lockheed Corporation, 1944.

Marson, Peter. The Constellation. Hutton, Eng.: Air Britain, Ltd., 1969. 70p.

____. Lockheed Constellation Series. Kent, Eng.: Air Britain, Ltd., 1982. 508p.

United States. Army Air Forces. Air Materiel Command. C-69, May 1941-February 1946. AMC Historical Study, no. 120. Wright Field, Ohio: USAAF Air Materiel Command, 1946. 28p.

Periodical/Anthology Literature:

"Advanced Mechanical Design Insures Performance and Safety in Constellation." Product Engineering, XVI (March 1945), 156-162.

Barr, Louis. "The Lockheed Constellation: A History." American Aviation Historical Society Journal, XXVIII (Fall-Winter 1983), 190-205; XXIX (Spring 1984), 44-55.

"Flying Shark: Lockheed's Constellation." Popular Mechanics, LXXIX (June 1943), 1-4.

Johnson, Clarence L. ("Kelly"). "Development of the Lockheed Constellation." Aeronautical Engineering Review, IV (July 1945), 5+.

Lindley, Denver. "First in the Sky: Constellation, Newest Thing in Transport Planes." Collier's, CIX (March 7, 1942), 21+.

"Lockheed Constellation." Flying, XXXVI (May 1945), 67-70.

Pierce, Paul. "Constellation, New Stratospheric Giant." Flying, XXXII (March 1943), 35+.

(Boeing) C-73

Bowers, Peter M. "The Boeing 247." The AOPA Pilot, XXIV (September 1981), 71-73.

Powell, Dennis M. "The Boeing 247." Air Pictorial, XXIV (July 1962), 194-197.

Seely, Victor D. "Boeing's Pacesetting 247." American Aviation Historical Society Journal, IX (Winter 1964), 239-270.

Watson, Dave. "Adaptable Annie: Boeing 247D." Flying, LIV (January 1954), 36+.

(Douglas) C-74 Globemaster
(see also C-124 Globemaster II)

Williams, Nicholas. "Globemaster: The Douglas C-74." American Aviation Historical Society Journal, XXV (Summer 1980), 82-106.

(Boeing) C-75 Stratoliner

Churchill, Edward. "The Stratoliner." Flying, XXXI (December 1942), 30+.

Powell, Dennis M. "The Veteran Stratoliners." Flight, LXVIII (March 1955), 32-35.

Scheppler, Robert H. "The Boeing Stratoliners." American Aviation Historical Society Journal, VIII (Spring 1963), 23-50.

Scheppler, Robert. "The Flying Whale." Air Classics, XII (May 1976), 26-35.

(Curtiss) C-76 Caravan

Blackmore, C. C. and Barbara Horner. C-76, July 1941-October 1943. ATSC Historical Study, no. 14. Wright Field, Ohio: USAAF Air Technical Service Command, 1944. 29p.

Boyne, Walter J. "The Basketcase Bummer: Curtiss C-76 Caravan." Airpower, IV (May 1974), 60-65.

"Curtiss Caravan." Aero Digest, XLII (January 1943), 186-188.

(Cessna) UC-78 Bobcat

"Wichita Bobcat." Scale Models, XII (April 1981), 184-186.

(Stinson) UC-81 Reliant

Lert, Peter. "Gullwing." Air Progress, XXXIX (March 1977), 42-46.

O'Hara, Bob. "Gullwing Stinson." Plane and Pilot, X (March 1974), 46-50.

(Fairchild) C-82 Packet

"Fairchild C-82 Packet." Aeronautical Engineering Review, III (December 1944), 149+.

"The 'Thing.'" Pegasus, XXVI (December 1957), 12-15.

UC-83 see L-5 Cub

C-84 see C-47/R4D Skytrain/Dakota

UC-86 see UC-61 Forwarder

(Consolidated) C-87 Liberator
(see also B-24 Liberator)

"The Consolidated C-87 Liberator Express." Canadian Aviation, XVI (July 1944), 87-94.

Mikesh, Robert C. "FDR and the 'Guess Where II.'" Air Classics, IX (April 1973), 32-37.

(Budd) C-93 Canastoga

Horner, Barbara. C-93. ATSC Historical Study, no. 15. Wright Field, Ohio: USAAF Air Technical Service Command, 1944. 18p.

"Structural Details of the Canastoga." Aero Digest, XLV (June 1, 1944), 80-83.

Watson, M. "Introducing the Budd RB-1 Canastoga Cargo Carrier and Troop Transport." Aero Digest, XLV (May 15, 1944), 75-77+.

UC-95 see L-2 Taylorcraft

(Boeing) C/KC-97 Stratofreighter/Stratotanker

"Available Jones." Flight, LXV (February 26, 1954), 235-236.

Beal, Thomas F. "Navigator Heritage: The KC-97L." Navigator, XXVII (Winter 1980), 15-19.

"Boeing Stratofreighters: Distinguished and Busy Progeny." U.S. Air Services, XXXIX (December 1954), 11-15.

Bywater, Murray A. "The KC-97 Gets 'Em There." Combat Crew, IX (August 1958), 22-25.

Calkins, Ken. "Available Jones." Boeing Magazine, XXIII (November 1953), 12-14.

Conklin, John D. "Those Squawking 'Talking Birds.'" Airman, VII (December 1963), 14-17.

Delaney, James L. "The KC-97L: Evolution of the Air Guard's Six-Engine Tanker." Aerospace Historian, XXIX (December 1982), 265-268.

Dockendorff, John R. "Old Stratofreighters Never Die." Combat Crew, XXVII (June-July 1977), 16-19, 16-18.

Geisenheyner, Stefan. "ANG KC-97Ls: Doing a Job for USAFE." Air Force and Space Digest, LII (November 1969), 66-69.

Kelsey, A. F. "Versatility Plus Performance: The Boeing Stratofreighter." Mechanical Engineering, LXXIII (March 1951), 234-235+.

Kennedy, Willis. "Odyssey of the Beast." Air Progress Aviation Review, VI (Summer 1982), 18-23.

McCarthy, Dan B. "The C-97." Air Classics, VIII (August 1972), 14-21, 58-59.

McGlasson, W. D. "Air Guard Airlift: Anywhere, Anytime in a C-97." National Guard, XXXIX (September 1985), 28-31.

Meurer, Fred. "Have a Drink on the Guard." Airman, XVII (March 1973), 34-37.
KC-97L.

Robinson, James W. "Regards to 'The Lady.'" Sergeants, XXI (December 1982), 14-15.

"Stratofreighter." Joint Services Recognition Journal, X (March 1955), 72-73.

Vaughan, David K. "Now They're All Gone." Combat Crew, XVIII (January 1967), 16-21.

(Boeing) C-98 Clipper

Bowers, Peter M. "The Boeing 314 Clipper." The AOPA Pilot, XXIII (November 1982), 87-89.

____. "The Great Clippers." Airpower, VII (November 1977), 20-35, 60-61.

Klaas, M. D. "Last of the Flying Clipper Ships." American Aviation Historical Society Journal, XIII (Summer 1968), 93-102.

(Northrop) UC-100 Gamma

Bowers, Peter M. "The Northrop Gamma." The AOPA Pilot, XXV (January 1984), 69-73.

(Consolidated) C-109 Liberator
(see also B-24/C-87 Liberator)

Kichell, Larry. "Flight to Chungking." Air Classics, XVIII (November 1982), 12-13.

Millikin, Eugene J. "Remembering the C-109." American Aviation Historical Society Journal, XXVI (Winter 1981), 272-278.

(Douglas) C-110/R3D

"The DC-5." Douglas Service Magazine, (March-April 1980), 30-35.

Holden, Henry A. "DC-5: The Forgotten Douglas." American Aviation Historical Society Journal, XXIX (Summer 1984), 96-101.

C-112 see C-54/R5O Skymaster

C-113 see C-46/R5D Commando

C-114/C-115 see C-54/R5O Skymaster

(Douglas) C-117/R4D-8 Skytrain II

Andrews, Hal. "Naval Aircraft: C-117D." Naval Aviation News, (November-December 1983), 16-17.

Douglas, Donald W. "Super DC-3." Aviation Week, L (January 31, 1949), 14-15.

Wiita, Douglas P. "Lessons From a Hummer." Aviation, V (February 1981), 6-12.

(Douglas) C-118/R6D Liftmaster

Burton, K. F. "Evolution for Cargo: DC-6A Liftmaster." Aero Digest, LIX (December 1949), 48-50+.

The Douglas DC-6 and DC-7 Series. Great Sampford, Eng.: Air Britain, Ltd., 1971. 90p.
C-118/R6D = DC-6.

"Flying the C-118." MATS Flyer, IV (January 1957), 18-21.

Hanuschek, Stanley. "Andes Run." Airman, VIII (December 1964), 42-45.

Karant, Max. "Return of the Six." Flying, XLII (June 1948), 30+.

Powell, Dennis M. "Big Six Decade: A Short History of the Douglas DC-6 Series." Flight, LXX (August 10, 1956), 222-224.

Smith, P. R. DC-6: Production List and Pictorial History. Twickenham, Eng.: DFR Marketing and Sales, 1984. 80p.

Williams, Nicholas. "425: An Example of Liftmaster Longevity." American Aviation Historical Society Journal, XXV (Fall 1980), 199-202.

(Fairchild) C-119/R4Q Flying Boxcar

Book/Document/Report Literature:

Frey, Royal. C-119, 1946-1953. AMC Historical Study, no. 247. Wright-Patterson AFB, Ohio: USAF Air Materiel Command, 1953. 67p.

Periodical/Anthology Literature:

Baer, H. S. "New Assault Transport Design Revealed." American Aviation, XVII (August 17, 1953), 19-20.

Boutelle, Robert S. "Performance of Packet Planes." Quartermaster Review, XXX (September-October 1950), 26-27, 125.

"Boxcar Town, U.S.A." Honeywell Flight Line, V (March 1954), 11-13.

Burney, J. C., Jr. "Wings for the Warrior." National Defense Transportation Journal, XI (March-April 1955), 38-42.

"C-119." Pegasus, XV (March 1950), 1-4.

"Fairchild C-119." Air Classics, VII (October 1970), 12-19.

"Fairchild C-119H." Aero Digest, LXV (August 1952), 24-32+.

"The H-Plane and Its Missions." Pegasus, XIX (July 1952), 16-17.

Hawthorne, Rudolph. "Fairchild C-119 Development." Aviation Age, XVIII (December 1952), 34-35+.

"The Loads of Charlie One One Nine." Pegasus, XIV (April 1951), 8-12.

McLarren, Robert. "New Design Brings Better Packet: C-119." Aviation Week, XLVII (December 8, 1947), 24+.

"Peacetime Tests Pay Off For Flying Boxcars in Combat." US Air Services, XXXVIII (January 1953), 18-22.

Trimble, Robert L. "Boxcar on New Tracks." Air Classics, XIII (December 1977), 58-61.

Waddell, L. S. "Phase-Out of Charlie One One Nine." Pegasus, XXIV (October 1955), 1-6.

C-120 see C-119/R4Q Flying Boxcar

(Lockheed) C-121/R7V-2 Super Constellation
(see also C-69/K70 Constellation)

Book/Document/Report Literature:

Ginter, Steven J. Lockheed C-121 Constellation. Naval Fighters Series, v. 8. Simi Valley, Calif.: Naval Fighters, 1983. 78p.

Periodical/Anthology Literature:

Andrews, Hal. "Naval Aircraft: Constellation." Naval Aviation News, (November 1982), 24-25.

Baxter, Gordon. "Curvacious Connie." Flying, CXIV (July 1987), 105-106.

"Connie Grows Up." MATS Flyer, III (August 1956), 24-25.

"Connies On the Move." MATS Flyer, IX (August 1962), 20-21.

Foster, Bob. "Last Flight to Tucson." National Guardsman, XXXII (June 1978), 8-10.

Hull, Tom. "Parting With Connie." Naval Aviation News, (November 1979), 34-37.

Lindquist, E. H. "Cadillac of the Skies." Air Classics, XXIII (July 1987), 14-23.

"Lockheed's R7V-2 Transport." BeeHive, XXX (July 1955), 32-33.

Mikesh, Robert C. "Lockheed C-121 Constellations in USAF and Civilian Service." American Aviation Historical Society Journal, XXV (Spring 1980), 10-18.

Moore, John A. "'Bataan.'" U.S. Army Aviation Digest, XVI (June 1970), 26-29.
Gen. MacArthur's C-121.

Schrader, Richard K. "'Bataan.'" Air Classics, XXIV (April 1988), 30-32.

Taylor, Henry A. "Constellation: Lockheed's 'Queen of the Skies.'" Air Enthusiast, no. 14 (1980), 29-42.

Underwood, Bob. "Last Flight of the Constellation." Naval Aviation News, (November 1982), 22-23+.

(Chase/Fairchild) C-123 Provider

Book/Document/Report Literature:

Bagwell, Margaret C. C-123, April 1945-September 1951. AMC Historical Study, no. 228. Wright-Patterson AFB, Ohio: USAF Air Materiel Command, 1952. 20p.

Periodical/Anthology Literature:

Bentley, W. C. "Operation Shakedown." Pegasus, XXV (July 1956), 6-11.

"Chase C-123B." Aero Digest, LXVI (June 1953), 28-32.

Epifano, Tony. "It's Up or Out After 30 Years for Aircraft, Too: Last Flight of the C-123 Provider." Officer, LXII (October 1986), 15-17.

____. "Last Flight of the C-123." Citizen Airman, XXXVIII (August 1986), 12-15.

George, James A. "Provider Used in South Vietnam Counterinsurgency Operations." Airman, VII (March 1963), 46-47.

Lee, B. S. "Assault Transport Order Goes to Chase." Aviation Week, LIII (October 23, 1950), 11-13.

Lefer, Henry. "New Chase Carrier Makes Debut." Aviation Week, LVIII (April 27, 1953), 15-16.

Lentz, Lawrence H. and Arthur Sorenson. "Patches: The Hole Story." Air Reservist, XXXII (September-October 1980), 14-15.

McGowan, Sam. "The Provider: The Fairchild C-123 Provider Series." Air Classics, XXIV (July 1988), 60-65.

Smith, Gerald. "Spray Away." Airman, XVIII (October 1974), 39-41.

Witze, Claude. "Logistics Pinch Turns C-123 to Cargo." Aviation Week and Space Technology, LXIV (June 11, 1956), 32-33.

____. "USAF Ready to Carry Walking Away With C-123." Aviation Week, LXIV (April 9, 1956), 28-31.

(Douglas) C-124 Globemaster II
(see also C-74 Globemaster)

Allen, Gene. "Shakey's Alive and Flying in the ANG." *MAC Flyer,* XXI (September 1974), 14-15.

"The C-124." *Air University Quarterly Review*, V (Winter 1951-1952), 82-85.

Cole, Dave. "Matron of the Lumbering Loads." *Airman,* XVI (April 1972), 43-44.

Davies, C. W. "Douglas C-124A Globemaster II." *Aero Digest,* LXV (November 1952), 26-32+.

Dorff, Richard W. "Old Shakey." *Airman,* X (May 1966), 18-21.

"Douglas C-124A Cargo Transport." *Interavia,* VI (December 1951), 679-682.

Eddy, Arthur. "The End of an Era: Old Shakey Bows Out." *MAC Flyer*, XXII (January 1975), 6-8.

Hines, Ray E. "Steering Old Shakey." *Aerospace Safety,* XXIII (August 1967), 24-25.

Kachner, Leonard. "Old Shakey's Last Flight to the World's Biggest Icebox." *Aerospace Historian*, XIV (Spring 1967), 29-31.
To the Arctic on October 8, 1962.

Koelling, Richard W. "Unsung Heros." *MAC Flyer*, XXXVI (February 1989), 24-25.

Martin, Robert P. "C-124 Passing Combat Test in Korea." *Aviation Week*, LVII (November 17, 1952), 62-68.

Purcell, Michael B. "Flying High." *Navigator,* XIX (January 1972), 26-27.

"USAF Cargo Carrier: The Globemaster II." *Skyways,* IX (February 1950), 28-29.

(Northrop) C-125 Raider

McLarren, Robert. "Northrop C-125 Based on Pioneer." *Aviation Week*, XLIX (November 8, 1948), 21-22+.

C-129 see C-47/R4D Skytrain/Dakota

(Lockheed) C/KC/MC/HC/LC-130 Hercules

Book/Document/Report Literature;

Caidin, Martin. *The Long Arm of America: The Story of the Mighty Hercules Air Assault Transport and Our Revolutionary Global Strike Force*. New York: E. P. Dutton, 1963. 369p.

Dabney, Joseph E. A Condensed History of the C-130. Marietta, Ga.: Public Information Office, Lockheed-Georgia Corporation, 1988. 10p.

____. Herk: Hero of the Skies. Lakemont, Ga: Copple House Books, 1979. 416p.

Donnelly, Mark S. and James E. Hill. Limitations to European C-130 Deployment. Thesis. Wright-Patterson AFB, Ohio: School of Engineering, Air Force Institute of Technology, 1986. 168p.

Drendel, Lou. C-130 Hercules in Action. Carrollton, Tex.: Squadron/Signal Publications, 1984. 50p.

Finlayson, Patillo A. Warner Robins Air Materiel Area's Weapons Systems: The C-130 Hercules--Logistical Support, 1958-1969. AFLC Historical Study, no. 486. Robins AFB, Ga.: Warner Robins Air Materiel Area, 1970. 220p.

Gaines, Mike. Hercules. Boston, Mass.: Jane's, 1984. 52p.

Harrington, Steven E. The C-130 Derivatives: Can They Perform the Tactical Airlift Mission? Research Paper. Maxwell AFB, Ala.: Air Command and Staff College, Air University, 1983. 30p.

Kinzey, Bert. Hercules. Blue Ridge Summit, Pa.: TAB, 1987. 64p.

Leader, Ray. C-130 Hercules. Color and Markings Series, v. 7. Blue Ridge Summit, Pa.: TAB, 1987. 64p.

McGowan, Sam. The C-130 Hercules: Tactical Airlift Missions, 1956-1975. Blue Ridge Summit, Pa.: TAB, 1988. 160p.

Mason, Francis K. Lockheed Hercules. New York: Sterling, 1985. 242p.

Morris, M. E. C-130: The Hercules. Novato, Calif.: Presidio Press, 1989.

Olausson, Lara. Lockheed Hercules Production List, 1954-1986. 5th ed. Lidkoping, Sweden, 1985. 108p.

Olson, Bennie A. C-130 Production and Support, 1952-1957. AMC Historical Study, no. 419. McClellan AFB, Calif.: Sacramento Air Materiel Area, 1958. 137p.

Pickarz, Eugene J. C-130 Hercules Acquisition Program: A Case Study of the Potential Impact of a System Level Warranty Application. Thesis. Wright-Patterson AFB, Ohio: School of Systems and Logistics, Air Force Institute of Technology, 1986. 111p.

Reed, Arthur. C-130 Hercules. Modern Combat Aircraft, no. 17. Osceola, Wis.: Motorbooks International, 1984. 72p.

St. John Turner, Paul. Lockheed C-130 Hercules. Aircraft in Profile, no. 223. Windsor, Eng.: Profile Publications, 1970. 24p.

Periodical/Anthology Literature:

Anderton, David A. "USAF Receives Highly-Versatile C-130." Aviation Week and Space Technology, LXV (December 3, 1956), 50-52+.

Andrews, Hal. "Naval Aircraft:Hercules." Naval Aviation News, (November 1971), 20-21.

Besecker, Raymond F. "C-130D Ski-Hercules." Navigator, XXVIII (Spring 1981), 19-21.

Brownfield, Orlen L. "The Legend of Hercules." MAC Flyer, XXIV (January 1977), 3-6.

"C-130 Hercules: Lockheed's Aerial Workhorse." Aerospace Historian, XXIV (September 1977), 158-162.

"C-130H." Air Progress, XLVIII (January 1986), 28-32.

Clark, Robert L. "The C-130." Aerospace Historian, XXVI (December 1979), 233-237.
Reprinted in Robin Higham and Carol Williams, eds., Flying Combat Aircraft of the USAAF-USAF.(Manhattan, Kans.: Sunflower University Press, 1981), pp.39-44.

Cook, John C. "Lockheed Hercules: The First 20 Years." Air Pictorial, XXXVI (August-September 1974), 302-305, 350-353.

Dabney, Joseph E. "Hercules History." Aircraft Illustrated, XII (September 1979), 435-441.

____. "The Labors of Hercules." Air Progress, XLVII (March 1985), 42+.

DeMar, Chuck. "The Corps' Flying Gas Station." Leatherneck, LXIII (April 1980), 46-49.

Drachlis, David B. "Getting Down and Dirty." Airman, XXIII (November 1979), 18-20.

Everett, Robert P. "A Dream Called 'Long Flight.'" Airman, XVI (July 1972), 2-8.
C-130 flight, Taiwan to Scott AFB, Illinois.

Frasher, Joan A. "Rescue of a Snowbird." Naval Aviation News, (May-June 1988), 14-15.
LC-130 # JDB-831

Gahren, Francis M. "Blackbirds of the Night: The MC-130E." Airlift, VI (Fall 1984), 8-9+.

Graham, Vickie M. "Legendary, Almost Beyond Belief." Airman, XXVIII (December 1984), 9-15.

Griswold, Wesley S. "62-Ton Plane Stops on a Dime." Popular Science, CLXXX (February 1962), 92-93+.

"Herculean Lockheed." RAF Flying Review, XII (May 1957), 34-37.

"Hercules 130 Story." Aircraft Engineering, LI (August 1979), 24-27.

"Hercules the Diehard." Naval Aviation News, (February 1982), 22-23.

"Here Come the 'H' Models." National Guard, XXXIII (August 1979), 18-20.

Howard, Byron L. "Herk: The First 30 Years." MAC Flyer, XXXI (September 1984), 3-5.

____. "The Legend Lives." MAC Flyer, XXVIII (December 1981), 16-17.

Hucker, Robert. "The Indestructable Herky Bird." Air Progress Aviation Review, IV (Summer 1980), 52-59.

Kosar, W. S. "The Hercules Unchained." U.S. Naval Institute Proceedings, CII (October 1976), 151-155.

____. "Research-Configured LC-130R Airplane." Antarctic Journal of the United States, XII (April 1977), 206-207.

Lambert, Mark. "Visit to Georgia." Flight, LXXXV (June 11, 1964), 979-987.

Lindquist, Peter W. "The C-130: Talented Tactical Transport." Air Force Magazine, LIX (November 1976), 69-73.

"Lockheed C-130A Hercules." Flight, LXX (September 14, 1956), 491-494.

"Lockheed Hercules: Still a Winner and Fighting to Stay That Way." Interavia, XXXII (November 1977), 1149-1151.

"Lockheed's Hercules: The Goodwill Aircraft." Interavia, XXVII (June 1972), 647-659.

McSurely, Alexander. "Air Force Orders C-130 Into Production." Aviation Week, LVII (October 6, 1952), 13-14.

Middlewood, Robert W. "The Hercules Story." Air Transportation, XXVI (February 1955), 14-16+.

Miller, Edward C. "Ski Hercules." Air Classics, XI (January 1975), 26-29.

"More Muscle for Rescue." Aerospace Safety, XXI (July 1965), 14-15. HC-130H.

Mullins, R. R. "New Tactical Role for KC-130." Marine Corps Gazette, LVII (February 1973), 49-50.

Powell, M. Phillip. "Mighty Hercules." Combat Arms, IV (November 1986), 30-33.

Prochoroff, Alan. "The Elusive 'Sitting Duck.'" Citizen Airman, XXXVIII (April 1986), 5-8.

"Report on the C-130E." MATS Flyer, IX (June 1962), 4-6.

Rosenberry, Donald L. "Aircrew Training System: C-130 Training Enters the Space Age." MAC Flyer, XXXIV (December 1987), 24-30.

Scarborough, Rowan. "What Will Replace the C-130?" Military Forum, IV (January-February 1988), 16+.

Scott, William B. "Combat Talon 2 Completes First Flight of Development Program." Aviation Week and Space Technology, CXXVIII (January 18, 1988), 36-37.
MC-130.

Sheffield, Dick. "Cleaning Up Their Act." Airman, XXII (April 1978), 38-41.

Solis, Onesimo, Jr. "One Herk (of a) Refueling." Combat Crew, XXXII (June 1982), 10-13.

Stanfield, Robert I. "C-130 Demonstrates Mission Flexibility." Aviation Week and Space Technology, LXVIII (May 19, 1958), 60-77.

Tracy, Joseph P. "Turbo Transport." Flying Safety, XII (June 1956), 16-21.

Volney, John. "The Labors of Hercules." Flying, LXII (January 1958), 28-30+.

Warwick, Graham. "Hercules: 25 Years of Tactical Airlift." Flight International, CXVI (July 14, 1979), 93-96.

Witterman, Benn H. "Romance With a Weenie Wagon." MAC Flyer, XVIII (July 1971), 16-20.

(Convair) C-131/R4Y Samaritan
(see also T-29 Flying Classroom)

Andrews, Hal. "Naval Aircraft: C-131." Naval Aviation News, (June 1979), 20-21.

Burgess, George D. "A Gold Watch for the Convair: Another of the Historic Prop Planes Retires From Active Service." MAC Flyer, XXII (March 1975), 27-28.

Collins, Helen F. "Farewell to 849." Naval Aviation News, (September 1979), 34-37.

Katzaman, Jim. "We Deliver--in the Clutch." Airman, XXVII (June 1983), 26-32.

Norton, John. "Going My Way?" Naval Aviation News, (July-August 1987), 26-27.

Simons, B. J. and P. J. Lynch. "The Convair YC-131C." Skyways, XIV (June 1955), 8-10, 36-37.

Williams, Nicholas M. "The Convair YC-131C Turboliners." American Aviation Historical Society Journal, XXV (Winter 1980), 251-259.

(Douglas) C-133 Cargomaster

Book/Document/Report Literature:

Maltars, Richard E. C-133 Cargomaster, 1953-1971. AFLC Historical Study, no. 494. Robins AFB, Ga.: History Office, Warner Robins Air Materiel Area, 1972. 217p.

Periodical/Anthology Literature:

Almond, Larry M. "Navigating the C-133." Navigator, XIX (January 1972), 22-24.

"C-133 Cargomaster." Flight, LXXIV (September 26, 1958), 519-523.

Cousins, R. D. "USAF Experience to Date With Turboprop Aircraft." Society of Automotive Engineers Journal, LXIV (January 1956), 21-23.

Craven, Harold L. "Queen of the Heavies." Airman, II (March-April 1958), 39-41.

"MATS Introduces Mr. Big." MATS Flyer, III (March 1956), 24-27.

"Prop Titan." Ryan Reporter, XVII (April 1956), 4-5+.

Stambler, Irwin. "Self-Sufficiency is Key to C-133A." American Aviation, XXVI (September 1956), 20-25.

Stone, Irving. "Douglas Rolls Out Turboprop C-133A Transport. " Aviation Week, LXIV (February 6, 1956), 26-27.

Sweeney, Robert. "C-133 Exhibits Stability, Controlability." Aviation Week and Space Technology, LXVIII (May 12, 1958), 70-83.

Witterman, Ben H. "Romance of the Weenie Wagon." MAC Flyer, XVIII (July 1971), 16-20.

Woods, Robert E. "Flying the Heavies: C-133." MATS Flyer, V (May 1958), 16-20.

(Boeing) C/KC-135 Stratolifter/Stratotanker

Book/Document/Report Literature:

Bassell, John C. History of the KC-135A Airplane, 1953-1958. AMC Historical Study, no. 296. Wright-Patterson AFB, Ohio: USAF Air Materiel Command, 1959. 86p.

Conley, Paul H. KC-135 Air Refueling in Southeast Asia. Research Paper. Maxwell AFB, Ala.: Air Command and Staff College, Air University, 1969. 99p.

Craig, Roger C. and Ira S. Paul, III. KC-135A Long-Haul Refueling. Research Paper. Maxwell AFB, Ala.: Air Command and Staff College, Air University, 1983. Unpaged.

Day, J. L. KC-135 Low-Altitude Tactics. Research Paper. Maxwell AFB, Ala.: Air Command and Staff College, Air University, 1988. 59p.

Deegan, William F., Jr. A Handbook for the Future KC-135 Aircraft Commander. Research Paper. Maxwell AFB, Ala.: Air Command and Staff College, Air University, 1982. 62p.

Dorr, Robert F. Boeing KC-135 Stratotanker. Modern Combat Aircraft, no. 17. Osceola, Wisc.: Motorbooks International, 1987. 112p.

Fanjoy, Richard O. "Combat Ready": A Handbook for the New KC-135 Copilot. Research Paper. Maxwell AFB, Ala.: Air Command and Staff College, Air University, 1981. 41p.

Fratt, Merlyn D., Jr. KC-135 Air Refueling in Southeast Asia: Support Requirements. Research Paper. Maxwell AFB, Ala.: Air Command and Staff College, Air University, 1969. 104p.

Hall, Samuel L. An Evaluation of C-135 Command Support Airlift. Research Paper. Maxwell AFB, Ala.: Air Command and Staff College, Air University, 1975. 52p.

Schiff, Barry J. Boeing 707. Fallbrook, Calif.: Aero Publishers, 1982. 180p.

Spaulding, George L. KC-135 Utilization: The Texaco Tanker. Research Paper. Maxwell AFB, Ala.: Air Command and Staff College, Air University, 1983. 39p.

Whittle, John A., et al. The Boeing 707 and 727. Great Sampford, Eng.: Air Britain, Ltd., 1973. 100p.

Periodical/Anthology Literature:

Allen, Ken. "Ride the Flying Boom." Airman, X (August 1966), 44-48.

Becker, George. "Global Weightlifters." Ryan Reporter, XXIV (January-February 1963), 6-11.

Blatter, Richard . "A New Lease on Life." Combat Crew, XXVII (August 1977), 4-7.

"Boeing KC-135A." RAF Flying Review, XVII (April 1962), 28-31.

Brown, Philip C. "KC-135 Stratotanker." In: Robin Higham and Carol Williams, eds. Flying Combat Aircraft of the USAAF-USAF. Ames, Iowa: Iowa State University Press, 1978. pp. 159-168.

Burrell, Harry R. "The Weight Lifter." Combat Crew, X (November 1959), 24-25.

Cokeley, R. G. "C-135 Short-Field Operations." MATS Flyer, X (April 1963), 20-25.

Cole, Dave. "Time of the Tigers. " Airman, XVI (October 1972), 6-10. The KC-135 in Southeast Asia.

Conley, Mark. "A Pilot's View of the KC-135R." Combat Crew, XXXVI (January 1986), 8-10.

Demers, W. A. "KC-135 Reengining: Revitalized Tankers are Helping to Plug a Widening Refueling Gap." Military Logistics Forum, IV (November-December 1987), 42+.

English, Malcolm. "Flight in a SAC KC-135 Tanker." Air Pictorial, XXLII (January 1980), 17-20.

Haugland, Vern. "KC-135 South of the Border." Air Force, XLI (January 1958), 70+.

Hornung, Rick. "Aerial Tankers Running Low." Military Logistics Forum, II (April 1986), 28-31+.

Hubbell, John G. "New Muscle For the Long Arm of SAC." Air Force, XL (July 1957), 48-53.

Kever, Robert S. "The SAC Express." Combat Crew, XIX (July 1968), 19-21.

Knutson, D. C. "Swept-Wing Savvy." Aerospace Safety, XIX (March 1963), 2-5.

Lambert, Mark. "Jet Tanker Time: The KC-135 and Its Boom." Flight, LXXXI (February 1, 1962), 169-173.

Leavenworth, Edgar C. "A Handful of Yoke." Combat Crew, XXI (December 1970), 8-11.

Lloyd, Alwyn T. "Plane of a Thousand Faces." Combat Crew, XXVII (September 1986), 4-7.

____. "30 Years Young." American Aviation Historical Society Journal, XXXII (Fall 1987), 172-181.

____. "Versatility Unlimited: The Boeing KC-135 Story." Air International, XIX (December 1980), 271-276, 291-295.

McArdle, Frank H. "The KC-135 in Southeast Asia." Air University Review, XIX (January-February 1968), 20-33.

McCarthy, Dan B. "Last Flight of the DASH-80." Air Classics, IX (August 1973), 28-35, 68.

McCullough, Bob. "Alert Tankers." Airman, XXI (July 1987), 28-30.

McDonald, Chuck. "To Be a Boomer." Profile, XXIV (March 1981), 12-15.

Murphy, Joseph S. "Air Force Go-Ahead on 707 Spurs New Jet Race." American Aviation, XIX (August 1, 1955), 21-22.

Pamentel, Laurence A. "A Rugged Bird." Combat Crew, XXXIII (October 1973), 28-30.

Peisher, Robert J. "What's Holding It Together: KC-135." Combat Crew, XXII (November 1971), 12-15.

Rea, Kenneth R. "Long Legs." Combat Crew, VIII (April 1958), 8-11.

Rogers, T. J. "First American Jet." Flying, LX (March 1957), 44-46+.

Ropelewski, Robert R. "CFM-56s Boost KC-35R Performance." Aviation Week and Space Technology, CXVIII (May 2, 1983), 42-46.

Scott, William B. "Special Mission Aircraft Aid SDI Sensor System Development." Aviation Week and Space Technology, CXXIX (October 3, 1988), 34-37.
NC-135 Argus.

Sonntag, Willie H. "Baptism of Fire." Combat Crew, IX (June 1959), 10-13.

Tregaskis, Richard. "Boeing's $15 Million Gamble." Collier's, CXXXIII (March 19, 1954), 70-71+.

Wilson, James R. "Tall Tail." Combat Crew, XV (March 1965), 14-15.

(Boeing) VC-137 Air Force One/Two

Benenson, Tom. "Flying the Presidential Jet." Plane and Pilot, XI (July 1975), 38-41.

Blair, Edwin T. "Check Ride." Airman, IV (March 1960), 24-25.

"Boeing VC-137A: The U.S. Government's VIP Transport." Interavia, XIV (June 1959), 690-691.

Gregory, William H. "707-320s Improve Nonstop Capability." Aviation Week and Space Technology, LXXI (December 7, 1959), 41+.

United States. Congress. House. Committee on Armed Services. Subcommittee on Research and Development. Air Force One Replacement Program: Hearings. 99th Cong., 1st sess. Washington, D.C.: GPO, 1986. 25p.

(Lockheed) C-140 Jetstar

"Air Force to Spur C-140 Replacement." Aviation Week and Space Technology, CXIV (April 27, 1981), 36-37.

Hornbuckle, Carl G. "Report on the Jetstar." MATS Flyer, X (May 1963), 4-7.

"Jetstar for MATS." RAF Flying Review, XVIII (January 1963), 30-33.

Lockheed-Georgia Company. Public Information Office. Jetstar. Marietta, Ga., 1987. 3p.

Stanfield, Robert I. "America's First Corporate Jet: Jetstar." Flying, LXVIII (March 1961), 24-27+.

(Lockheed) C-141 Starlifter

Book/Document/Report Literature:

Arnold, Palmer G. C-141 Pilot's Handbook, Research Paper. Maxwell AFB, Ala.: Air Command and Staff College, Air University, 1980. 38p.

Kraus, Walter L. and Jose M. Matteson. C-141 Starlifter. Scott AFB, Ill.: History Office, Military Airlift Command, 1973. 110p.

Lockheed-Georgia Company. Public Information Office. C-141 Starlifter. Marietta, Ga., 1987. 3p.

Martin, Harold H. Starlifter: The C-141--Lockheed's High-Speed Flying Truck. Brattleboro, Vt.: Stephen Greene, 1972. 190p.

Periodical/Anthology Literature:

Adkins, L. A., Jr. "C-141 Reliability Flight Test Program." Quality Control, XXIII (February 1967), 370-373.

Ashlock, James R. "C-141 Flight Tests Continue on Schedule." Aviation Week and Space Technology, LXXXI (August 31, 1964), 38-41+.

Black, Michael P. "Flying the Starlifter." MAC Flyer, XXVI (February 1979), 4-7.

Brewer, Alton P. H., Jr. "C-141 Starlifter." In: Robin Higham and Carol Williams, eds. Flying Combat Aircraft of the USAAF-USAF, Ames, Iowa: Iowa State University Press, 1978. pp. 101-112.

Burton, Paul. "Proving Out the Starlifter." BeeHive, XL (Spring 1967), 14-17.

"The C-141 Starlifter." American Aviation, XXVIII (April 1965), 18-31.

Cleveland, F. A. and R. D. Gilson. "Development Highlights of the C-141 Starlifter." Journal of Aircraft, II (July-August 1965), 278-287.

Coble, Donald W. "Introducing the Starlifter." Airman, VII (September 1963), 24-27.

"Curtain Calls for the Starlifter." MAC Flyer, XII (May 1966), 9-13.

Downs, Robert B. "The C-141 Takes Shape." MATS Flyer, IX (January 1962), 4-7.

Gilley, Jack G. "The Big Push." Aerospace Safety, XX (January 1964), 4-7.

____. "Up Front in the C-141A." Combat Crew, XV (February 1965), 26-27.

Haid, William. "C-141B Stretch Program: A Case Study of the Relationship Between the Military and Defense Contractors." Air Force Journal of Logistics, X (Fall 1986), 23-26; XI (Winter 1987), 17-20.

Holder, William G. "Stretching It." Airman, XXIV (June 1980), 39-43.

Huntington, Dean. "Up Front in the Stretched Starlifter." MAC Flyer, XXVIII (June 1981), 8-10.

Lambert, Mark. "Visit to Georgia." Flight, LXXXV (June 11, 1964), 978-985.

Lawrence, James J. "Readiness: C-141 Participation in 'Red Flag.'" Aerospace Safety, XXXIV (January 1978), 1-5.

"The Lockheed C-141 Starlifter." Air Pictorial, XXIX (April 1967), 107-111.

"Lockheed's Big Lifter." Flying Review International, XIX (January 1964), 33-37, 45.

"Lockheed's Stretched Starlifter." Flight International, CVIII (October 9, 1975), 525-526.

McLarty, R. Douglas. "The Flying Fruit Basket." Airman, XVIII (November 1974), 36-40.

Pocock, Charles L. "C-141 Operations in 'Coffin Corner.'" MAC Flyer, XVII (December 1970), 18-20.

____. "Wild Ride in a Big Bird." MAC Flyer, XVII (June 1970), 16-18.

Sapp, R. A. "Bigger is Better: Stretching the C-141 Starlifter." Lockheed Horizons, (October 1985), 38-44.

Service, Brian M. "Transatlantic Flight in a Starlifter." Air Pictorial, XXIX (April 1967), 110-111.

Slaybaugh, T. J. "Shakey's Successor." Aerospace Safety, XXI (February 1965), 10-12.

Spillers, William H. "Starlifter Progress." MATS Flyer, XI (October 1964), 20-23.

Sturm, Ted R. "The Starlifter Turns Ten." Airman, XVIII (October 1974), 42-47.

Swearengen, John F. "C-141 Starlifter: Tomorrow's Workhorse." Aerospace Safety, XVIII (November 1962), 1-4.

Wilkelmson, Ron. "How They Do It at the Factory." MAC Flyer, XXVIII (October 1981), 16-19.

(Waco) CG-4 Glider

Mrazek, James E. Fighting Gliders of World War II. New York: St. Martin's Press, 1977. 207p.

Newcomb, Harold and Russell W. Schmidt. "On Whispering Wings." Airman, XXXII (June 1988), 10-15.

Nicholas, William H. "Gliders: Silent Weapons of the Sky." National Geographic Magazine, LXXXVI (August 1944), 149-160.

"Pilots Expendable." Air Classics, III (May 1967), 43-60.

Richmond, Stu. "Waco CG-4A Glider." RC Modeller, XIV (May 1977), 32-37.

Thomas, Arthur J. "CG-4A Glider." In: Robin Higham and Abigail T. Siddall, eds. Flying Combat Aircraft of the USAAF-USAF. Ames, Iowa: Iowa State University Press, 1975. pp. 108-118.

____. "Waco's Flying Box." In: Michael O'Leary, ed. Air Progress Pilot Reports. Canoga Park, Calif.: Challenge Publications, 1979. pp. 92-97.

United States. Army Air Forces. Training Division. Pilot Training Manual for Glider CG-4A. AAF Manual 51-130-1. Washington, D.C., n.d.

GK-1 see UC-61 Forwarder

GV see C-130 Hercules

JR/RR see C-4 Ford Trimotor

JRB/SNB see C-45 Expeditor

(Martin) JRM Mars

Andrews, Hal. "Naval Aircraft: Mars." Naval Aviation News, (April 1980), 20-21.

Caidin, Martin. "The Navy's Old Lady." AirTech, IV (June 1944), 24-27+.

Dupuy, R. Ernest. "Mars Takes Wings." Forum, CI (April 1939), 228-232.

Edwards, Ron. "Martin Mars Flying Boat." Air Classics, XVII (November 1981), 14-26, 68-71.

Serrell, A. "Mars: No Bettah da Kine." U.S. Naval Institute Proceedings, LXXXIII (August 1957), 856-861.

Shloss, Leon W. "The Secret Mission of the Mars." In: Norman Carlisle, ed. The Air Forces Reader. Indianapolis, Ind.: Bobbs-Merrill, 1944. pp. 239-246.

"The U.S. Navy's Big Four." Esso Air World, VIII (March-April 1956), 136-138.

(Sikorsky) JRS

Andrews, Hal. "Naval Aircraft: JRS-1." Naval Aviation News, (March-April 1987), 14-15.

Salo, Mauno. "The Sikorsky S-43." American Aviation Historical Society Journal, XXVI (Fall 1981), 178-196.

KB-29 see B-29 Superfortress

KB-47 see B-47 Stratojet

(Boeing) KB-50
(see also B-50 Superfortress)

Brown, Charles D. "Revamped Refueler." Flying Safety, XIV (July 1958), 16-19.

RA see C-2 Fokker Trimotor

R2D see C-32/C-33

R3D see C-110

(Convair) R3Y Tradewind

"Convair Tradewind." Interavia, IX (February 1954), 91-92.

Poling, George E. "On the Trail of the Tradewind." Airpower, VIII (May 1978), 10-18, 22-24, 62.

"Tradewind Turboprop." Flight, LXVII (February 4, 1955), 139-141.

R4C see C-30 Condor

R4D see C-47 Skytrain/Dakota

R4Q see C-119 Flying Boxcar

R4Y see C-131 Samaritan

R5C see C-46 Commando

R5D see C-54 Skymaster

R5O see C-56 Lodestar

R6D see C-118 Liftmaster

R7O see C-69 Constellation

R7V see C-121 Super Constellation

(Sikorsky) RS/C-6 Guardian

Andrews, Hal. "Naval Aircraft: RS." Naval Aviation News, (April 1982), 24-25.

"Sikorsky Guardian." Aviation, XXIV (January 16, 1928), 148-150.

(Fokker) T-2

Book/Document/Report Literature:

Casey, Louis S. The First Nonstop Coast-to-Coast Flight and the Historic T-2 Airplane. Annals of Flight, v. 1, no. 1. Washington, D.C.: Smithsonian Institution Press, 1964. 90p.

Periodical/Anthology Literature:

"Army Fokker Crosses Continent in 27 Hours." Aviation, XIV (May 14, 1923), 524-526.

Casey, Lou. "Flight of the T-2." National Aeronautics, I (Spring 1973), 52-57.

Craven, Harold L. "They Hedge-Hopped to Glory." Transportation Proceedings, III (September 1969), 18-20.

"Epoch-Making Flight." US Air Services, VIII (June 1923), 9-13.

"Famous Flights: The Nonstop Flight Across America." Popular Aviation, I (October 1927), 19-21, 64.

"History's Longest Hop Across America." Literary Digest, LXXVII (May 19, 1923), 38-44.

Kelly, Oakley G. and John A. Macready. "The 2,000-Mile Flight of the Army Fokker T-2: Official Report on the Flight From San Diego to Fort Benjamin Harrison." Aviation, XIV (January 1, 1923), 8-11.

Macready, John A. "The First Nonstop Flight Across America." National Geographic Magazine, XLVI (July 1924), 1-8.

Morgan, Len. "Kelly, Macready, and the T-2." Air Force, XXXIX (August 1956), 290-292+.

Scrivner, John H. "The Impossible Has Happened." Aerospace Historian, XIV (Summer 1967), 103-109.

"The United States Army Giant Monoplane T-2." Aeronautical Digest, I (December 1922), 281+.

(Bell-Boeing) V-22 Osprey

Book/Document/Report Literature:

Bell Helicopter Textron. Public Relations Department. The V-22 Osprey. Fort Worth, Tex., 1987. 10p.

Boeing Helicopter Company. Public Relations Department. The Bell-Boeing V-22 Osprey Tilt-Rotor Aircraft Program. Philadelphia, Pa., 1987. 22p.

United States. Congress. House. Committee on Science and Technology. Subcommittee on Transportation, Aviation, and Materials. Joint Service Advanced Vertical Wing (JVX) Program: Hearings, 98th Cong., 1st sess. Washington, D.C.: GPO, 1983. 118p.

____. Government Accounting Office. National Security and International Affairs Division. DOD Acquisition: Case Study of the Navy V-22 Osprey Joint Vertical Lift Aircraft Program. Washington, D.C., 1986. 17p.

Periodical/Anthology Literature:

Andrews, Hal. "V-22 Osprey: An Inside Look." Naval Aviation News, (January-April 1989), 12-13, 12-14.

Booren, Stephen D. "The Marines' MV-22A: Key Element in a Deadly Trio." Rotor and Wing International, XXII (June 1988), 28-30.

Brahney, James H. "V-22 Tiltrotor: Manufacturing Technology at Work." Aerospace Engineering, VIII (March 1988), 8-13.

Braybrook, Roy M. "Tiltrotor Technology and the Bell-Boeing V-22 Osprey." Pacific Defense Reporter, XII (September 1985), 53-55.

Bulban, Erwin J. "Tiltrotor Tool-Up: Bell Prepares to Give Wing to the V-22." Rotor and Wing International, XX (July 1986), 46-47, 64.

____. "V-22 Testing: Flexing the Osprey's Wings." Rotor and Wing International, XXIII (January 1989), 56-58.

Cagini, John D. "V-22 Osprey: Revolution is in the Air." Aviation Mechanics Journal, XIX (December 1988), 18-25.

Carr, Fred. "Osprey." Marines, XIV (February 1985), 14-17.

Crawford, C. C., Jr. "JVX: What an Opportunity." Vertiflite, XXVIII (July-August 1982), 34-39.

DeMeis, Richard. "AT3 and the V-22." Aerospace America, XXVI (June 1988), 29-31.

Fink, Donald E. "V-22 Osprey Development Contract Tests New Procurement Policy." Aviation Week and Space Technology, CXXII (June 3, 1985), 220, 223-227.

Gavaghan, Helen. "Tilting at New Aviation Markets." New Scientist, CXVI (October 1, 1987), 38-43.

Grimes, Vincent P. "Osprey: A Look to the Future." National Defense, LXXII (July-August 1987), 35-40.

Gumble, Bruce. "V-22 Training: Breaking New Ground." Military Forum, IV (March 1988), 43+.

Hanifen, Timothy C. "Air Combat Maneuvering for the MV-22A." Marine Corps Gazette, LXXI (May 1987), 84-91.

____. "Arming the Osprey." Marine Corps Gazette, LXXI (July 1987), 64-68.

____. "Escorting the Assault Force." Marine Corps Gazette, LXXI (March 1987), 48-51.

____. "Osprey--Force Multiplier." Marine Corps Gazette, LXX (February 1986), 22-23.

Harvey, David. "Anatomy of a Program." Sea Power, XXVII (July 1984), 16-18+.

____. "SV-22 and the Growing List of Tiltrotor Missions." Rotor and Wing International, XXI (October 1987), 76-77+.

Jacobs, Keith. "Osprey--Tilt-Rotor Takes Off." International Combat Arms, VI (July 1988), 30-33.

Lambert, Mark. "JVX: The World's First Production Tilt-Rotor." Interavia, XXXVIII (July 1983), 755-757.

____. "V-22 Osprey: Seeing is Believing." Interavia, XLIII (July 1988), 674-675.

____. "V-22 Osprey: The Aircraft for All Seasons." Interavia, XL (December 1985), 1321-1323.

Mashman, Joe. "The Best of Both Worlds." Professional Pilot, XXI (September 1987), 48-53.

Moxon, Julian. "V-22 Osprey: Changing the Way Man Flies." Flight International, CXXXIII (May 14, 1988), 22-30.

____. "V-22 Osprey: Triservice Workhorse." Flight International, CXXX (August 30, 1986), 154-157.

O'Rourke, Gerald. "Tilt-Rotor Straight Talk." The Hook, XV (Winter 1987), 66-73.

Pochhacker, Christian. "Bell-Boeing New Tiltrotor Osprey, V-22." Defence Update International, no. 89 (July 1988), 4-11.

Rhodes, Jeffrey P. "The Amazing Osprey." Air Force Magazine, LXX (January 1987), 80-83.

"Roll-Out: First V-22." Aviation Week and Space Technology, CXXVIII (May 23, 1988), 43-82.

"Tilt-Rotor." Air Progress, XLVIII (August 1986), 32-33+.

Tracy, R. J. "The V-22: Preparing for Full-Scale Development." Vertiflite, XXXI (November-December 1985), 27-29.

"USMC's V-22 Rolls Out." Armed Forces Journal International, CXXV (June 1988), 50-53.

Walsh, Edward J. "Forward Tilt for Navy/USMC Aviation." Sea Power, XXXI (June 1988), 57-58.

Warwick, Graham. "JVX: Tiltrotor Sets Out to Succeed." Flight International, CXXIV (November 5, 1983), 1214-1220.

E. Trainer Aircraft

(Curtiss) A-1

Andrews, Hal. "Naval Aircraft: Navy No. A-1." Naval Aviation News, (May 1976), 20-21.

Bowers, Peter M. "Antique Plane Tests a Modern Pilot." Flying, LVIII (June 1956), 26-27+.

Eddy, Myron. "The U.S. Navy's First Airplane." Flying, LXX (February 1962), 54-55+.

Germeraad, D. P. "Airborne With the Curtiss A-1 Replica." Aerospace Engineering, XX (December 1961), 20-21+.

Trimble, Robert L. "The Navy's First: The Story of Glenn Curtiss' A-1." Air Trails, III (Fall 1978), 24-34.
Reprinted in Air Classics, XV (April 1979), 78-88.

AT-4/AT-5 see PC-1

(North American) AT-6/SNJ Texan/Harvard

Book/Document/Report Literature:

Kohn, Leo J. The Story of a Texan. Appleton, Wis.: Aviat Publications, 1978. 40p.

Morgan, Len. The AT-6 Texan. Famous Aircraft Series. New York: Arco, 1985. 64p.

Ohlrich, Walter and Jeffrey L. Ethell. The Incredible T-6 Pilot-Maker. Osceola, Wis.: Specialty Press, 1983. 144p.

Periodical/Anthology Literature:

Baxter, Gordon. "AT-6 to Osh." Flying, CXV (November 1988), 100-102.

Botting, Fred. "T-6: Beginning of the End." Air Training, IV (August 1954), 25-27.

Bowers, Peter M. "The Thundering Texan." Wings, XVI (October and December 1986), 10-21, 30+.

Cole, Wayne S. "AT-6 Texan." In: Robin Higham and Carol Williams, eds. Flying Combat Aircraft of the USAAF-USAF. Ames, Iowa: Iowa State University Press, 1978. pp. 19-26.

Davisson, Budd. "Pilot-Maker: We Fly the Texan and Find How the Classic Trainer Lives Up to Its Name." Air Progress, XLVII (March 1985), 70-74+.

Ethell, Jeffrey L. "The T-6: Ready When Needed." Air Force Magazine, LXV (April 1982), 72-75.

____. "The Wonderful Six." Air Force Magazine, LXIV (January 1981), 72-78.

Harrison, Ben. "It's a Warbird." Air Classics, XII (August 1976), 34-45.

Kitchell, Larry. "A Texan Bites the Dust." Air Classics, XIX (July 1983), 14-15, 74-75.

Lichster, Lothar. "Texan's Last Stand." Air Classics, II (January-February 1965), 38-39+.
With the Civil Air Patrol.

Morgan, Len. "A Harvard Reunion." Flying, CX (January 1983), 36-39.

____. "Once a Harvard Man, Always a Harvard Man." Flying, CXVI (February 1989), 82-85.

O'Leary, Michael. "T-6 Potpourri." Air Classics, IX (June 1973), 53-59, 77.

Scanlan, Hugh. "Handling the Harvard." Aeroplane Monthly, VIII (July 1980), 377-381.

Scott, Phil. "Strong, Noisy Type." Flying, CXV (May 1988), 60-65.

Stone, Tim. "SNJ." Air Progress, XXXIX (June 1977), 60-62.

Weeghman, Richard B. "Pilot Report: North American T-6." Flying, LXXXIV (June 1969), 60-65.

(Beech) AT-7/SNB Navigator
(see also C-45 Expeditor)

"Beech AT-7 Navigator." Aero Digest, LII (March 1946), 82-85+.

O'Malley, B. P. "The SNB Bugsmasher." Air Classics, VIII (January 1972), 34-38.

AT-8 see UC-78 Bobcat

(Curtiss) AT-9 Jeep

Boyne, Walter J. "Air Jeep." Wings, IV (February 1974), 18-26, 29.

AT-11 see AT-7 Navigator

AT-12 see P-35 Guardsman

(Boeing) AT-15

"The Boeing AT-15 Bomber Crew Trainer." Aviation, XLI (December 1942), 263+.

AT-16 see AT-6/SNJ Texan/Harvard

AT-17 see UC-78 Bobcat

AT-18 see A-29/C-63 Hudson

AT-19 see UC-81 Reliant

(Avro) AT-20 Anson

Hall, Alan W. The Avro Anson. London, Eng.: Almark, 1972. 64p.

Middleton, Don. "Avro Anson." Aeroplane Monthly, VIII (April 1980), 187-193.

"Old Lady in a New Dress: The Avro Anson V, Its Evolution From a Prewar Light Bomber to an Aviation Trainer." Aviation, XLIII (November 1943), 226+.

(Fairchild) AT-21 Gunner

Ackerman, Beatrice, et al. AT-21, February 1941-July 1945. ATSC Historical Study, no. 55. Wright Field, Ohio: USAAF Air Technical Service Command, 1945. 71p.

Mitchell, Kent A. "The Saga of the Fairchild AT-21." American Aviation Historical Society Journal, XXXII (Fall 1987), 162-171.

"Plastic Gunnery Trainer." Air Progress, III (July 1943), 20-25.

AT-22 see B-24 Liberator

AT-23 see B-26 Marauder

AT-24 see B-25 Mitchell

(Boeing) B and W

"Boeing's First." Aeroplane Monthly, III (June 1975), 289-290.

Morrow, James A. "The Boeing B and W." American Aviation Historical Society Journal, XI (Spring 1966), 38-46.

Searles, Robert A. "Reflections: Boeing's Biplane Beginnings." Business and Commercial Aviation, LVIII (June 1986), 42-50.

BT-3 see PT-9

(Seversky) BT-8

"Seversky Basic Trainer." Aviation, XXXIV (February 1935), 70-71.

(North American) BT-9/BT-10

Gault, Owen and Vital Ferry. "Empire Builder: The Story of the North American BT-9." Air Classics, XII (October 1976), 40-48, 90-92.

"North American NA-16." Aviation, XXXV (July 1936), 38+.

(Fleetwing) BT-12 Sophomore

Ackerman, Beatrice, et al. XBT-12/BT-12, October 1938-December 1944. ATSC Historical Study, no. 24. Wright Field, Ohio: USAAF Air Technical Service Command, 1944. 20p.

Guishard, C. H. "Fleetwing's BT-12." Aero Digest, XLI (December 1942), 186-187.

Knox, J. A. and J. J. Boericke. "Designing for Easy Maintenance: Fleetwing's BT-12 Training Plane." Aviation, XLI (December 1942), 108-112.

(Vultee) BT-13 Valiant

Cooke, Robert C. "Design Analysis of the Valiant, BT-13." Automotive and Aviation Industries, XCII (January 15, 1945), 26-29+.

"Design Details of the Valiant, BT-13." Aero Digest, XLIII (January 1945), 76-91.

Hester, Larry. "Good Valiant Vibes." Flying, CXIV (April 1987), 92+.

Schnepf, Ed. "The Princely Valiant." Air Progress, XL (May 1978), 56+.

Smith, Gene. "Vibrator." Air Progress, XLIX (November 1987), 78-82.

Strunk, Al and Robin Higham. "The Vultee BT-13A." *Aerospace Historian*, XXIV (December 1977), 212-216.
Reprinted in Robin Higham and Carol Williams, eds., *Flying Combat Aircraft of the USAAF-USAF* (Manhattan, Kans.:Sunflower University Press, 1981), pp. 81-86.

"The Vultee Valiant SNV." *AirTech*, IV (June 1944), 18-23.

BT-14 see BT-9

BT-15/BT-16 see BT-13 Valiant

(Curtiss) JN-4 Jenny

Book/Document/Report Literature:

Bowers, Peter M. *Curtiss JN-4*. Aircraft in Profile, no. 37. Windsor, Eng.: Profile Publications, 1965. 13p.

Casari, Robert B. *The Curtiss Jennies*. Chillicothe, Ohio: Military Aircraft Publications, 1975. 152p.

Curtiss Aeroplane and Motor Company. *The Curtiss JN*. Buffalo, N.Y., 1920. 8p.

____. *The Curtiss Standard JN-4B Tractor Handbook*. Buffalo, N. Y., 1917. 32p.

____.____. Buffalo, N.Y., 1918; Glendale, Calif.: Aviation Book Co., 1971. 72p.

Lincke, Jack R. *Jenny Was No Lady: The Story of the JN-4D*. New York: W. W. Norton, 1970. 288p.

Periodical/Anthology Literature:

Baxter, Gordon. "A Hard Act to Follow." *Flying*, CVIII (November 1981), 70+.

Bowers, Peter M. "Legendary Jenny." *Aviation Quarterly*, II (Fall 1976), 404-452.

Browne, Jay. "Jenny as a Lady--Sometimes." *Aviation Quarterly*, IV (Winter 1978), 66-75.

"Curtiss Model JN-4B Tractor." *Aerial Age Weekly*, V (April 26, 1917), 186-187.

Davisson, Bud. "A Lady Named Jenny." *Air Progress*, XLI (December 1979), 54-61.

Gilbert, John. "The Jenny." *Flying*, LXXXIV (March 1969), 52-55.

"Instructions for Assembling Curtiss Standard JN-4B Aeroplanes." *Aerial Age Weekly*, V (August 20-27, 1917), 822-824, 870-872.

Kucera, J. M. "Jenny Has Flown." Aeroplane Monthly, IV (August 1976), 416-421.

Lamb, Dean I. "Suggestions for the Fliers of the JN-4B." Air Service Journal, II (May 9, 1918), 657-658.

Molson, Kenneth M. "The JN-4." American Aviation Historical Society Journal, XVII (Winter 1972), 225-241; XVIII (Summer 1973), 84-95.

"Notes on the Curtiss JN Aircraft Family." Cross and Cockade Journal, IV (Winter 1963), 335-342.

Phillips, F. G. "An Ode to Jenny." Aerospace Historian, XXIV (December 1977), 217-221.

Rust, Kenn C. "Old Jennies Never Die." American Aviation Historical Society Journal, VIII (Fall 1963), 202-205.

Turner, Bill. "Jenny: A 60-Year Old Glamor Girl." Sport Aviation, XXVI (March 1977), 47-52.

(Curtiss) N2C Fledgling

"Curtiss Fledgling." Airway Age, X (June 1929), 860+.

Mock, Robert M. "Curtiss Fledgling Can Be Fitted for All-Round Service as Land-Plane or Seaplane." Aviation, XXIV (April 9, 1928), 886-887+.

(Consolidated) N2Y Fleet

"Construction Details of Consolidated Fleet Trainers." Aero Digest, XXVII (August 1935), 50-51.

(N.A.F.) N3N Yellow Peril

Andersen, Brian. "Yellow Peril." Aviation Quarterly, VI (Summer 1980), 212-223.

(Boeing) NB-1

Van Deurs, George. "Fly Before Buy: The Boeing NB-1 Occasionally Went Into Flat Spins, Whirling Like a Maple Seed." Aerospace Historian, XXI (Fall 1974), 139-143.

NE-1 see L-4 Cub

NJ see BT-9

(Stearman) NS

Levy, Deed. "The Origins of a Great Navy Training Plane: The NS/NS-2." Naval Aviation Museum Foundation, V (Fall 1985), 10-15, 18.

(Stearman) PT-13 Kaydet

Davisson, Bud. "Different Stearman." Air Progress, XXXIV (February 1974), 46-51.

Hammett, Tom. "Stearman." Flying, CXI (September 1984), 102-107.

Lowe, Thomas E. "China Stearmans." American Aviation Historical Society Journal, XXI (Summer 1976), 150-152.

Mayborn, Mitch. "Stearman's Ageless Wonder." Air Classics, VIII (June 1972), 32-40, 63.

Szurovy, Geja G. "'Take the Kid Up and Turn Him Upside Down!" Private Pilot, XXI (February 1986), 22-29.

(Ryan) PT-16

Bassett, James R. "Aristocrats Join the Army." Air Classics, IX (July 1973), 54-63.

____. The Ryan PT-16/PT-20: Aristocrats Join the Army. Research Report. Maxwell AFB, Ala.: Air War College, Air University, 1973. 15p.

Carpenter. Dorr B. and Mitch Mayborn. Ryan Guidebook. Dallas, Tex.: Flying Enterprise Publications, 1975. 120p.

Mayborn, Mitch. Ryan PT/ST Series. Aircraft in Profile, no. 158. Windsor, Eng.: Profile Publications, 1972. 24p.

(Stearman) PT-17 Kaydet
(see also PT-13 Kaydet)

Baxter, Gordon. "Grand Piano." Flying, CXV (September 1988), 92-93.

Bowers, Peter M. "Evolution of the Stearman PT-17." Air Classics, I (Winter 1963-1964), 28-33.

Gabrielle, Douglas. "Day of the Stearman." Air Progress, XXXIX (September 1977), 58-63.

Park, Edwards. "Reunion With a PT-17." Flying, CII (June 1978), 48-54.

PT-18 see PT-13/PT-17 Kaydet

(Fairchild) PT-19 Cornell

Bowers, Peter M. "Fairchild's Fair Children." Air Classics, IV (June 1968), 21-25.

Thieblot, A. "Engineering Fairchild's PT-19." Aero Digest, XXXVII (October 1940), 130-132, 134+.

PT-20 to PT-22 see PT-16

PT-23/PT-26 see PT-19 Cornell

PT-27 see PT-13/17 Kaydet

(Standard) SJ Standard

"An American Training Machine." The Aeroplane, XIII (August 22, 1917), 520-522.

"The Standard Aeroplane Corporation Model J. Training Tractor." Aviation and Aeronautical Engineering, II (April 1, 1917), 216-217.

"The Standard Model J-R." Aviation and Aeronautical Engineering, III (August 1, 1917), 28-29.

SNB see C-45 Expeditor/AT-7 Navigator

(Curtiss) SNC-1

"Curtiss SNC-1 Trainer." Aero Digest, XXXIX (July 1941), 184+; XL (August 1941), 105+.

SNJ see AT-6/SNJ Texan/Harvard

SNV see BT-13 Valiant

T1A see T2V

(North American) T-2 Buckeye

Allietta, Dick. "Bravo Buckeye." Skyline, XXV (January 1967), 40-47.

Andrews, Hal. "Naval Aircraft: Buckeye." Naval Aviation News, (January 1971), 20-21.

Bulban, Erwin J. "T-2J Designed as Primary-Through-Advanced Trainer." Aviation Week and Space Technology, LXVI (May 13, 1957), 52-53+.

Buerger, John. "Buckeyes Aboard: The Navy's Newest Jet Trainer." Skyline, XVII (Summer 1959), 36-41.

____. "Stand By for Jets." Skyline, XVI (Spring 1958), 32-35.

Butter, Kirk. "The T-2 Buckeye and the VT Squadrons." Air Classics, IX (April 1973), 48-55, 82.

Ginter, Steven J. North American Rockwell T-2 Buckeye. Naval Fighters Series, no. 15. Simi Valley, Calif.: Naval Fighters, 1987. 62p.

"Rockwell Buckeye." Air Enthusiast, V (October 1973), 162-169.

(Lockheed) T-2V Sea Star

"High Speed Trainer." Ordnance, XL (July-August 1955), 112-113.

"Lockheed T-2V Jet Trainer." Aviation Age, XXV (May 1956), 20-25.

"Lockheed's Sea Star." RAF Flying Review, XIII (February 1958), 26-29.

T-6 see AT-6/SNJ Texan/Harvard

T-7 see UC-45 Expeditor/AT-7 Navigator

T-11 see UC-45 Expeditor/AT-7 Navigator

T-17 see PT-17 Kaydet

T-19 see PT-19 Cornell

(North American) T-28 Trojan

Andrews, Hal. "Naval Aircraft: T-28 Trojan." Naval Aviation News, (September 1975), 20-21.

Clarke, Richard W. "Last of the Big Time Radials." Air Classics, IX (February 1973), 32-39.

Davisson, Bud. "Charlie Trojan." Air Progress, XLIII (October 1981), 74+.

Ginter, Steven J. North American T-28 Trojan. Naval Fighters Series, no. 5. Simi Valley, Calif.: Naval Fighters, 1981. 66p.

Larsen, Jim. "Trojan." Air Classics, XV (March 1979), 44-49.

"North American T-28B." Aero Digest, LXVIII (May 1954), 31-33+.

O'Hara, Bob. "Chitty-Chitty-Bang-Bang: The T-28." Air Classics, VIII (August 1972), 46-53.

Papineau, Ned. "T-28: Last of the Round-Engine Trainers." Air Progress, XXXIX (April 1977), 30+.

"Pilot's Report: T-28." Skyways, IX (December 1950), 14-15, 40.

Trimble, Robert L. "Naval Trojans." Air Combat, VIII (May 1980), 62-72.

Wheeler, Howard. "T-28 Spans the Generation Gap." Naval Aviation News, (July 1982), 22-23.

(Convair) T-29 Flying Classroom
(see also C-131/R4Y Samaritan)

Ginter, Steven J. and Nick Williams. Convair T-29 Flying Classroom. Naval Fighters Series, no. 14. Simi Valley, Calif.: Naval Fighters, 1987. 114p.

Jouse, Thomas H. and Mike R. Martini. "Farewell to a Lady." Navigator, XXII (Fall 1975), 26-29.

Self, Mary. Convair 250/T-29/T-32, May 1948-January 1949. AMC Historical Study. Wright-Patterson AFB, Ohio: USAF Air Materiel Command, 1949. 4p.

"T-29 Flying Classroom." Flying Safety, X (January 1954), 14-16.

Zimmerman, John. "The T-29." In: Robin Higham and Carol Williams, eds. Flying Combat Aircraft of the USAAF-USAF. Manhattan, Kans.: Sunflower University Press, 1981. pp. 91-96.

(Fairchild) T-31/NQ-1

"Fairchild's T-31 Basic Trainer." Aviation Age, XIV (December 1950), 22-23.

Fenwick, Any C. T-31, May 1946-November 1949. AMC Historical Study, no. 205. Wright-Patterson AFB, Ohio: USAF Air Materiel Command, 1950. 76p.

Mitchell, Kent. "The Fairchild XNQ-1/T-31 Trainer." American Aviation Historical Society Journal, XXII (Spring 1977), 48-55.

T-32 see T-29 Flying Classroom

(Lockheed) T-33 Silver Star
(see also F-80 Shooting Star/TF-80C)

Book/Document/Report Literature:

Poston, Freddie L. The T-33 as a Tactical Fighter for Limited Warfare in Southeast Asia. Research Paper. Maxwell AFB, Ala.: Air Command and Staff College, Air University, 1962. 25p.

Periodical/Anthology Literature:

Beavers, Jim. "Every Man a Tiger." Air Force Magazine, LX (January 1977), 45-49.

Collins, Richard L. "T-Bird." Flying, CXIII (December 1986), 74-77+.

Conley, Douglas C. "U.S. Air Force Series T-33A Aircraft." Aerospace Historian, XXIV (September 1977), 168-172.

Crandell, Susan. "T-Cat." Flying, CIII (October 1978), 57-59+.

Deal, Duane W. "The T-Bird: Try It, You'll Like It." Interceptor, XX (January 1978), 6-11.

Ethell, Jeffrey L. "Jet Pilot." Air Progress, XLVI (December 1984), 42+.

Fuchs, Ron. "Those Terrific T-Birds." Airman, XXVIII (January 1984), 21-26.

Hall, G. Warren. "Research and Development History of the U.S. Air Force Stability T-33." American Aviation Historical Society Journal, XIX (Winter 1974), 270-274.

Levier, Tony and Sammy Mason. "Spinning the 'T.'" Interceptor, X (February 1968), 8-11.

O'Leary, Michael. "Fang T-Bird Phase Out." Air Classics, XXIV (August 1988), 40-45.
Removal from Florida ANG.

Thompson, Bill. "Twilight of the T-Bird." Airman, XXXI (June 1987), 23-26.

Vandyk, Anthony. "T-33 Trains Jet Pilots of 19 Nations." American Aviation, XIX (May 21, 1956), 47-48.

Wilson, George W. "Hail, Hail, the Gangstart's Here." Aerospace Safety, XVII (June 1961), 6-18.

(Beech) T-34 Mentor

Andrews, Hal. "Naval Aircraft: Mentor." Naval Aviation News, (January 1973), 20-21.

____. "Naval Aircraft: T-34C Mentor." Naval Aviation News, (April 1978), 23-24.

____. "Naval Aircraft: T-34C Turbo-Mentor." Naval Aviation News, (July 1982), 24-25.

"Beech Mentor." RAF Flying Review, XII (November 1956), 34-37.

"Beech T-34A Mentor." Aero Digest, LXIX (August 1954), 30-36.

Davisson, Bud. "T-34C." Air Progress, XXXVII (August 1975), 54+.

Drendel, Lou. "Magnificent Mentor: A Pilot Reviews,His Favorite Aircraft." Air Progress Aviation Review, V (Fall 1981), 50-69.

Greenwood, Jim. "New for the Navy." Flight Magazine, XLIV (December 1955), 50-51, 66, 74-76.

Lert, Peter. "Turbine Mentor." Air Progress, XLIII (December 1981), 24-31.

____. "Warbird From Wichita: T-34." Air Progress, XXXVII (August 1975), 58-64.

Moll, Nigel. "T-34C-1 Turbo-Proposition." Flying, CXI (July 1984), 36+.

Munson, Russell. "The T-34C." Flying, CV (September 1979), 76-81.

Schemmer, Benjamin F. "I Flew the T-34C and It Didn't Crash." Armed Forces Journal International, CXV (July 1978), 36-37+.

Smith, Gene. "Thunderstorm." Air Progress, XLII (March 1980), 32+.

Stone, Robert R. "The T-34C: A Thing of Beauty." Approach, XXI (September 1975), 1-3.

"T-34 Charlie: A Better Mentor." Air International, X (May 1976), 221-230.

(Temco) T-35 Buckaroo

Davisson, Bud. "T-35: The Race Doesn't Always Go to the Swift." Air Progress, XXV (July 1974), 40-45.

Tacks, D. R. "Temco T-35." AeroDigest, LXIII (December 1951), 20-27.

"Temco Trainer Designed for Easy Production and Maintenance." Aviation Age, XX (December 1953), 68-71.

(Cessna) T-37 Tweety Bird
(see also A-37 Dragonfly)

Book/Document/Report Literature:

Bagwell, Margaret C. History of the T-37 Weapon System, January 1952-April 1956. AMC Historical Study, no. 286. Wright-Patterson AFB, Ohio: USAF Air Material Command, 1957. 53p.

Fucci, Daniel T. and Jimmy T. White. A Handbook for the T-37 Flight Commander. Research Paper. Maxwell AFB, Ala.: Air Command and Staff College, Air University, 1983. 38p.

Periodical/Anthology Literature:

Alexander, Louis. "The Air Force's New Cessna T-37 Jet Trainer." Air Force, XLI (January 1958), 62-64.

Bunting, Richard. "The Tweety Bird and Me." Airman, VII (December 1963), 10-12.

"The Cessna T-37 Tweety Bird." Flying Review International, XIX (December 1963), 32-36.

"Cessna's XT-37 Makes First Flight." Skyways, XIII (November 1954), 37-38.

Davis, Jim. "It Really Does Use Monkeys and Mirrors." Aerospace Safety, XXXVI (May 1980), 12-15.

Davisson, Bud. "T-37: Three-Ton Dog Whistle." Air Progress, XXXVIII (October 1976), 20-24, 84.

Duddy, Brian J. "New Life for the T-37." Flying Safety, XLIII (November 1987), 5-7.

Hagan, Bob. "Knee High to a T-37." Flying Safety, XII (September 1956), 20-24.

Katzaman, Jim. "Aces High." Airman, XXIX (February 1985), 42-47.

Kell, Jerry. "The T-37 Story." USAF Instructor's Journal, I (January 1964), 62-65.

Piper, Danny and Dan McCauley. "The Trail of the T-37: Where It All Began for 30,000 Air Force Pilots." Air Force Magazine, LVIII (April 1975), 65-69.

Prochoroff, Alan. "To Fly: Oh-oh, What a Feeling!" Airman, XXVII (June 1983), 8-16.

Rose, Eugene A. "What a Tweet!" Navigator, XXV (Summer 1978), 16-19.

Stanfield, Robert I. "Evaluation Shows T-37A is Fast, Simple, Compact." Aviation Week and Space Technology, LXVI (April 1, 1957), 48-49+.

(Northrop) T-38 Talon
(see also F-5 Freedom Fighter/Tiger II)

Book/Document/Report Literature:

Brady, Roger A. and Paul F. Bridges, Jr. A Survival Course for Talon Drivers: A Handbook for New T-38 Instructor Pilots. Research Paper. Maxwell AFB, Ala.: Air Command and Staff College, Air University, 1982. 87p.

Northrop Corporation. Public Relations Department. T-38 Highlights. Los Angeles, Calif., 1987. 3p.

Periodical/Anthology Literature:

"Anniversary of a Classic." Air Classics, XVII (August 1981), 50-57.

Buck, Arthur W. "The T-38." Air University Quarterly Review, XIII (Fall 1961), 22-29.

Carson, Don. "Teaching Tactics to TAC 'Migs.'" Air Force Magazine, LVII (March 1974), 44-47.

Collins, Richard L. "Air Force Preflight." Flying, CXII (November 1985), 17-18.

____. "Inside a Military Operations Area Looking Out From the Sharp End of a T-38." Flying, CXII (November 1985), 76-78+.

Connors, Slim. "T-38 Instructor Pilot." Air Progress Aviation Review, V (June 1981), 38-41, 70-72.

____. "The Satisfactions of a T-38 Instructor Pilot." Air Force Magazine, LIX (January 1981), 58-62.

Croft, David R. and Robert W. Wickman. "I Fly the T-38!" Air Force Magazine, LIV (August 1971), 36-41.

Ethell, Jeffrey L. "Flying the T-38 Talon." Air Classics, X (December 1974), 46-51, 76-77.

Harms, Johnny N. "Talons for the Eaglets." Aerospace Safety, XVII (July 1961), 2-4.

Henrich, Stephen G. "T-38 Performance." USAF Fighter Weapons Review, (Spring 1978), 22-25.

Holm, Skip and Michael O'Leary. "Supersonic!" Air Progress, XLVII (May 1985), 42+.

Martinez, Ed. "Slipping the Surly Bonds." Airman, XIV (May 1970), 4-11.

Merington, Larry. "The Mean Machine." Air Progress Aviation Review, V (Fall 1981), 16-25.

Miller, E. M. "This Trainer Can Go, Go, Go." Air Force, XLIV (March 1961), 99-102.

"The Northrop T-38 Supersonic Trainer." Interavia, XVI (April 1961), 465-468.

Power, Thomas R. "The Little White Mice That Fly." Airman, XXIII (September 1979), 17-20.

Scott, William B. "NASA T-38 Trainers May Be Replaced." Aviation Week and Space Technology, CXVII (July 12, 1982), 52-53+.

Sweeney, Richard L. "Flying the Talon." Interavia, XVI (April 1961), 469-470.

____. "Northrop's T-38 Talon." Flying, LXVIII (March 1961), 2-3+.

Tilton, Jim. "The Talented Talon." Airman, XVIII (September 1974), 30-33.

Westervelt, Durward G. "Supersonic Bridge to Aerospace." Airman, V (June 1961), 9-11.

(North American) T-39 Sabreliner

Book/Document/Report Literature:

United States. Government Accounting Office. National Security and International Affairs Division. Analysis of Air Force Decision to Replace T-39 Aircraft. Washington, D.C., 1986. 10p.

Periodical/Anthology Literature:

Andrews, Hal. "Naval Aircraft: Sabreliner." Naval Aviation News, (April 1978), 20-21.

Blount, Earl. "Sabreliner." Skyline, XVII (Winter 1958), 50-55.

Campbell, J. K. "A Little Jewel." Combat Crew, XII (November 1961), 16-18.

Dimon, Richard B. "Bogie, Bogie--Navy Style: Venerable Game of Hide and Seek is Serious Business for T-39D Saberliners in Radar Intercept Officer Training Program at Glynco NAS." Skyline, XXII (March 1964), 34-41.

____. "Sabreliner: Service With a Smile." Skyline, XXIV (April 1966), 42-47.

Fulmer, Harold G. "Flying Tips on the Sabreliner." Combat Crew, XV (August 1964), 12-14.

Lauren, Ward. "Certified Airworthy." Skyline, XVIII (February 1960), 44-49.

"Little Jet With Long Legs." Skyline, XXII (January 1964), 38-43.

Olcott, John W. and Richard N. Aarons. "Rockwell Sabreliner." Business and Commercial Aviation, XLVI (June 1980), 45-51.

Rogers, Bill. "Sabreliner." Flying, LXXVIII (May 1966), 40-47.

"Sabrelift." MAC Flyer, XXII (December 1975), 10-13.

Solomon, Howard. "So You're Going to Fly the T-39." Combat Crew, XX (June 1969), 11-13.

"Versatile Twin." RAF Flying Review XVII (March 1962), 26-29.

T-40 see C-140 Jet Star

(Cessna) T-41 Mescalero

Book/Document/Report Literature:

Clarke, Charles ("Bill"). The Cessna 172. Blue Ridge Summit, Pa.: TAB, 1987. 293p.

Periodical/Anthology Literature:

Bueker, R. C. "The Small Plane is Back." Airman, IX (September 1965), 43-44.

Collins, Richard L. "Skyhawk: Attuned to the Times." Flying, CVII (November 1980), 24-29.

Cook, LeRoy. "Skyhawk Celebrates 30 Years of Service." Private Pilot, XX (October 1985), 46-51.

Phillips, Ed. "Skyhawk." Aviation Mechanics Journal, XIX (February-March 1989), 40-53, 35+.

Smith, Ken. "Cessna 172: Survival of the Fittest." Plane and Pilot, XIV (November 1978), 58-62.

Stead, Dave. "Tame the Mescalero." Airman, XVII (November 1973), 36-39.

Sutherland, Don. "The Mescalero Connection." Airman, XXXII (September 1988), 34-38.

Wilkinson, Stephen. "Is This the World's Best Airplane?: Cessna Skyhawk 172." Flying, LXXXVIII (April 1971), 38-40.

Zack, Marv. "Those Vintage 172s." Air Progress, XLII (March 1980), 39+.

(Beech) T-42 Baron

Collins, Richard L. "Pilot Report: Basic Baron." Flying, XCIII (July 1973), 62-69+.

Lert, Peter. "Classic Light Twin." Air Progress, XLIV (March 1982), 74+.

McClellan, J. Mac. "Pilot Report: Baron." Flying, CXI (June 1984), 58+.

Schlaeger, G. J. "Pilot Report: Beechcraft Model 55 Baron." Flying, LXVIII (March 1961), 40-41+.

(Boeing) T-43 Flying Classroom

Billick, Donald M. "The Future is Now." Navigator, XXI (January 1974), 5-9.

"Boeing's Military Twin." Air Enthusiast, V (September 1973), 193-194.

Lerner, Vance. "The T-43A Navigator Trainer." Aerospace Safety, XXIX (December 1973), 8-9.

Minton, David H. Boeing 737. Aero Series, v. 37. Blue Ridge Summit, Pa.: TAB, 1989. 112p.

O'Lone, Richard G. "T-43A Designed to Spur Navigator Training." Aviation Week and Space Technology, XCVIII (March 12, 1973), 19-20.

Slocum, Jerald. "Baby Boomer: Boeing 737." Flying, CXV (March 1988), 70-77.

Smith, P. R. Boeing 737. Air Portfolio, no. 1. New York: Jane's, 1986. 64p.

"T-43A Airborne Navigator Trainer." Navigator, XIX (March 1972), 14-20.

Ulsamer, Edgar. "UNTS: The Last Word in Navigator Training." Air Force Magazine, LVI (August 1973), 56-59.
T-43's Undergraduate Navigator Training System.

(Beech) T-44/U-21 King Air
(see also C-6 King Air)

Andrews, Hal. "Naval Aircraft: T-44." Naval Aviation News, (September 1981), 20-21.

Brechner, Beryl. "Royal Ride." Flying, CVI (February 1980), 58-61+.

Dobson, James R. "A Detailed History of the U-21." U.S. Army Aviation Digest, XXVIII (June 1982), 14-19.

Lert, Peter. "King Air F90." Air Progress, XLIV (April 1982), 10-17.

____ and Michael O'Leary. "Quest for Performance." Air Progress, XLV (August 1983), 30-37.

McClellan, J. Mac. "Long Live the King." Flying, CIX (June 1982), 24-27.

Winters, Curtis J. "The T-44A and Contract Maintenance." U.S. Naval Institute Proceedings, CVI (January 1980), 107-109.

(McDonnell-Douglas/BAe) T-45 Goshawk

Andrews, Hal. "The Goshawk Flies." Naval Aviation News, (July-August 1988), 18-20.

Braybrook, Roy M. "The Case for the Hawk." Asian Defence Journal, no. 4 (April 1982), 48-50+.

Clostermann, Jacques. "IDR Flies the Hawk." International Defence Review, no. 11 (November 1987), 1481-1483.

Gaines, Mike. "T-45: Tailhook Trainer." Flight International, CXXXIII (March 12, 1988), 20-25.

"Hawks in the Air." Defence Update International, no. 77 (November 1986), 43-49.

Higman, Susan. "Inflatable Scenario for T-45TS: First Production Goshawk in October 1989." Sea Power, XXXI (August 1988), 25-26.

Housman, Damian. "The Navy's New Hawk." International Combat Arms, IV (May 1986), 50-55+.

Hooton, Edward. "Hawks Join the U.S. Navy." NATO's Sixteen Nations, XXIX (November-December 1985), 78-79.

Smith, Bruce A. "Douglas Begins Flight Tests of Navy T-45A Jet Trainer." Aviation Week and Space Technology, CXXVIII (May 2, 1988), 36-39.

Stewart, Van N. "The Road to Kingsville: An Update on the T-45 Training System." Wings of Gold, X (Summer 1985), 13-17.

Weatherly, Frank. "Next Generation Trainer." Flying Safety, XXXVIII (March 1982), 16-17.

(Fairchild) T-46 Thunderpup

Berry, F. Clifton, Jr. "T-46A: The U.S. Air Force's Next Trainer." International Defence Review, XVII (September 1984), 1361-1362.

Braybrook, Roy M. "Tweety Bird Replacement." Air International, XXVIII (June 1985), 273-277, 279-280.

Callegari, William. "Designed to Be Stalled." Astronautics and Aeronautics, XXI (September 1983), 60-68.

DeMeis, Richard. "Thunderpup Caught in Pentagon Wringer." Aerospace America, XXV (February 1987), 20-23.

Schlitz, William P. "T-46A: The New Primary Trainer." Air Force Magazine, LXVII (August 1984), 82-86.

Shawler, Wendell H. "T-46A Flight Test Results." Cockpit, (July-September 1986), 4-9.

Wanstall, Brian. "Good for 30 Years?: The U.S. Air Force's Future Trainer." Interavia, XXXVIII (March 1983), 227-229.

(Cessna) T-47 Citation II

"Cessna T-47A Begins Operations as a Naval Navigation Trainer." Aviation Week and Space Technology, CXXII (January 14, 1985), 25-26.

Kennedy, Norval. "Citation Attack." Professional Pilot, XVIII (August 1984), 72-73.

Lowery, John. "Flightcheck: Cessna Citation." Professional Pilot, XVIII (November 1984), 106-113.

McClellan, J. Mac. "Cessna T-47: The Navy's Super Citation." Flying, CXV (December 1988), 70-77.

____. "Second Coming." Flying, CXIII (July 1986), 30-35.

Moll, Nigel. "Bogeyman." Flying, CXV (December 1988), 78-80.

(Huff-Daland) TA-2

"The Huff-Daland Biplane." Flight, XIV (March 16, 1922), 158-160.

"The Huff-Daland Thick-Wing Biplane." Aviation, XII (February 6, 1922), 161-163.

United States. Army Air Service. Performance Test of Huff-Daland T-2, Equipped With Curtiss OX-5 Engine. USAS Information Circular, v. 5, no. 407. Washington, D.C.: GPO, 1923. 4p.

____. ____. Performance Test of Huff-Daland T-2, Equipped With Laurence R. Engine. USAS Information Circular, v. 5, no. 411. Washington, D.C.: GPO, 1923. 5p.

(Douglas) TA-4 Skyhawk
(see also A-4 Skyhawk)

Adkinson, John L. A History of the Use of the TA-4F Aircraft as a Tactical Air Co-Ordinator (Airborne) in the War in Southeast Asia. Research Report. Maxwell AFB, Ala.: Air War College, Air University, 1977. 78p.

Belisle, Ken. "Tandem Hawk." Scale Modeller, X (January 1975), 34-40.

(Huff-Daland) TA-6

"Huff-Daland Model TA-6 Advanced Training Plane." Aviation, XV (October 1, 1923), 408-409.

(McDonnell-Douglas) TAV-8 Harrier
(see also AV-8 Harrier)

Greeley, Brendan M., Jr. "Aviation Week Pilot Report: Improved VTOL Performance of TAV-8B Adds Realism to Attack Force Training." Aviation Week and Space Technology, CXXVII (August 3, 1987), 68-76.

"A Jump Jet Made For Two." Air Enthusiast, I (1971), 233+.

TB-25 see B-25 Mitchell

(McDonnell-Douglas) TF-15 Eagle
(see also F-15 Eagle)

Singleton, I. Jerry. "Flying the 'Bald' Eagle." TAC Attack, XVII (December 1977), 4-9.

(Lockheed) TF-80C/TV Shooting Star
(see also F-80 Shooting Star/T-33 Silver Star)

"First Jet Trainers." Aero Digest, LVI (June 1948), 38-40.

(McDonnell-Douglas) TF/A-18 Hornet
(see also F/A-18 Hornet)

Bauer, William D. "Employment of the Two-Seat TF/A-18 Airplane." Naval War College Review, XXXIII (March-April 1980), 103-107.

(North American) TF-51D Mustang
(see also P-51 Mustang)

Lert, Peter. "Last of the TFs: We Fly and Report on the Last Flying TF-51D Mustang." Air Classics, XIV (October 1978), 34-39, 82-83.

(Temco) TT-1 Pinto

Haddaway, George E. "Jet For Beginners." Flight Magazine, XLVIII (May 1959), 42+.

TV see TF-80C/TV Shooting Star

Wright Types

Book/Document/Report Literature:

Goodall, Michael H. The Wright Aircraft. London, Eng.: Gentry Books, 1972. 148p.

Wolko, Howard S., ed. Wright Flyer: An Engineering Perspective. Washington, D.C.: Smithsonian Institution Press, 1987. 106p.

Wright, Orville. Early History of the Airplane. Dayton, Ohio: Wright Airplane Company, 1922. 24p.

Wright Airplane Company. Wright Flyers. New York: Premier Press, 1912. 16p.

Periodical/Anthology Literature:

Alford, Leon P. "The Wright Aeroplane: A Noteworthy Invention." American Machinist, XXXI (October 1, 1908), 473-478.

Brown, Harold H. "Some Construction Details of the Wright Aeroplane." Aeronautics, III (October 1908), 14-15, 44.

Claudy, Carl H. "The Army's First Airplane." Aero Digest, XI (December 1927), 636-640.

___. "Our First Army Flying Machine." Technical World, XII (October 1909), 222-231.

"Construction of the Wright Aeroplane." Scientific American, XCIX (September 26, 1908), 208-210.

Dienstbach, Carl. "The Perfect Flying Machine." American Aeronaut, I (January 1908), 3-11.

Drew, Andrew. "Learning to Fly a Wright Biplane." Aero, II (August 1911), 385-386, 410, 442.

Gibbs-Smith, Charles H. "The World's First Practical Airplane." NCR World, (Fall 1978), 5-6.

Hardie, George. "The First Successful Homebuilt Airplane." Sport Aviation, XXVII (December 1978), 36-37.

Harrington, Floyd. "Signal Corps No. 1." Soldiers, XXX (March 1975), 6-7.

Hildenbrandt, Alfred. "The Wright Brothers' Flying Machine." American Magazine of Aeronautics, II (January 1908), 13-16.

Humphreys, Frederic E. "The Wright Flyer and Its Possible Uses in War." Professional Memoirs, II (March-April 1910), 99-107.
Reprinted in the Journal of the United States Artillery, XXXIII (March-April 1911), 144-147.

Lanchester, F. W. "The Wright and Voisin Types of Flying Machine: A Comparison." Aeronautical Journal, XIII (January 1909), 4-12.

Loening, Grover C. "The New Wright Aeroboat Type C." Aeronautics, XIV (June 15, 1914), 170-171.

"The Model E. Wright Biplane." Flight, VI (January 3, 1914), 18-19.

"New Wright Model E." Aeronautics, XIII (October 1913), 140-141.

Roosevelt, Theodore. "Roosevelt Administration in 1907 Gave the United States Army the First Aeroplane Ever Used for Military Purpose." Aerial Age, III (May 15, 1916), 261-263.

"Secret of the Wright Air-Ship." Literary Digest, XXXVI (June 13, 1908), 861-862.

Worman, Charles G. "Signal Corps No. 1." Military Collector and Historian, (Spring 1971), 1-12.

Wright, Orville and Wilbur. "Wright Brothers' Aeroplane." Century Magazine, LXXVI (September 1908), 641-650.
Reprinted under Orville's name alone in US Air Services, VIII (December 1923), 26-32.

"The Wright Biplane." Flight, II (March 12, 1910), 174-177.

"The Wright Biplane Model B." Aeronautics, IX (September 1911), 93-100.

"Wright Biplane Model B and a Historical Sketch of Its Development." Scientific American Supplement, LXXII (December 9, 1911), 380-382.

"Wright Flyer: Design Analysis." Aero Digest, LXVII (July 1953), 28-37.

F. Prototype Aircraft

(Douglas) XA2D-1 Skyshark

Karstens, Nick. "Surplus to Requirements." Wings, XI (April 1981), 24-41.

(North American) XA2J-1 Savage

Redman, Rod. "Turbine Savage: North American's XA2J-1." Air Classics, X (July 1974), 20-23.

(Northrop) XA-9A

Blair, Edwin T. "Cutting Out the Guesswork." Airman, XVII (July 1973), 24-30.

Geddes, J. Philip. "The XA-9A: Northrop's AX Entry." Interavia, XXVII (August 1972), 866-868.

"The Northrop XA-9A Close Support Aircraft." International Defence Review, V (August 1972), 370-371.

(Beech) XA-38 Grizzly

"Flying Cannon: Unusual Beech XA-38 Flying Destroyer." Air Classics, V (October 1968), 10-13.

Trimble, Robert L. "Beech XA-38 Grizzly." Air Classics, XIX (August 1983), 16-25.

(Vultee) XA-41

United States. Army Air Forces. Air Technical Service Command. XA-41, April 1942-September 1945. ATSC Historical Study, no. 49. Wright Field, Ohio: USAAF Air Technical Service Command, 1945. 15p.

"The Unusual Vultee Attack Fighter, XA-41." Air Classics, VII (May 1971), 8-19.

(Douglas) XA-42/XB-42

Snodgrass, Raymond. XA-42, XB-42, XB-42A, May 1943-August 1947. AMC Historical Study, no. 174. Wright-Patterson AFB, Ohio: USAF Air Materiel Command, 1948. 26p.

(Boeing) XB-15

"Boeing B-15, Granddaddy of the B-17." Air Classics, III (July 1966), 12-15.

Bouche, Georges G. "Granddaddy: The XB-15." Aerospace Historian, XXVI (September 1979), 171-181.

(Douglas) XB-19

"Army to Convert XB-19A to Transport." Aviation News, III (March 12, 1945), 10-11.

"B-19, Air Fleet Leviathan." Modern Plastics, XVIII (August 1941), 29-31+.

"B-19, Army's Newest Super Bomber." Life, VIII (April 8, 1940), 28-29.

"Building the Douglas B-19." Aviation, XXXIX (October 1940), 58-59.

"Engineering Problems of Assembling the Douglas B-19 Bomber." Aero Digest, XXXVII (December 1940), 130-131.

Klemin, Alexander. "Brobdingriagian: Douglas B-19 Bomber." Scientific American, CLXIV (April 1941), 214-215.

McSurely, Andrew. "Data on XB-19 Installations and Army's Big Plane Program." Aviation News, II (December 21, 1944), 9-10.

O'Hara, Bob. "Douglas B-19: Pioneer Heavy Bomber." Air Classics, II (March-April 1965), 8-11+.

Strompl, G. A. "Douglas Builds the World's Largest Airplane." Machine Design, XLVI (July 1940), 146-151; XLVII (December 1940), 113-115.

Trimble, Robert L. "Bombing Behemoth." Air Classics, XVII (October 1981), 28-41.

Wagner, Ray. "The Douglas XB-19." Air Pictorial, XXIII (August 1961), 232-235.

"The World's Biggest Bomber." Popular Mechanics, LXXIV (December 1940), 814-815.

(North American) XB-21 Dragon

"North American Dragon Bomber." Aero Digest, XXX (February 1937), 82-83.

(North American) XB-28

Daniels, C. M. "Born to Lose: North American XB-28, America's Secret Stratosphere Bomber." Airpower, V (May 1975), 36-39.

Trimble, Robert L. "America's Mystery Bomber." Air Classics, XVII (September 1981), 16-26.

Tuttle, Jim. "The North American XB-28." American Aviation Historical Society Journal, XXIV (Summer 1979), 82-89.

(Martin) XB-33

Case, H. W. XB-33, B-33, October 1940-November 1943. ATSC Historical Study, no. 50. Wright Field, Ohio: USAAF Air Technical Service Command, 1945. 35p.

XB-35 see YB-49

(Boeing) XB-38
(see also B-17 Flying Fortress)

Dean, Jack. "Forgotten Fortress." Airpower, X (September 1980), 10-19.

(Boeing) XB-40
(see also B-17 Flying Fortress)

Birdsall, Stephen. "The Destroyer Escorts." Air Power Historian, XII (July 1965), 92-94.

Freeman, Roger A. "The Fighter Fortresses." Air Pictorial, XXIII (June 1961), 172-173+.

United States. Army Air Forces. Air Technical Service Command. XB-40. ATSC Historical Study, no. 12. Wright Field, Ohio: USAAF Air Technical Service Command, 1944. 19p.

(Consolidated) XB-41
(see also B-24 Liberator)

United States. Army Air Forces. Air Technical Serice Command. XB-41. ATSC Historical Study, no. 13. Wright Field, Ohio: USAAF Air Technical Service Command, 1944. 15p.

XB-42 see XA-42/XB-42

(Douglas) XB-42/XB-43 Mixmaster

Boyne, Walter J. "Douglas XB-42/42A/43." Airpower, III (September 1973), 6-15, 67.

Gunston, William T. ("Bill"). "Douglas Mixmasters." Aeroplane Monthly, III (September 1975), 426-432.

Keiller, A. C. "The Douglas Pushers." Air Classics, VIII (November 1972), 51-55.

Wood, C. C. "Design Development of the Douglas XB-42." Aviation, XLVI (February 1947), 37-42.

(Convair) XB-46

Boyne, Walter J. "Convair's Needle-Nosed Orphan: The One and Only B-46 Jet Bomber." Airpower, VI (September 1976), 8-19, 66.

(Martin) XB-48

Boyne, Walter J. "XB-48: Martin's Pipe Dream." Airpower, V (September 1975), 8-21.

Miller, Martin J. XB-48, August 1944-January 1950. AMC Historical Study, no. 202. Wright-Patterson AFB, Ohio: USAF Air Materiel Command, 1950. 79p.

(Martin) XB-51

Boyne, Walter J. "Attack: XB-51." *Airpower,* VIII (July 1978), 20-39.

Sternick, Robert. "Martin's XB-51." *Air Classics,* IX (February 1973), 18-23, 62.

"XB-51, Jet Bomber." *Skyways,* IX (January 1950), 18-19.

(North American) XB-70 Valkyrie

Book/Document/Report Literature:

Mills, Arthur J. *The Requirement for the B-70.* Research Report. Maxwell AFB, Ala.: Air War College, Air University, 1961. 68p.

Pace, Steve. *North American Valkyrie, XB-70.* Aero Series, no. 30. Fallbrook, Calif.: Aero Publishers, 1984. 104p.

Rees, Ed. *Manned Missile: The Story of the B-70.* New York: Duell, Sloan and Pearce, 1960. 182p.

Sherman, Dennis M. *The National Security Act, a Blueprint for the Congressional Role in Weapons Development: A Case Study of the B-70 Bomber.* PhD. Dissertation. Madison, Wisc.: University of Wisconsin at Madison, 1978.

Periodical/Anthology Literature:

Blair, Edwin T. "Footprints of a Space Monster." *Airman,* VII (February 1963), 14-16.

____. "The Valkyrie." *Airman,* IV (June 1960), 2-7.

Carter, L. J. "McNamara Legacy--a Revealing Case History: Death of the B-70." *Science,* CLIX (February 23, 1968), 859-863.

"Flight Diary of the Valkyrie." *Flying Review International,* XXI (May 1966), 511-512.

Foxworth, Thomas G. "North American XB-70 Valkyrie." *Historical Aviation Album,* nos. 7-8 (1969-1970), 76-87, 164-175.

Gunston, William T. ("Bill"). "XB-70 Valkyrie: The Bomber With the Mostest." *Aeroplane Monthly,* III (April 1975), 168-184.

Herron, Ed. "XB-70A." *Skyline,* XXIII (March 1965), 2-13.

Lauren, Ward. "Riding the Shock Wave." *Skyline,* XIX (January 1961), 20-27.

Miller, Ed Mack. "The Gutting of the Valkyrie." *Air Force,* XLIII (January 196), 29-33.

Pace, Steve. "Triplesonic Twosome." *Wings,* XVI (February 1986), 10-35.

Pike, Iain. "B-70." Flight, LXXXV (June 25, 1964), 1055-1062; LXXXVI (July 2, 1964), 18-24.

Plattner, C. M. "USAF Surveys XB-70 Program Future." Aviation Week and Space Technology, LXXXIV (June 20, 1966), 26-30.

____. "XB-70A's Research Role Will Contribute to Mach 3 Flight Data for SST Development." Aviation Week and Space Technology, LXXX (May 18, 1964), 26-30.

Poole, Gordon L. "The Big White One." Airman, VIII (December 1964), 6-10.

Rees, Ed. "Furor Over Fantastic Plane: The B-70 Bomber." Life, XLIX (October 17, 1960), 124-130+.
Abridged in Reader's Digest, LXXVIII (January 1961), 83-88 as "Make Way for the B-70."

Schistedt, Albert, Jr. "America's Most Amazing Airplane, the B-70." Science Digest, XLVIII (December 1960), 1-5.

Stimson, Thomas E., Jr. "The Controversial B-70: Military Monster or Supersonic Transport?" Popular Mechanics, CXIX (June 1963), 67-71+.

"Valkyrie: The Bomber That Never Was." Flying Review International, XX (January 1965), 22-25, 42.

Wheeler, Keith. "Billion-Dollar Baby: The Birth of the XB-70." Reader's Digest, LXXXVII (July 1965), 128-133.

____. "The Full Story of the 2.8 Seconds That Killed the XB-70." Life, LXI (November 11, 1966), 126-143.

____. "XB-70: Building the World's Weirdest Airplane." Life, LVIII (January 15, 1965), 74-80+.

"XB-70A Roll-Out." Machine Design, XXXVI (June 8, 1964), 162-167.

(Seversky) XBT

"Seversky Trainer: New XBT." Aviation, XXXV (October 1936), 44-45.

(Northrop) XBT2C-1/2

Andrews, Hal. "Naval Aircraft: XBT2C-1." Naval Aviation News, (September 1979), 20-21.

___. "Naval Aircraft: XBT2C-2." Naval Aviation News, (November-December 1987), 16-17.

(Fairchild) XC-31

"Fairchild Military Cargo Transport." Aviation, XXXIII (December 1934), 410-412.

"Fairchild XC-31 Military Cargo Transport." Aero Digest, XXV (December 1934), 44-45.

(Lockheed) XC-35
(see also UC-36/37 Electra)

Holcomb, Mal. "Crossing the High Frontier." Airpower, XI (September 1981), 37-43.

Wright, Theodore P. "They Take the High Road: Collier Trophy Award for 1938 to U.S. Army Air Corps for Designing, Building, and Equipping First Successful Stratospheric Airplane, XC-35." Collier's, CII (September 24, 1938), 14+.

(Convair) XC-99
(see also B-36 Peacemaker)

"Convair's Improved C-99." Aero Digest, LX (January 1950), 13+.

"Engineering Facts About the World's Largest Landplane: Air Force Transport, the XC-99." Automotive Industries, XCVIII (February 15, 1948), 42-43+.

"Flying Freight Train: The Air Force's XC-99." Popular Science, CLII (February 1948), 142-143.

Gibson, Jim. "The First Behemoth." Airman, XII (July 1968), 28-29.

Gunston, William T. ("Bill"). "Capacious Convair." Aeroplane Monthly, VI (February 1978), 8-19, 66.

Harper, Jennifer. "Kelly's Durable Relic." Air Force Magazine, LXIV (September 1981), 166-167.

"The XC-99: First Global Freighter." Air University Quarterly Review, IV (Fall 1950), 90-92.

XC-105 see XB-15

(Fairchild) XC-120 Packplane
(see also C-119 Flying Boxcar)

"The Fairchild XC-120 Packplane." Interavia, V (November 1950), 560-563.

(LTV-Hiller-Ryan) XC-142

Anderton, David A. "XC-142 Detail Design Effort Hits Peak." Aviation Week and Space Technology, LXXVIII (January 14, 1963), 60-61+.

Everett, Robert P. "A New Way to Fly." Airman, XI (July 1967), 10-13.

Fricker, John. "Bridging the VTOL Gap." Flying Review International, XX (November 1964), 21+.

Madison, Stuart G. "Progress With the XC-142A." Interavia, XXI (January 1966), 52-53.

____. "Tilt-Wing Transport." Aerospace Safety, XXI (August 1965), 20+.

Schoeni, Arthur L. "XC-142A: The Tri-Service Turboprop Tiltwing Transport." Air Classics, XV (May 1979), 38-49, 94-96.

Sunday, Terry L. "Tri-Service Tiltwing: The Amazing Story of Vought's XC-142, Pioneer of VTOL Flight." Airpower, XIV (July 1984), 10-23.

Williams, D. H. "V/STOL = XC-142A." Ryan Reporter, XXIII (July-August 1962), 22-25.

"XC-142A Airborne." Ryan Reporter, XXVI (Winter 1964), 36-39.

(Convair) XF/YF2Y-1 Sea Dart

Allen, Richard S. "Sea Dart." Wings, XIII (August 1983), 38+.

Boyne, Walter J. "Sea-Ski Fighter: The Story of the Convair Sea Dart." Air Pictorial, XXXIII (November 1971), 408-410.

Everhart, M. L. "Convair Sea Dart Flies." Aero Digest, LXVII (September 1953), 20-21.

Gunston, William T. ("Bill"). "Hydro-Ski Fighter." Aeroplane Monthly, III (December 1975), 634-639.

Hull, Tom. "Sea Dart." Naval Aviation News, (January 1981), 40-45.

Johnsen, Frederick A. "The Sea Dart." Air Classics, VIII (November 1971), 22-25.

Jones, Merrill. "Flight Test at San Diego." Honeywell Flight Line, VI (June 1955), 1-6.

Long, B. J. "Sea Dart: U.S. Navy XF2Y-1 and YF2Y-1 Experimental Supersonic Seaplanes." American Aviation Historical Society Journal, XXIV (Spring 1979), 2-12.

(Grumman) XF-3F

Seybel, Roger. "Three-Time Prototype." Wings, XIX (February 1989), 30+.

(Grumman) XF5F-1 Skyrocket

Gault, Owen. "Grumman's Tiger Twins: The Skyrocket and the Tigercat." Air Classics, IX (August 1973), 22-27.

(Vought) XF5U-1 Flying Flapjack

Allen, Richard S. "The First Flying Saucer." Airpower, XVII (May 1987), 44+.

Andrews, Hal. "Naval Aircraft: Flying Flapjack." Naval Aviation News, (June 1973), 20-21.

Janasen, Katie. "When Pancakes Flew." Smithsonian Air and Space, III (April-May 1988), 94-95.

Schoeni, Arthur L. "Flying Flapjack." Air Classics, IV (September 1967), 10-23.

____. "The Flying Flapjack." Air Classics, XI (August 1975), 16-25, 72-77.

____. "Flying Pancakes." Aeroplane Monthly, III (November 1975), 566-571.

(Curtiss) XF7B

Andrews, Hal. "Naval Aircraft: XF7B." Naval Aviation News, (November 1980), 20-21.

(Boeing) XF8-B1

Koehnen, Richard C. "Boeing's 5-in-1 Fighter: XF8-B1, the Ultimate Piston-Engined Jack-of-All-Trades." Airpower, V (July 1975), 26-39, 49-55.

Marlowe, Philip. "Boeing's Forgotten Fighter." Air Classics, XI (April 1975), 48-53.

(Grumman) XF10F Jaguar

DeMeis, Richard. "No Room to Swing a Cat." Wings, VI (August 1976), 30-48.

(Hughes) XF-11

Burton, Charles. "Howard Hughes and the 10,000 Foot Split-S." Air Classics, XVIII (August 1982), 14-25.

____. "XF-11: The Plane That Almost Killed Howard Hughes." Air Classics, XIX (April 1983), 62-71, 80.

Boyne, Walter J. "The Most Elusive Hughes: The Mystery D-2/F-11 Reconnaissance-Bomber." Wings, VII (June 1977), 44-55.

____. "The Phantom Eye." Tactical Air Reconnaissance Digest, IV (August 1970), 18-23.

"Hughes XF-11 to Test Photo Plane Worth." Aviation Week, XLVII (October 13, 1947), 23-24.

United States. Army Air Forces. Air Materiel Command. Hughes D-2, D-5, F-11: December 1939-August 1946, AMC Historical Study, no. 123. Wright Field, Ohio: USAAF Air Materiel Command, 1946. 74p.

(Republic) XF-12 Rainbow

Brabham, Lowery L. "Flying the XF-12 Rainbow." Air Classics, XI (November 1975), 26-29.
Reprinted in Michael O'Leary, ed. Air Progress Pilot Reports (Canoga Park, Calif.: Challenge Publications, 1979), pp. 80-85.

____. "XR-12 Rainbow." In: Robin Higham and Abigail T. Siddall, eds. Flying Combat Aircraft of the USAAF-USAF. Ames, Iowa: Iowa State University Press, 1975. pp. 154-159.

Hendrix, Lin. "The Republic XF-12 Rainbow." American Aviation Historical Society Journal, XXI (Winter 1976), 282-285.

McLarren, Robert. F-12 Based on Fighter Experience: Republic's Photo Plane." Aviation Week, XLVII (November 10, 1947), 28-30.

Martin, Robert E. "Republic's Rainbow." Air Classics, V (August 1969), 38-43.

(Northrop) XF-15 Reporter

Wright, Jay E. "The First and Last Northrop F-15A." American Aviation Historical Society Journal, XIV (Fall 1969), 186-188.

(Curtiss) XF15C-1

Andrews, Hal. "End of the Line: The Last Curtiss Navy Fighter." Air Enthusiast, III (March 1976), 98-109.

____. "Naval Aircraft: Curtiss XF15C-1." Naval Aviation News, (May 1978), 20-21.

XF-19 see F-117A Stealth

(McDonnell) XF-85 Goblin

"Parasite Jet Fighter: McDonnell XF-85." Mechanical Engineering, LXX (August 1948), 686-687.

Powers, Richard D. "Monstro and the Goblins." American Aviation Historical Society Journal, XVIII (Fall 1973), 146-159.

Smith, Richard K. "An Escort Appended: The Story of the McDonnell XF-85 Goblin." Flying Review International, XXII (December 1967), 1060-1062.

Sunday, Terry L. "Built-in Escort." Airpower, XV (January 1985), 10-21.

"XF-85: Bug on a Trapeze." Flying, XLIV (April 1949), 30-31+.

(Curtiss) XF-87 Blackhawk

"Curtiss XF-87 Blackhawk." Air Pictorial, XV (July 1953), 212-213.

Schultz, Helen. XF-87, 1945-1949. AMC Historical Study, no. 204. Wright-Patterson AFB, Ohio: USAF Air Materiel Command, 1950. 103p.

(McDonnell) XF-88 Voodoo

Anderton, David A. "Fighter Preview: McDonnell XF-88." Aviation Week, LIII (September 4, 1950), 27-29+.

Bristow, Dean. "The XF-88." Talon, III (January 1958), 12-13.

Gunston, William T. ("Bill"). "Experimental Voodoo." Aeroplane Monthly, X (July 1982), 354-357.

Moran, Sebastian. "Voodoo." Air Classics, XI (March 1975), 66-69.

(Lockheed) XF-90

Anderton, David A. "Lockheed XF-90 is Transonic Contender." Aviation Week, LIII (December 25, 1950), 27-29.

Bodie, Warren M. "Penetrator Fighter: Lockheed XF-90." Airpower, XI (March 1981), 20-37.

Kohn, Leo J. "XF-90: Lockheed's Penetration Fighter." Air Classics, X (June 1974), 16-18+.

"Lockheed XF-90 Penetration Fighter." Aero Digest, LIX (July 1949), 22+.

McLarren, Robert. "XF-90 Newest USAF Transonic Fighter." Aviation Week, L (May 23, 1949), 12-14.

"Strike Fighter: XF-90." Skyways, IX (May 1950), 14-15.

(Republic) XF-91

"Republic XF-91 Interceptor Flies on Rocket Power." US Air Services, XXXVIII (February 1953), 9-10.

Sunday, Terry L. "Thunderceptor." Airpower, XVI (September 1986), 40-52.

"Supersonic Rocket-Powered Flight Made by Republic XF-91." American Helicopter, XXIX (February 1953), 3-4.

(Consolidated) XF-92

Downie, D. "The Delta Wing." Skyways, XI (May 1952), 22-23, 55-56.

"Jet Interceptor: XF-92." Skyways, IX (March 1950), 14-15.

Morrow, Ardath. XF-92, 1945-1949. AMC Historical Study, no. 190. Wright-Patterson AFB, Ohio: USAF Air Materiel Command, 1949. 48p.

(Republic) XF-103

Cushman, Robert H. "XF-103 Demise Clouds Dual-Cycle Future." Aviation Week and Space Technology, LXVIII (September 9, 1957), 101+.

Holder, William G. "Lost Century Fighters." Air Classics, IX (July 1973), 46-53, 78-79, 82.

"The Lost Centurys." TAC Attack, XII (November 1972), 16-19.

Pace, Steve. "Supersonic Cavaliers." Airpower, XVI (November 1986), 24-43.

"Trisonic Titanium Republic." Air Enthusiast, VII (1978), 198-213.

XF-107 see YF-107

(Bell) XF-109

"Bell XF-109." The Aeroplane, LXXXVIII (January 21, 1955), 67-68.

Holder, William G. "Lost Century Fighters." Air Classics, IX (July 1973), 46-53, 78-79, 82.

Pace, Steve. "Supersonic Cavaliers." Airpower, XVI (November 1986), 24-43.

Roos, Frederick W. "Will the Real F-109 Please Stand Up." American Aviation Historical Society Journal, XXII (Spring 1977), 34-35.

XF-110 see F-4 Phantom II

(Hall) XFH-1

Neville, L. E. "The Hall Shipboard Fighter: Wasp-Powered XFH-1." Aviation, XXVII (November 30, 1929), 1067-1072.

(Bell) XFL-1 Airabonita
(see also P-39 Airacobra)

Andrews, Hal. "Naval Aircraft: Airabonita." Naval Aviation News, (December 1980), 24-25.

(Ryan) XFR-1 Fireball

Andrews, Hal. "Naval Aircraft: Fireball." Naval Aviation News, (July 1980), 20-21.

Gunston, William T. ("Bill"). "Fireball: Ryan's Prop/Jet Fighter." Aeroplane Monthly, III (August 1975), 370-375.

Hersey, Stephen. "Fireball Flight Demonstrations Officially Unveil Navy Jet Plane." Aviation News, IV (November 1, 1945), 8-9.

"Jet and Conventional Engines Combined in Ryan FR-1 Fireball." Aviation, XLIV (November 1945), 171+.

McDowell, Ernest R. "Ryan's Fireball." American Aviation Historical Society Journal, VIII (Winter 1963), 231-247.

Marlowe, Philip. "One Turning, One Burning: The Ryan Fireball." Air Classics, X (August 1974), 64-73.

"Ryan Fireball: The U.S. Navy's First Jet Fighter." Popular Science, CXLVIII (January 1946), 96-97.

Salmon, B. T. "Ryan FR-1 Fireball: Composite-Powered Aircraft." Aero Digest, LIV (March 1947), 66-68+.

Lockheed XFV/Convair XFY Pogoplane

Allen, Richard S. "Bolt Upright." Airpower, XVI (January 1986), 44+.

Baer, Harry S., Jr. "Five Firms Active on VTO Fighter Projects." American Aviation, XVII (March 1954), 13-14.

Casari, Robert. "Fighter Revolution." Flying, LV (August 1954), 32-33, 64.

Coughlin, William J. "Convair VTO Fighter Prepares for First Free-Flight Tests." Aviation Week, LX (June 21, 1954), 26-28.

____. "Navy Shows Off Newest Aircraft: Convair XFY-1." Aviation Week, LXI (November 15, 1954), 18-19.

"The First VTO." Joint Services Recognition Journal, X (April 1954), 86-87.
XFY-1.

"Hummingbird Fighters." Aircraft, XXXII (May 1954), 34-36.

"Navy Takes Wraps Off Convair and Lockheed Vertical-Takeoff Fighters." Aviation Week, LX (March 22, 1954), 14-18.

Ryan, Cornelius. "These Fighters Take Off Straight Up." Collier's, CXXXIII (April 2, 1954), 42-47.

Suggs, Dalton B. "Hanger Flying the Convair Pogo." Aviation Age, XXIV (November 1955), 150-155.

Taylor, John W. R. "America's VTO Fighters." Air Power, II (Autumn 1954), 3-10.

"Vertical Risers." Interavia, IX (May 1954), 295-297.

(Rockwell International) XFV-12

Covault, Craig. "XFV-12 Ground Tests 90% Completed." Aviation Week and Space Technology, CVII (September 12, 1977), 48-51.

Robinson, Clarence A., Jr. "XFV-12 May Spur Navy VTOL Family." Aviation Week and Space Technology, XCVIII (April 16, 1973), 12-17.

(Huff-Daland) XHB-1 Cyclops

Boyne, Walter J. "Huff-Daland in 3D." Wings, VI (December 1977), 55-63.

"The Huff-Daland Cyclops." Flight, XIX (December 22, 1927), 158+.

"Huff-Daland Cyclops Described." US Air Services, XI (October 1926), 41-43.

"Mystery Bomber: The Cyclops." Scientific American, CXXXIV (April 1926), 262+.

(Hughes) XHK Spruce Goose/Hercules

Book/Document/Report Literature:

Barton, Charles. Howard Hughes and His Flying Boat. Fallbrook, Calif.: Aero Publishers, 1982. 272p.

Hatfield, David. Howard Hughes' H-4 Hercules. Los Angeles, Calif.: JWC Publications, 1972. 104p.
Reprinted five years later in a 112-page edition by Northrop University, Inglewood, California.

McDonald, John J. Howard Hughes and the Spruce Goose. Blue Ridge Summit, Pa.: TAB, 1981. 160p.

Odekirk, Glenn E. Spruce Goose HK Hercules. Los Angeles, Calif.: Frank Alcanter, Inc., 1982. 126p.

Periodical/Anthology Literature:

Bangs, Scholer. "Hughes Answers Senate Probe With First Flying Boat Flight." Aviation Week, XLVII (November 10, 1947), 11-12.

Colby, Doug. "Spruce Goose: Genius Unveiled." Plane and Pilot, XVIII (December 1982), 14-18.

Dobson, Robert R. "First Eyewitness Report: Howard Hughes' Spruce Goose." Plane and Pilot, XV (December 1979), 40-45.

Foster, Mark S. "Flying Lumber Yard: Henry J. Kaiser, Howard Hughes, and the Famous Spruce Goose Caper." Aerospace Historian, XXXIII (June 1986), 96-103.

Gibbons, Robert H. "The Forgotton Behemoth." Air Classics, VIII (November 1972), 10-15.

"Goose on Display." Air Classics, XIX (November 1983), 33-35.

Liang, Gerald. "Home for the Hercules." Air Classics, XVIII (June 1982), 38-41, 80-81.

McIlvenna, Don E. "The Hughes Superplane and the Second Front Climax During World War II." Southern California Quarterly, (Winter 1975), 371-382.

Nixon, Stuart and John E. Terpatra. "Howard Hughes and His Great White Whale." Air Line Pilot, XLIX (July-August 1980), 6-13, 6-15.

Roberts, Lois J. "Howard Hughes' Efforts to Preserve the Spruce Goose." American Aviation Historical Society Journal, XXX (Winter 1985), 294-306.

Schnepf, Ed. "The Spruce Goose is Alive and Well." Air Progress, XXXIX (August 1977), 14-16.

Treadway, Bruce and Gerald Liang. "Hughes' Flying Boat." Air Classics, XVII (February 1981), 41-53.

(Douglas) XO2D-1

Karstens, Nicholas V. and Robert E. Williams. "The Douglas XO2D-1." American Aviation Historical Society Journal, XXIX (Fall 1984), 228-237.

(Curtiss) XO-18 Falcon

Bowers, Peter M. "The Life and Times of the Curtiss XO-18." American Aviation Historical Society Journal, XXXII (Spring 1987), 42-47.

(Consolidated) XP4Y-1 Corregidor

Hanson, Chuck. "The Consolidated Model 31/XP4Y-1 Corregidor Flying Boat." American Aviation Historical Society Journal, XXVII (Summer 1982), 136-147.

(Convair) XP5Y-1

"Convair XP5Y-1." Air Pictorial, XV (April 1953), 113-114.

(Curtiss) XP-6A Hawk

Jefferies, Walter M. "XP-6A." American Aviation Historical Society Journal, XIV (Fall 1969), 169-172.

(Martin) XP6M-1 Seamaster

"And Now--the Seamaster." Flight, LXVII (January 14, 1955), 36-37.

Andrews, Hal. "Naval Aircraft: Seamaster." Naval Aviation News, (June 1974), 22-23.

Decker, J. L. "Seamaster: Its Development and Some Considerations From the Accidents." Aeronautical Engineering Review, XVI (November 1957), 58-62.

Gunston, William T. ("Bill"). "Martin Seamaster: The Eventful Story of the U.S. Navy's Last Flying Boat." Aeroplane Monthly, XI (November 1974), 948-953.

"Martin XP6M-1 Seamaster is First Multi-Jet Attack Seaplane." Aero Digest, LXX (February 1955), 94+.

Murphy, William L. "The Seamaster Remembered." Naval Aviation News, (December 1981), 31-33.

Vicek, Adolph, Jr. "Producing the First Seamaster." Flight, LXIX (March 30, 1956), 350-352.

White, E. M. "Martin Seamaster." Air Pictorial, XVII (March 1955), 72-73.

(Curtiss) XP-21 Demon

"Lightweight Demon." RAF Flying Review, XXVII (August 1962), 36-37.

(Curtiss) XP-31

Boyne, Walter J. "Mini-Shrike." Airpower, IV (March 1974), 8-18.

(Bell) XP-39E

Hall, C. L. "Final Report on XP-39E Airplane." American Aviation Historical Society Journal, XXVII (Spring 1982), 54-63.

(Seversky) XP-41
(see also P-35 Guardsman)

"XP-41: Granddaddy of the P-47 Thunderbolt." Wings, XI (October 1981), 32-36.

(Curtiss) XP-42
(see also P-36 Hawk)

Beauchamp, Gerry. "XP-42: Perennial Test-Bed." American Aviation Historical Society Journal, XIX (Summer 1974), 82-92.

(Curtiss) XP-46 Hawk
(see also P-40 Warhawk)

Beauchamp, Gerry. "Plucked Hawks: The Curtiss XP-46 and XP-46A." American Aviation Historical Society Journal, XXVI (Spring 1981), 26-29.

(Lockheed) XP-49
(see also P-38 Lightning)

Condon, Jean. XP-49. ATSC Historical Study, no. 16. Wright Field, Ohio: USAAF Air Technical Service Command, 1944. 18p.

(Grumman) XP-50

O'Leary, Michael. "Grumman's XP-50." Air Classics, XXIV (June 1988), 14-19, 71-73.

XP-53 see XP-46 Hawk

(Vultee) XP-54

Gunston, William T. ("Bill"). "Prototype Pursuit: XP-54." Aeroplane Monthly, VII (October 1979), 550-553.

Matson, Kay, Amy C. Fenwick, and C. Gilbert. XP-54. ATSC Historical Study, no. 17. Wright Field, Ohio: USAAF Air Technical Service Command, 1944. 22p.

(Curtiss) XP-55 Ascender

Gunston, William T. ("Bill"). "Prototype Pursuit: XP-55 Ascender." Aeroplane Monthly, VII (November 1979), 580-583.

"XP-55, Ascender Fighter Plane." Aero Digest, XLVIII (March 15, 1945), 93-94.

(Northrop) XP-56

Gunston, William T. ("Bill"). "Prototype Pursuit: XP-56." Aeroplane Monthly, VII (December 1979), 636-638.

Sunday, Terry L. "Black Bullet." Wings, XVIII (December 1988), 10-19.

United States. Army Air Forces. Air Materiel Command. XP-56, July 1939-January 1946. AMC Historical Study, no. 121. Wright Field, Ohio: USAAF Air Materiel Command, 1946. 26p.

(Lockheed) XP-58 Chain Lightning

Bodie, Warren M. "Double or Nothing." Wings, X (June-July 1980), 18-23, 46, 8-21.

Gunston, William T. ("Bill"). "Prototype Pursuit: XP-58 Chain Lightning." Aeroplane Monthly, VIII (January 1980), 51-53.

Holder, William G. "The XP-58 Escort Lightning." TAC Attack, XIV (September 1974), 18-19.

Smith, P. B. "Final Report on the Procurement, Inspection, Testing, and Acceptance of Two Lockheed XP-58 Airplanes." American Aviation Historical Society Journal, XXVII (Summer 1982), 120-135.

United States. Army Air Forces. Air Technical Service Command. XP-58, February 1939-November 1944. ATSC Historical Study, no. 53. Wright Field, Ohio: USAAF Air Technical Service Command, 1945. 17p.

(Curtiss) XP-60

Bagwell, Margaret C. XP-60. ATSC Historical Study, no. 18. Wright Field, Ohio: USAAF Air Technical Service Command, 1944. 23p.

Wagner, Ray. "The Curtiss P-60 Series." Air Pictorial, XXIII (April 1961), 106-109.

(Curtiss) XP-62

Gunston, William T. ("Bill"). "Prototype Pursuit: XP-62." Aeroplane Monthly, VIII (February 1980), 78-80.

Porter, Eugene and C. Gilbert. XP-62. ATSC Historical Study, no. 19. Wright Field, Ohio: USAAF Air Technical Service Command, 1944. 18p.

XP-65 see XP-50

(McDonnell) XP-67 Bat

Gunston, William T. ("Bill"). "Prototype Pursuit: XP-67 Bat." Aeroplane Monthly, VIII (February 1980), 78-80.

James, Derek N. "McDonnell XP-67." Aircraft Illustrated, VIII (October 1975), 402-407.

McLaren, David R. "The XP-67 Bat." Air Classics, VIII (August 1972), 10-13.

United States. Army Air Forces. Air Materiel Command. XP-67, May 1940-January 1946. AMC Historical Study, no. 122. Wright Field, Ohio: USAAF Air Materiel Command, 1946. 25p.

(Curtiss) XP-71

Bagwell, Margaret C. XP-71. ATSC Historical Study, no. 20. Wright Field, Ohio: USAAF Air Technical Service Command, 1944. 23p.

(Republic) XP-72
(see also P-47 Thunderbolt)

Gunston, William T. ("Bill"). "Prototype Pursuit: XP-72." Aeroplane Monthly, VIII (March 1980), 134-135.

(General Motors) XP-75 Eagle

Fenwick, Amy C. XP-75, P-75. ATSC Historical Study, no. 21. Wright Field, Ohio: USAAF Air Technical Service Command, 1944. 47p.

____. XP-75, P-75, 1944-1947 (Supplement). AMC Historical Study, no. 176. Wright-Patterson AFB, Ohio: USAF Air Materiel Command, 1948. 47p.

Holder, William G. "Body by Fisher: The XP-75." TAC Attack, XIV (October 1974), 16-17.

Holley, Irving B., Jr. "A Detroit Dream of Massed-Produced Fighter Aircraft: The XP-75 Fiasco." Technology and Culture, XXVIII (July 1987), 578-639.

Schrader, Richard K. "The Fighter From General Motors." Air Progress Warbirds Illustrated, VIII (January-February 1989), 6-9.

Searles, Robert A. "Reflections: GM's 'Composite' Fighter." Business and Commercial Aviation, LVIII (January 1986), 30-35.

Trimble, Robert L. "America's Mystery Fighter." Air Classics, XVI (July 1980), 56-65.

XP-76 see XP-39E

(Bell) XP-77

Berliner, Don. "Bell XP-77." Model Aviation, VIII (July 1982), 41-44.

"Case History of the XP-77 Airplane Project." American Aviation Historical Society Journal, XXVI (Winter 1981), 311-313.

Colchogoff, George D. "Final Report on the XP-77 Airplane." American Aviation Historical Society Journal, XXVI (Winter 1981), 314-327.

Holder, William G. "Lightweight Fighter: World War II Style." TAC Attack, XV (May 1975), 26-27.

Matson, Kay, C. Gilbert, and Amy C. Fenwick. XP-77, 1941-1944. ATSC Historical Study, no. 22. Wright Field, Ohio: USAAF Air Technical Service Command, 1944. 20p.

O'Leary, Michael. "XP-77: First of the Lightweight Fighters." Air Classics, X (March 1974), 27-31.

XP-78 see P-51 Mustang

(Northrop) XP-79

"Flying Wind Chopper: XP-79." Science Illustrated, III (August 1940), 44-45.

(Convair) XP-81

Sunday, Terry L. "Convair's Flying Compromise." Wings, XVII (October 1987), 36+.

"XP-81, First Jet-Propjet Model." Aero Digest, LII (March 1946), 80-81.

(Bell) XP-83

Koehnen, Richard C. "Bell's No-Name Fighter." Airpower, XII (January 1982), 18-29.

(Curtiss) XP-87 Blackhawk

"AAF Unveils New XP-87 Fighter." Aviation Week, XLVII (September 15, 1947), 11-12.

Boyne, Walter J. "A Fighter for All-Weather: Curtiss XP-87 Blackhawk." Wings, V (February 1975), 30-40, 58.

(Boeing) XPBB-1 Sea Ranger

Bowers, Peter M. "The Lone Ranger." Wings, XIV (December 1984), 8-29.

(Sikorsky) XPBS-1

Sullivan, Dick and Jay Sullivan. "Forgotten Dreadnought: Sikorsky's XPBS-1 Flying Boat." Airpower, VII (January 1977), 50-61.

(Berliner-Joyce) XPS-1

"First Retractable Fighter." Air Classics, III (March 1967), 20-21+.

(Hall) XPTBH-2

"Aluminum Flash: Experimental Hall XPTBH-2." Air Classics, IV (April 1968), 48-53.

XR-12 see XF-12 Rainbow

(Lockheed) XRO-1

Andrews, Hal. "Naval Aircraft: XRO-1." Naval Aviation News, (July-August 1985), 16-17.

(Brewster) XSBA-1

Stark, Tom. "Brewster XSBA-1." Model Airplane News, XCVIII (February 1979), 25-27, 97.

(Douglas) XTB2D-1

Andrews, Hal. "Naval Aircraft: XTB2D-1." Naval Aviation News, (January 1982), 20-21.

(Grumman) XTB2F-1

Klein, Bernard C. F. "Grumman 55 (XTB2F-1)." World War II Journal, I (January 1989), 20-21.

(Vought) XTBU Seawolf

Andrews, Hal. "Naval Aircraft: XTBU." Naval Aviation News, (March 1980), 20-21.

(Vultee) YA-19

"Vultee Attack Bomber." Aero Digest, XXVIII (January 1936), 74-77.

"Vultee Attack Bomber." Aviation, XXXV (January 1936), 31-33; XXXVII (January 1938), 32-33, 71.

(Keystone) YB-7

Westburg, Peter W. "B-7: The Gull-Wing Bomber." Air Classics, XI (June 1975), 18-25, 66-73.

(Boeing) YB-9

Hucker, Robert. "Boeing's Death Angel." Air Classics, XVII (February 1981), 12-21, 78-80.

(Northrop) YB-49/XB-35 Flying Wing
(see also B-2 Stealth)

Book/Document/Report Literature:

Baker, Francis J. The Death of the Flying Wing: The Real Reason Behind the Cancellation of Northrop Aircraft's RB-49. PhD. Dissertation. Claremont, Calif.: Claremont Graduate School, 1984.

Coleman, Theodore, with Robert Wenham. Jack Northrop and the Flying Wing. New York: Paragon Press, 1988. 278p.

The Flying Wings of Northrop. Appleton, Wis.: Aviat Publications, 1975. 100p.

Lippisch, Alexander. The Delta Wing: History and Development. Ames, Iowa: Iowa State University Press, 1981. 126p.

Matson, Kay and Eugene Porter. XB-35, November 1941-July 1946. AMC Historical Study, no. 119. Wright Field, Ohio: USAAF Air Materiel Command, 1946. 30p.

Morrow, Ardath. YB-35, YB-49, May 1946-May 1949. AMC Historical Study, no. 200. Wright-Patterson AFB, Ohio: USAF Air Materiel Command, 1950. 69p.

Woodridge, E. T. Winged Wonders: The Story of the Flying Wing. Washington, D.C.: Smithsonian Institution Press, 1983. 230p.

Periodical/Anthology Literature:

Allen, Richard S. "Northrop's Flying Wing." Wings, XVI (February 1986), 36+.

Bangs, Scholer. "Design Details of the Northrop XB-35." Aviation, XLVI (April 1947), 37-41.

Biddle, W. "The Saga of the Flying Wing: The Man and the Machine." Discover, VII (February 1986), 24-25.

"Flying Wing: Special Section." Aviation, XLV (August 1946), 55-68.

Gunston, William T. ("Bill"). "All-Wing Northrops." Aeroplane Monthly, II (January-February 1974), 442-449, 482-489.

Haught, Dave. "Northrop XB-35." Model Aviation, XIII (August 1987), 64-66, 163-164, 170.

"Inside the Flying Wing." Popular Science, CL (January 1947), 95-97.

"Jet Wing Takes Off: YB-49." Life, XXIII (November 3, 1947), 30-39.

"Looking Into the XB-35." Popular Science, CL (May 1947), 116-117.

Northrop, Jack K. "Development of the All-Wing Aircraft." Royal Aeronautical Society Journal, LI (June 1947), 481-509.

____. "Northrop XB-35 Flying Wing Superbomber." Aviation, XLV (August 1946), 55-61.

"Northrop's Winged Wonders." Air Classics, III (May 1967), 60+.

Sears, William R. "Flying Wing Could Stealthily Reappear." Aerospace America, XXV (July 1987), 16-19.

Shloss, Leon W. and Alexander R. Boone. "100-Ton Flying Wing Outstrips B-29." Popular Science, CXLVIII (May 1946), 78-80.

Stanley, Max. "I Fly the Wing." Flying, XXXIX (September 1946), 34-35+.

Taylor, F. J. "Jack Northrop and His Flying Wing." Reader's Digest, L (March 1947), 29-32.

"10,000-Mile Flying Wing." Popular Mechanics, LXXXV (June 1946), 129-131.

"U.S. Shows Off New Superplane." Life, XXI (July 8, 1946), 23-25.

Walton, Russ. "Jack Northrop and His Wonderful Wing." Flying, XLII (February 1948), 19-21+.

(Convair) YB-60
(see also B-36 Peacemaker)

"New Big Boy: Convair YB-60." Aviation Week, LV (July 30, 1951), 16-17.

Schrader, Richard K. "Battle of the Giants." Air Classics, XXIV (March 1988), 14-19, 72.

(Boeing) YC-14

"Boeing Unveils First YC-14 Short Field Transport Aircraft." Defense Transportation Journal, XXXII (August 1976), 53-54.

Elson, Benjamin M. "USAF YC-14 Prototypes Begin New Flight Test Phase." Aviation Week and Space Technology, CV (November 29, 1976), 64-65.

Hadden, F. G. "YC-14B Prototype Testing." Society of Experimental Test Pilots Technical Review, XIII (Fall 1977), 128-136.

McPherson, R. L. "YC-14 Flight Test Program." Society of Experimental Test Pilots Technical Review, XIII (Fall 1977), 112-127.

North, David M. "YC-14 Designed to Cut Pilot Workload." Aviation Week and Space Technology, CVII (July 18, 1977), 41+.

O'Lone, Richard G. "First Flight Launches YC-14 Testing." Aviation Week and Space Technology, CV (August 16, 1976), 22-23.

"YC-14: All Blow and No Puff." Air International, XI (November 1976), 227-232.

(McDonnell-Douglas) YC-15

Allward, Maurice. "McDonnell-Douglas YC-15." Design Profile, no. 29. Air Pictorial, XXXVIII (December 1976), 484-487.

Chopping, Douglas. "Flying the YC-15." International Defence Review, X (August 1977), 672-675.

Field, Hugh. "YC-15 in the Air." Flight International, CXII (July 23, 1977), 269-272.

Lane, J. P. "YC-15 Development and Test Highlights." Society of Experimental Test Pilots Technical Review, XIII (Fall 1977), 85-111.

North, David M. "YC-15 Holds Few Transitional Problems." Aviation Week and Space Technology, CV (October 4, 1976), 40-43.

Schlitz, William P. "YC-15 STOL." Air Force Magazine, LVIII (October 1975), 16-23.

YC-34 see C-32

YC-108 see B-17 Flying Fortress

YC-116 see C-54 Skymaster

(Chase) YC-122 Cargo Glider/Avituc

Angstadt, I. R. "Is This the Prototype?" Aero Digest, LVIII (May 1949), 42-43+.

____. "Larger Avituc." Aero Digest, LIX (September 1949), 74+.

Smyser, A. E., Jr. "Cargo Plane Evolves From Glider." Aviation Week, XLVIII (January 5, 1948), 23-24+.

(Northrop) YC-125 Raider

O'Leary, Michael. "Trimotor Raider: The USAF's Most Unlikely Transport." Air Classics, X (July 1974), 67-73.

(Stroukoff) YC-134 Pantobase
(see also C-123 Provider)

Hawkes, Robert. "YC-134 Marks Change in Flow System." Aviation Week and Space Technology, LXV (December 24, 1956), 34+.

(Lockheed) YF-12/A-11 Blackhawk
(see also SR-71 Blackbird)

Book/Document/Report Literature:

James and Associates. YF-12 Experiments Symposium. NASA-CP-2054. Edwards, Calif.: NASA Dryden Research Center, 1978. 286p.

Tibbetts, Larry N. The F-12 Interceptor: Case Study of a Loser. Research Report. Maxwell AFB, Ala.: Air War College, Air University, 1968. 87p.

Periodical/Anthology Literature:

"America's Most Discussed Military Aircraft: The YF-12A." Interavia, XIX (December 1964), 1806-1807.

Bergman, Jules. "Our Watchdog on the Edge of Space: The YF-12A Interceptor." Reader's Digest, LXXXV (December 1964), 56-60.

Butz, J. S., Jr. "A-11: Born in the Skunk Works, Reared in Secret, It Blazes New Heights in Aircraft Performance." Air Force and Space Digest, XLVII (April 1964), 33-37.

"A Giant Step." Interceptor, VII (January 1965), 8-11.

Gunston, William T. ("Bill"). "A-11: Some Thoughts on a 2,000-mph Aircraft." Flight, LXXXV (March 12, 1964), 377-379.

Humphries, Orin L. "Blackhawk: America's Secret Guardian." Airpower, IX (May 1979), 10-29, 54.

Johnson, Clarence L. ("Kelly"). "Some Development Aspects of the YF-12A Interceptor Aircraft." Journal of Aircraft, VII (July-August 1970), 355-359.

"Johnson's Masterpiece: The YF-12A." Flying Review International, XX (January 1965), 16-21.

McDonald, Allen K. "The YF-12A Interceptor Weapon System." Air University Review, XVI (March-April 1965), 13-17.

Miller, Richmond L., Jr. "Flight Testing the F-12 Series Aircraft." Journal of Aircraft, XII (September 1975), 695-698.

"NASA and the YF-12." Interceptor, XII (April 1970), 20-23.

Plattner, C. M. "YF-12A Performance in F-108 Category." Aviation Week and Space Technology, LXXXI (October 19, 1964), 54-63.

Scholin, Allan R. "Fastest in the Air: USAF's YF-12A." Air Force and Space Digest International, I (July 1965), 36-37.

"Sheehan, Biran. "The Astounding YF-12A." Airman, IX (November 1965), 16-17.

Stone, Irving. "New Roles for the YF-12A." Air Force and Space Digest, LII (September 1969), 16+.

Witze, Claude. "Less Than the Whole Truth." Air Force and Space Digest, XLVII (April 1964), 16+.

(Northrop) YF-17 Cobra

Allward, Maurice. "Northrop's YF-17." Design Profile, no. 22. Air Pictorial, XXXVIII (April 1976), 129-134.

Fink, Donald E. "YF-17 Evolved From Previous Data Base." Aviation Week and Space Technology, C (April 15, 1974), 46-47+.

Geddes, J. Philip. "The Lightweight, High-Performance YF-17." Interavia, XXVIII (December 1973), 1315-1318.

____. "Northrop's YF-17 Makes Its Bow." Interavia, XXIX (June 1974), 538-539.

Godfrey, W. H. "Northrop's New Fighter Generation: YF-17." Air Enthusiast, VI (March 1974), 105-108, 146.

"New Fighter for Europe?" Air International, XIX (September 1980), 126-130, 143.

"Northrop YF-17 Technology-Demonstrator Aircraft." Aircraft Illustrated, VII (July 1974), 268-270.

Ulsamer, Edgar. "The Lightweight Fighter Halts the Cost Spiral." Air Force Magazine, LVI (October 1973), 64-67.

"YF-17 Makes 61-Minute Maiden Flight." Aviation Week and Space Technology, C (June 17, 1974), 20-21.

(Lockheed) YF-22/(Northrop) YF-23 Advanced Tactical Fighter

Book/Document/Report Literature:

Braybrook, Roy M. Supersonic Fighter Developments. San Leandro, Calif.: Haynes Publications, 1987. 225p.

Sweetman, Bill. Advanced Fighter Technology. Osceola, Wis.: Motorbooks International, 1987. 150p.

United States. Government Accounting Office. National Security and International Affairs Division. Case Study of the Air Force Advanced Tactical Fighter Program. Washington, D.C., 1986. 16p.

Periodical/Anthology Literature:

"Advanced Fighter Technology." Aviation Week and Space Technology, CXXIV (June 23, 1986), 47-175.

"The Advanced Tactical Fighter." Defense Science and Electronics, IV (May 1985), 14-16+.

"ATF Schemes Cut to Two." Machine Design, LIX (January 8, 1987), 12+.

Braybrook, Roy M. "The Shape of Things to Come." Air International, XXVII (September 1984), 143-147, 149.

Brown, S. F. "21st Century Superfighters." Popular Science, CCXXIX (October 1986), 80-86+.

Canan, James W. "ATF: Hot and Stealthy." Air Force Magazine, LXX (April 1987), 59-63+.

Dornheim, Michael A. "Lockheed Devoting Key Resources to Advanced Tactical Fighter." Aviation Week and Space Technology, CXXIII (July 22, 1985), 143-151.

Dudney, Robert S. "The ATF and Its Friends." Air Force Magazine, LXXII (January 1989), 46-53.

Hoey, Brian. "Fighters of Tomorrow." Airman, XXIX (August 1985), 16-20.

Kitfield, James. "ATF: Planning by New Rules." Military Logistics Forum, III (September 1986), 13-14+.

Lake, Julian S. "ATF: Out of the Shadows." Defense Science, VII (May 1988), 35-36+.

Moxon, Julian and Graham Warwick. "Advanced Tactical Fighter." Flight International, CXXVI (October 20, 1984), 1048-1049, 1055-1056, 1060-1061.

Pace, Steve. "Fighter on the Fringe." Wings, XVII (April 1987), 22-37.

Piccirillo, Albert C. "Advanced Tactical Fighter: Design Goals and Technical Challenges." Aerospace America, XXII (November 1984), 74-79.

Rawles, James W. "ATF Program Faces Tough Challenges." Defense Electronics, XIX (September 1987), 61-62+.

Sweetman, Bill. "Advanced Tactical Fighter: Holding for Launch." Interavia, XL (June 1985), 605-607.

____. "A New Approach for USAF's ATF." Interavia, XLII (January 1987), 49-51.

Welling, William. "ATF: A Look at Expected Innovations." Military Technology, XII (July 1988), 85-86+.

YF-93/YF-94 see F-86 Sabre Jet

(North American) YF-107

Allen, Francis J. "Forgotten Fighter." Air Classics, XX (April 1984), 34-39, 78.

Holder, William G. "Lost Century Fighters." Air Classics, IX (July 1973), 46-53, 78-79, 82.

"The Lost Centurys." TAC Attack, XII (November 1972), 16-19.

Mizcroft, Baker. "The Super Super Sabre." Air Classics, X (August 1974), 30-39, 54-57.

Pace, Steve. "Supersonic Cavaliers." Airpower, XVI (November 1986), 24-43.

(Bell) YFM Airacuda

"Bell XFM-1: New Ship of Radical Design for the Air Corps." Aviation, XXXVI (September 1937), 49-50.

"A Classic Revisited." Wings, VII (June 1977), 28-35.

Hucker, Robert. "Bell's Radical Bomber-Destroyer." Air Classics, XVI (April 1980), 26-28, 78-79.

Schrader, Richard K. "Airacuda." Air Classics, XXIV (February 1988), 46-47, 82.

Thomas, Merle. "Airacuda: The Killer That Never Was." Air Classics, II (August-September 1965), 18-21+.

"Twin-Engined Bell XFM-1 Army Fighter." Aero Digest, XXXI (August 1937), 25-26.

(Douglas) YO-31C

Westburg, Peter W. "Douglas Observation Monoplanes: The YO-31C." Air Classics, XII (December 1976), 18-25, 64-67.

(Curtiss) YP-23

Westburg, Peter W. "The Last Curtiss Biplane Fighter." Air Classics, XII (February 1976), 56-73.

(Curtiss) YP-37 Hawk
(see also P-36 Hawk)

Beauchamp, Gerry. "Futuristic Hawks: A Story About the Curtiss YP-37 Aircraft." American Aviation Historical Society Journal, XXII (Spring 1977), 12-25.

O'Leary, Michael. "Variation on a Theme." Air Classics, XXII (December 1986), 62-65.

(North American) YP-64
(see also T-6 Texan)

Gann, Harry K. "The North American P-64." American Aviation Historical Society Journal, XXIII (Fall 1978), 296-299.

Petersen, Norm. "The P-64." Sport Aviation, XXXI (February 1982), 34-38.

(Waco) YPF-7

Smith, Gene. "Tailspin Tommy's Aircraft." Air Progress, XLVIII (November 1986), 38+.

(Waco) YPT-14

"Berliner, Don. "Waco 'D.'" Model Aviation, XIII (May 1987), 58-64, 176.

Jefferies, Walter M. "Waco's Military 'Ds.'" American Aviation Historical Society Journal, IX (Summer 1964), 134-135.

"Waco YPT-14 Primary Trainer." Aero Digest, XXXVII (August 1940), 91-92.

(Ryan) YPT-25

"New Ryan PT-25 Trainer Built of Non-Strategic Materials." Aero Digest, XLI (October 1942), 201-202.

III. Rotary-Wing Aircraft

BAPTIZED in the fire of the CBI late in World War II, American rotary-wing aircraft began to come of age during the Korean Conflict and matured in the four decades since. This section examines certain of the literature as it relates to rotary-wing aircraft (generally meaning helicopters, with a nod toward the autogiro), but unlike the last part, is not subdivided by type. Aircraft here are, however, also ordered not by manufacturer, but by numerical designation(s). Although this arrangement is somewhat non-traditional/non-chronological, it offers the advantage of a single order and thus quick access. Here again, details on specific electronic suits/maintenance packages/armaments specifically designed for a given aircraft are entered with that machine.

(Bell) AH-1 Cobra/Sea Cobra

Book/Document/Report Literature:

Bell Helicopter Textron. Public Relations Department. AH-1W Super Cobra. Fort Worth, Tex., 1986. 6p.

____. The Modernized AH-1S. Fort Worth, Tex., 1983. 21p.

Peoples, Kenneth. Bell AH-1 Cobra Variants. Datagraph, no. 4. Arlington, Tex.: Aerofax, Inc., 1988. 64p.

Richardson, Doug. AH-1 Cobra. Modern Fighting Aircraft, no. 13. New York: Arco, 1987. 63p.

United States. Army. Headquarters. The Attack Helicopter Team Handbook. FM 17-50-1. Washington, D.C.: GPO, 1983. 88p.

Periodical/Anthology Literature:

"AH-1G in Vietnam by Unit, January 1, 1972." American Aviation Historical Society Journal, XX (Fall 1975), 168-169.

"The AH-1J Sea Cobra." Approach, XVII (February 1972), 10-13.

Allen, Travis M. "Toprotor." Marine Corps Gazette, LXXII (January 1988), 29-30.

Andrews, Hal. "Naval Aircraft: Cobra." Naval Aviation News, (April 1979), 20-21.

Brown, David A. Hueycobra Offers High Maneuverability." Aviation Week and Space Technology, LXXXIV (June 13, 1966), 54-55+.

"Cobra: More Dangerous Than Ever." Defence Update International, no. 35 (1983), 47-53.
AH-1S.

"Cobra: NATO's Anti-Tank Spearhead." Flight International, CXIII (June 13, 1978), 1678-1680.

Cohill, George L. "Modernized Cobra Fire Control System." U.S. Army Aviation Digest, XXV (October 1979), 5-13.

Cook, John C., Jr. "The Attack Helicopter." U.S. Naval Institute Proceedings, XCVIII (October 1972), 97-98.
AH-1J.

Drenz, Charles F. "The Cobra/TOW Helicopter." National Defense, LX (November-December 1975), 202-203.

Edwards, Richard T. "The Tank Killers." Army, XXIX (February 1979), 28-31.

"Fangs for the Cobra." Air Classics, VIII (January 1972), 56-60, 64.

Funk, David L. "Let's Do Something With the Cobra." Armor, LXXXIV (July-August 1975), 16-21.

Gavin, Franc. "Fanged Cobra Gunship." International Combat Arms, IV (May 1986), 62-66.

Harrison, Alan L. "AH-1G Transition." U.S. Army Aviation Digest, XVI (February 1970), 7-9.

Harvey, David. "Bell AH-1W: 'Off the Shelf' R & D Builds a Super Cobra." Rotor and Wing International, XX (June 1986), 48-50.

_____. "Super Cobra Fills the Gap for the U.S. Marines." Rotor and Wing International, XVIII (December 1984), 48-53.
AH-1T.

"Hueycobra Attacks Viet Cong Targets: Photographs." Aviation Week and Space Technology, LXXXVIII (May 20, 1968), 64-67.

Humston, D. E. "Cobra Intercept." U.S. Naval Institute Proceedings, CXIV (January 1988), 97-98.

Kalista, C. J. "The Huey Cobra Joins the Army." U.S. Army Aviation Digest, XV (April 1966), 13-20.

Klass, Philip J. "Army Tests Nocturnal Hueycobra." Aviation Week and Space Technology, XCIII (October 5, 1970), 57-59.

Lambert, Mark. "The Bell AH-1G Hueycobra." Flight International, XCII (July 27, 1967), 154-156.

_____. "The U.S. Army's Cobra Companies." International Defence Review, XIV (September 1981), 1179-1181.

McDouglal, Winn. "Aerial Field Artillery for the Corps." Field Artillery Journal, XLII (March-April 1974), 54-58.

Meller, R. "The TOW/Cobra Antitank Helicopter: More Mobile Firepower for the U.S. Army in Europe." International Defence Review, VIII (December 1975), 882-886.

Mullen, Jack L. "Training the Cobra Pilot/Gunner Team." U.S. Army Aviation Digest, XXI (November 1975), 1-4.

Negrette, Arthur J. "The AH-1W: Tougher Cobra With a Deadlier Strike." Rotor and Wing International, XXII (September 1988), 44-49.

O'Grady, George. "AH-1S Cobra." Aerophile, II (October 1979), 22-35.

Reisch, M. C. "Marine AH-1T: Seven Tons of Leatherneck Fury." Defense Systems Review and Military Communications, II (September 1984), 45+.

Ropelewski, Robert R. "Marine/Bell AH-1W Demonstrates Improved Performance." Aviation Week and Space Technology, CXXV (August 25, 1986), 52-53+.

St. Louis, Robert P. "Modernized Cobra." U.S. Army Aviation Digest, XXIV (January-March 1978), 2-7, 19-25, 8-13.

_____. "Modernized Cobra Update." U.S. Army Aviation Digest, XXV (October 1979), 24-25.

Sprinkle, James D. "The Hueycobra in Vietnam." American Aviation Historical Society Journal, XX (Fall 1975), 162-170.

_____. "The Hueycobra: Its Origins." American Aviation Historical Society Journal, XX (Spring 1975), 30-36.

Steenblik, Jan and Manny Robio. "Cobras in the Treetops." Boy's Life, LXX (June 1980), 20-23.

Stoops, C. B. "New Punch with TOW Cobras." Marine Corps Gazette, LXIV (December 1980), 21-24.

Zambone, Joe. "Cobra in the Combined Arms Team." National Guard, XXXIX (September 1985), 14-16+.

(Bell) UH-1 Iroquois/Huey

Book/Document/Report Literature:

Bell Helicopter Textron. Public Relations Department. History of the UH-1 Huey Series. Fort Worth, Tex., 1987. 6p.

Chant, Christopher F. Bell UH-1. Super Profile Series. Newbury Park, Calif.: Interbooks, Inc., 1985. 56p.

Drendel, Lou. Huey. Carrollton, Tex.: Squadron/Signal Publications, 1983. 64p.

Mutza, Wayne. UH-1 Huey in Action. Carrollton, Tex.: Squadron/Signal Publications, 1986. 50p.

Scutts, Jerry. UH-1 Iroquois/AH-1 Hueycobra. Osceola, Wis.: Motorbooks International, 1984. 112p.

Periodial/Anthology Literature:

Andrews, Hal. "Naval Aircraft: Huey." Naval Aviation News, (February 1980), 20-21.

"The Bell UH-1D Transport." Interavia, XVIII (September 1963), 1331-1334.

Bruggink, Gerard. "Ditching the Huey." U.S. Army Aviation Digest, XIII (May 1967), 36-46.

Craig, Constant P. "Requiem for a Lightweight." Marine Corps Gazette, LXXI (October 1987), 48-51.
UH-1N.

Eppinger, Dale L. "Be a Hover-Lover." Combat Crew, XXII (January 1971), 4-7.

"Getting to Know the Iroquois." U.S. Army Aviation Digest, VI (June 1960), 25-27.

Grady, Nanse. "It Looks Like a Huey." U.S. Army Aviation Digest, XXIV (December 1978), 28-29.

Greene, John H. "Peaceful Injun." U.S. Army Aviation Digest, VIII (November 1962), 31-33.

"HU-1A Iroquois Designed for Frontline Service." U.S. Army Aviation Digest, V (June 1959), 11-15.

Jensen, David. "UH-1H Upgrades: Can Huey's Be Made to Fly Forever?" Rotor and Wing International, XXII (July 1988), 32-33.

Kelly, Mike. "Hueys in Vietnam." U.S. Army Aviation Digest, XXVII (June 1981), 10-14.

Krisman, Michael J. "'Can Do' Helicopter." Army Digest, XXI (July 1966), 12-16.

Lauterbach, John W. "Introducing the Twin Huey." U.S. Army Aviation Digest, XVI (August 1970), 22-27.
UH-1N.

Merryman, James N. "Aerial Artillery: UH-1B/XM-3." Field Artilleryman, XXVI (April 1963), 2-5.

"New Metal Birds: The Bell Company's HU-1A, the World's First Offensive Helicopter." Newsweek, LX (October 29, 1962), 37-38.

Nolan, Charles and Henry Roll. "Do You Really Know the UH-1?" U.S. Army Aviation Digest, XV (July 1969), 18-20.

Scholin, Allan R. "U.S. Army Pilots Flying the Bell UH-1D Iroquois Have Set Ten World Helicopter Records." Air Force and Space Digest International, I (January 1965), 50+.

Scoggin, Vernon C. "Expanding Huey's Mission." Marine Corps Gazette, LXXII (April 1988), 58-59.

Skarich, Anton J. "Triple-Nickel Winds at the World." U.S. Army Aviation Digest, XVI (January 1970), 2-5.
UH-1N.

Smith, Elton J. "Know Your Iroquois." U.S. Army Aviation Digest, VIII (April 1962), 8-10.

Stanfield, Robert I. "Minimum Effort Required to Pilot Bell's UH-1 Iroquois." Aviation Week and Space Technology, LXIX (December 8, 1958), 76-77+.

Tragesser, John N. "Utility Tactical Transport, the Armed UH-1B." Field Artilleryman, XXX (June 1964), 3-7.

Vogels, David. "'Grandma,' the Vietnam Veteran Flies Again." U.S.Army Aviation Digest, XXVI (January 1980), 20-21.

HNS-1 see R/H-4/HNS-1

HOR-1 see H/HH/OH/UH-43

HRP-1 see H/CH/HH/SH-21

(Bell) HSL-1

Andrews, Hal. "Naval Aircraft: HSL-1." Naval Aviation News, (September 1980), 20-21.

HTE-1 see H/HTE-1/OH-23

HO2S see R/H-5/HO2S/HO3S

HSS-2 see H/CH/HH/RH/SH/UH-3/HSS-2

(Kaman) SH/UH-2 Seasprite

Andrews, Hal. "Naval Aircraft: Seasprite." Naval Aviation News, (June 1971), 20-21; (February 1983), 24-25.

Brown, David A. "Kaman UH-2A Demonstrates All-Weather Capability, Maneuverability for Navy Roles." Aviation Week and Space Technology, LXXVIII (April 8, 1963), 80-81+.

Galdorisi, George. "SH-2F LAMPS Weapon System Trainer: The Tactical Wave of the Future." U.S. Naval Institute Proceedings, CVI (March 1980), 129-131.

Hammond, R. E. and Pat Tierney. "The LAMPShip Team." U.S. Naval Institute Proceedings, CIV (March 1978), 154-158.

Myer, Robert J. "UH-2 Open Sea Salvage." Approach, XIV (November 1968), 14-16.

"Seasprite Joins the Fleet." Ryan Reporter, XXIV (January-February 1963), 18-21.

"Seasprite Revived." Naval Aviation News, (February 1983), 22-23.

(Sikorsky) H/CH/HH/RH/SH/VH-3 and HSS-2 Sea King/Jolly Green Giant

Book/Document/Report Literature:

Chant, Christopher F. Sea King. Super Profile Series. Newbury Park, Calif.: Interbook, Inc., 1986. 56p.

Periodical/Anthology Literature:

Andrews, Hal. "Naval Aircraft: Sea King." Naval Aviation News, (July 1972), 20-21.

Bader, John T. "SH-3H: Sub Killer." Naval Aviation News, (November 1982), 8-12.

Bass, A. C. "Pilot Report: Sikorsky HSS-2." Flying, LXVIII (February 1961), 34-35+.

Champlin, G. F. "Antisubmarine Sikorsky." American Helicopter, LII (September 1958), 9+.

"Coming Cargo Copter: The CH-3C." MATS Flyer, X (April 1963), 4-7.

Damonte, Carlton R. "Caring for 'Big Charlie.'" Aerospace Maintenance Safety, XX (September 1965), 6-10.

Jackson, William G. "Army One: On the Presidential Run." Army Digest, XXV (June 1970), 13-15.
VH-3.

Little, Patrick J. "New SAR Sea Kings." Aircraft Illustrated, XII (March 1979), 114-118.

Lutton, Thomas C. "SAR: A New Dimension." U.S. Naval Institute Proceedings, XCVI (February 1970), 114-115.

McClung, Frank. "Commanders of the Copter Crossing." MAC Flyer, XIX (December 1967), 9-11.
June 1, 1967 transatlantic flight of two USAF HH-3Es.

Mills, Richard L. "Flying the CH-3C." MATS Flyer, XII (June 1965), 23+.

Murray, Michael K. "Sea King: ASW Workhorse." U.S. Naval Institute Proceedings, CVII (June 1981), 99-100.

Powers, Dave. "The Ritual of the Sub Hunt." Flying, CXIV (April 1987), 56-58+.

Rawlings, John D. R. "The Jolly Green Giant." Airfix Magazine, XXII (March 1981), 304-307.

"The Sea King Antisubmarine Helicopter." Interavia, XXIV (January 1969), 70-72.

"Sitka HH-3F Crews Hone New SAR Skills." Rotor and Wing International, XXIII (February 1989), 59-60.

Stevens, Gary L. "The Transition From C-141s to Helicopters." Aerospace Historian, XXIV (June 1977), 70-77.

"A Tri-Service Helicopter." Interavia, XIX (April 1964), 511-513.

"USAF HH-3s Make Record Helicopter Atlantic Flight." Aviation Week and Space Technology, LXXXVII (August 14, 1967), 36-38.

York, Alan D. "Unknown Quantity." MAC Flyer, XXVII (August 1980), 19-21.

HO3S see R/H-5/HO2S/HO3S

(Sikorsky) R/H-4 and HNS-1

Carter, Rowland. "Helicopter for War." Flying, XXXII (March 1944), 38+.

Foley, Dennis L. "Ode to the R-4." Canadian Aviation Historical Society Journal, XVIII (Winter 1980), 112-115, 124.

Glines, Carroll V. "The Skyhook." Air Force Magazine, LXXI (July 1988), 104-113.

Schrader, Richard K. "Sikorsky's World War II Helos." Air Classics, XXIII (May 1987), 16-19.

OH-4 see OH-58 Kiowa

(Sikorsky) R/H-5 and HO2S/HO3S Dragonfly

Andrews, Hal. "Naval Aircraft: HO3S." Naval Aviation News, (January 1977), 20-21.

Kaplan, Howard M. "Flight of the Dragonfly." Army, XXXVII (February 1987), 41-43.

(Hughes/McDonnell-Douglas) OH-6 Cayuse

Book/Document/Report Literature:

Porter, Donald J. The McDonnell-Douglas OH-6A Helicopter. Aero Series, v. 38. Blue Ridge Summit, Pa.: TAB, 1989. 96p.

Periodical/Anthology Literature:

"Army Guard Builds, Tests a Modified OH-6A." Rotor and Wing International, XXII (December 1988), 67-68.

Blodget, Robert. "Hughes 500: Light, Turbine-Powered Helicopter." Flying, LXXX (February 1967), 32-33+.

Green, David. "Pilot Report: A Tale About Flying the NOTAR." Rotor and Wing International, XXI (August 1987), 38-41.

"Hughes OH-6A." Air Pictorial, XXVII (December 1965), 425-427.

"Hughes OH-6A." Flight, LXXXIX (February 24, 1966), 316-318.

Kernstock, Nicholas C. "NOTAR Reduces Pilot Workload, Improves Response in OH-6A." Aviation Week and Space Technology, CXXX (February 13, 1989), 44-47.

Merritt, Donald E. S. "OH-6A Cayuse Helicopter: A Proven Battlefield Warrior." National Guard, XXXVIII (November 1984), 28-30.

Scholin, Allan R. "23 World Records for Rotary-Winged Aircraft Have Been Claimed by the U.S. Army's Hughes OH-6A Light Observation Helicopter." Air Force and Space Digest International, II (June 1966), 44-45.

Smith, Barry D. "How Helicopters Shake the Bushes for Illegal Aliens." Rotor and Wing International, XXIII (February 1989), 38-40.

Weeghman, Richard B. "Hughes' Utility Machine." Flying, XCVI (June 1975), 105-108.

UH-12B see H/HO-23/HTE-1 Raven

(Bell) H/UH/OH-13 and HTL Sioux

Baugh, Wilford A., Jr. "Night Flying Hints for the Sioux." U.S. Army Aviation Digest, V (April 1959), 24+.

Bulban, Erwin J. "Bell Demonstrates H-13 Tactical Uses." Aviation Week and Space Technology, LXXVI (April 9, 1962), 111-113.

Christian, William P. "History of OH-13 Mishaps." U.S. Army Aviation Digest, XVI (May 1970), 52-55.

Franklin, Robert B., Jr. "Turbocharger OH-13S." U.S. Army Aviation Digest, XV (April 1969), 22-23.

Gardner, William H. "Utilization of the TH-13." U.S. Army Aviation Digest, XV (January 1969), 4-5.

Thompson, David E. "The Sioux Signs Off." Air Pictorial, XL (May 1978), 175+.

Weichael, Henry, Jr. "H-13 Flight Rules of Thumb." U.S. Army Aviation Digest, I (April 1955), 28-31.

(Sikorsky) H/HH/CH/UH-19 Chickasaw

Benham, E. M. "Across by Helicopter." BeeHive, XXVII (Fall 1952), 23-27.

Colucci, Frank. "S-55: Chopper Extraordinary." Wings, XI (April 1981), 42-51.

"Sikorsky H-19." MAC Flyer, XXVIII (March 1981), 16-17.

"Sikorsky's in the Army Now." BeeHive, XXVIII (Fall 1952), 16-17.

(Boeing-Vertol) H/CH/HH/SH-21 and HRP-1 Shawnee/Workhorse

Boeing Vertol Company. Public Relations Department. CH-21 Series. Philadelphia, Pa., 1987. 3p.

____.____. HRP Series. Philadelphia, Pa., 1987. 3p.

Coleman, Frank. "Vertol H-21C Sets New Distance Record." American Helicopter, XLIV (September 1956), 12+.

McMahon, F. K. "Army Workhorse." U.S. Army Aviation Digest, I (September 1955), 5-14.

Spencer, Edward A. "Skis for the Shawnee." U.S. Army Aviation Digest, VI (November 1960), 27-30.

Treadway, Robert N. "The H-21C in Artillery Employment." Field Artilleryman, XXIII (July 1962), 42-48.

Vance, W. E. "Coast-to-Coast in an H-21." U.S. Army Aviation Digest, II (October 1956), 10-14.

(Hiller) H/CH/HO-23 and HTE-1 Raven

Fricker, John. "Flying the Hiller 12B." The Aeroplane, XC (April 13, 1956), 252-255.

Lambert, Mark. "Hiller 12E in the Air." Flight, LXXVIII (October 14, 1960), 616-617.

(Boeing-Vertol) H/UH-25 and HUP-1 Retriever

Boeing-Vertol Company. Public Relations Department. HUP/UH-25 Retriever Series. Philadelphia, Pa., 1987. 3p.

Thoreson, Richard M. "The Deuce." Naval Aviation Museum Foundation, V (Fall 1985), 43-46.

(Sikorsky) H/SH/UH/VH-34 and HUS-1 Chocktaw/Seahorse

Anderton, David A. "Service Experience Leads to T-58 Changes." Aviation Week and Space Technology, LXXVIII (April 29, 1963), 48-49+.

Andrews, Hal. "Naval Aircraft: H-34." Naval Aviation News, (September-October 1986), 16-17.

Cassidy, D. F. "Mountain Operations With the H-34." U.S. Army Aviation Digest, III (March 1957), 5-9.

Nelson, Herbert A. "Bigger Payloads for the HUS." Marine Corps Gazette, XLV (April 1961), 52-56.

"Sikorsky S-58." Flight, LXVIII (November 11, 1955), 737-738.

(Cierva-Pitcairn) PA-36 Autogiro

Book/Document/Report Literature:

Autogiro Company of America. The Autogiro. Philadelphia, Pa.: Pitcairn-Cierva Autogiro Company of America, 1930. Unpaged.

Brooks, Peter W. Cierva Autogiros: The Development of Rotary-Wing Flight. Washington, D.C.: Smithsonian Institution Press, 1988. 416p.

De La Cierva, Juan, with Don Rose. Wings of Tomorrow: The Story of the Autogiro. New York: Brewer, Warren and Putnam, 1931. 300p.

Townson, George. Autogiro: The Story of the Windmill Plane. Blue Ridge Summit, Pa.: TAB, 1985. 225p.

United States. Congress. House. Committee on Military Affairs: Development of the Autogiro and Rotary-Wing Aircraft: Hearings. 75th Cong., 3rd sess. Washington, D.C.: GPO, 1938. 104p.

Periodical/Anthology Literature:

Ely, Louis B. "The Autogiro: An Observation Post." Field Artillery Journal, XXIX (May 1939), 202-214.

Gunther, Carl. "The First Whirly Aboard." The Hook, XIII (Fall 1985), 10-13.

____. "Pitcairn's Marvelous Autogiro." Private Pilot, XV (September 1980), 38-45.

Hawkes, Frank M. "Flying the Autogiro." U.S. Air Services, XV (December 1930), 36-37.

Kingston-Smith, Frank. "Mr. Pitcairn's Autogiros." Airpower, XIII (March 1983), 28+.

LaPage, W. L. "The Autogiro Analyzed." Society of Automotive Engineers Journal, XXVII (September 1930), 257-262.

Lewis, H. Latane, II. "The Military Value of Autogiros." U.S. Air Services, XVII (October 1932), 36-37.

McNair, Leslie J. "And the New Autogiro." Field Artillery Journal, XXVII (January 1937), 5-17.

Nelson, William. "The Autogiro as a Military Craft." U.S. Naval Institute Proceedings, CXLI (October 1929), 290-292, 336.

____. "The Pitcairn Autogiro." Aviation, XXVI (June 22, 1929), 2187-2189.

Wetherill, Richard W. "The Autogiro Goes to Work." National Aeronautic Magazine, X (May 1932), 6-12.

(Sikorsky) H/CH-37 and HR2S-1 Mojave

Andrews, Hal. "Naval Aircraft: HR2S-1/H-37." Naval Aviation News, (January-February 1988), 18-19.

Elzy, Arthur F. "Sikorsky HR2S-1 Transport Helicopter." Fortitudine: Bulletin of the Marine Corps Historical Program, XVIII (Fall 1988), 19-20.

Wicker, Rush R. "CH-37 Mojave--Workhorse of Korea." U.S. Army Aviation Digest, XIV (July 1968), 32-34.

Witze, Claude. "Marines Put HR2S in Operational Status." Aviation Week and Space Technology, LXVI (April 8, 1957), 62-63+.

(Kaman) H/HH/OH/UH-43 and HOK-1 Huskie

Bass, A. C. "Pilot Report: Kaman's H-43B Huskie--USAF's Crash-Rescue Helicopter." Flying, LXIX (July 1961), 30-31+.

Bulban, Erwin J. "H-43B Tailored to USAF Rescue Concept." Aviation Week and Space Technology, LXIX (December 22, 1958), 71+.

Coughlin, Robert D. "H-43B Huskie." Interceptor, II (July 1960), 8-11.

Gardner, Bob. "Here's What's New in Rescue." MAC Flyer, XX (September 1973), 16-18.

Kirkland, R. G. "Flying the H-43B Kaman Huskie." MATS Flyer, IX (February 1962), 6-8.

Newton, W. A. "H-43B Test Pilot Report." MATS Flyer, IX (June 1962), 14-15.

____. "The Lifesaver." Combat Crew, XII (September 1961), 8-11.

Searles, Robert A. "Reflections: The Helicopter Gets Jet Power." Business and Commercial Aviation, LXII (February 1988), 111-112.

Sheehan, Brian. "Rescue's New Look." Airman, III (September 1959), 7-8.

Shershun, Carroll S. "Never Fear, Pedro's Here." Airman, XIII (August 1969), 10-13.

Stanfield, Robert I. "H-43A Possesses Stick and Rudder Qualities." Aviation Week and Space Technology, LXX (February 23, 1959), 231-233+.

(Boeing-Vertol) H/CH/HH/RH/UH-46 Sea Knight

Book/Document/Report Literature:

Boeing-Vertol Company. Public Relations Department. Helicopter Flying Qualities-Characteristics: CH-46E. 4 vols. Warminster, Pa.: Naval Air Development Center, 1983.

Periodical/Anthology Literature:

Andrews, Hal. "Naval Aircraft: Sea Knight." Naval Aviation News, (February 1978), 20-21.

Bright, Jeff. "500 Feet and Zero Knots." Approach, XXXIII (December 1987), 12-15.

Gregory, William H. "Vertol Flies CH-46A Assault Helicopter." Aviation Week and Space Technology, LXXVII (October 22, 1962), 30-32.

Harvey, David. "NARF: Resolving the Sea Knight's Midlife Crisis." Rotor and Wing International, XX (December 1986), 44-45.

Hooper, W. Ewan. "Heavyweights From Vertol." Flight International, XCIV (December 12, 1968), 975-976.

"Marine CH-46A Designed for Ship, Field Operations." Aviation Week and Space Technology, LXXVII (November 5, 1962), 108-109.

"Marines Seek Increased Speed, Range." Aviation Week and Space Technology, CIII (September 29, 1975), 62-64.

Mawhinney, William A. "H-46 Emergency Floatation System." Approach, XXXIV (July 1988), 30-31.

Moody, Joe D. "Medium Helicopter Assets." Marine Corps Gazette, LXV (May 1981), 50-55.

Pursley, Charles R. "Vertical Replenishment and the UH-46A." U.S. Naval Institute Proceedings, XCII (November 1966), 182-185.

Sheldon, Kim. "H-46: Workhorse of the Fleet." Naval Aviation News, (November-December 1988), 21-23.

Shenfeld, Gary. "The Navy's Flyable Fludic Helo." Data, XIII (October 1968), 28+.

(Boeing-Vertol) CH-47 Chinook

Book/Document/Report Literature:

Anderton, David A. Boeing Helicopters: CH-47 Chinook. Minigraph, no. 27. Arlington, Tex.: Aerofax, Inc., 1988. 48p.

Blender, G. L., et al. Airworthiness and Flight Characteristics Test of the CH-47D Helicopter. USAAEFA-82-07. Edwards AFB, Calif.: Army Aviation Engineering Flight Activity, 1984. 243p.

United States. Government Accounting Office. Procurement and Systems Acquisition Division. Can the Army's $2.8 Billion Program to Modernize the CH-47 Helicopter Be Improved? Washington, D.C., 1978. 28p.

Periodical/Anthology Literature:

Baldwin, Truxtun R. "The Improved Chinook." U.S. Army Aviation Digest, XIII (December 1967), 22-27.

Boody, Peter. "Volusia Triangle." Flying, CVI (June 1980), 43-48.

Burroughs, Leonard H. "Chinook Deployability." U.S. Army Aviation Digest, XI (April 1965), 12-15.

"Chinook--Boeing-Vertol's Versatile Chopper." Flying Review International, XIX (February 1964), 34-38+.

Daly, Richard W. "By Hook or By Crook." Military Engineer, LXI (July-August 1969), 251-255.

Eaton, Loren D. "Armed Chinooks." U.S. Army Aviation Digest, XII (July 1966), 26-29.

Ellis, Charles W. "Boeing-Vertol's Heavy-Lift Master." Defense and Foreign Affairs Digest, VI (August 1978), 20+.

Fink, Donald E. "CH-47D Offers Upgraded Performance." Aviation Week and Space Technology, CXIX (July 18, 1983), 54-56+.

Gallup, John F. "Camouflage and Dispersion of Chinooks." U.S. Army Aviation Digest, XI (November 1965), 19-21.

Gerry, John C. "The Chinook: Advanced Battlefield Mobility Vehicle." U.S. Army Aviation Digest, VIII (August 1962), 3-7.

Harvey, David. "'Big D' From Philadelphia: Boeing-Vertol's Gun-Toting Chinook." Rotor and Wing International, XX (July 1986), 49-51.

"Here Comes the CH-47B." U.S. Army Aviation Digest, XIII (August 1967), 12-13.

Hesson, James M. "The Chinook Goes Modern." National Defense, LX (November-December 1975), 204-205.

____. "Why?: CH-47D." U.S. Army Aviation Digest, XXV (September 1979), 2-7.

MacMahon, F. K. "The Chinook: Service Transport Helicopter." Interavia, XVII (July 1962), 866-868.

Mohr, Donald A. "Guns A Go-Go." U.S. Army Aviation Digest, XXII (June 1976), 8-13.

Nelms, Douglas W. "Updating the Big Bird." Army, XXXI (May 1981), 50-51+.

O'Connor, J. C. "Progress Report: CH-47 Modernization Program." Society of Experimental Test Pilots Technical Review, XVI (Spring 1981), 54-64.

"Old Chinooks to Be Upgraded to Like-New CH-47D Models." Aviation Week and Space Technology, CXXIX (October 31, 1988), 63-67.

Patrick, Dennis, et al. "CH-47 Operations in Central and South America." U.S. Army Aviation Digest, XXVIII (September 1982), 9-11.

Rawlings, Morris D. "Armed CH-47A Helicopter Employment." U.S. Army Aviation Digest, XII (August 1966), 26-28.

____. "Tote That Bale--Efficiently." Army, XIX (July 1969), 38-42.

Sutton, Geoff. "CH-47 Modernization Program: On Schedule and Within Budget." Defense Management Journal, XVIII (Spring 1982), 26-33.

Thompson, James B. "Self-Deployment of CH-47 Medium-Lift Helicopters to Europe." U.S. Army Aviation Digest, XXV (August 1979), 6-7.

"Vertol Announces Details of Army's Chinook." American Helicopter, XIV (April 1960), 60+.

"The Vertol CH-47 Chinook." Interavia, XIX (April 1964), 574-577.

Wilson, Leonard R. "The Chinook: A Personal View." U.S. Army Aviation Digest, X (April 1964), 18-20.

(Sikorsky) HH-52

Linn, Don. "Sikorsky's HH-52: The Coast Guard's Flying Boat." Air Classics Quarterly Review, VI (Fall 1979), 38-44.

(Sikorsky) H/CH/HH/HM/RH-53 Stallion/Sea Dragon

"Aerial Minesweeping." Defence Update International, no. 76 (October 1986), 53+.

Andrews, Hal. "Naval Aircraft: Sea Stallion." Naval Aviation News, (May 1973), 20-21.

____. "Naval Aircraft: Super Stallion." Naval Aviation News, (January-February 1984), 16-17.

Brownlow, Cecil. "HH-53B Readied for SEA Role as Interim CARA: Combat Aircrew Rescue Aircraft." Aviation Week and Space Technology, LXXXVII (July 3, 1967), 64-65+.

Coleman, Herbert J. "Navy Supports Continuation of CH-53E." Aviation Week and Space Technology, CVI (March 7, 1977), 23-24.

Colucci, Frank. "Sea Draggers: The Airborne Mine Countermeasure Force." Sea Power, XXVI (May 1983), 50-52+.
HM-53.

Cutler, Katie. "Slinging It." Airman, XXIII (February 1979), 18-23.

Davis, H. R., II. "CH-53s: A Real Lift for the Ground Commander." Marine Corps Gazette, LXV (May 1981), 20-22.

Davis, R. M. "Helicopter Minesweeping Operations." International Defence Review, XI (March 1978), 385-388.

Everett, Robert P. "The Big Ugly Fat Fellow." Airman, XIII (March 1969), 4-7.
HH-53.

Gaines, Mike. "We Fly With the Super Jolly Greens." Flight International, CXVII (March 8, 1980), 751-756.

Grimes, Vincent. "Super Stallion." International Combat Arms, VI (July 1988), 42-45+.

Grossman, Stanley L. "Computerizing the Helicopter." Marine Corps Gazette, LXXI (October 1987), 43-47.

Groves, Donald. "Helos Revolutionize Minesweeping." Our Navy, LXVII (March 1972), 2-5.

Harrison, Kirby. "Airborne Sweeping." Naval Aviation News, (March 1982), 16-23.

Harvey, David. "Navy/Marines Corral a Super Breed of Workhorse." Rotor and Wing International, XXI (May 1987), 40-43+.

____. "Pave Low III Rotorcraft: The USAF's Night Stalkers." Rotor and Wing International, XXIII (February 1989), 48-50.
MH-53H.

Kettering, A. J. "The Evolution of Heavy Lift." Marine Corps Gazette, LVII (May 1973), 20-24.

"Keystone Outfits the President's CH-53D." Rotor and Wing International, XXII (February 1988), 53-54.

Kleppsattel, F. M. "CH-53E Super Stallion." Marine Corps Gazette, LVI (May 1972), 43-46.

Krueger, S. P. "The Super Stallion." Vertiflite, XXIX (March-April 1983), 30-32, 41.

Lightfoot, R. B. "The Heavy Assault Transport Helicopter." Journal of the Royal Aeronautical Society, LXX (October 1966), 914-922.
CH-53A.

Manning, Harry. "Countering the Naval Mine Menace." International Combat Arms, VI (November 1988), 42-47+.

Pugh, Craig. "New Eyes for the Giant." Airman, XXIII (December 1979), 30-33.

Shillito, Robert G. "The Ten-Year Hover." Marine Corps Gazette, LXX (September 1986), 61-65.
CH-53E.

"The Sikorsky CH-53A." Interavia, XXIII (September 1968), 1091-1093.

"Weapon System Series: The MH-53E Super Stallion." NATO's Sixteen Nations, XXIX (June-July 1984), 86-88.

Wheeler, Howard A. "CH-53A: Heavy Hauling in the Mediterranean." Naval Aviation News, (November-December 1983), 8-9.

(Sikorsky) CH-54 Tarhe/Skycrane

Blackwell, Brendan. "HLH." U.S. Army Aviation Digest, XIX (August 1973), 8-16.

Dodd, Joe. "Skycrane." Marine Corps Gazette, XLIV (April 1960), 54-55.

Jacobs, Norman P. "Big Lift by Flying Crane." Army Digest, XXIII (January 1968), 24-26.

McVey, Curtis. "Arctic Test of the 'Crane." U.S. Army Aviation Digest, XV (November 1969), 2-6.

"The Old, the New, and the Memories." National Guard, XXXIII (May 1979), 12-15.

Pearson, J. F. "Chopper They Call When the Chips are Down: The Sikorsky Skycrane." Popular Mechanics, CXLVII (February 1977), 84+.

"The Skycrane Shows Its Stuff." BeeHive, XXXVIII (Fall 1963), 8-10.

"Skycranes Wrapped for Posterity." Rotor and Wing International, XXIII (January 1989), 67-69.

Tierney, Richard K. "The Flying Crane." U.S. Army Aviation Digest, X (November 1964), 12-15.

Williams, Charles. "Sikorsky Flying Crane." American Helicopter, LIV (March 1959), 11+.

(Hughes) H/TH-55 Osage

Endicott, Al. "Last Flight of the TH-55." U.S. Army Aviation Digest, XXXIV (August 1988), 22-23.

(Lockheed) AH-56 Cheyenne

"AH-56A Demonstrates Maneuverability." Aviation Week and Space Technology, LXXXVII (December 25, 1967), 64-66.

"A Case History: Cheyenne." Armed Forces Journal, CVII (November 22, 1969), 14-15.

Dillard, Robert J. "The Army Cheyenne: Designed to Survive." U.S. Army Aviation Digest, XVII (May 1971), 2-7.

Fitch, John B. "Call In the Gunships." Army Digest, XXVII (April 1971), 32-35.

Hessman, James D. "The Army Gets a New Weapon System." Armed Forces Journal, CVI (December 14, 1968), 7+.

Hieronymus, W. S. "AH-56 Faces New Army Tests." Aviation Week and Space Technology, XCIII (November 16, 1970), 58-59+.

"An Indian in Warpaint: The Aggressive Cheyenne." Flying Review International, XXII (July 1967), 715+.

Kluever, Emil E. "The Cheyenne." U.S. Army Aviation Digest, XIV (January 1968), 2-4.

"The Lockheed AH-56A." Interavia, XXII (October 1967), 1562-1563.

Martin, Norman W. "Armament for the Cheyenne." Ordnance, LIII (March-April 1969), 502-505.

Myers, F. W. and Cyril F. Horton. "Advanced Attack Helicopter: A Case Study." Defense Management Journal, X (September 1974), 36-38.

Nihart, Brooke. "Cheyenne Dead After Lingering Illness." Armed Forces Journal, CX (September 1972), 60-61.

Ropelewski, Robert R. "Army Completing AH-56A Tests." Aviation Week and Space Technology, XCVI (May 22, 1972), 55-59.

Tuck, William R. "The AH-56 in Support of Tank Operations." Armor, LXXVII (November-December 1968), 3-10.

Ulsamer, Edgar E. "Here Comes the Rigid Rotors." Air Force and Space Digest, LI (February 1968), 52-55.

(Bell) TH-57 Sea Ranger
(see also OH-58 Kiowa)

Andrews, Hal. "Naval Aircraft: TH-57 Sea Ranger." Naval Aviation News, (March 1981), 20-21.

(Bell) OH-58 Kiowa/Aeroscout

Book/Document/Report Literature:

Bell Helicopter Textron. Public Relations Department. U.S. Army OH-58D (AHIP). Fort Worth, Tex., 1987. 3p.

____. ____. OH-58D: U.S. Army Scout With Eye-Opening Technology. Fort Worth, Tex., 1986. 5p.

Bender, G. L. and E. L. Mitchell. Preliminary Airworthiness Evaluation of the OH-58D Helicopter. Edwards, Calif.: Army Aviation Engineering Flight Activity, 1984. 221p.

Peth, Stephen B. Crew Configuration and Task Saturation in Army OH-58 Helicopters. Research Paper. Maxwell AFB, Ala.: Air Command and Staff College, Air University, 1981. 47p.

United States. Government Accounting Office. Mission Analysis and Acquisition Division. Army Helicopter Improvement Program's Future May Depend on Success in Controlling Cost. Washington, D.C., 1983. 19p.

____. ____. National Security and International Affairs Division. Changing Roles and Priorities for the Army's OH-58D Helicopter Program. Washington, D.C., 1988. 9p.

Periodical/Anthology Literature:

"Bell OH-4A." Flight, LXXXVI (August 6, 1964), 210-212.

Bulban, Erwin J. "The AHIP Advanced Scout: Near Ready to Go on Patrol." Rotor and Wing International, XIX (April 1985), 74-77, 91.

Chappell, George W. "The Deep Battle Surgeon." Infantry, LXXIX (January-February 1989), 35-36.

DeMeis, Richard. "Designing the Military Helicopter: Bell's OH-58 Scout." Aerospace America, XXII (September 1984), 88-90, 92.

Fairweather, Robert S. "Arming the Aeroscout." U.S. Army Aviation Digest, XXVIII (February 1985), 45-47.

____. "Kiowa Joins Up." U.S. Army Aviation Digest, XV (November 1969), 26-27.

Foster, W. "The AHIP: Eyes and Ears of the Battlefield Commander." Vertiflite, XXIX (November-December 1983), 23-25.

Harvey, David. "A Comeback Story: The OH-58D Aeroscout." Rotor and Wing International, XXII (February 1988), 44-46.

____. "How Aerial-Scout Testing Helped Sight the Army's Needs." Rotor and Wing International, XXII (December 1988), 64-65, 76.

Mancuso, Vincent P. "The Improved OH-58." U.S. Army Aviation Digest, XXV (November 1979), 42-46.

Merritt, Donald E. S. "Remembrance and Reality: The OH-58 Story." U.S. Army Aviation Digest, XXIX (October 1983), 39-44.

Mordoff, Keith. "U.S. Army Gains Experience With OH-58D in Germany." Aviation Week and Space Technology, CXXVIII (February 8, 1988), 36-39.

Morris, Jerry J. and Donald E. S. Merritt. "Wire Strikes: The OH-58 Strikes Back." U.S. Army Aviation Digest, XXVII (December 1981), 24-27.

Oaks, Adam P. "OH-58D: The New Eye on the Battlefield." Field Artillery, (October 1988), 41-42.

Olcott, John W. "The Making of a Classic." Business and Commercial Aviation, LIX (July 1986), 42-47.

Ropelewski, Robert R. "Mast-Mounted Sight, New Rotor Boost Effectiveness of OH-58D." Aviation Week and Space Technology, CXXV (September 8, 1986), 66-67+.

Shifrin, Carole A. "Bell Expects Deployment, Test Results Will Assure Funding of OH-58D." Aviation Week and Space Technology, CXXVIII (February 8, 1988), 43-46.

Tisdale, Dan, Jr. "OH-58 Kiowa Power Margin." U.S. Army Aviation Digest, XVII (June 1971), 9-11.

Warwick, Graham. "Texas Ranger." Flight International, CXIX (April 1, 1981), 1113-1114.

Weeghman, Richard B. "For Whom Doth Bell Toil?" Flying, LXXVIII (May 1966), 102-107.

(Sikorsky) H/EH/HH/SH/UH/VH/MH-60 Hawk

Book/Document/Report Literature:

Blackwell, Brendan P. Blackhawk Around the World: Plan for the First Helicopter Circumnavigation of the Globe. Research Paper. Maxwell AFB, Ala.: Air Command and Staff College, Air University, 1979. 81p.

Mullen, G. Michael. Choppers Grounded: The Supply-Demand Problem. Washington, D.C.: National Defense University Press, 1988. 80p.

Smith, W. W. and Richard T. Christensen. UH-60A Aviator Qualification Course: Student Handbook for Blackhawk. Fort Rucker, Ala.: Army Aviation Center, 1984. 336p.

Periodical/Anthology Literature:

Bedard, Paul. "LAMPS III: The Navy's Bright Light." Military Logistics Forum, III (November-December 1986), 22-24.
SH-60.

"Blackhawk." Defence Update International, no. 86 (February 1988), 28-38.

"The Blackhawks are Coming." Air International, XVI (April 1979), 163-169.

Blackwell, Brendan P. "Blackhawk Around the World." U.S. Army Aviation Digest, XXVII (July-August 1981), 38-43, 20-25.

Brown, David A. "Blackhawk Exhibits Speed, Agility." Aviation Week and Space Technology, XCIV (February 15, 1971), 38-43.

Butler, John K. "UTTAS." Armor, LXXXVI (May-June 1971), 19-20.

Clouse, Gale E., Jr. "HH-60D Night Hawk." MAC Flyer, XXX (October 1983), 16-19.

Costello, Gene. "Blackhawk." U.S. Army Aviation Digest, XXIV (February 1978), 34-39.

Ferguson, James D. "UTTAS: Birth of a Huey Replacement." Flight International, CX (August 21, 1976), 449-454.

Fink, Donald E. "Workload for Pilot Reduced in SH-60B." Aviation Week and Space Technology, CXIII (December 1, 1980), 111+.

Galdorisi, George. "SH-60F: New Capabilities for the Battle Group." U.S. Naval Institute Proceedings, CXIII (February 1987), 88-91.

____. "Turning on LAMPS III." U.S. Naval Institute Proceedings, CXII (October 1986), 147-149.

Gilson, Charles. "Some Operational Aspects of the UH-60A Blackhawk." International Defence Review, XIII (July 1980), 1067-1074.

Grange, David E., Jr. "The Infantryman's Helicopter in the 1980s." U.S. Army Aviation Digest, XXVI (April 1980), 2-3.

Green, David. "Pilot Report: Air Force's Sikorsky HH-60 Night Hawk." Rotor and Wing International, XXI (August 1987), 42-45, 57.

Harvey, David. "Blackhawks on Custom's Patrol: Holding the Line on Drug Smuggling." Rotor and Wing International, XX (March 1986), 26-27, 71.

____. "CV-Helo: The U.S. Navy's New Inner-Zone Goalie." Rotor and Wing International, XXI (October 1987), 70-73.
SH-60F.

____. "A Deadly Duo for Subs: Aboard Ship with LAMPS III." Rotor and Wing International, XXI (December 1987), 46-48+.
SH-60B.

____. "The HH-60: Navy CSAR Gets a Shot in the Arm." Rotor and Wing International, XXII (August 1988), 42-43, 58.
CSAR = Combat Search and Rescue.

____. "LAMPS III: Persian Gulf Action Highlights System Efficiency." Defense Science, VII (August 1988), 28+.

____. "The Preparation: The Helicopter Bred for Assault." Defense and Foreign Affairs, VII (August 1979), 4-7.
UH-60.

____. "The USAF's HH-60D: Combat SAR for the 1990s." Rotor and Wing International, XVIII (March 1984), 44-49.

____. "The U.S. President's New Doomsday Helicopter." Rotor and Wing International, XXII (February 1988), 42-43, 65.
VH-60.

___. "UTTAS: The Technology of Helicopter Survival." Defense and Foreign Affairs Digest, VI (August 1978), 4-6+.

Houser, George. "LAMPS Mk. III." Military Electronics/Countermeasures, IV (October 1978), 34+.

Kandebo, Stanley W. "Army, Sikorsky Assess Plans for Next-Generation Blackhawk." Aviation Week and Space Technology, CXXVIII (March 21, 1988), 52-53.

___. "Manufacturer Details Configurations of Navy/Coast Guard SH-60F Derivatives." Aviation Week and Space Technology, CXXVI (May 18, 1987), 96-97.

Kefford, B. "The UH-60A Blackhawk: A World-Wide Force Multiplier." Aircraft Engineering, LIV (October 1982), 9-15.

Larson, George C. "A Legacy for the Night Hawk." Air Force Magazine, LXVII (February 1984), 102-104.

Lopez, Ramon. "Blackhawks Enter Production." International Defence Review, XI (August 1978), 1269-1274.

___. "UH-60A Blackhawk in Production." Interavia, XXXIII (November 1978), 1111-1116.

Manningham, Dan. "LAMPS III." U.S. Naval Institute Proceedings, CIV (March 1978), 159-161.

___. "SH-60 Seahawk: A Pilot's Report." U.S. Naval Institute Proceedings, CVIII (January 1982), 98-104.

Mullen, George M. "Choppers Grounded." U.S. Army Aviation Digest, XXXIV (September-October 1988), 2-14, 8-14.

O'Connor, J. C. "UTTAS Testing." Society of Experimental Test Pilots Technical Review, XIII (Fall 1977), 153-162.

Purtell, J. M. "Light Airborne Multi-Purpose System." Vertiflite, XXVII (September-October 1981), 10-13.
LAMPS SH-60.

Rausa, Rosario. "LAMPS Mk. III." Naval Aviation News, (June 1980), 8-19.

Reardon, Mark. "All the Chips at Sea." Seapower, XXV (July 1982), 24-30.

"Seahawk--Bird of Prey." Surface Warfare, V (April 1980), 6-8.

"SH-60 Seahawk." Air Progress, XLV (July 1983), 55+.

"Sikorsky Hawks." Air International, XXVII (July 1984), 9-16, 45-46.

"Sikorsky MH-60K Design Stresses Commonality With Existing Aircraft." Aviation Week and Space Technology, CXXVIII (April 11, 1988), 48+.

"Sikorsky SH-60B: The U.S. Navy's LAMPS Mk. III." Flight International, CXIII (February 25, 1978), 510-511.

Sundaram, Gowri S. "LAMPS Mk. III: The Long Arm of Shipborne ASW." International Defence Review, XVI (March 1983), 287-290.

Thomas, Wayne. "We Fly the Army's New Super Copter: UH-60A Blackhawk." Mechanics Illustrated, LXXIV (July 1978), 46-47+.

Turner, William E. "The UH-60 Blackhawk: Posturing for the Future." Vertiflite, XXXIII (November-December 1987), 22-25.

Walker, Richard B. "Fielding the Blackhawk." Army Logistics, IX (November-December 1977), 24-25.

Wetmore, Warren C. "Reliability Emphasized in UH-60 Design." Aviation Week and Space Technology, CVII (April 11, 1977), 44-47.

____. "SH-60B to Fill Several Navy Roles." Aviation Week and Space Technology, CVIII (April 3, 1978), 40-41.

____. "UH-60 Termed Low Technical Risk." Aviation Week and Space Technology, CVI (April 4, 1977), 59-63.

Winkel, Raymond H. and Dan Manningham. "LAMPS: The Ship System With Wings." U.S. Naval Institute Proceedings, CVI (March 1980), 114-117.

Zanow, William L. "Airmobility: The Fielding of the UH-60A Blackhawk." Army Aviation, XXIX (September 15, 1980), 28-30.

Zeringue, Brian. "Army Hawks at Sea." Soldiers, XLII (October 1987), 16-18.

(Hughes/McDonnell-Douglas) AH-64 Apache.

Book/Document/Report Literature:

Bender, G. L., et al. Airworthiness and Flight Characteristics Test: YAH-64 Advanced Attack Helicopter--Part I. Edwards AFB, Calif.: Directorate for Development and Qualification, Army Aviation Engineering Flight Activity, 1981. 181p.

Colucci, Frank. McDonnell-Douglas Apache. Aero Series, v. 33. Blue Ridge Summit, Pa.: TAB, 1988. 96p.

David, R. G. Fielding the AH-64: Logistics Lessons Learned. Research Paper. Maxwell AFB, Ala.: Air Command and Staff College, Air University, 1988. 42p.

Gunston, William T. ("Bill"). AH-64 Apache. Combat Aircraft Series, no. 6. Harrisburg, Pa.: Stackpole Books, 1986. 48p.

Monson, Lyle. McDonnell-Douglas (Hughes) AH-64 Apache. Minigraph, no. 18. Arlington, Tex.: Aerofax, Inc., 1987. 48p.

Picasso, B. D., et al. Airworthiness and Flight Characteristics Test: YAH-64 Advanced Attack Helicopter--Part II. Edwards AFB, Calif.: Directorate for Development and Qualification, Army Aviation Engineering Flight Activity, 1982. 152p.

Richardson, Doug. AH-64 Apache. Modern Fighting Aircraft, no. 12. New York: Arco, 1987. 64p.

Rorke, James B. Apache for the Battlefield of Today and the 21st Century. Mesa, Ariz.: McDonnell-Douglas Helicopter Company, 1987. 13p.

Periodical/Anthology Literature:

Adshead, Robin. "Apache in Europe." Armed Forces, VII (February 1988), 72-75.

"Apache (AH-64) Tank Killer: Two Helicopters in One." Defense Systems Review and Military Communications, II (September 1984), 39+.

"Apache Warpath Hugs 'Nap of the Earth.'" Officer, LVIII (December 1982), 17-19.

Austin, S. A. and John A. Moore. "AH-64: A Total System for Battle." U.S. Army Aviation Digest, XXXIII (October 1987), 18-29.

Bauer, Daniel R. "Tank Killer." Armor, LXXXVI (May-June 1977), 8-10.

Brown, William J. and Jeffrey V. Van Rope. "AH-64: A Total System for Battle." U.S. Army Aviation Digest, XXXIV (February 1988), 20-29.

Browne, Edward M. "The Advanced Attack Helicopter." U.S. Army Aviation Digest, XXVII (May 1981), 4-8.

____. "The AH-64 Apache and Its Combat System." Military Technology, VI (October 1982), 148-152.

Brunelle, Ron and Philip A. Mooney. "AH-64: A Total System for Battle." U.S. Army Aviation Digest, XXXIII (November 1987), 16-23; XXXIV (January and April 1988), 24-36, 20-26.

Camacho, Thomas M. "Apaches Land in Tar Heel Country." National Guard, XLII (February 1988), 30-32+.

Coughlin, James. "Making It Tough for Tanks." Defense Electronics, XVIII (August 1986), 51-52+.

Cox, Frank. "Tough Bird." Soldiers, XLI (December 1986), 13-16.

____. "War Bird." Soldiers, XLIII (March 1988), 13-15.

Drenz, Charles F. "AH-64 Apache." U.S. Army Aviation Digest, XXXI (December 1985), 2-8.

____. "AH-64A Apache: Ready for the Battlefield." Vertiflite, XXIX (November-December 1983), 20-23.

Ennis, Ned B. "Putting the Deep in Deep Attack." Army, XXXVIII (June 1988), 44-47.

Fink, Donald E. "Few YAH-64 Design Changes Expected." Aviation Week and Space Technology, CVI (January 10, 1977), 83-84.

___. "Hughes AAH Entry Shuns Innovations." Aviation Week and Space Technology, XCIX (November 19, 1973), 47-49.

___. "Hughes Starts Phase Two on YAH-64." Aviation Week and Space Technology, CVI (January 3, 1977), 17-19.

___. "YAH-64 Design Leans on Vietnam Experience." Aviation Week and Space Technology, CIII (August 18, 1975), 42-47.

Fulton, Richard P. "Training the Trainers." U.S. Army Aviation Digest, XXXI (November 1985), 36-44.

Funk, David L. "The Advanced Attack Helicopter Challenge." Armor, LXXXVII (January-February 1978), 16-22.

Geddes, J. Philip. "AH-64: The U.S. Army's Advanced Attack Helicopter." International Defence Review, XI (August 1978), 1247-1252.

___. "Hughes Helicopters' AH-64 in Advanced Development." Interavia, XXXIII (August 1978), 728-730.

___. "Hughes YAH-64 Advanced Attack Helicopter." Interavia, XXX (September 1975), 972-975.

Geisenheyner, Stefan. " In Search of a Tactical Concept: The Attack Helicopter." Asian Defence Journal, no. 10 (October 1983), 30-32+.

Green, David L. "Flying the Army's Latest Warrior: The Hughes AH-64 Apache." Rotor and Wing International, XIX (April 1985), 48-58, 64-69.

Grett, Stanley E. "Advanced Attack Helicopter." U.S. Army Aviation Digest, XXV (April 1979), 14-19.

Harrison, Ben L. "The Bigger-Than-Life AH-64A Apache: Awesome, Mean--Now What Do We Do With It?" Army, XXXVIII (January 1988), 10-12+.

Harvey, David. "Here Come the Apaches." Rotor and Wing International, XVIII (January 1984), 72-77.

___. "Zap! Kaboom! Look Out for Laser-Toting Apaches." Rotor and Wing International, XXII (May 1988), 36-39.

Hirsch, Norman B. "Helicopterbore Anti-Armour Warfare Illustrated by AH-64." Armada International, VII (March-April 1983), 6-8+.

____, et al. "AH-64: A Total System for Battle." U.S. Army Aviation Digest, XXXII (July, September-October 1986), 3-9, 36-45, 36-44; XXXIII (January 1987), 6-17.

"Hughes AH-64: The Advanced Attack Helicopter." Flight International, CXIII (February 25, 1978), 512-517.

"Indians From Arizona." Air International, XXVI (May 1984), 224-228, 248-252.

Johnson, Robert L., Jr. "AH-64 and OT 11 Tactical Lessons Learned." U.S. Army Aviation Digest, XXXI (January 1985), 2-12.

Kerr, John N. "Army's YAH-64: Survivable Tank Killer." U.S. Army Aviation Digest, XXIII (November 1977), 4-5+.

Lambert, Mark. "Apache on the Warpath: The Hughes AH-64 in Production at Last." Interavia, XXXVIII (May 1983), 436-438.

____. "Flying the AH-64 Apache." Interavia, XXXVIII (August 1983), 833-836.
Reprinted in International Defence Review, XVI (September 1983), 1241-1245.

Lord, K. M. "Tank-Killing Copter Sees in the Dark: YAH-64 Advanced Attack Helicopter." Popular Science, CCXV (November 1979), 137+.

McConnell, Lewis J. "AH-64: A Total System for Battle." U.S. Army Aviation Digest, XXXIV (November 1988), 38-44.

"McDonnell-Douglas Team Studies Development of Advanced Apache." Aviation Week and Space Technology, CXXVIII (January 25, 1988), 45-46.

McGinty, Kenneth R. and Gregory A. Brockman. "Cross Forward Line of Own Troops With AH-64 Apache." U.S. Army Aviation Digest, XXXIII (January 1987), 2-5.

____. "Scouts Out With the AH-64 Apache." U.S. Army Aviation Digest, XXXIII (September 1987), 16-20.

McGlasson, W. D. "AH-64: The Attack Helicopter of the Future." National Guard, XXXVII (October 1983), 22-25.

Meyer, Deborah G. "Apache's Agility and Smooth Flight." Armed Forces Journal International, CXXI (October 1983), 76+.

Moore, John A. and Wendell W. Shivers. "AH-64: A Total System for Battle." U.S. Army Aviation Digest, XXXIV (May 1988), 18-30.

Neims, Douglas W. "The Awesome Apache." Army, XXXVII (January 1987), 38-47.

Philips, George. "AH-64: A Total System for Battle." U.S. Army Aviation Digest, XXXIV (December 1988), 44-49.

"Riding High With the Apache." Defense Electronics, XVI (December 1984), 66-68.

Roop, Troy J. "Keeping the Apache Flying and Fighting." U.S. Army Aviation Digest, XXVIII (March 1982), 38-41.

Ropelewski, Robert R. "Apache to Provide Night/Bad Weather Capability." Aviation Week and Space Technology, CXVII (September 13, 1982), 42-43+.

Rorke, James B. "AH-64 Apache--For the Battlefield of Today and the 21st Century." Defence Update International, no. 84 (December 1987), 8-17+.

Saint, Crosbie E. and Walter H. Yates, Jr. "Attack Helicopter Operations in the AirLand Battle." Military Review, LXVIII (June-July, October 1988), 2-15, 2-9, 2-10.
Three articles reprinted from Army Aviation considering, in order, Close Operations, Deep Operations, and Rear Operations.

Scott, William B. "AH-64A Apaches Represent Blend of Electronic, Airframe Technologies." Aviation Week and Space Technology, CXXVII (July 20, 1987), 72-73+.

Shivers, Wendell W. and Jeffrey W. Van Rope. "AH-64: A Total System for Battle." U.S. Army Aviation Digest, XXXIV (March 1988), 14-26.

Starr, David W. "AH-64 Update." U.S. Army Aviation Digest, XXXI (March 1985), 38-39.

Thompson, Leroy. "Heavy-Duty Apache." Air Progress Aviation Review, VI (Winter 1982), 42-45.

Wanstall, Brian. "Wars Don't Stop at Night." Interavia, XXXVII (August 1982), 773-775.

Yarlett, William, et al. "AH-64 Apache Combat Mission Simulator." U.S. Army Aviation Digest, XXXI (March-May, August-September 1985), 40-44, 40-44, 38-43, 40-45, 40-44.

(Aerospatiale) HH-65 Dolphin

"Aerospatiale Dauphin." Interavia, XXX (August 1975), 867-869.

"Coast Guard Nears Helicopter Decision." Aviation Week and Space Technology, CIX (July 3, 1978), 101+.

Manningham, Dan. "A Decade of Dauphins." Business and Commercial Aviation, LIV (January 1984), 43-48.

North, David M. "Coast Guard Helicopter Buy Awaited." Aviation Week and Space Technology, CVI (May 16, 1977), 18-19.

"USCG Selects Dolphin 2 Helicopter." Interavia, XXXIV (August 1979), 733-734.

Wyatt, D. E. "HH-65A Dolphin Emergency Exits." On Scene, (Spring 1986), 7-9.

(McDonnell-Douglas/Bell and Boeing/Sikorsky) LHX

Book/Document/Report Literature:

Harris, Wesley L., et al. U.S. Army's LHX Program. Washington, D.C.: Army Science Board, 1984. 50p.

Hubbard, R. D. and R. L. Tomaine. Influence of Operational Requirements on LHX (Light Helicopter Family) Concept Formulation. St. Louis, Mo.: Army Aviation Systems Command, 1987. 8p.

Lovett, Michael L. Light Combat Helicopter. Research Paper. Maxwell AFB, Ala.: Air Command and Staff College, Air University, 1984. 43p.

United States. Army. Army Training and Doctrine Command. Light Helicopter Family Tradeoff Analysis. 10 vols. Fort Monroe, Va., 1985.

____. Department of Defense. Defense Science Board. Report on LHX Requirements. Washington, D.C., 1986. 54p.

____. Government Accounting Office. National Security and International Affairs Division. Case Study of the Army Light Helicopter Program. Washington, D.C., 1986. 22p.

Periodical/Anthology Literature:

Ambrose, James R. "Getting Airborne With the Army's New Light Helicopter Program." Army, XXXV (August 1985), 39-42+.

Anderson, Ronald K. "LHX and the Streamlining Initiative." Program Manager, XIV (March-April 1985), 11-13.

____. "LHX: Integrated, Survivable." Journal of Electronic Defense, X (April 1987), 73-74+.

____. "LHX Targets Dollars as Well as Bogeys." Aerospace America, XXIV (July 1986), 24-27.

____. "Light Helicopter Family (LHX) Program Overview." U.S. Army Aviation Digest, XXXIII (January 1987), 38-41.

____. "Tomorrow's Fleet: The Light Helicopter Family." Vertiflite, XXXI (November-December 1985), 42-46.

Artis, Donald R., Jr. "LHX: Keeping Pace With Technology." U.S. Army Aviation Digest, XXVIII (December 1982), 9-12.

Brittingham, Michael L. "LHX: Sweet Dream and Flying Machine." U.S. Army Aviation Digest, XXVIII (December 1982), 13-16.

Cannet, M. "Reflections on the LHX Program." International Defence Review, XX (January 1987), 41+.

Capaccio, Tony. "Army LHX Hovers on the Brink." Defense Electronics, XIX (October 1987), 93+.

Cockerham, Samuel G. "Design Frontier: LHX Combat Helicopter." Defense and Foreign Affairs, XII (June-July 1984), 24-26.

Cribbins, Joseph F. "Why Not an LHX?" U.S. Army Aviation Digest, XXIX (January 1983), 38-39.

Dornheim, Michael A. "GAO Criticizes LHX Program, Claiming R & D Work Insufficient." Aviation Week and Space Technology, CXXX (January 30, 1989), 22-24.

____. "McDonnell-Douglas/Bell Team Introduces Its LHX Concept." Aviation Week and Space Technology, CXXX (March 6, 1989), 56-59.

Fairweather, Robert S., Jr. "LHX: Army Aviation's Entry Into the Future." U.S. Army Aviation Digest, XXIX (January 1983), 34-37.

Feaster, Lewis. "LHX: Helicopter of the Future." U.S. Army Aviation Digest, XXVII (May 1981), 9-11.

Gaines, Mike. "LHX: A Giant Step." Flight International, CXXVII (June 1, 1985), 167-171.

Harvey, David. "Bell Tilts Its Technological Thrust Toward LHX." Rotor and Wing International, XVIII (September 1984), 52-57.

____. "First Team: The Boeing-Sikorsky Game Plan." Rotor and Wing International, XX (November 1986), 42, 44, 70.

____. "LHX: The Helicopter Program of the Century." Rotor and Wing International, XVIII (January 1984), 56-71.

____. "Superteam: Bell-McDonnell-Douglas Playing Catchup?" Rotor and Wing International, XX (November 1986), 43, 45, 71.

____. "U.S. Army Ready to Roll With LHX?" Rotor and Wing International, XX (January 1986), 48-51.

____. "Words from Quad A: The Army Wants LHX--Badly." Rotor and Wing International, XVIII (July 1984), 48-53.

Haystead, John. "Gearing Up for LHX." Journal of Electronic Defense, VIII (October 1985), 41-42+.

Kernstock, Nicholas C. "McDonnell-Douglas/Bell LHX Design Will Use 680 High-Agility Rotor." Aviation Week and Space Technology, CXXX (March 6, 1989), 60-61.

Lambert, Mark. "Is the 680 Rotor Bell's LHX Candidate?" Interavia, XL (September 1985), 981-982.

____. "LHX: Not Just Another Helicopter." Interavia, XL (May 1985), 448-450.

Leibstone, Marvin. "LHX Requirements, Capabilities, Acquisition." Military Technology, XI (January 1987), 85-87.

"LHX." Defence Update International, no. 41 (1983), 20-23.

Lopez, Ramon. "LHX: Helicopter Program of the Century." International Defence Review, XVII (May 1984), 585-588.

____. and Mark Lambert. "LHX: The U.S. Army Wants, 5,000, Industry Needs the Business." Interavia, XXXVIII (September 1983), 972-974.

Ludvigaen, Eric C. "New Light Helicopter is Goal of Massive Army LHX Program." Army, XXXIV (March 1984), 38-42+.

McNair, Carl H., Jr. "LHX: The Conceptual Formulation of Why, What, and When." U.S. Army Aviation Digest, XXVIII (December 1982), 2-5.

Nordeen, Lon O. "LHX: The U.S. Army's Rotorcraft for the Battlefield of the 21st Century." National Defense, LXIX (March 1985), 36-39.

Parker, Ellis D. "Vision of LHX." U.S. Army Aviation Digest, XXXII (December 1986), 2-5.

Ropelewski, Robert R. "Army Seeks Light Helicopter Bids." Aviation Week and Space Technology, CXIX (August 29, 1983), 16-17.

Schrage, D. P. "The LHX Preliminary Design Process." Vertiflite, XXIX (November-December 1983), 14-17.

Shifrin, Carole A. "Army May Use LHX Technology to Upgrade Current Production Fleet." Aviation Week and Space Technology, CXXVIII (April 25, 1988), 23-24.

____. "Manufacturers Believe Need Remains for New Advanced Army Helicopter." Aviation Week and Space Technology, CXXVIII (January 18, 1988), 21+.

Singley, G. T. "Light Helicopter Family (LHX): The U.S. Army's Future Light Rotorcraft Fleet." Vertiflight, XXX (November-December 1984), 18-22.

Siuru, William D., Jr. and John D. Busick. "LHX." Military Review, LXVI (December 1986), 26-31.

Turk, Paul. "LHX: The U.S. Army Prepares For a Giant Leap Forward." Interavia, XXXIX (December 1984), 11-13+.

Warwick, Graham. "Advanced Helicopter Concepts Compete." Flight International, CXXI (March 13, 1982), 615-616, 625-626.

Wilson, Michael. "LHX: The Smallest With the Mostest." International Defence Review, XVII (September 1984), 38-40.

(Hughes) XH-17 Flying Crane

Prouty, R. W. "The Amazing Hughes XH-17 Flying Crane." Rotor and Wing International, XIV (August 1980), 46-48, 66.

(Bell) XH-40
(see also UH-1 Huey)

Weichsel, Hans, Jr. "XH-40 Test Program." U.S. Army Aviation Digest, III (May 1957), 5-10.

"XH-40." U.S. Army Aviation Digest, II (February 1956), 15-18.

(Lockheed) XH-51 Aerogyro

"The Aerogyro: Lockheed's Rigid-Rotor Concept." Interavia, XVIII (September 1963), 1338-1340.

Hibbard, H. L. and R. R. Heppe. "Lockheed Favors the Rigid-Rotor Helicopter." Interavia, XVII (June 1962), 727-729.

Reed, W. S. "First Photos Show Lockheed's Helicopter." Aviation Week and Space Technology, LXXV (July 21, 1961), 26-27.

Wilson, G. C. "Lockheed to Market Rigid-Rotor XH-51A." Aviation Week and Space Technology, LXXXI (December 28, 1964), 44-45.

"XH-51 Reaches 160-mph in Test Flights." Aviation Week and Space Technology, LXXVIII (May 13, 1963), 90-91.

Zuments, George. "How Low Can You Go?" U.S. Army Aviation Digest, XV (May 1969), 26-31.

(Sikorsky) XH-59

Fink, Donald E. "Advancing Blade Flight Data Gathered." Aviation Week and Space Technology, CXVI (March 29, 1982), 42-44+.

Linden, A. W. "XH-59A ABC Aircraft Flight Tests at Fort Rucker, Alabama." Aircraft Engineering, LIV (December 1982), 14-18.

(Piasecki) YH-16 Transporter

Berry, M. "Piasecki Unveils YH-16 Transporter." American Helicopter, XXXII (October 1953), 12, 16-17.

Boeing Vertol Company. Public Relations Department. YH-16 Series. Philadelphia, Pa., 1987. 3p.

Hawthorne, Randolph. "Piasecki's Bid for 'Copter Market." Aviation Age, XXIV (July 1955), 26-30.

McSurely, Andrew. "Piasecki Unveils Giant Transport Copter." Aviation Week, LIX (September 21, 1953), 16-18.

"Piasecki's Largest." The Aeroplane, LXXXVIII (June 10, 1955), 776-777.

(Hiller) YH-32 Hornet

"Ramjet Helicopter." Military Review, XXXVII (April 1957), 63-64.

(Cessna) YH-41 Seneca

"Cessna Delivers First YH-41 Helicopter to the U.S. Army." American Helicopter, XLVIII (November 1957), 13-14.

Coleman, Frank. "Cessna's New CH-1C." American Helicopter, LIV (March 1959), 9+.

Searles, Robert A. "Reflections: Cessna's First (and Last) Helicopter." Business and Commercial Aviation, LIX (September 1986), 41-47.

Appendix I: List of Journals Consulted

Each of the 334 journals/magazines listed contributed at least 1 article to this compilation.

AEI Defense Review
Aerial Age
Aerial Age Weekly
Aero
Aero Album
Aero Digest
Aeronautical Digest
Aeronautical Engineering Review
Aeronautical Journal
Aeronautics
Aerophile
The Aeroplane
Aeroplane Monthly
Aerospace Accident and Maintenance Review
Aerospace America
Aerospace Engineering
Aerospace Historian
Aerospace International
Aerospace Maintenance Safety
Aerospace Medicine
Aerospace Safety
Air Classics
Air Classics Quarterly Review
Air Combat
Air Combat Quarterly Review
Air Enthusiast
Air Extra
Air Force
Air Force and Space Digest
Air Force and Space Digest International
Air Force Magazine
Air Force Journal of Logistics
Air International
Air Line Pilot
Air News
Air Op
Air Pictorial

Air Power Historian
Air Power Journal
Air Progress
Air Progress Aviation Review
Air Reservist
Air Service Journal
Air Trails
Air Training
Air Transportation
Air University Quarterly Review
Air University Review
Aircraft
Aircraft Age
Aircraft Engineering
Aircraft Illustrated
Aircraft Journal
Airfax Magazine
Airlift
Airlift Operations Review
Airman
Airpower
AirTech
Airview
All Hands
American Aeronaut
American Aviation
American Aviation Historical Society Journal
American Helicopter
American History Illustrated
American Machinist
American Magazine
American Magazine of Aeronautics
American Mercury
American Politics Quarterly
Analog
Antarctic Journal of the United States
The AOPA Pilot
Approach
Armada International
Armed Forces Journal
Armed Forces Journal International
Armor
Army
Army Digest
Army Information Digest
Army Logistics
Army Quarterly and Defence Review
Army Research, Development, and Acquisition
Asian Defence Journal
Astronautics and Aeronautics
Atlanta
Atlantic
Automotive and Aviation Industries
Automotive Engineering
Automotive Engineers
Automotive Industries
Aviation
Aviation Age
Aviation and Aeronautical Engineering
Aviation News
Aviation Op
Aviation Quarterly
Aviation/Space
Aviation Week
Aviation Week and Space Technology
Barrons
BeeHive
Boeing Magazine
Boy's Life
Bulletin of the Atomic Scientists
Business and Commercial Aviation
Business Week

California

Canadian Aeronautics and Space Journal

Canadian Aviation

Canadian Aviation Historical Society Journal

Car and Driver

Chronicle of War: Vietnam

Citizen Airman

Coast Guard Magazine

Cockpit

Collier's

Combat Crew

Combat Forces Journal

Common Cause Magazine

Commonweal

Concepts

Congressional Digest

Congressional Record

Coronet

Countermeasures

Cross and Cockade Journal

Current History

Data

Defence Update International

Defense and Foreign Affairs

Defense Electronics

Defense Management Journal

Defense Monitor

Defense Science and Electronics

Defense Systems Management Review

Design News

Discover

Douglas Service Magazine

Electrical Review

Electronic Business

Electronic Design

Electronic Warfare

Electronics

Encore

Engineer

Exxon Air World

FAA Aviation News

Far Eastern Economic Review

Field Artillery Journal

Field Artilleryman

Flight

Flight International

Flight Magazine

Flight Operations

Flying

Flying Review International

Flying Safety

Forbes

Fortitudine

Fortune

Freeman

Grumman Aerospace Horizons

Harper's Magazine

High Technology

Honeywell Flightline

The Hook

IEEE Spectrum

Industrial Finishing

Industry Week

Infantry

Interavia

Interceptor

International Antique Airplane Digest

International Defence Review

International Security

Iron Age

Joint Services Recognition Journal

Journal of Aeronautical Science

Journal of Aircraft

Journal of Aviation Medicine

Journal of Defense and Diplomacy

Journal of Electronic Defense

Journal of Engineering for Gas Turbines and Power

Journal of the United States Artillery

Leatherneck

Life

Literary Digest

Logistics Forum

Look

Los Angeles

The MAC Flyer

Machine Design

Machinist

Maintenance

Marine Corps Gazette

Marines

Marketing and Media Decisions

The MATS Flyer

Mech

Mechanical Engineering

Mechanics Illustrated

Michigan History

Military Collector and Historian

Military Electronics/Countermeasures

Military Engineer

Military Forum

Military Intelligence

Military Logistics Forum

Military Review

Military Technology

Missiles and Rockets

Model Airplane News

Model Aviation

Model Builder

Modern Plastics

Ms

National Aeronautic Magazine

National Aeronautics

National Defense

National Defense Transportation Journal

National Geographic Magazine

National Guard

National Guardsman

National Journal

National Review

Nation's Business

NATO Journal

NATO's Fifteen Nations

NATO's Sixteen Nations

Naval Aviation Museum Foundation

Naval Aviation News

Naval War College Review

Navigation

Navigator

Navy

New Republic

New Scientist

New York Times Magazine

New Yorker

Newsweek

OAG Frequent Flyer

Officer

On Scene

Ordnance

Pacific Defense Reporter

Pegasus

Plane and Pilot

Popular Aviation

Popular Mechanics

Popular Science

Private Pilot

Proceedings of the Institute of Mechanical Engineers

Product Engineering

Professional Memoirs

Professional Pilot

Profile

Program Manager

Public Policy

Quality Control

Quartermaster Review

RAF Flying Review

RC Modeler

Reader's Digest

Rolls Royce Magazine

Rotor and Wing International

Royal Aeronautical Society Journal

Royal Air Forces Quarterly

Ryan Reporter

Saturday Evening Post

Saturday Review

Scale Modeller

Scale Models

Science

Science Digest

Scientific American

Sea Classics

Seapower

Sergeants

Shell Aviation News

Shipmate

Sibley Journal

Signal

Skyline

Skyways

Slipstream

Smithsonian

Smithsonian Air and Space

Society of Automotive Engineers Journal

Society of Experimental Test Pilots Technical Review

Soldier of Fortune

Soldiers

Space/Aeronautics

Space World

Sperryscope

Sport Aviation

Spur

Steel

Strategic Review

Surface Warfare

TAC Attack

Tactical Air Reconnaissance Digest

Tactical Air Warfare Center Quarterly Report

Talon

Technical World

Technology and Culture

Technology Review

Texas Monthly

Time

Tool Production

Translog

Transportation Proceedings

True

USAF Fighter Weapons News Letter

USAF Fighter Weapons Review

USAF Instructors Journal

U.S. Army Aviation Digest

U.S. Naval Institute Proceedings

U.S. News and World Report

Vectors

Vertiflite

Vietnam

Vital Speeches of the Day

Washington Monthly

Western Aerospace

Western Flying

Wings

Wings of Gold

World War Enthusiast

World War I Aero

World War II Journal

World War II Magazine

Appendix II: Late Entries

The citations below came to the compiler's attention between the first submission of the manuscript and the final correction of proofs. Included are many 1989 and some 1990 titles as well as those representing a large contribution of work made available by Challenge Publications. To ease use of this unindexed section, entries are arranged by numerical designations of the different individual aircraft models.

Abbreviations

AAHSJ = American Aviation Historical Association Journal

AWST = Aviation Week and Space Technology

USNIP = U.S. Naval Institute Proceedings

A-1

Cornelius, Art. "Skyraider." Air Classics, VI (November-December 1969), 32-41.

Dorr, Robert F. Douglas A-1 Skyraider. Osceola, Wis.: Motorbooks International, 1989. 199p.

Lichster, Lothar. "SPAD." Air Classics, II (May-June 1965), 66+.

A-3D

Blackburn, Tom. "The A3D Joins the Fleet." The Hook, XVII (Winter 1989), 48-49.

Morgan, Rich. "The A3D: The Sanford Whales." The Hook, XVII (Winter 1989), 38-47.

A3J/A-5

Grove, Michael and Jay Miller. North American Rockwell A3J/A-5 Vigilante. Aerofax Minigraph, no. 9. Arlington, Tex.: Aerofax, Inc., 1989. 56p.

A-4

Beech, Eric and Tim Hall. "Skyhawk Soars Again." Flight International, CXXXIV (December 10, 1988), 22-27.

Chesneau, Roger J. "Warship Wings, no. 3: McDonnell-Douglas A-4 Skyhawk." Warship, no. 24 (Fall 1982), 300-301.

Kinzey, Bert. A-4 Skyhawk in Detail and Scale. Detail and Scale Series, v. 32. Blue Ridge Summit, P..: TAB, 1989. 72p.

O'Leary, Michael. "Last of the Skyhawks." In: Michael O'Leary, ed. USAF-USN Jet Fighters, 1950-1965. Canoga Park, Calif.: Challenge Publications, 1986. pp. 8-15.

A-7

"A-7 Corsair." TAC Attack, XXIX (August 1989), 16-17.

Brown, David A. "LTV Begins Flight Tests of Reengined A-7F for Air Guard Mission." AWST, CXXXI (December 4, 1989), 19-21.

Holder, William G. A-7 Corsair II. Aero Series, v. 39. Blue Ridge Summit, Pa.: TAB, 1990. 96p.

Kinzey, Bert and Roy Leader. A-7 Corsair II. Colors and Markings Series, v. 15. Blue Ridge Summit, Pa.: TAB, 1990. 64p.

Phillips, Edward H. "USAF, LTV Probe Wing Cracks That Grounded Air Guard A-7s." AWST, CXXX (May 29, 1989), 29-30.

Van der Veen, Linn. "A-7." Flying Safety, XLV (January 1989), 2-4.

A-10

Mason, Eldred P. "Down and Dirty." Airpower, XIX (July 1989), 20-43+.

Nordeen, Lon. "Close Air Support: A Demanding, Dangerous, and Controversial Mission." National Defense, LXXIX (July-August 1989), 26-30.

Van der Veen, Linn. "A-10." Flying Safety, XLV (January 1989), 5-8.

A-20

Clayton, Mark. "'Gloria.'" Aerospace Historian, XXXV (Winter 1988), 235-242.

"Jungle Douglas." In: Michael O'Leary, ed. Search for the Ghostly Warriors of World War II. Canoga Park, Calif.: Challenge Publications, 1989. pp. 84-89.

Westel, Freeman. "Cry Havoc and Loose the Birds of War." Air Classics, III (January-February 1966), 42-47+.

A-24 see SBD

A-26

Dean, Jack. "Mother was an A-26." Air Classics, III (September-October 1986), 6-15.

Jamison, Theodore R. "Nightmare in the Korean Hills." AAHSJ, XXXIV (Summer 1989), 82-93.

Ryan, Frank. "War Classic." Air Classics, I (Summer 1964), 8-11.

A-28/A-29

Andrews, Hal. "Naval Aircraft: PBO-1 Hudson." Naval Aviation News, (March-April 1990), 16-17.

AC-130

"AF Readies AC-130U Gunship for First Flight." AWST, CXXXII (February 19, 1990), 64-65.

AJ

Scarborough, William E. "The North American AJ Savage." The Hook, XVII (Fall-Winter 1989), 28-43, 16-27.

AV-8

Anderson, Joseph T. "AV-8B Night Attack." Marine Corps Gazette, LXXIII (May 1989), 26-29.

Bally, J. "Short Take-Off and Vertical Landing: A Close Look at an Historic Technical Breakthrough." Armada International, XIII (October-November 1989), 46+.

Butcher, Richard. "AV-8B Enters Service." In: Michael O'Leary, ed. Naval Airpower. Canoga Park, Calif.: Challenge Publications, 1985. pp. 14-23.

Calvert, Denis J. "Cherry Point Harriers." In: Michael O'Leary, ed. Devil Dogs of the Air. Canoga Park, Calif.: Challenge Publications, 1987. pp. 19-23, 94-99.

Falconer, Jonathan. "Vectoring for Success." Armed Forces, VIII (July 1989), 310-313.

Jackson, Robert. Sea Harrier and AV-8B. Weapons and Warfare Series. New York: Sterling, 1989. 159p.

"Marines Go Operational With AV-8B Super Harrier." Air Combat, XIV (May 1986), 10+.

Most, Michael. "Evolution of a Design: The AV-8B Harrier II." Aviation Mechanics Journal, XIX (March 1989), 18-27.

"The Unstoppable Super Harrier." In: Michael O'Leary, ed. USMC Airpower in Action. Canoga Park, Calif.: Challenge Publications, 1989. pp. 26-53.

B-1

"AF Grounds B-1B Fleet Following Wing Cycling Problem." AWST, CXXX (April 3, 1989), 21-22.

Holder, William G. B-1 Bomber. 2nd ed. Aero Series, v. 35. Blue Ridge Summit, Pa.: TAB, 1988. 104p.

Lubkin, Yale Jay. "Congress Sounds Off on the B-1B EW Suite." Defense Science, VII (December 1989), 14+.

"Probe Faults Commander, Pilot in B-1B Crash at Ellsworth AFB." AWST, CXXX (March 13, 1989), 71-72.

Scott, William. "B-1B Stability Enhancement Tests Verify Low-Level Mission Capability." AWST, CXXX (June 28, 1989), 34-35.

U.S. Congress. Senate. Committee on Armed Services. Subcommittee on Strategic Forces and Nuclear Deterrence. B-1B Defensive Avionics System: Hearings. 100th Cong., 2nd sess. Washington, D.C.: GPO, 1988. 42p.

_____. Government Accounting Office. National Security and International Affairs Division. Strategic Bombers: B-1B Cost and Performance Remain Uncertain. Washington, D.C.: GPO, 1989. 60p.

Waddell, Milton H. "B-1B." Flying Safety, XLV (February 1989), 2-3.

B-2

"B-2 Bomber Takes to the Air." Armada International, XIII (October-November 1989), 84-85.

"B-2 First Flight." AWST, CXXXI (July 24, 1989), 13, 22-28.

"B-2 Taxi Tests." AWST, CXXXI (July 17, 1989), 16-23.

Bond, D. F. "AF Will Buy Low-Rate B-2s Under Fixed-Prince Incentive Contracts." AWST, CXXXI (August 7, 1989), 26-27.

_____. "USAF Says B-2's Range Exceeds B-1B's With Varied Payloads, Flight Profiles." AWST, CXXXI (October 23, 1989), 30-31.

Dornheim, Michael A. "Initial Test Flight Reveals New Details About Design Characteristics of B-2." AWST, CXXXI (July 31, 1989), 85-89.

Eshel, Tamir. "Stealth Aircraft Comes of Age." Defence Update International, no. 94 (March 1989), 22-37.

Garrison, P. "Now You See It, Now You Don't." Flying, CXVI (November 1989), 128+.

Gilmartin, Patricia A. "Full B-2 Cost Put at $155 Billion as Congressional Opposition Mounts." AWST, CXXXII (April 2, 1990), 22-23.

Gilmore, C. P. "B-2: What Stealth's First Flight Reveals." Popular Science, CXXXII (October 1989), 45-49, 90.

Graham, Vickie M. "Debut!" Airman, XXXIII (August 1989), 18-25.

Hoekema, D. A. "The B-2: Winning Weapon of the Last War." Christian Century, CVI (August 30, 1989), 772-773.

Lepingwell, John W. R. "Soviet Strategic Air Defense and the Stealth Challenge." International Security, XIV (Fall 1989), 64-100.

Moore, Tim. B-2 First Test Flight." Combat Crew, XXXIX (September 1989), 4-5.

Morrocco, John D. "Bipartisan Opposition to B-2 Grows Despite Release of New Information." AWST, CXXXI (July 17, 1989), 22-23.

_____. "SAC Chief Indicates Scaled Back B-2 Bomber Procurement Possible." AWST, CXXXII (March 12, 1990), 22-23.

O'Leary, Michael. "Stealth Unveiled." Air Combat, XVII (March-April 1989), 54-57.

Randolph, Bernard P. "The B-2 Bomber." Vital Speeches of the Day, LV (June 1, 1989), 494-497.

Rhodes, Jeffrey P. "On Stealthy Wings." Air Force Magazine, LXXII (February 1989), 43-49.

Richardson, Doug. "Is Stealth Misleading?" Interavia, XLIV (October 1989), 968-970.

_____. Stealth: Deception, Evasion, and Concealment in the Air. New York: Crown, 1989. 192p.

Schnepf, Edwin, ed. Stealth: Northrop's B-2 Stealth Bomber and America's Stealth Force. Canoga Park, Calif.: Challenge Publications, 1989. 66p.

Scott, William B. "Success of B-2 Program Linked to Decision to Build, Deploy B-1B." AWST, CXXXII (February 5, 1990), 71+.

_____. and Michael A. Dornheim. "Post-Flight Review Indicates Airworthiness of B-2 Design." AWST, CXXXI (July 24, 1989), 22-25.

Smith, Bruce A. "B-2 Peak Production Delays Drive Up Program Costs." AWST, CXXXI (July 24, 1989), 26-27.

Stringer, Jerry. "New Facts About the New Bomber: Lifting the B-2 Veil." Airman, XXXIII (June 1989), 22-23.

Sweetman, Bill. "B-2 Bomber for the 21st Century." Interavia, XLIV (January 1989), 22-26.

_____. Stealth Bomber. Osceola, Wis.: Motorbooks International, 1989. 176p.

Varni, Gerard. "Stealth Bomber." International Combat Arms, VII (September 1989), 42-51.

Watts, Barry. "The Coming Age of Stealth." In: U.S. Commission on Integrated Long-Term Strategy, Working Group on Technology. Technology for National Security. Washington, D.C.: GPO, 1988. pp. 87-93.

B-10

Boyne, Walter J. "B-10: Martin's Soldier." Air Classics, II (May-June 1965), 32-37.

B-17

Birdsall, Steve. "Hell's Angels: A Galaxy of B-17 Flying Fortresses in Combat." Air Classics, V (July-August 1968), 34-53.

Birdsong, George. Stormy Weather. Dublin, Calif.: Hambleden Publishing Co., 1989.
Recollections of the pilot of the 91st BG's "Delta Rebel #2," the first Fortress to complete 25 ETO missions.

Bowers, Peter M. "The Saga of the 'Shoo Shoo Baby.'" Airpower, XIX (May 1989), 12-25.

Doherty, Robert E. "Last Flight of the 'Mischief Maker II.'" Air Classics, XXV (May 1989), 14-19, 81.

Fields, Kenneth W. "New Guinea Fort." In: Michael O'Leary, ed. Search for the Ghostly Warriors of World War II. Canoga Park, Calif.: Challenge Publications, 1989. pp. 18-23.

Gibson, Everett K. "'Texas Raiders.'" Warbirds International, VII (November-December 1988), 34-41.

"Let's Save the 'Swamp Ghost.'" Warbirds International, VIII (November-December 1989), 28-31.

Mitchell, Mitch. "A Once and Future Queen." Warbirds International, VII (September-October 1988), 50-53, 71.

O'Leary, Michael. "The First Forts." In: Michael O'Leary, ed. Victorious Eagles. Canoga Park, Calif.: Challenge Publications, 1989. pp. 20-41, 81-93, 98-99.

Peltzer, Mike. "B-17 Survivors." AAHSJ, XXXIV (Spring 1989), 12-15.

Schnepf, Edwin, ed. The Saga of the B-17 Flying Fortress. Canoga Park, Calif.: Challenge Publications, 1990. 114p.

Thorp, Walter N. "Baby Reborn." Air Classics, XXV (March 1989), 34-45. "Shoo Shoo Baby."

_____. "Salute to the Baby." Warbirds International, VIII (July-August 1989), 46-53.

B-24

Alford, Bob. "The Search for 'Nothing Sacred.'" Warbirds International, VIII (May-June 1989), 6-11.

Boghosian, S. Samuel. "Mystery of the Missing Liberators." Warbirds International, VIII (September-October 1989), 66-70, 72, 76.

Bowman, Martin W. B-24, 1939-1945. 2nd ed. New York: Sterling, 1989. 192p.

Carigan, William. Ad Lib: Flying the B-24 Liberator in World War II. Manhattan, Kans.: Sunflower University Press, 1988. 96p.

Carlin, Robert. "Twin-Tailed Time Bomb." Air Classics, V (July-August 1968), 16-25.

Cundiff, Michael J. "Ten Knights in a Bar Room": Missing in Action in the South Pacific, 1943. Ames, Iowa: Iowa State University Press, 1989. 144p.

Dorr, Robert F. "B-24 Liberator: The Mostest." Air Power History, XXXVII (Spring 1990), 27-31.

Holder, William J. "Different Faces--Different Jobs." Aviation Mechanics Journal, XIX (May and July 1989), 24-31, 34-40.

Nash, Colleen A. "Grand Old Liberator." Air Force Magazine, LXXII (September 1989), 134+.

O'Leary, Michael. B-24 Liberator. Canoga Park, Calif.: Challenge Publications, 1989. 122p.

____. "Restoration of the Decade." Air Classics, XXVI (January 1990), 40-53, 76.
B-24J "All American."

"Salute to the Mighty Liberator: Pictorial Survey of America's Most-Produced World War II Bomber." Air Classics, XXV (June 1989), 62-73.

Sloat, Chuck. "The Unknown Liberator." Warbirds International, VII (May-June 1988), 16-25.

Smith, Gene. "Flying the Mighty Liberator." Warbirds International, VIII (September-October 1989), 38-47, 78-79.
"Delectable Doris."

B-25

Dean, Jack. "Gun Ship." Air Classics, III (November-December 1966), 24-29.

Sylva, Dave. "B-25s Over Tokyo." Flying Safety, XLV (June 1989), 16-19.

B-26

Pennington, Robert A. "Lost Marauder." In: Michael O'Leary, ed. Search for the Ghostly Warriors of World War II. Canoga Park, Calif.: Challenge Publications, 1989. pp. 90-97.

Ryan, Frank. "Wild and Wooly Widow Maker." Air Classics, I (Fall 1964), 16-21.

B-29

Scott, P. "Restoration Work." Flying, CXVI (November 1989), 66-68+.
"Enola Gay."

B-52

Berlan, G. J. Forward Offense: Preparing the B-52 for Conventional Warfare. Maxwell AFB, Ala.: Airpower Research Institute, Air University, 1989. 114p.

Francillon, Rene J. B-52: Ageing Buffs, Youthful Crews. Osceola, Wis.: Motorbooks International, 1989. 128p.

Harford, J. D. B-52 Aircraft on the Central Front. Study Project. Carlisle Barracks, Pa.: Army War College, 1989. 55p.

Harrell, A. Lee. "Weaseling in the Buff." Airpower Journal, III (Spring 1989), 37-42.

Waddell, Milton H. "B-52." Flying Safety, XLV (February 1989), 4-5.

B-57

Delve, Ken. Canberra and B-57. Warbirds Illustrated, no. 51. New York: Sterling, 1989. 64p.

Jackson, Robert. Canberra: The Operational Record. Washington, D.C.: Smithsonian Institution Press, 1989. 168p.

BT

Andrews, Hal. "Naval Aircraft: Douglas BT." Naval Aviation News, (September-October 1989), 16-17.

BT-13

Ewing, Dwight. "The Saga of Ol' Snarly." Air Classics, VI (March-April 1969), 38-43.

C-1A

Andrews, Hal. "Naval Aircraft: C-1A/TF Trader." Naval Aviation News, (May-June 1989), 14-15.

C-2(1)

"'Question Mark's' Anniversary." Air Classics, XXV (May 1989), 38-39.

Renfro, Robert B. "'Question Mark's' Flight." AAHSJ, XXXIV (Summer 1989), 136-143.

C-2(2)

"Improved Greyhound." In: Michael O'Leary, ed. Navair. Canoga Park, Calif.: Challenge Publications, 1986. pp. 82-85.

C-5

"C-5B First Flight." In: Michael O'Leary, ed. Military Airpower Review, VI (1986), 10+.

Wilkinson, Stephen. "Big." Smithsonian Air and Space, IV (August-September 1989), 28-39.

C/KC-10

"Crash Course for Cool Heads." Airman, XXXIII (October 1989), 10-14.

Wanstall, Brian. "Tankers Boost Combat Credability." Interavia, XLIV (June 1989), 559+.

C-12(1)

Brown, D. J. "Lockheed Vega." Air Classics, III (November-December 1966), 30-35.

C-17(2)

Ladnier, Ron R. "C-17 Aircrew Training." Airlift, XI (Spring 1989), 1-5.

McGowan, Sam. "Airlift and the New C-17." National Defense, LXXIX (July-August 1989), 40-43.

Scarborough, Rowan. "New Hurdles for the C-17." Military Forum, V (May 1989), 17+.

Shellner, Earl. "Bridging the Airlift Gap: The C-17's Timely Arrival." Airman, XXXIII (August 1989), 16-17.

_____. "Giving Airlift a Boost." Airman, XXXIII (August 1989), 10-16.

C-23A

O'Leary, Michael. "Sherpa." In: Michael O'Leary, ed. Air Force in Action. Canoga Park, Calif.: Challenge Publications, 1987. pp. 58-61, 76-77.

_____._____. In: Michael O'Leary, ed. NATO's Deadly Warplanes. Canoga Park, Calif.: Challenge Publications, 1987. pp. 80-85.

C-26A

Shifrin, Carole A. "Fairchild Delivers First Two C-26As to Air National Guard." AWST, CXXX (April 24, 1989), 36-39.

C-45

Phillips, Ed. "Twin Beech: The Early Years." Aviation Mechanics Journal, XIX (August 1989), 24-31.

Sloat, Chuck. "Salute to the Bug Smasher." Warbirds International, VII (January-February 1988), 48-56.

C-47

Johnsen, Frederick A. "McChord's Gooney." In: Michael O'Leary, ed. Search for the Ghostly Warriors of World War II. Canoga Park, Calif.: Challenge Publications, 1989. pp. 24-29, 98.

C-69

O'Hara, Bob. "Connie." Air Classics, V (September 1968), 25-43.

C-75

O'Hara, Bob. "The Very Round Robin." Air Classics, III (May-June 1966), 42-47.

C-98

Klass, M. D. "The Incredible Clippers." Air Classics, V (September-October 1968), 10-24.

C-119

Mitchell, Kent A. "The One and Only Fairchild C-119H Flying Van." AAHSJ, XXXIV (Spring 1989), 46-49.

C-123

Dean, Jack. "Payload." Wings, XIX (October 1989), 36+.

C-124

McKinney, Sherman R. "The Vicious Wind." Flying Safety, XLV (March 1989), 20-23.

C/KC-130

Hunter, James R. "The Medriders: Refueling in Flight." USNIP, CXVI (February 1990), 49-51.

Kolcum, E. H. "Lockheed Weighs Investment Risks of Developing New C-130 Version." AWST, CXXXI (November 27, 1989), 45+.

McGowan, Sam. "Blind Bat." Air Force Magazine, LXXII (July 1989), 82-85.

Mayhew, Mark. "C-130." Flying Safety, XLV (February 1989), 10-12.

Morris, M. E. C-130: The Hercules. San Rafael, Calif.: Presidio, 1989. 128p.

Scott, William B. "Future C-130 Version May Incorporate STOL Features and Autonomous Capability." AWST, CXXIX (March 29, 1989), 34-35.

Sweetman, Bill. "Air Support for Special Operations." Interavia, XLIV (February 1989), 133-136.

C/KC-135

Laming, Tim. Flying Tankers. Osceola, Wis.: Motorbooks International, 1989. 124p.

Wall, James L. "C/KC-135." Flying Safety, XLV (February 1989), 13-15.

C-141

Wittel, Taylor. "C-141 Structural Design Criteria and Severe Missions." MAC Flyer, XXXII (February 1990), 20-23.

CG-4

Gault, Owen. "Pilots Expendable." Air Classics, III (September-October 1966), 43-59.

DF

O'Hara, Bob. "Douglas' Seagoing Gooney Bird: The DF Flying Boat." Air Classics, VI (May-June 1969), 58+.

E-2

Sullivan, Mark D. "Hawkeye Mishaps: Facts and Fallacies." Approach, XXXIV (February 1983), 8-9.

E-3

Holmes, Jerry. "NATO's Own: AWACS." NATO's Sixteen Nations, XXXIV (1989), 62-65.

E-6A

Dornheim, Michael A. "E-6A Tacamo Tail Damage Prompts Delivery Delay." AWST, CXXX (April 17, 1989), 28-29.

Frasher, Joan A. "Turnkey Training for the E-6A Hermes." Naval Aviation News, (March-April 1989), 12-13.

EA-6

Jenkins, Denis R. Grumman EA-6A Intruder/EA-6B Prowler. Arlington, Tex.: Aerofax, Inc., 1989. 57p.

Lloyd, Alwyn T. "Golden Wings Among the Silver." Combat Crew, XXXIX (June 1989), 8-11.

Kandebo, Stephen W. "Grumman Prepares for Flight Tests of Advanced Capability EA-6B." AWST, CXXXI (September 25, 1989), 41+.

EF-111

"EF-111 Tactical Jammer." In: Michael O'Leary, ed. Military Airpower Review, II (1984), 10+.

"Tactical Jamming Raven." In: Michael O'Leary, ed. NATO's Deadly Warplanes. Canoga Park, Calif.: Challenge Publications, 1987. pp. 54-59.

ES-3

Nordwall, B. D. "Tests to Verify Airworthiness of New ES-3A Configuration." AWST, CXXXI (September 25, 1989), 36-37+.

F3D

Amody, Francis J. "Skyknights, Nightmares, and MIGs." AAHSJ, XXXIV (Winter 1989), 307-313.

F-4

Dorr, Robert F. McDonnell F-4 Phantom. Aero Series, v. 36. Blue Ridge Summit, Pa.: TAB, 1989. 104p.

Evans, Julius L. "F-4 MIG Killer." Naval Aviation News, (May-June 1989), 32-33.

Johnson, Jerome L. "F/RF-4." Flying Safety, XLV (January 1989), 9-12.

Kinzey, Bert. Colors and Markings of the F-4E Phantom II. Colors and Markings Series, v. 13. Blue Ridge Summit, Pa.: TAB, 1989. 64p.

Mersky, Peter. "Phantom Farewell." Naval History, II (Summer 1988), 37-45.

Shaw, Robbie. F-4 Phantom: Guardian of the Free World. Osceola, Wis.: Motorbooks International, 1989. 144p.

F-4B

Ryan, Frank. "Boeing's BIPE: Classic F4B Navy Fighter." Air Classics, III (November-December 1966), 54+.

F4F

"The Bethpage Cat." In: Michael O'Leary, ed. Hawks Over the Pacific. Canoga Park, Calif.: Challenge Publications, 1989. pp. 24-37, 86-93.

F4U

Dean, Jack. "Pirate of the Pacific." Airpower, XIX (September 1989), 10-29.

O'Leary, Michael. "Corsair Graveyard." Warbirds International, IX (January-February 1990), 6-12.

Parchman, John C. "Pursuing the Bent-Wing Bird." Warbirds International, VIII (March-April 1989), 10-13.

F-5

Larke, Graham. "F-5." Flying Safety, XLV (January 1989), 13-15.

F-5

O'Leary, Michael. "Lockheed's Photo Fighters." In: Michael O'Leary, ed. Spy Planes. Canoga Park, Calif.: Challenge Publications, 1988. pp. 54-63.

F6F

Gamble, Dale and Tim Adams. "America's Answer to the Japanese Zero." All Hands, no. 810 (August 1984), 16-17.

F7U

Moore, Jack. "Surviving the Cutlass." The Hook, XVII (Fall 1989), 20-27.

F-8

Holliday, Joe. "DeHavilland's Wooden Wonder." Air Classics, V (January-February 1968), 54-61+.

F8U

Ginter, Steve. Vought's F-8 Crusader, Part III: Marine Fighter Squadrons. Simi Valley, Calif.: Naval Fighters, 1989. 66p.

Kinzey, Bert. F-8 Crusader in Detail and Scale. Detail and Scale Series, no. 31. Blue Ridge Summit, Pa.: TAB, 1989. 72p.

Schoeni, Arthur. "Crusader F8U." In: Michael O'Leary, ed. USAF-USN Jet Fighters, 1950-1965. Canoga Park, Calif.: Challenge Publications, 1986. pp. 38-47.

F9F

Tuttle, Bruce N. "F9F Panther." In: Michael O'Leary, ed. USAF-USN Jet Fighters, 1950-1965. Blue Ridge Summit, Pa.: TAB, 1986. pp. 74-81.

F11F

Trimble, Robert L. "Grumman F11F Tiger." In: Michael O'Leary, ed. USAF-USN Jet Fighters, 1950-1965. Canoga Park, Calif.: Challenge Publications, 1986. pp. 28-33, 80-81.

F-14

Chesneau, Roger J. "Warship Wings, no. 6: Grumman F-14 Tomcat." Warship, no. 27 (Summer 1983), 186-187.

"Gaddafi's Nightmare." In: Michael O'Leary, ed. Navair. Canoga Park, Calif.: Challenge Publications, 1986. pp. 10-26.

Howard, Gregg. "F-14 Crew Co-ordination in Out-of-Control Flight." Approach, XXXIV (January 1989), 8-9.

Lundquist, E. H. "Tomcats Come Aboard." In: Michael O'Leary, ed. Killer Elite. Canoga Park, Calif.: Challenge Publications, 1987. pp. 40-47.

Morrocco, John D. "Delays in Navy ATF Program Bolster Case for F-14D." AWST, CXXXI (September 4, 1989), 30-31.

_____. "Tomcat 21 Will Include Technologies to Reduce F-14's Radar Signature." AWST, CXXX (May 22, 1989), 30-31.

F-15

Hill, Martin V. "F-15." Flying Safety, XLV (January 1989), 16-19.

Kandebo, Stanley W. "Modified F-15B to Demonstrate STOL, Maneuver Capability." AWST, CXXX (May 29, 1989), 44-47.

Morrocco, John D. "ATF Challenged by F-15XX, New Single-Engine ATF." AWST, CXXXII (February 19, 1990), 16-17.

Smith, Gene. "Hard to Beat." In: Michael O'Leary, ed. Killer Elite. Canoga Park, Calif.: Challenge Publications, 1987. pp. 48-55, 68-69.

Varni, Gerard. "F-15E Dual-Role Fighter--Twice as Nice." International Combat Arms, VII (September 1989), 12-19.

F-16

Bond, D. F. "Budget Cuts, Cost Growth Cloud F-16 Plans for 1990s." AWST, CXXXI (December 4, 1989), 18-19.

_____. "Close Air Support Move in Congress Might Disrupt Production of F-16." AWST, CXXXI (November 13, 1989), 22-24.

Cole, Tim. "Bird-of-Prey." Popular Mechanics, CLXVI (July 1989), 70-72, 118.

Ethell, Jeffrey L. "Airborne in the Electric Jet." In: Michael O'Leary, ed. Killer Elite. Canoga Park, Calif.: Challenge Publications, 1987. pp. 20-25.

Scott, William B. "F-16C/D High Angle-of-Attack Instability Overcome by Flight Control Modifications." AWST, CXXXI (November 13, 1989), 61+.

Shifrin, Carole A. "TAC Demonstration Bolsters Support for F-16 in CAS Role." AWST, CXXX (April 17, 1989), 49-50.

Thornton, John P. "F-16 Air Defense Fighter." USAF Fighter Weapons Review, XXXVII (Spring 1989), 30-33.

United States. Office of the Secretary of Defense, Director, Operational Test and Evaluation. Operational Test Plan Concept for Evaluation of Close Air Support Aircraft: Summary. Washington, D.C., 1989. 130p.

"U.S. 7th Air Force Shifts to All F-16 Fighter Force." AWST, CXXX (June 12, 1989), 229-230.

F/A-18

Braybrook, Roy M. "MIG-29 vs. F/A-18: Keep Out of Dogfights!" Pacific Defense Reporter, XV (February 1989), 50-52.

DeVeer, Stephen A. "Mishap Mission to Mexico." Approach, XXXIV (February 1989), 18-21.

Donahue, Paul F. "The F/A-18D for the A-6E: A Good Deal That Warrants a Closer Look." Marine Corps Gazette, LXXIII (July 1989), 26-27.

"Double-Edged Sword." In: Michael O'Leary, ed. Navair. Canoga Park, Calif.: Challenge Publications, 1986. pp. 27-35.

"F/A-18 Hornet: Airpower for the Navy." In: Michael O'Leary, ed. Military Airpower Review, III (1984), 10+.

"F/A-18 Hornet: The Multi-Mission Strike Fighter." Asian Defence Journal, no. 12 (December 1988), 68-71+.

Harris, Dick. "Flying the F/A-18 Hornet." Boy's Life, LXXVII (November 1987), 38+.

Jopling, H. W. "Fabulous F/A-18: Fact or Fiction?" Marine Corps Gazette, LXXIII (July 1989), 25-26.

Krupp, Dennis T. and David J. Rash. "F/A-18D Hornet: Strike Fighter for the Future." Marine Corps Gazette, LXXIII (July 1989), 22-25.

Macak, William C. "F/A-18 and the Armed Reconnaissance Mission." Marine Corps Gazette, LXXIV (March 1990), 26-27.

Noorda, Tim. "Low-Altitude Training in the Hornet." The Hook, XVII (Winter 1989), 102-105.

Rollinger, Martin G. "F/A-18 All-Weather Close Air Support." Marine Corps Gazette, LXXIV (March 1990), 28-29.

Smith, Gene. "Every Boy Should Have One." In: Michael O'Leary, ed. Contemporary Warplanes. Canoga Park, Calif.: Challenge Publications, 1986. pp. 6-15.

_____. "Hornet Victorious." In: Michael O'Leary, ed. Killer Elite. Canoga Park, Calif.: Challenge Publications, 1987. pp. 56-83.

"Strike Hornet." In: Michael O'Leary, ed. USMC Airpower in Action. Canoga Park, Calif.: Challenge Publications, 1989. pp. 54-63, 84-89.

F-21A

"Kfirs for the Navy." In: Michael O'Leary, ed. Navair. Canoga Park, Calif.: Challenge Publications, 1986. pp. 60-65.

Lowe, Malcolm. "Israel's Kfir Jet Fighter: The Lion Cub Roars." International Combat Arms, VII (May 1989), 51-54.

"Navy's F-21A Kfir." In: Michael O'Leary, ed. Military Airpower Review, VI (1986), 10+.

F-80

Schrader, Richard K. "Shooting Star." Air Classics, XXV (April 1989), 38-47.

F-86

Ervice, Greg and Mitch Mitchell. "Song of the Sabre." Warbirds International, VIII (July-August 1989), 18-23.

O'Leary, Michael. "The Navy's Sabres." Air Combat, XVII (May-June 1989), 34-41.
FJ Fury.

F-100

Kinzey, Bert. F-100 Super Sabre in Detail and Scale. Detail and Scale Series, v. 33. Blue Ridge Summit, Pa.: TAB, 1989. 64p.

Menard, David W. F-100 Super Sabre. Colors and Markings Series, v. 14. Blue Ridge Summit, Pa.: TAB, 1990. 64p.

F-101

Grant, Robert S. "Black Voodoo." Warbirds International, VII (May-June 1988), 6-9.

F-104

Lanier, L. H. "NF-104 Starfighter." In: Michael O'Leary, ed. USAF-USN Jet Fighters, 1950-1965. Canoga Park, Calif.: Challenge Publications, 1986. pp. 48-65.

F-106

DiChristina, Mariette. "Forward Flap Lifts Fighter." Popular Science, CCXXXV (July 1989), 83-84.

Phillips, Edward H. "NASA Langley Expands Flight Envelope of F-106B Equipped With Vortex Flaps." AWST, CXXX (June 5, 1989), 22-23.

F-111

"F-111 Swing-Wing Bomber in Action." Defence Update International, no. 92 (December 1988), 30-42.

Henderson, B. W. "USAF, Rockwell Begin Work on F-111D/F Avionics Upgrade." AWST, CXXXII (January 29, 1990), 51+.

Scott, William B. "One Digital Computer to Replace Six Flight Control Units in F-111." AWST, CXXX (March 6, 1989), 51-53.

Thornborough, Anthony M. F-111 Aardvark. Warbirds Photofax. New York: Sterling, 1989. 48p.

Titus, Nathan T. "F-111." Flying Safety, XLV (January 1989), 26-28.

F-117A

Dornheim, Michael A. "USAF Continues to Increase Daytime Flights of F-117A Stealth Aircraft." AWST, CXXXI (July 3, 1989), 24-25.

____. "USAF/Lockheed F-117A Has High-Wing Sweep, But Low-Wing Loading." AWST, CXXX (May 1, 1989), 26-27.

Lake, Julian S. "Stealth: The F-117A Fighter." Defense Science, VIII (January 1989), 24-26.

Scott, William B. "USAF Expands Use of F-117A, Adds More Daytime Flights." AWST, CXXX (May 1, 1989), 24-25.

"Stealth Fighter Photographed Over Nevada." Popular Mechanics, CLXVI (August 1989), 15-18+.

Sweetman, Bill. "Wobbly Goblin: Not Perfect, But First." Interavia, XLVI (January 1989), 28-30.

FB

Andrews, Hal. "Naval Aircraft: FB." Naval Aviation News, (March-April 1989), 16-17.

AH-1

Grimes, Vincent. "Supercobra." International Combat Arms, VII (July 1989), 41-45.

Verier, Mike. Cobra: Tank Killer Supreme. Osceola, Wis.: Motorbooks International, 1989. 126p.

UH-1

McLean, Herbert E. "A Flight on the Huey 'Memory Machine.'" Rotor and Wing International, XXIV (February 1990), 60-63.

Sprinkle, James D. "Helatkltron Three: The Seawolves." AAHSJ, XXXIII (Winter 1988), 290-310.

SH-2

Kandebo, Stanley W. "Navy Plans to Roll Out Upgraded SH-2 Seasprite Helicopter in October." AWST, CXXX (June 19, 1989), 143-147.

Ruth, Michael S. "LAMPS Mk. I: Brighter Than Ever." Surface Warfare, IX (July-August 1984), 12-13.

Schrader, Richard K. "Super Seasprite." Air Combat, XVII (May-June 1989), 6-10+.

H-5

Thornton, John W. "Rotorhead." The Hook, XVII (Summer 1989), 64-65.

OH-6

Porter, Donald J. The McDonnell-Douglas OH-6A Helicopter. Aero Series, v. 38. Blue Ridge Summit, Pa.: TAB, 1989. 96p.

H-46

Kocks, Kathleen. "Cherry Point: The Marines' Leading SAR Helicopter Unit." Rotor and Wing International, XXIII (August 1989), 42-43.

H-47

Harvey, David. "Outfitting the MH-47E For Its Special Operations Mission." Rotor and Wing International, XXIII (October 1989), 60-61, 74.

Murray, Frank. "CH-47 Chinook 'Hands Off' Flying." U.S. Army Aviation Digest, XXVI (May 1989), 28-29.

H-52

Burgess, Rick. "Sea Guard Retires at 26 Plus." Naval Aviation News, (January-February 1990), 10-11.

H-53

Harvey, David. "Pave Low III Rotorcraft: The USAF's Night Stalkers." Rotor and Wing International, XXIII (February 1989), 48-50.

Stamp, Joy. "Pave Low III-E Helos at Woodbridge." Rotor and Wing International, XXIII (September 1989), 40-42.

"Super Stallions of the Warhose (HMH-465) Squadron." In: Michael O'Leary, ed. USMC Airpower in Action. Canoga Park, Calif.: Challenge Publications, 1989. pp. 6-13, 76-79.

OH-58

"Armed OH-58D Readies for Army Induction." Rotor and Wing International, XXIII (March 1989), 54-55.

"OH-58D Advanced Scout." Defence Update International, no. 96 (May 1989), 20-29.

H-60

Cripps, David B., et al. Airworthiness and Flight Characteristics Evaluation of the EH-60A (Quick Fix) Helicopter: Final Report. USAAEFA-87-07. Edwards AFB, Calif.: Army Aviation Engineering Flight Activity, 1988. 98p.

Galdorisi, George. "LAMPS III: How to Procure a Winner." USNIP, CXV (January 1989), 93-97.
SH-60.

"High Altitude Night Hawk." In: Michael O'Leary, ed. Contemporary Warplanes. Canoga Park, Calif.: Challenge Publications, 1986. pp. 16-17.
HH-60A.

AH-64

Bond, D. F. "U.S. Army Approves Initial Design Phase of Longbow Radar System for AH-64." AWST, CXXXI (July 31, 1989), 27-28.

DeLucien, Albert. "Fielding the AH-64 Apache." U.S. Army Aviation Digest, XXXIV (July-August 1989), 54-55.

Ferrell, Mark and Scott Reynolds. "Apache Thunder: The Advanced JAAT Test." U.S. Army Aviation Digest, XXXIV (May 1989), 2-10.
JAAT = Joint Air Attack Team.

Harvey, David. "Is the Army Way the Way to Fly a Combat Helicopter?" Rotor and Wing International, XXIII (April 1989), 62-63, 86.

Manuel, Philip K. "Aviation Logistics: History of Apache Maintenance Training." U.S. Army Aviation Digest, XXXIV (June 1989), 15-17.

Sack, Friedrich W. "Using Helicopters to Fight Tanks." Armada International, XIII (February-March 1989), 36-38.

H-65

McClellan, J. Mac. "Day of the Dauphin." Flying, CXVI (September 1989), 76-81.

J2F

Trimble, Robert L. "The Duck That Roared." In: Michael O'Leary, ed. Devil Dogs of the Air. Canoga Park, Calif.: Challenge Publications, 1987. pp. 74-85.

JN-4

Baxter, Gordon. "Jenny in a Barn." Flying, CXVI (August 1989), 130+.

JRM

O'Leary, Michael. "The Largest Warbird." Warbirds International, VII (September-October 1988), 54-63.

LHX

Harvey, David. "How LHX Will Do the Job." Rotor and Wing International, XXIII (April 1989), 58-61.

Rhodes, Jeffrey P. "LHX on the Horizon." Air Force Magazine, LXXII (April 1989), 60-65.

Nieuport

Bruce, J. M. Nieuport Aircraft of World War I. Vintage Warbirds, no. 10. New York: Sterling, 1989. 64p.

NLB-1

Lincoln, Marshall. "The Barling Bomber." Air Classics, II (September-October 1965), 28-33.

O-1

Wagner, Richard P. "One Flew Over the Ammo Dump." Soldier of Fortune, XIV (April 1989), 65-68.

O-5

Lincoln, Marshall. "They Conquered the World." Air Classics, III (January-April 1966), 62+, 4-9+.

O-47

Thomas, Merle. "O-Bird." Air Classics, II (November-December 1965), 46-51.

OS2U

Edington, Robert. "Kingfisher." Air Classics, II (November-December 1965), 34-39.

Schoeni, Arthur L. "Floatplane Hero." In: Michael O'Leary, ed. Hawks Over the Pacific. Canoga Park, Calif.: Challenge Publications, 1989. pp. 6-23, 72-79.

OV-1

Love, Terry. OV-1 Mohawk in Action. Carrollton, Tex.: Squadron/Signal Publications, 1989. 50p.

Rux, William A. "Tchepone Navy." U.S. Army Aviation Digest, XXXIV (June 1989), 22-27.

OV-10

Scott, William B. "Upgrades of OV-10s Will Extend Aircraft's Service Life Into 1990s." AWST, CXXX (July 10, 1989), 40-41+.

Williams, Steve. "OV-10 Aerial Observer." In: Michael O'Leary, ed. Devil Dogs of the Air. Canoga Park, Calif.: Challenge Publications, 1987. pp. 26-27.

P-3

Nash, Trevor. "ASW Operations with PATWING 11." Armed Forces, VIII (May 1989), 220-224.

Roscoe, Jack F. and David L. Hall. "Sharpening the Sword of Orion." USNIP, CXV (June 1989), 93-95.

Sullivan, Kerry. "Track and Attack State of the Art Antisub Warfare." International Combat Arms, VII (May 1989), 38-41.

P5M

Trimble, Robert L. "Martin's Fighting Marlin." In: Michael O'Leary, ed. Seaplanes at War. Canoga Park, Calif.: Challenge Publications, 1988. pp. 24-29, 89-90.

P6M

Allen, Francis J. "Poseidon's Giant: The Story of the Martin P6M Sea-Master." AAHSJ, XXXIV (Summer 1989), 144-153.

P-30

Dean, Jack. "The Pursuit With Six Names." Air Classics, IV (March-April 1967), 6-13.

P-36

Boyne, Walter J. "Hawks for Hire." Air Classics, VI (November-December 1969), 6-23.

Hyer, Charles R. "Curtiss P-36 in Squadron Service." AAHSJ, XXXIII (Winter 1988), 242-257.

P-38

Abberger, T. J. "P-38 Combat Log." In: Michael O'Leary, ed. Victorious Eagles 2. Canoga Park, Calif.: Challenge Publications, 1990. pp. 24-43, 74-80, 90-98.

Veronico, Nick. "The Last Lightnings." Warbirds International, VIII (September-October 1989), 22-25.

P-39

Furler, E. H. "P-400: Dog of War." In: Michael O'Leary, ed. Fabulous Flops. Canoga Park, Calif.: Challenge Publications, 1990. pp. 88-95.

O'Leary, Michael. "Cobra in the Clouds." In: Michael O'Leary, ed. Victorious Eagles. Canoga Park, Calif.: Challenge Publications, 1989. pp. 42-57, 74-80.

_____. "Cobra With a Cannon." Warbirds International, VII (September-October 1988), 14-31, 78-89.

P-40

Ethell, Jeffrey L. "NWM and the Kittyhawk." Warbirds International, VIII (March-April 1989), 50-69.
NWM = National Warplane Museum, Genesco, New York.

O'Leary, Michael. "P-40." Warbirds International, VII (July-August 1988), 6-27, 76-87.

_____. "Surplus." Warbirds International, VIII (July-August 1989), 54-69.

Schnepf, Edwin, ed. Warhawks Against the Rising Sun: The War of the Curtiss P-40 Fighting the Imperial Japanese Army and Navy Air Forces. Canoga Park, Calif.: Challenge Publications, 1989. 98p.

P-47

Hess, William N. P-47 Thunderbolt. Warbirds Photofax. New York: Sterling, 1989. 48p.

P-51

Dean, Jack. "Mustang." Air Classics, II (November-December 1965), 27-33.

Hatch, Gardner N. P-51 Mustang. Padukah, Ky.: Turner Publishing Co., 1988. 144p.

"Mustang Men." In: Michael O'Leary, ed. Victorious Eagles 2. Canoga Park, Calif.: Challenge Publications, 1990. pp. 44-47.

P-63

Furler, E. F., Jr. "Bill's Ultimate Cobra." In: Michael O'Leary, ed. Victorious Eagles 2. Canoga Park, Calif.: Challenge Publications, 1990. pp. 54-83.

O'Leary, Michael. "Last Kingcobra." Warbirds International, V (Winter 1986), 32-37.

PA-36

Aellen, Richard. "The Autogiro and Its Legacy." Smithsonian Air and Space, IV (December 1989-January 1990), 52-59.

PBY

Beebe, William A., 2nd. "World War II Catalina Delivered Safely to Cherry Point." Fortitudine, XIX (Winter 1989-1990), 24, 18-19.

DiLeo, Joseph and David J. Moore. "Cat on Display." Warbirds International, VIII (March-April 1989), 18-21.

Haller, Mike and Julie Leonard. "The Queen Comes Home." Warbirds International, VII (January-February 1988), 28-37.

Spencer, Ted M. "Queen of Dago Lake." Warbirds International, V (Winter 1986), 40-47.

"Those Killer Cats." In: Michael O'Leary, ed. Seaplanes at War. Canoga Park, Calif.: Challenge Publications, 1988. pp. 36-43.

PB2Y

Andrews, Hal. "Naval Aircraft: Coronado." Naval Aviation News, (November-December 1989), 22-23.

PN-9

Asher, Don. "The Seaplane That Sailed to Glory." Air Classics, II (November-December 1965), 14-17.

PT

O'Hara, Bob. "Fairchild's Fair Children." Air Classics, IV (September-October 1967), 20-23.

PT-17

Andrews, Hal. "Naval Aircraft: N2S/PT-17." Naval Aviation News, (January-February 1990), 16-17.

PT-19

Sloat, Chuck. "A Fairchild Restoration." Warbirds International, VIII (May-June 1989), 22-28.

PV-2

Veronico, Nick and John F. Whyte. "Gun-Nosed Harpoon." Warbirds International, VIII (March-April 1989), 30-37.

RB-45

Myers, Howard S., Jr. "We Flew Tornados." In: Michael O'Leary, ed. Spy Planes. Canoga Park, Calif.: Challenge Publications, 1988. pp. 26-31.

R3Y-2

O'Leary, Michael. "Tradewind." In: Michael O'Leary, ed. Seaplanes at War. Canoga Park, Calif.: Challenge Publications, 1988. pp. 66-67.

S-3A/B

Chesneau, Roger J. "Warship Wings, no. 2: The Lockheed S-3A Viking." Warship, no. 23 (Summer 1982), 178-180.

Nash, Trevor. "Viking at Sea." Armed Forces, VIII (June 1989), 268-269.

"S-3B Viking." In: Michael O'Leary, ed. Navair. Canoga Park, Calif.: Challenge Publications, 1986. pp. 52-56.

SBD

"Dauntless From Douglas." In: Edwin Schnepf, ed. Hawks Over the Pacific 2. Canoga Park, Calif.: Challenge Publications, 1990. pp. 16-35, 64-83.

O'Leary, Michael. "A Banshee for the Army." In: Michael O'Leary, ed. Victorious Eagles 2. Canoga Park, Calif.: Challenge Publications, 1990. pp. 48-53, 82-83.

_____. "Dauntless Does It." Warbirds International, VI (Winter 1987), 32-37.

SE-5A

Connors, John F. SE-5A in Action. Aircraft, no. 69. Carrollton, Tex.: Squadron/Signal Publications, 1985. 50p.

Hardesty, Bergen. "SE-5A Warbird." Air Classics, III (January-February 1966), 38-41.

Sopwith

Stasinos, Nick. "Hun Killer." Air Classics, IV (September-October 1967), 54-65.
Camel.

SPAD

Connors, John F. SPAD Fighters in Action. Aircraft, no. 93. Carrollton, Tex.: Squadron/Signal Publications, 1989. 50p.

SR-71

"Blackbird: It's Still the Best!" In: Michael O'Leary, ed. Spy Planes. Canoga Park, Calif.: Challenge Publications, 1988. pp. 32-47.

Brodie, Lennie. "Can the U.S. Afford to Retire the SR-71 Blackbird?" Defense Electronics, XXI (September 1989), 57-58+.

Haggerty, Michael J. "Bye, Bye Blackbird." Airman, XXXIV (March 1990), 10-12.

O'Leary, Michael. Black Magic. Osceola, Wis.: Motorbooks International, 1989. 144p.

Thornborough, Anthony M. Lockheed Blackbirds. Osceola, Wis.: Motorbooks International, 1989. 64p.

"U.S. Reconnaissance Weakened by SR-71 Program Termination." AWST, CXXXII (January 22, 1990), 38-44.

T-6

Davis, Larry. T-6 Texan in Action. Carrollton, Tex.: Squadron/Signal Publications, 1989. 58p.

O'Hara, Bob. "The Flying Legend: NAA's Great T-6 Texan." Air Classics, IV (March-April 1967), 20-25.

T-38

Gonzales, Laurence. "Toward Pure Energy." Harper's Magazine, CCLXXIX (August 1989), 72-76.

"Terrific Talon." In: Michael O'Leary, ed. Air Force in Action. Canoga Park, Calif.: Challenge Publications, 1987. pp. 6-11.

T-41

Phillips, Ed. "Skyhawk." Aviation Mechanics Journal, XIX (March 1989), 40-47.

T-45

"Assembly of T-45A Trainer Gets Underway." Sea Power, XXXII (February 1989), 41-44.

Phillips, E. H. "Team Correcting Deficiencies in Navy's T-45A Trainer." AWST, CXXXI (October 30, 1989), 46-48.

"T-45 Training System." In: Michael O'Leary, ed. USMC Airpower in Action. Canoga Park, Calif.: Challenge Publications, 1989. pp. 14-17.

T-47

Andrews, Hal. "Naval Aircraft: T-47 Citation." Naval Aviation News, (July-August 1989), 16-17.

Phillips, E. H. "USAF Tanker-Transport Training to Stress Low and High Altitude Mission Profiles." AWST, CXXX (May 1, 1989), 51-52.

TBD

Doll, Thomas. "And Then There Were None: Story of the TBD Devastator." Air Classics, II (March-April 1965), 32-37+.

McCullough, Anson. "Torpedo Bomber Devastator." Air Classics, V (January-February 1968), 28-43.

TT-1

Smith, Gene. "The U.S. Navy's Forgotten Trainer." Warbirds International, VIII (May-June 1989), 34-37, 76-82.

U-2

Bissell, Richard M., Jr. "Origins of the U-2: Interview." Air Power History, XXXVI (Winter 1989), 15-24.

O'Leary, Michael. "NASA's High Flyers." Air Combat, XVII (May-June 1989), 42-50.

_____. "U-2 Goes for the Gold." Air Classics, XXV (September 1989), 48-54.

Pocock, Chris. Dragon Lady: The History of the U-2. Osceola, Wis.: Motorbooks International, 1989. 214p.

Roberts, Fred. "Dragon Lady." In: Michael O'Leary, ed. Spy Planes. Canoga Park, Calif.: Challenge Publications, 1988. pp. 6-15.

Shellner, Earl. "The Dragon Lady." Airman, XXXIII (March 1989), 36-38.

Shumway, Todd. "Tradition of Tactical Reconnaissance." Defense Electronics, XXI (September 1989), 63-66.

"White Bird." In: Michael O'Leary, ed. Spy Planes. Canoga Park, Calif.: Challenge Publications, 1988. pp. 48-53.

UC-43

Berry, Peter. Beechcraft Staggerwing. Blue Ridge Summit, Pa.: TAB, 1990. 160p.

Smith, Gene. "The Embassy Staggerwing." Warbirds International, VII (January-February 1988), 6-13.

UPF-7

Karstens, Nicholas. "WACO UPF-7 Scale Drawings." AAHSJ, XXXIV (Winter 1989), 314-316.

V-22

Bassett, Edward W. "Allison's T406 Engine: Powering the Tiltrotor." Rotor and Wing International, XXIII (March 1989), 38-40.

Hanley, Art. "Tilt-Rotor V-22 Osprey." Wings, XIX (June 1989), 30+.

Harvey, David. "V-22's First Flight: Short, But Very Sweet." Rotor and Wing International, XXIII (May 1989), 52-53.

Jaszka, Paul R. "V-22 Gets Ready to Tilt Its Rotor." Defense Electronics, XX (December 1988), 80-85+.

Jenkins, Chris. "V-22 Osprey." NATO's Sixteen Nations, XXXIII (November 1988), 83-86.

Lawrence, William S. "The V-22: 'Am I Gonna Like It?'" Vertiflite, XXXIV (November-December 1988), 51-55.

Leibstone, Marvin and Ezio Bonsignore. "Osprey Loses U.S. Funds." Military Technology, XIII (June 1989), 51-52.

Martin, Stanley and Allen H. Schoen. "V-22 Osprey Enabling Technologies." Vertiflite, XXXIV (November-December 1988), 46-50.

O'Shea, Bernard. "V-22 Osprey Tilt Rotor Brings New Dimension to Amphibious Warfare." Pacific Defense Reporter, XV (December 1988-January 1989), 98-101.

"The Rotor Osprey." In: Michael O'Leary, ed. USMC Airpower in Action. Canoga Park, Calif.: Challenge Publications, 1989. pp. 20-25, 90.

Shifrin, Carole A. "Bell-Boeing V-22 Tilt-Rotor Prototype Makes First Flight." AWST, CXXX (March 27, 1989), 20-21.

_____. "V-22 Prepared for Further Expansion of Flight Envelope." AWST, CXXX (April 10, 1989), 34-40.

VT-13

Smith, Gene. "A Valiant Restoration." Warbirds International, VII (May-June 1988), 58-71.

VZ-5FA

Mitchell, Kent A. "The One and Only Fairchild VZ-5FA Fledgling." AAHSJ, XXXIV (Winter 1989), 258-261.
Experimental U.S. Army STOL assault aircraft.

Wright

Allward, Maurice. "They Followed the Flyer." In: John W. R. Taylor, ed. Aircraft Seventy. New York: Arco, 1969. pp. 49-54.

X-4

O'Leary, Michael. "Northrop X-4 Restoration." *Warbirds International*, V (May-June 1988), 34-37, 80-81.

X-14

Wagstaff, Sunny L. "X-14B Retired." *Warbirds International*, VIII (July-August 1989), 28-29.

X-15

"X-15 Plus 30 Years." *NASA Activities*, XX (May 1989), 4-5.

X-29

Ahmadi, Ali-Reza. *A Correlation Study of X-29A Aircraft and Associated Analytical Developments*. DE Thesis. Lawrence: University of Kansas, 1988.

Gera, Joseph, *et al*. "Real-Time Comparison of X-29A Flight Data and Simulation Data." *Journal of Aircraft*, XXVI (February 1989), 117-123.

Hicks, Jack W. and Neil W. Matheny. "Flight Tests Confirm X-29 Technologies." *Exxon Air World*, XLI (Spring 1989), 18-21.

Scott, William B. "High AOA Characteristics of X-29 Exceed Predictions." *AWST*, CXXXII (March 6, 1990), 68-69.
AOA = Angles of Attack.

X-30

Covault, Craig. "Bush Administration Speeds NASP Review, Spurred by Soviet, German Competition." *AWST*, CXXX (June 5, 1989), 24-25.

_____. "White House Acts to Reverse Aero-Space Plane Cancellation." *AWST*, CXXX (April 24, 1989), 20-21.

Heppenheimer, T. A. "2015: The Age of Spaceplanes." *Final Frontier*, II (May-June 1989), 36-39.

Kandebo, Stanley W. "Researchers Explore Slush Hydrogen as Fuel for NASP." *AWST*, CXXX (June 26, 1989), 37-38.

Kolcum, Edward H. "Contractors Pursue NASP Technology Despite Possible Funding Cutbacks." *AWST*, CXXX (May 22, 1989), 96-97.

United States. Defense Science Board. Task Force the the National Aerospace Plane. *Final Report*. Washington, D.C.: GPO, 1988. 90p.

Welling, William. "Spaceplane." *International Combat Arms*, VII (January 1989), 66-73+.

XB-38

"XB-38: Fortress With a Difference." In: Michael O'Leary, ed. Fabulous Flops. Canoga Park, Calif.: Challenge Publications, 1990. pp. 58-63.

XB-40

"Fighter Fortress." In: Michael O'Leary, ed. Fabulous Flops. Canoga Park, Calif.: Challenge Publications, 1990. pp. 18-23.

XB-42/XB-43

"America's First Jet Bomber." In: Michael O'Leary, ed. Fabulous Flops. Canoga Park, Calif.: Challenge Publications, 1990. pp. 78-87.

Schrader, Richard K. "Douglas XB-43." Air Classics, XXV (July 1989), 14-21+.

_____. "Mixmaster." Air Classics, XXV (June 1989), 16-25.

XC-108

Thompson, Scott. "Return of the XC-108." Warbirds International, VII (March-April 1988), 6-13.

XCG-7/XCG-8

Schrader, Richard K. "Gliders From Douglas." Air Classics, XXV (May 1989), 18-23, 62-63, 80-81.

XF5F-1

Lucabaugh, Dave and Bob Martin. "The Scarce Skyrocket." AAHSJ, XXXIV (Summer 1989), 96-107.

XF5U-1

McCullough, Anson. "The Flying Flapjack: Vought's Unusual XF5U." Air Classics, IV (January-February 1967), 8-15.

XF-11

Boyne, Walter J. "The Phantom Eye: How Howard Hughes Developed a Spy Plane." Air Classics, VI (November-December 1967), 52+.

XF-90

"Deep Penetration." In: Michael O'Leary, ed. Fabulous Flops. Canoga Park, Calif.: Challenge Publications, 1990. pp. 72-77.

XF-91

Rochfort, Gary L. "The Republic XF-91 Thunderceptor." AAHSJ, XXXIV (Spring 1989), 16-31.

XFV-1

Johnson, Frank. "Vertical Riser." In: Michael O'Leary, ed. Fabulous Flops. Canoga Park, Calif.: Challenge Publications, 1990. pp. 30-39, 96-98.

XHK

Westal, Freeman. "The Plywood Ghost." Air Classics, III (July-August 1966), 10-14.

XOSE-1/2

Schrader, Richard K. "End of an Era: The Story of the Navy's Last Floatplane." In: Michael O'Leary, ed. Fabulous Flops. Canoga Park, Calif.: Challenge Publications, 1990. pp. 18-23.

XP-75

Schrader, Richard K. "The Fighter From General Motors." Warbirds International, VIII (January-February 1989), 6-8, 81.

_____. "Dayton's Eagle." Warbirds International, VIII (July-August 1989), 30-31.

XPBB-1

"Lone Ranger." In: Michael O'Leary, ed. Hawks Over the Pacific 2. Canoga Park, Calif.: Challenge Publications, 1990. pp. 40-41.

YF-22/YF-23

Morrocco, John D. "Budget Pressures Forcing USAF to Ease Requirements for ATF." AWST, CXXXI (August 28, 1989), 23-25.

Model/Manufacturer/ Popular Name Index

IN those cases where in the alphabetical listing below models have the same numerical designation, they are chronologically separated with roman numerals in parentheses. Thus, for example, B-2(1) refers to the 1930s Curtiss Condor II while B-2(2) denotes Northrop's new Stealth Bomber. Cross references are provided between derivatives/Popular Names and basic airframes in those cases where the former are not referenced. All warbirds are alphabetically entered under the name of their manufacturer, although in certain cases, companies merged into other firms are referenced under the larger organization (e.g., Hiller see Fairchild), with appropriate cross references between the two.